An Introduction to

PUBLISHED BY THE PRESS SYNDICATE OF THE UNIVERSITY OF CAMBRIDGE
The Pitt Building, Trumpington Street, Cambridge, United Kingdom

CAMBRIDGE UNIVERSITY PRESS
The Edinburgh Building, Cambridge CB2 2RU, UK
40 West 20th Street, New York, NY 10011-4211, USA
477 Williamstown Road, Port Melbourne, VIC 3207, Australia
Ruiz de Alarcón 13, 28014 Madrid, Spain
Dock House, The Waterfront, Cape Town 8001, South Africa

http://www.cambridge.org

First published 2003

Printed in the United Kingdom at the University Press, Cambridge

Typeface CMR 10/13 pt *System* LaTeX 2ε [TB]

A catalogue record for this book is available from the British Library

Library of Congress Cataloguing in Publication data

ISBN 0 521 81026 4 hardback
ISBN 0 521 00794 1 paperback

Contents

	Preface	*page* vii
1	**Solution of equations by iteration**	1
1.1	Introduction	1
1.2	Simple iteration	2
1.3	Iterative solution of equations	17
1.4	Relaxation and Newton's method	19
1.5	The secant method	25
1.6	The bisection method	28
1.7	Global behaviour	29
1.8	Notes	32
	Exercises	35
2	**Solution of systems of linear equations**	39
2.1	Introduction	39
2.2	Gaussian elimination	44
2.3	LU factorisation	48
2.4	Pivoting	52
2.5	Solution of systems of equations	55
2.6	Computational work	56
2.7	Norms and condition numbers	58
2.8	Hilbert matrix	72
2.9	Least squares method	74
2.10	Notes	79
	Exercises	82
3	**Special matrices**	87
3.1	Introduction	87
3.2	Symmetric positive definite matrices	87
3.3	Tridiagonal and band matrices	93

3.4	Monotone matrices	98
3.5	Notes	101
	Exercises	102
4	**Simultaneous nonlinear equations**	104
4.1	Introduction	104
4.2	Simultaneous iteration	106
4.3	Relaxation and Newton's method	116
4.4	Global convergence	123
4.5	Notes	124
	Exercises	126
5	**Eigenvalues and eigenvectors of a symmetric matrix**	133
5.1	Introduction	133
5.2	The characteristic polynomial	137
5.3	Jacobi's method	137
5.4	The Gerschgorin theorems	145
5.5	Householder's method	150
5.6	Eigenvalues of a tridiagonal matrix	156
5.7	The QR algorithm	162
5.7.1	The QR factorisation revisited	162
5.7.2	The definition of the QR algorithm	164
5.8	Inverse iteration for the eigenvectors	166
5.9	The Rayleigh quotient	170
5.10	Perturbation analysis	172
5.11	Notes	174
	Exercises	175
6	**Polynomial interpolation**	179
6.1	Introduction	179
6.2	Lagrange interpolation	180
6.3	Convergence	185
6.4	Hermite interpolation	187
6.5	Differentiation	191
6.6	Notes	194
	Exercises	195
7	**Numerical integration – I**	200
7.1	Introduction	200
7.2	Newton–Cotes formulae	201
7.3	Error estimates	204
7.4	The Runge phenomenon revisited	208
7.5	Composite formulae	209

7.6 The Euler–Maclaurin expansion 211
7.7 Extrapolation methods 215
7.8 Notes 219
Exercises 220

8 Polynomial approximation in the ∞-norm 224
8.1 Introduction 224
8.2 Normed linear spaces 224
8.3 Best approximation in the ∞-norm 228
8.4 Chebyshev polynomials 241
8.5 Interpolation 244
8.6 Notes 247
Exercises 248

9 Approximation in the 2-norm 252
9.1 Introduction 252
9.2 Inner product spaces 253
9.3 Best approximation in the 2-norm 256
9.4 Orthogonal polynomials 259
9.5 Comparisons 270
9.6 Notes 272
Exercises 273

10 Numerical integration – II 277
10.1 Introduction 277
10.2 Construction of Gauss quadrature rules 277
10.3 Direct construction 280
10.4 Error estimation for Gauss quadrature 282
10.5 Composite Gauss formulae 285
10.6 Radau and Lobatto quadrature 287
10.7 Note 288
Exercises 288

11 Piecewise polynomial approximation 292
11.1 Introduction 292
11.2 Linear interpolating splines 293
11.3 Basis functions for the linear spline 297
11.4 Cubic splines 298
11.5 Hermite cubic splines 300
11.6 Basis functions for cubic splines 302
11.7 Notes 306
Exercises 307

12 Initial value problems for ODEs 310
12.1 Introduction 310
12.2 One-step methods 317
12.3 Consistency and convergence 321
12.4 An implicit one-step method 324
12.5 Runge–Kutta methods 325
12.6 Linear multistep methods 329
12.7 Zero-stability 331
12.8 Consistency 337
12.9 Dahlquist's theorems 340
12.10 Systems of equations 341
12.11 Stiff systems 343
12.12 Implicit Runge–Kutta methods 349
12.13 Notes 353
Exercises 355

13 Boundary value problems for ODEs 361
13.1 Introduction 361
13.2 A model problem 361
13.3 Error analysis 364
13.4 Boundary conditions involving a derivative 367
13.5 The general self-adjoint problem 370
13.6 The Sturm–Liouville eigenvalue problem 373
13.7 The shooting method 375
13.8 Notes 380
Exercises 381

14 The finite element method 385
14.1 Introduction: the model problem 385
14.2 Rayleigh–Ritz and Galerkin principles 388
14.3 Formulation of the finite element method 391
14.4 Error analysis of the finite element method 397
14.5 *A posteriori* error analysis by duality 403
14.6 Notes 412
Exercises 414

***Appendix A* An overview of results from real analysis** 419

***Appendix B* WWW-resources** 423

Bibliography 424
Index 429

Preface

This book has grown out of printed notes which accompanied lectures given by ourselves and our colleagues over many years to undergraduate mathematicians at Oxford. During those years the contents and the arrangement of the lectures have changed substantially, and this book has a wider scope than is currently taught. It contains mathematics which, in an ideal world, would be part of the equipment of any well-educated mathematician.

Numerical analysis is the branch of mathematics concerned with the theoretical foundations of numerical algorithms for the solution of problems arising in scientific applications. The subject addresses a variety of questions ranging from the approximation of functions and integrals to the approximate solution of algebraic, transcendental, differential and integral equations, with particular emphasis on the stability, accuracy, efficiency and reliability of numerical algorithms. The purpose of this book is to provide an elementary introduction into this active and exciting field, and is aimed at students in the second year of a university mathematics course.

The book addresses a wide range of numerical problems in algebra and analysis. Chapter 2 deals with the solution of systems of linear equations, a process which can be completed in a finite number of arithmetical operations. In the rest of the book the solution of a problem is sought as the limit of an infinite sequence; in that sense the output of the numerical algorithm is an 'approximate' solution. This need not, however, mean any relaxation of the usual standards of rigorous analysis. The idea of convergence of a sequence of real numbers (x_n) to a real number ξ is very familiar: given any positive value of ε there exists a positive integer N_0 such that $|x_n - \xi| < \varepsilon$ for all n such that $n > N_0$. In such a situation one can obtain as accurate an approximation to ξ as required by calculating sufficiently many members of the sequence, or just one member, sufficiently far along. A 'pure mathematician' would prefer the exact answer, ξ, but the sorts of guaranteed accurate approximations which will be discussed here are entirely satisfactory in real-life applications.

Numerical analysis brings two new ideas to the usual discussion of convergence of sequences. First, we need, not just the existence of N_0, but a good estimate of how large it is; and it may be too large for practical calculations. Second, rather than being asked for the limit of a given sequence, we are usually given the existence of the limit ξ (or its approximate location on the real line) and then have to construct a sequence which converges to it. If the rate of convergence is slow, so that the value of N_0 is large, we must then try to construct a better sequence, one that converges to ξ more rapidly. These ideas have direct applications in the solution of a single nonlinear equation in Chapter 1, the solution of systems of nonlinear equations in Chapter 4 and the calculation of the eigenvalues and eigenvectors of a matrix in Chapter 5.

The next six chapters are concerned with polynomial approximation, and show how, in various ways, we can construct a polynomial which approximates, as accurately as required, a given continuous function. These ideas have an obvious application in the evaluation of integrals, where we calculate the integral of the approximating polynomial instead of the integral of the given function.

Finally, Chapters 12 to 14 deal with the numerical solution of ordinary differential equations, with Chapter 14 presenting the fundamentals of the finite element method. The results of Chapter 14 can be readily extended to linear second-order partial differential equations.

We have tried to make the coverage as complete as is consistent with remaining quite elementary. The limitations of size are most obvious in Chapter 12 on the solution of initial value problems for ordinary differential equations. This is an area where a number of excellent books are available, at least one of which is published in two weighty volumes. Chapter 12 does not describe or analyse anything approaching all the available methods, but we hope we have included some of those in most common use.

There is a selection of Exercises at the end of each chapter. All these exercises are theoretical; students are urged to apply all the methods described to some simple examples to see what happens. A few of the exercises will be found to require some heavy algebraic manipulation; these have been included because we assume that readers will have access to some computer algebra system such as Maple or Mathematica, which then make the algebraic work almost trivial. Those involved in teaching courses based on this book may obtain copies of LaTeX files containing solutions to these exercises by applying to the publisher by email (`solutions@cambridge.org`). Although the material presented in this book does not presuppose the reader's acquaintance with mathematical software packages, the importance of these cannot be overemphasised. In Appendix B, a brief set of pointers is provided to relevant software repositories.

Our treatment is intended to maintain a reasonably high standard of rigour, with many theorems and formal proofs. The main prerequisite is therefore some familiarity with elementary real analysis. Appendix A lists the standard theorems (labelled **Theorem A.1**, **A.2**, ..., **A7**) which are used in the book, together with proofs of one or two of them which might be less familiar. Some knowledge of basic matrix algebra is assumed. We have also used some elementary ideas from the theory of normed linear spaces in a number of places; complete definitions and examples are given. Some prior knowledge of these areas would be helpful, although not essential.

The chart below indicates how the chapters of the book are interrelated. They show, in particular, how Chapters 1 to 5 form a largely self-contained unit, as do Chapters 6 to 10.

Roadmap of the book

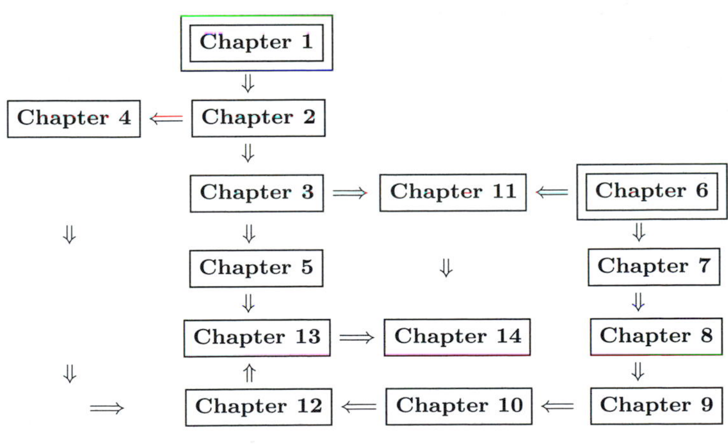

We have included some historical notes throughout the book. As well as hoping to stimulate an interest in the development of the subject, these notes show how wide a historical range even this elementary book covers. Many of the methods were developed by the great mathematicians of the seventeenth and eighteenth centuries, including Newton, Euler and Gauss, but what is usually known as Gaussian elimination for the solution of systems of linear equations was known to the Chinese two thousand years ago. At the other end of the historical scale, the analysis of the eigenvalue problem, and the numerical solution of differential equations, are much more recent, and are due to mathematicians who are still very much alive. Many of our historical notes are based on the excellent biographical database at the history of mathematics website

`http://www-history.mcs.st-andrews.ac.uk/history/`

We have tried to eradicate as many typographical errors from the text as possible; however, we are mindful that some may have escaped our attention. We plan to post any typos reported to us on

`http://web.comlab.ox.ac.uk/oucl/work/endre.suli/index.html`

We wish to express our gratitude to Professor Bill Morton for setting us off on this *tour de force*, to David Tranah and Jonathan Walthoe at Cambridge University Press for encouraging us to persist with the project, and to the staff of the Press for not only improving the appearance of the book and eliminating a number of typographical errors, but also for correcting and improving some of our mathematics. We also wish to thank our colleagues at the Oxford University Computing Laboratory, particularly Nick Trefethen, Mike Giles and Andy Wathen, for keeping our spirits up, and to Paul Houston at the Department of Mathematics and Computer Science of the University of Leicester for his help with the final example in the book.

Above all, we are grateful to our families for their patience, support and understanding: this book is dedicated to them.

ES & DFM *Oxford, May 2003.*

1

Solution of equations by iteration

1.1 Introduction

Equations of various kinds arise in a range of physical applications and a substantial body of mathematical research is devoted to their study. Some equations are rather simple: in the early days of our mathematical education we all encountered the single *linear* equation $ax+b=0$, where a and b are real numbers and $a \neq 0$, whose solution is given by the formula $x = -b/a$. Many equations, however, are *nonlinear:* a simple example is $Ax^2 + Bx + C = 0$, involving a quadratic polynomial with real coefficients A, B, C, and $A \neq 0$. The two solutions to this equation, labelled x_1 and x_2, are found in terms of the coefficients of the polynomial from the familiar formulae

$$x_1 = \frac{-B + \sqrt{B^2 - 4AC}}{2A}, \qquad x_2 = \frac{-B - \sqrt{B^2 - 4AC}}{2A}. \tag{1.1}$$

It is less likely that you have seen the more intricate formulae for the solution of cubic and quartic polynomial equations due to the Italian mathematicians Scipione del Ferro (1465–1526) and Lodovico Ferrari (1522–1565), respectively, which were published by Girolamo Cardano (1501–1576) in 1545 in his *Artis magnae sive de regulis algebraicis liber unus.* In any case, if you have been led to believe that similar expressions involving radicals (roots of sums of products of coefficients) will supply the solution to any polynomial equation, then you should brace yourself for a surprise: no such closed formula exists for a general polynomial equation of degree n when $n \geq 5$. It transpires that for each $n \geq 5$ there exists a polynomial equation of degree n with integer coefficients which cannot be solved in terms of radicals;[1] such is, for example, $x^5 - 4x - 2 = 0$.

Since there is no general formula for the solution of polynomial equations, no general formula will exist for the solution of an arbitrary nonlinear equation of the form $f(x) = 0$ where f is a continuous real-valued function. How can we then decide whether or not such an equation possesses a solution in the set of real numbers, and how can we find a solution?

The present chapter is devoted to the study of these questions. Our goal is to develop simple numerical methods for the approximate solution of the equation $f(x) = 0$ where f is a real-valued function, defined and continuous on a bounded and closed interval of the real line. Methods of the kind discussed here are iterative in nature and produce sequences of real numbers which, in favourable circumstances, converge to the required solution.

1.2 Simple iteration

Suppose that f is a real-valued function, defined and continuous on a bounded closed interval $[a, b]$ of the real line. It will be tacitly assumed throughout the chapter that $a < b$, so that the open interval (a, b) is nonempty. We wish to find a *real number* $\xi \in [a, b]$ such that $f(\xi) = 0$. If such ξ exists, it is called a **solution** to the equation $f(x) = 0$.

Even some relatively simple equations may fail to have a solution in the set of real numbers. Consider, for example,

$$f\colon x \mapsto x^2 + 1\,.$$

Clearly $f(x) = 0$ has no solution in any interval $[a, b]$ of the real line. Indeed, according to (1.1), the quadratic polynomial x^2+1 has two roots: $x_1 = \sqrt{-1} = \imath$ and $x_2 = -\sqrt{-1} = -\imath$. However, these belong to the set of imaginary numbers and are therefore excluded by our definition of solution which only admits *real* numbers. In order to avoid difficulties of this kind, we begin by exploring the existence of solutions to the equation $f(x) = 0$ in the set of real numbers. Our first result in this direction is rather simple.

[1] This result was proved in 1824 by the Norwegian mathematician Niels Henrik Abel (1802–1829), and was further refined in the work of Evariste Galois (1811–1832) who clarified the circumstances in which a closed formula may exist for the solution of a polynomial equation of degree n in terms of radicals.

Theorem 1.1 *Let f be a real-valued function, defined and continuous on a bounded closed interval $[a, b]$ of the real line. Assume, further, that $f(a)f(b) \leq 0$; then, there exists ξ in $[a, b]$ such that $f(\xi) = 0$.*

Proof If $f(a) = 0$ or $f(b) = 0$, then $\xi = a$ or $\xi = b$, respectively, and the proof is complete. Now, suppose that $f(a)f(b) \neq 0$. Then, $f(a)f(b) < 0$; in other words, 0 belongs to the open interval whose endpoints are $f(a)$ and $f(b)$. By the Intermediate Value Theorem (Theorem A.1), there exists ξ in the open interval (a, b) such that $f(\xi) = 0$. □

To paraphrase Theorem 1.1, if a continuous function f has opposite signs at the endpoints of the interval $[a, b]$, then the equation $f(x) = 0$ has a solution in (a, b). The converse statement is, of course, false. Consider, for example, a continuous function defined on $[a, b]$ which changes sign in the open interval (a, b) an even number of times, with $f(a)f(b) \neq 0$; then, $f(a)f(b) > 0$ even though $f(x) = 0$ has solutions inside $[a, b]$. Of course, in the latter case, there exist an even number of subintervals of (a, b) at the endpoints of each of which f *does* have opposite signs. However, finding such subintervals may not always be easy.

To illustrate this last point, consider the rather pathological function

$$f\colon\, x \mapsto \frac{1}{2} - \frac{1}{1 + M|x - 1.05|}\,, \tag{1.2}$$

depicted in Figure 1.1 for x in the closed interval $[0.8, 1.8]$ and $M = 200$. The solutions $x_1 = 1.05 - (1/M)$ and $x_2 = 1.05 + (1/M)$ to the equation $f(x) = 0$ are only a distance $2/M$ apart and, for large and positive M, locating them computationally will be a challenging task.

Remark 1.1 *If you have access to the mathematical software package Maple, plot the function f by typing*

```
plot(1/2-1/(1+200*abs(x-1.05)), x=0.8..1.8, y=-0.5..0.6);
```

at the Maple command line, and then repeat this experiment by choosing $M = 2000$, 20000, 200000, 2000000, *and* 20000000 *in place of the number* 200. *What do you observe? For the last two values of M, replot the function f for x in the subinterval* $[1.04999, 1.05001]$. ◇

An alternative sufficient condition for the existence of a solution to the equation $f(x) = 0$ is arrived at by rewriting it in the equivalent form $x - g(x) = 0$ where g is a certain real-valued function, defined

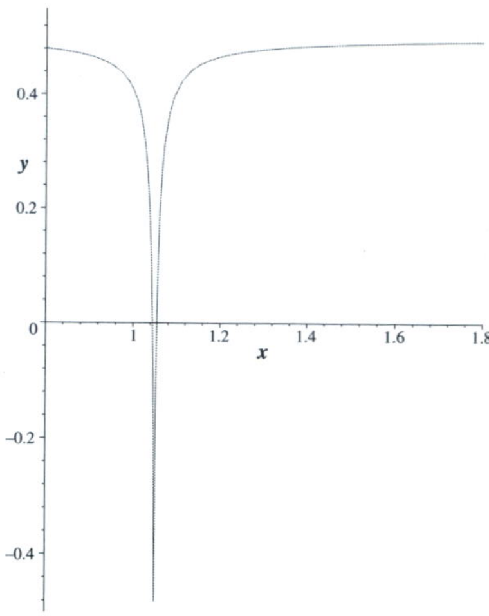

Fig. 1.1. Graph of the function $f\colon x \mapsto \frac{1}{2} - \frac{1}{1+200|x-1.05|}$ for $x \in [0.8, 1.8]$.

and continuous on $[a, b]$; the choice of g and its relationship with f will be clarified below through examples. Upon such a transformation the problem of solving the equation $f(x) = 0$ is converted into one of finding ξ such that $\xi - g(\xi) = 0$.

Theorem 1.2 (Brouwer's Fixed Point Theorem) *Suppose that g is a real-valued function, defined and continuous on a bounded closed interval $[a, b]$ of the real line, and let $g(x) \in [a, b]$ for all $x \in [a, b]$. Then, there exists ξ in $[a, b]$ such that $\xi = g(\xi)$; the real number ξ is called a* **fixed point of the function** *g.*

Proof Let $f(x) = x - g(x)$. Then, $f(a) = a - g(a) \leq 0$ since $g(a) \in [a, b]$ and $f(b) = b - g(b) \geq 0$ since $g(b) \in [a, b]$. Consequently, $f(a)f(b) \leq 0$, with f defined and continuous on the closed interval $[a, b]$. By Theorem 1.1 there exists $\xi \in [a, b]$ such that $0 = f(\xi) = \xi - g(\xi)$. $\square$

Figure 1.2 depicts the graph of a function $x \mapsto g(x)$, defined and continuous on a closed interval $[a, b]$ of the real line, such that $g(x)$ belongs to $[a, b]$ for all x in $[a, b]$. The function g has three fixed points in the interval $[a, b]$: the x-coordinates of the three points of intersection of the graph of g with the straight line $y = x$.

Of course, any equation of the form $f(x) = 0$ can be rewritten in the

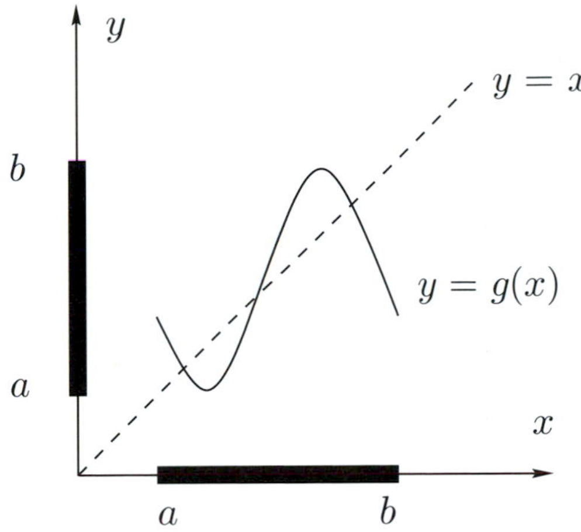

Fig. 1.2. Graph of a function g, defined and continuous on the interval $[a,b]$, which maps $[a,b]$ into itself; g has three fixed points in $[a,b]$: the x-coordinates of the three points of intersection of the graph of g with $y = x$.

equivalent form of $x = g(x)$ by letting $g(x) = x + f(x)$. While there is no guarantee that the function g, so defined, will satisfy the conditions of Theorem 1.2, there are many alternative ways of transforming $f(x) = 0$ into $x = g(x)$, and we only have to find one such rearrangement with g continuous on $[a,b]$ and such that $g(x) \in [a,b]$ for all $x \in [a,b]$. Sounds simple? Fine. Take a look at the following example.

Example 1.1 *Consider the function f defined by $f(x) = \mathrm{e}^x - 2x - 1$ for $x \in [1,2]$. Clearly, $f(1) < 0$ and $f(2) > 0$. Thus we deduce from Theorem 1.1 the existence of ξ in $[1,2]$ such that $f(\xi) = 0$.*

In order to relate this example to Theorem 1.2, let us rewrite the equation $f(x) = 0$ in the equivalent form $x - g(x) = 0$, where the function g is defined on the interval $[1,2]$ by $g(x) = \ln(2x+1)$; here (and throughout the book) ln means $\log_{\mathrm{e}}$. As $g(1) \in [1,2]$, $g(2) \in [1,2]$ and g is monotonic increasing, it follows that $g(x) \in [1,2]$ for all $x \in [1,2]$, showing that g satisfies the conditions of Theorem 1.2. Thus, again, we deduce the existence of $\xi \in [1,2]$ such that $\xi - g(\xi) = 0$ or, equivalently, $f(\xi) = 0$.

We could have also rewritten our equation as $x = (\mathrm{e}^x - 1)/2$. However, the associated function $g\colon x \mapsto (\mathrm{e}^x - 1)/2$ does not map the interval $[1,2]$ into itself, so Theorem 1.2 cannot then be applied. ◇

Although the ability to verify the existence of a solution to the equation $f(x) = 0$ is important, none of what has been said so far provides a *method* for solving this equation. The following definition is a first step in this direction: it will lead to the construction of an algorithm for computing an approximation to the fixed point ξ of the function g, and will thereby supply an approximate solution to the equivalent equation $f(x) = 0$.

Definition 1.1 *Suppose that g is a real-valued function, defined and continuous on a bounded closed interval $[a, b]$ of the real line, and assume that $g(x) \in [a, b]$ for all $x \in [a, b]$. Given that $x_0 \in [a, b]$, the recursion defined by*

$$x_{k+1} = g(x_k)\,, \qquad k = 0, 1, 2, \ldots\,, \tag{1.3}$$

is called a **simple iteration***; the numbers x_k, $k \geq 0$, are referred to as* **iterates***.*

If the sequence (x_k) defined by (1.3) converges, the limit must be a fixed point in $[a, b]$ of the function g, since g is continuous on the closed interval $[a, b]$. Indeed, writing $\xi = \lim_{k\to\infty} x_k$, we have that $\xi \in [a, b]$ and

$$\xi = \lim_{k\to\infty} x_{k+1} = \lim_{k\to\infty} g(x_k) = g\left(\lim_{k\to\infty} x_k\right) = g(\xi)\,, \tag{1.4}$$

where the second equality follows from (1.3) and the third equality is a consequence of the continuity of g.

A sufficient condition for the convergence of the sequence (x_k) is provided by our next result which represents a refinement of Brouwer's Fixed Point Theorem, under the additional assumption that the mapping g is a contraction.

Definition 1.2 (Contraction) *Suppose that g is a real-valued function, defined and continuous on a bounded closed interval $[a, b]$ of the real line. Then, g is said to be a* **contraction** *on $[a, b]$ if there exists a constant L such that $0 < L < 1$ and*

$$|g(x) - g(y)| \leq L|x - y| \quad \forall\, x, y \in [a, b]\,. \tag{1.5}$$

Remark 1.2 *The terminology 'contraction' stems from the fact that when (1.5) holds with $0 < L < 1$, the distance $|\,g(x) - g(y)\,|$ between the images of the points x, y is (at least $1/L$ times)* smaller *than the distance*

$|x-y|$ between x and y. More generally, when L is any positive real number, (1.5) is referred to as a **Lipschitz condition**.[1]

Armed with Definition 1.2, we are now ready to state the main result of this section.

Theorem 1.3 (Contraction Mapping Theorem) *Let g be a real-valued function, defined and continuous on a bounded closed interval $[a, b]$ of the real line, and assume that $g(x) \in [a, b]$ for all $x \in [a, b]$. Suppose, further, that g is a contraction on $[a, b]$. Then, g has a unique fixed point ξ in the interval $[a, b]$. Moreover, the sequence (x_k) defined by (1.3) converges to ξ as $k \to \infty$ for any starting value x_0 in $[a, b]$.*

Proof The existence of a fixed point ξ for g is a consequence of Theorem 1.2. The uniqueness of this fixed point follows from (1.5) by contradiction: for suppose that g has a second fixed point, η, in $[a, b]$. Then,

$$|\xi - \eta| = |g(\xi) - g(\eta)| \leq L|\xi - \eta|,$$

i.e., $(1-L)|\xi - \eta| \leq 0$. As $1 - L > 0$, we deduce that $\eta = \xi$.

Let x_0 be any element of $[a, b]$ and consider the sequence (x_k) defined by (1.3). We shall prove that (x_k) converges to the fixed point ξ. According to (1.5) we have that

$$|x_k - \xi| = |g(x_{k-1}) - g(\xi)| \leq L|x_{k-1} - \xi|, \qquad k \geq 1,$$

from which we then deduce by induction that

$$|x_k - \xi| \leq L^k |x_0 - \xi|, \qquad k \geq 1. \tag{1.6}$$

As $L \in (0, 1)$, it follows that $\lim_{k\to\infty} L^k = 0$, and hence we conclude that $\lim_{k\to\infty} |x_k - \xi| = 0$. □

Let us illustrate the Contraction Mapping Theorem by an example.

Example 1.2 *Consider the equation $f(x) = 0$ on the interval $[1, 2]$ with $f(x) = \mathrm{e}^x - 2x - 1$, as in Example 1.1. Recall from Example 1.1 that this equation has a solution, ξ, in the interval $[1, 2]$, and ξ is a fixed point of the function g defined on $[1, 2]$ by $g(x) = \ln(2x + 1)$.*

[1] Rudolf Otto Sigismund Lipschitz (14 May 1832, Königsberg, Prussia (now Kaliningrad, Russia) – 7 October 1903, Bonn, Germany) made important contributions to number theory, the theory of Bessel functions and Fourier series, the theory of ordinary and partial differential equations, and to analytical mechanics and potential theory.

Table 1.1. *The sequence* (x_k) *defined by (1.8).*

k	x_k
0	1.000000
1	1.098612
2	1.162283
3	1.201339
4	1.224563
5	1.238121
6	1.245952
7	1.250447
8	1.253018
9	1.254486
10	1.255323
11	1.255800

Now, the function g is defined and continuous on the interval $[1, 2]$, and g is differentiable on $(1, 2)$. Thus, by the Mean Value Theorem (Theorem A.3), for any x, y in $[1, 2]$ we have that

$$|\,g(x) - g(y)\,| = |\,g'(\eta)(x - y)\,| = |g'(\eta)|\,|\,x - y\,| \tag{1.7}$$

for some η that lies between x and y and is therefore in the interval $[1, 2]$. Further, $g'(x) = 2/(2x + 1)$ and $g''(x) = -4/(2x + 1)^2$. As $g''(x) < 0$ for all x in $[1, 2]$, g' is monotonic decreasing on $[1, 2]$. Hence $g'(1) \geq g'(\eta) \geq g'(2)$, *i.e.*, $g'(\eta) \in [2/5, 2/3]$. Thus we deduce from (1.7) that

$$|\,g(x) - g(y)\,| \leq L|\,x - y\,| \qquad \forall x, y \in [1, 2]\,,$$

with $L = 2/3$. According to the Contraction Mapping Theorem, the sequence (x_k) defined by the simple iteration

$$x_{k+1} = \ln(2x_k + 1)\,, \qquad k = 0, 1, 2, \ldots, \tag{1.8}$$

converges to ξ for any starting value x_0 in $[1, 2]$. Let us choose $x_0 = 1$, for example, and compute the next 11 iterates, say. The results are shown in Table 1.1. Even though we have carried six decimal digits, after 11 iterations only the first two decimal digits of the iterates x_k appear to have settled; thus it seems likely that $\xi = 1.26$ to two decimal digits. ◇

You may now wonder how many iterations we should perform in (1.8) to ensure that all six decimals have converged to their correct values. In order to answer this question, we need to carry out some analysis.

Theorem 1.4 *Consider the simple iteration (1.3) where the function g satisfies the hypotheses of the Contraction Mapping Theorem on the bounded closed interval $[a,b]$. Given $x_0 \in [a,b]$ and a certain tolerance $\varepsilon > 0$, let $k_0(\varepsilon)$ denote the smallest positive integer such that x_k is no more than ε away from the (unknown) fixed point ξ, i.e., $|x_k - \xi| \le \varepsilon$, for all $k \ge k_0(\varepsilon)$. Then,*

$$k_0(\varepsilon) \le \left[\frac{\ln|x_1 - x_0| - \ln(\varepsilon(1-L))}{\ln(1/L)}\right] + 1, \tag{1.9}$$

where, for a real number x, $[x]$ signifies the largest integer less than or equal to x.

Proof From (1.6) in the proof of Theorem 1.3 we know that

$$|x_k - \xi| \le L^k |x_0 - \xi|, \quad k \ge 1.$$

Using this result with $k = 1$, we obtain

$$\begin{aligned} |x_0 - \xi| &= |x_0 - x_1 + x_1 - \xi| \\ &\le |x_0 - x_1| + |x_1 - \xi| \\ &\le |x_0 - x_1| + L|x_0 - \xi|. \end{aligned}$$

Hence

$$|x_0 - \xi| \le \frac{1}{1-L}|x_0 - x_1|.$$

By substituting this into (1.6) we get

$$|x_k - \xi| \le \frac{L^k}{1-L}|x_1 - x_0|. \tag{1.10}$$

Thus, in particular, $|x_k - \xi| \le \varepsilon$ provided that

$$L^k \frac{1}{1-L}|x_1 - x_0| \le \varepsilon.$$

On taking the (natural) logarithm of each side in the last inequality, we find that $|x_k - \xi| \le \varepsilon$ for all k such that

$$k \ge \frac{\ln|x_1 - x_0| - \ln(\varepsilon(1-L))}{\ln(1/L)}.$$

Therefore, the smallest integer $k_0(\varepsilon)$ such that $|x_k - \xi| \le \varepsilon$ for all

$k \geq k_0(\varepsilon)$ cannot exceed the expression on the right-hand side of the inequality (1.9). □

This result provides an upper bound on the maximum number of iterations required to ensure that the error between the kth iterate x_k and the (unknown) fixed point ξ is below the prescribed tolerance ε. Note, in particular, from (1.9), that if L is close to 1, then $k_0(\varepsilon)$ may be quite large for any fixed ε. We shall revisit this point later on in the chapter.

Example 1.3 *Now we can return to Example 1.2 to answer the question posed there about the maximum number of iterations required, with starting value $x_0 = 1$, to ensure that the last iterate computed is correct to six decimal digits.*

Letting $\varepsilon = 0.5 \times 10^{-6}$ and recalling from Example 1.2 that $L = 2/3$, the formula (1.9) yields $k_0(\varepsilon) \leq [32.778918] + 1$, so we have that $k_0(\varepsilon) \leq 33$. In fact, 33 is a somewhat pessimistic overestimate of the number of iterations required: computing the iterates x_k successively shows that already x_{25} is correct to six decimal digits, giving $\xi = 1.256431$. ◇

Condition (1.5) can be rewritten in the following equivalent form:

$$\left| \frac{g(x) - g(y)}{x - y} \right| \leq L \qquad \forall\, x, y \in [a, b]\,, \quad x \neq y\,,$$

with $L \in (0, 1)$, which can, in turn, be rephrased by saying that the absolute value of the slope of the function g does not exceed $L \in (0, 1)$. Assuming that g is a differentiable function on the open interval (a, b), the Mean Value Theorem (Theorem A.3) tells us that

$$\frac{g(x) - g(y)}{x - y} = g'(\eta)$$

for some η that lies between x and y and is therefore contained in the interval (a, b).

We shall therefore adopt the following assumption that is somewhat stronger than (1.5) but is easier to verify in practice:

$$\begin{aligned} &g \text{ is differentiable on } (a, b) \text{ and} \\ &\exists L \in (0, 1) \text{ such that } |g'(x)| \leq L \text{ for all } x \in (a, b)\,. \end{aligned} \tag{1.11}$$

Consequently, Theorem 1.3 still holds when (1.5) is replaced by (1.11).

We note that the requirement in (1.11) that g be differentiable is indeed more demanding than the Lipschitz condition (1.5): for example, $g(x) = |x|$ satisfies the Lipschitz condition on any closed interval of the real line, with $L = 1$, yet g is not differentiable at $x = 0$.[1]

Next we discuss a local version of the Contraction Mapping Theorem, where (1.11) is only assumed in a neighbourhood of the fixed point ξ rather than over the entire interval $[a, b]$.

Theorem 1.5 *Suppose that g is a real-valued function, defined and continuous on a bounded closed interval $[a, b]$ of the real line, and assume that $g(x) \in [a, b]$ for all $x \in [a, b]$. Let $\xi = g(\xi) \in [a, b]$ be a fixed point of g (whose existence is ensured by Theorem 1.2), and assume that g has a continuous derivative in some neighbourhood of ξ with $|g'(\xi)| < 1$. Then, the sequence (x_k) defined by $x_{k+1} = g(x_k)$, $k \geq 0$, converges to ξ as $k \to \infty$, provided that x_0 is sufficiently close to ξ.*

Proof By hypothesis, there exists $h > 0$ such that g' is continuous in the interval $[\xi - h, \xi + h]$. Since $|g'(\xi)| < 1$ we can find a smaller interval $I_\delta = [\xi - \delta, \xi + \delta]$, where $0 < \delta \leq h$, such that $|g'(x)| \leq L$ in this interval, with $L < 1$. To do so, take $L = \frac{1}{2}(1 + |g'(\xi)|)$ and then choose $\delta \leq h$ such that

$$|g'(x) - g'(\xi)| \leq \tfrac{1}{2}(1 - |g'(\xi)|)$$

for all x in I_δ; this is possible since g' is continuous at ξ. Hence,

$$|g'(x)| \leq |g'(x) - g'(\xi)| + |g'(\xi)| \leq \tfrac{1}{2}(1 - |g'(\xi)|) + |g'(\xi)| = L$$

for all $x \in I_\delta$. Now, suppose that x_k lies in the interval I_δ. Then,

$$x_{k+1} - \xi = g(x_k) - \xi = g(x_k) - g(\xi) = (x_k - \xi)g'(\eta_k)$$

by the Mean Value Theorem (Theorem A.3), where η_k lies between x_k and ξ, and therefore also belongs to I_δ. Hence $|g'(\eta_k)| \leq L$, and

$$|x_{k+1} - \xi| \leq L|x_k - \xi|\,. \tag{1.12}$$

This shows that x_{k+1} also lies in I_δ, and a simple argument by induction shows that if x_0 belongs to I_δ, then all x_k, $k \geq 0$, are in I_δ, and also

$$|x_k - \xi| \leq L^k |x_0 - \xi|\,, \qquad k \geq 0\,. \tag{1.13}$$

Since $0 < L < 1$ this implies that the sequence (x_k) converges to ξ. □

[1] If you are familiar with the concept of Lebesgue measure, you will find the following result, known as **Rademacher's Theorem**, revealing. *A function f satisfying the Lipschitz condition (1.5) on an interval $[a, b]$ is differentiable on $[a, b]$, except, perhaps, at the points of a subset of zero Lebesgue measure.*

If the conditions of Theorem 1.5 are satisfied in the vicinity of a fixed point ξ, then the sequence (x_k) defined by the iteration $x_{k+1} = g(x_k)$, $k \geq 0$, will converge to ξ for any starting value x_0 that is sufficiently close to ξ. If, on the other hand, the conditions of Theorem 1.5 are violated, there is no guarantee that any sequence (x_k) defined by the iteration $x_{k+1} = g(x_k)$, $k \geq 0$, will converge to the fixed point ξ for any starting value x_0 near ξ. In order to distinguish between these two cases, we introduce the following definition.

Definition 1.3 *Suppose that g is a real-valued function, defined and continuous on the bounded closed interval $[a, b]$, such that $g(x) \in [a, b]$ for all $x \in [a, b]$, and let ξ denote a fixed point of g. We say that ξ is a* **stable fixed point** *of g, if the sequence (x_k) defined by the iteration $x_{k+1} = g(x_k)$, $k \geq 0$, converges to ξ whenever the starting value x_0 is sufficiently close to ξ. Conversely, if no sequence (x_k) defined by this iteration converges to ξ for any starting value x_0 close to ξ, except for $x_0 = \xi$, then we say that ξ is an* **unstable fixed point** *of g.*

We note that, with this definition, a fixed point may be neither stable nor unstable (see Exercise 2).

As will be demonstrated below in Example 1.5, even some very simple functions may possess both stable and unstable fixed points. Theorem 1.5 shows that if g' is continuous in a neighbourhood of ξ, then the condition $|g'(\xi)| < 1$ is sufficient to ensure that ξ is a stable fixed point. The case of an unstable fixed point will be considered later, in Theorem 1.6.

Now, assuming that ξ is a stable fixed point of g, we may also be interested in the speed at which the sequence (x_k) defined by the iteration $x_{k+1} = g(x_k)$, $k \geq 0$, converges to ξ. Under the hypotheses of Theorem 1.5, it follows from the proof of that theorem that

$$\lim_{k\to\infty} \frac{|x_{k+1} - \xi|}{|x_k - \xi|} = \lim_{k\to\infty} \left| \frac{g(x_k) - g(\xi)}{x_k - \xi} \right| = |g'(\xi)| \,. \tag{1.14}$$

Consequently, we can regard $|g'(\xi)| \in (0, 1)$ as a measure of the speed of convergence of the sequence (x_k) to the fixed point ξ.

Definition 1.4 *Suppose that $\xi = \lim_{k\to\infty} x_k$. We say that the sequence (x_k) converges to ξ* **at least linearly** *if there exist a sequence (ε_k) of positive real numbers converging to 0, and $\mu \in (0, 1)$, such that*

$$|x_k - \xi| \leq \varepsilon_k \,, \quad k = 0, 1, 2, \ldots \,, \qquad \text{and} \qquad \lim_{k\to\infty} \frac{\varepsilon_{k+1}}{\varepsilon_k} = \mu \,. \tag{1.15}$$

If (1.15) holds with $\mu = 0$, then the sequence (x_k) is said to converge to ξ **superlinearly**.

If (1.15) holds with $\mu \in (0,1)$ and $\varepsilon_k = |x_k - \xi|$, $k = 0,1,2,\ldots$, then (x_k) is said to converge to ξ **linearly**, *and the number $\rho = -\log_{10}\mu$ is then called the* **asymptotic rate of convergence** *of the sequence. If (1.15) holds with $\mu = 1$ and $\varepsilon_k = |x_k - \xi|$, $k = 0,1,2,\ldots$, the rate of convergence is slower than linear and we say that the sequence converges to ξ* **sublinearly**.

The words 'at least' in this definition refer to the fact that we only have inequality in $|x_k - \xi| \le \varepsilon_k$, which may be all that can be ascertained in practice. Thus, it is really the sequence of bounds ε_k that converges linearly.

For a linearly convergent sequence the asymptotic rate of convergence ρ measures the number of correct decimal digits gained in one iteration; in particular, the number of iterations required in order to gain one more correct decimal digit is at most $[1/\rho] + 1$. Here $[1/\rho]$ denotes the largest integer that is less than or equal to $1/\rho$.

Under the hypotheses of Theorem 1.5, the equalities (1.14) will hold with $\mu = |g'(\xi)| \in [0,1)$, and therefore the sequence (x_k) generated by the simple iteration will converge to the fixed point ξ linearly or superlinearly.

Example 1.4 *Given that α is a fixed positive real number, consider the function g defined on the interval $[0,1]$ by*

$$g(x) = \begin{cases} 2^{-\left\{1+(\log_2(1/x))^{1/\alpha}\right\}^{\alpha}} & \text{for } 0 < x \le 1\,, \\ 0 & \text{for } x = 0\,. \end{cases}$$

As $\lim_{x\to 0+} g(x) = 0$, the function g is continuous on $[0,1]$. Moreover, g is strictly monotonic increasing on $[0,1]$ and $g(x) \in [0,1/2] \subset [0,1]$ for all x in $[0,1]$. We note that $\xi = 0$ is a fixed point of g (cf. Figure 1.3).

Consider the sequence (x_k) defined by $x_{k+1} = g(x_k)$, $k \ge 0$, with $x_0 = 1$. It is a simple matter to show by induction that $x_k = 2^{-k^{\alpha}}$, $k \ge 0$. Thus we deduce that (x_k) converges to $\xi = 0$ as $k \to \infty$. Since

$$\lim_{k\to\infty}\left|\frac{x_{k+1}}{x_k}\right| = \mu = \begin{cases} 1 & \text{for } 0 < \alpha < 1\,, \\ \frac{1}{2} & \text{for } \alpha = 1\,, \\ 0 & \text{for } \alpha > 1\,, \end{cases}$$

we conclude that for $\alpha \in (0,1)$ the sequence (x_k) converges to $\xi = 0$ sublinearly. For $\alpha = 1$ it converges to $\xi = 0$ linearly with asymptotic rate

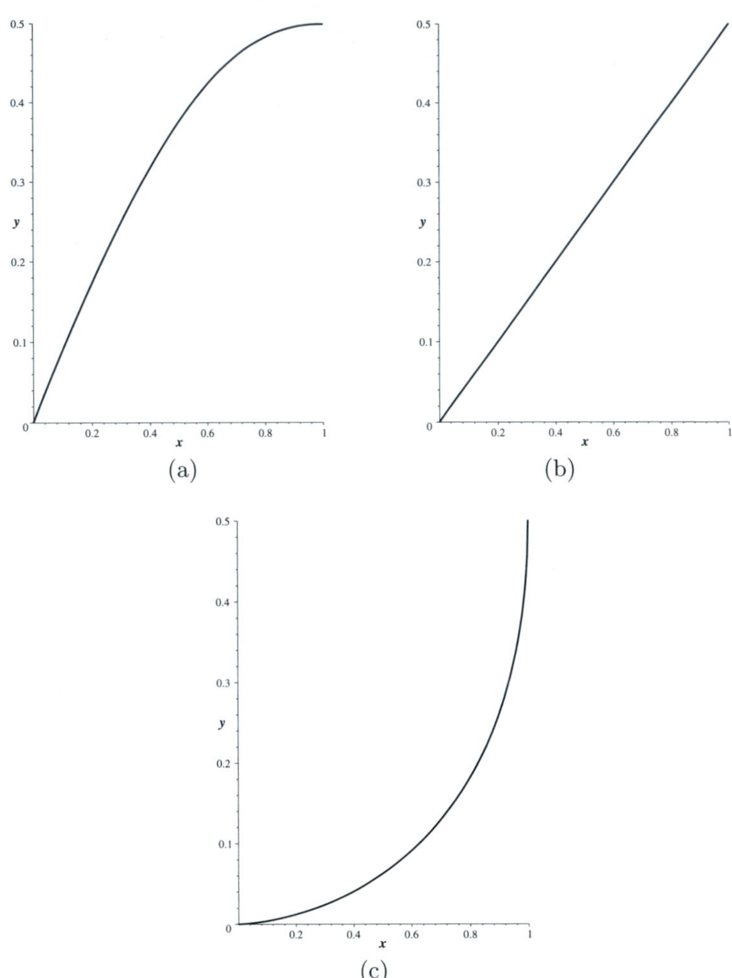

Fig. 1.3. Graph of the function g from Example 1.4 on the interval $x \in [0,1]$ for (a) $\alpha = 1/2$, (b) $\alpha = 1$, (c) $\alpha = 2$.

$\rho = -\log_{10} \mu = \log_{10} 2$. When $\alpha > 1$, the sequence converges to the fixed point $\xi = 0$ superlinearly. The same conclusions could have been reached by showing (through tedious differentiation) that $\lim_{x \to 0+} g'(x) = \mu$, with μ as defined above for the various values of the parameter α. $\diamond$

For a linearly convergent simple iteration $x_{k+1} = g(x_k)$, where g' is continuous in a neighbourhood of the fixed point ξ and $0 < |g'(\xi)| < 1$, Definition 1.4 and (1.14) imply that the asymptotic rate of convergence of the sequence (x_k) is $\rho = -\log_{10}|g'(\xi)|$. Evidently, a small value of $|g'(\xi)|$ corresponds to a large positive value of ρ and will result in more rapid convergence, while if $|g'(\xi)| < 1$ but $|g'(\xi)|$ is very close to 1, ρ will be a small positive number and the sequence will converge very slowly.[1]

Next, we discuss the behaviour of the iteration (1.3) in the vicinity of an *unstable fixed point* ξ. If $|g'(\xi)| > 1$, then the sequence (x_k) defined by (1.3) does not converge to ξ from any starting value $x_0 \neq \xi$; the next theorem gives a rigorous proof of this fact.

Theorem 1.6 *Suppose that $\xi = g(\xi)$, where the function g has a continuous derivative in some neighbourhood of ξ, and let $|g'(\xi)| > 1$. Then, the sequence (x_k) defined by $x_{k+1} = g(x_k)$, $k \geq 0$, does not converge to ξ from any starting value x_0, $x_0 \neq \xi$.*

Proof Suppose that $x_0 \neq \xi$. As in the proof of Theorem 1.5, we can see that there is an interval $I_\delta = [\xi-\delta, \xi+\delta]$, $\delta > 0$, in which $|g'(x)| \geq L > 1$ for some constant L. If x_k lies in this interval, then

$$|x_{k+1} - \xi| = |g(x_k) - g(\xi)| = |(x_k - \xi)\, g'(\eta_k)| \geq L|x_k - \xi|\,,$$

for some η_k between x_k and ξ. If x_{k+1} lies in I_δ the same argument shows that

$$|x_{k+2} - \xi| \geq L|x_{k+1} - \xi| \geq L^2|x_k - \xi|\,,$$

and so on. Evidently, after a finite number of steps some member of the sequence $x_{k+1}, x_{k+2}, x_{k+3}, \dots$ must be outside the interval I_δ, since $L > 1$. Hence there can be no value of $k_0 = k_0(\delta)$ such that $|x_k - \xi| \leq \delta$ for all $k \geq k_0$, and the sequence therefore does not converge to ξ. □

Example 1.5 *In this example we explore the simple iteration (1.3) for g defined by*

$$g(x) = \tfrac{1}{2}(x^2 + c)$$

where $c \in \mathbb{R}$ is a fixed constant.

The fixed points of the function g are the solutions of the quadratic equation $x^2 - 2x + c = 0$, which are $1 \pm \surd(1 - c)$. If $c > 1$ there are no solutions (in the set $\mathbb{R}$ of real numbers, that is!), if $c = 1$ there is one solution in $\mathbb{R}$, and if $c < 1$ there are two.

[1] Thus $0 < \rho \ll 1$ corresponds to slow linear convergence and $\rho \gg 1$ to fast linear convergence. It is for this reason that we defined the asymptotic rate of convergence ρ, for a linearly convergent sequence, as $-\log_{10}\mu$ (or $-\log_{10}|g'(\xi)|$) rather than μ (or $|g'(\xi)|$).

Suppose now that $c < 1$; we denote the solutions by $\xi_1 = 1 - \surd(1-c)$ and $\xi_2 = 1+\surd(1-c)$, so that $\xi_1 < 1 < \xi_2$. We see at once that $g'(x) = x$, so the fixed point ξ_2 is unstable, but that the fixed point ξ_1 is stable provided that $-3 < c < 1$. In fact, it is easy to see that the sequence (x_k) defined by the iteration $x_{k+1} = g(x_k)$, $k \geq 0$, will converge to ξ_1 if the starting value x_0 satisfies $-\xi_2 < x_0 < \xi_2$. (See Exercise 1.) If c is close to 1, $g'(\xi_1)$ will also be close to 1 and convergence will be slow. When $c = 0$, $\xi_1 = 0$ so that convergence is superlinear. This is an example of quadratic convergence which we shall meet later. ◇

The purpose of our next example is to illustrate the concept of asymptotic rate of convergence. According to Definition 1.4, the asymptotic rate of convergence of a sequence describes the relative closeness of successive terms in the sequence to the limit ξ as $k \to \infty$. Of course, for small values of k the sequence may behave in quite a different way, and since in practical computation we are interested in approximating the limit of the sequence by using just a small number of terms, the asymptotic rate of convergence may sometimes give a misleading impression.

Example 1.6 *In this example we study the convergence of the sequences (u_k) and (v_k) defined by*

$$u_{k+1} = g_1(u_k), \qquad k = 0, 1, 2, \ldots, \qquad u_0 = 1\,,$$
$$v_{k+1} = g_2(v_k), \qquad k = 0, 1, 2, \ldots, \qquad v_0 = 1\,,$$

where

$$g_1(x) = 0.99x \qquad \textit{and} \qquad g_2(x) = \frac{x}{(1 + x^{1/10})^{10}}\,.$$

Each of the two functions has a fixed point at $\xi = 0$, and we easily find that $g_1'(0) = 0.99$, $g_2'(0) = 1$. Hence the sequence (u_k) is linearly convergent to zero with asymptotic rate of convergence $\rho = -\log_{10} 0.99 \approx 0.004$, while Theorem 1.5 does not apply to the sequence (v_k). It is quite easy to show by induction that $v_k = (k+1)^{-10}$, so the sequence (v_k) also converges to zero, but since $\lim_{k\to\infty}(v_{k+1}/v_k) = 1$ the convergence is sublinear. This means that, in the limit, (u_k) will converge faster than (v_k). However, this is not what happens for small k, as Table 1.2 shows very clearly.

The sequence (v_k) has converged to zero correct to 6 decimal digits when $k = 4$, and to 10 decimal digits when $k = 10$, at which stage u_k

Table 1.2. *The sequences (u_k) and (v_k) in Example 1.6.*

k	u_k	v_k
0	1.000000	1.000000
1	0.990000	0.000977
2	0.980100	0.000017
3	0.970299	0.000001
4	0.960596	0.000000
5	0.950990	0.000000
6	0.941480	0.000000
7	0.932065	0.000000
8	0.922745	0.000000
9	0.913517	0.000000
10	0.904382	0.000000

is still larger than 0.9. Although (u_k) eventually converges faster than v_k, we find that $u_k = (0.99)^k$ becomes smaller than $v_k = (k+1)^{-10}$ when

$$k > \frac{10}{\ln(1/0.99)} \ln(k+1)\,.$$

This first happens when $k = 9067$, at which point u_k and v_k are both roughly 10^{-40}. In this rather extreme example the concept of asymptotic rate of convergence is not useful, since for any practical purposes (v_k) converges faster than (u_k). $\diamond$

1.3 Iterative solution of equations

In this section we apply the idea of simple iteration to the solution of equations. Given a real-valued continuous function f, we wish to construct a sequence (x_k), using iteration, which converges to a solution of $f(x) = 0$. We begin with an example where it is easy to derive various such sequences; in the next section we shall describe a more general approach.

Example 1.7 *Consider the problem of determining the solutions of the equation $f(x) = 0$, where $f\colon x \mapsto \mathrm{e}^x - x - 2$.*

Since $f'(x) = \mathrm{e}^x - 1$ the function f is monotonic increasing for positive x and monotonic decreasing for negative values of x. Moreover,

$$\left.\begin{aligned} f(1) &= \mathrm{e} - 3 < 0\,, \\ f(2) &= \mathrm{e}^2 - 4 > 0\,, \\ f(-1) &= \mathrm{e}^{-1} - 1 < 0\,, \\ f(-2) &= \mathrm{e}^{-2} > 0\,. \end{aligned}\right\} \tag{1.16}$$

Hence the equation $f(x) = 0$ has exactly one positive solution, which lies in the interval $(1, 2)$, and exactly one negative solution, which lies in the interval $(-2, -1)$. This is illustrated in Figure 1.4, which shows the graphs of the functions $x \mapsto \mathrm{e}^x$ and $x \mapsto x + 2$ on the same axes. We shall write ξ_1 for the positive solution and ξ_2 for the negative solution.

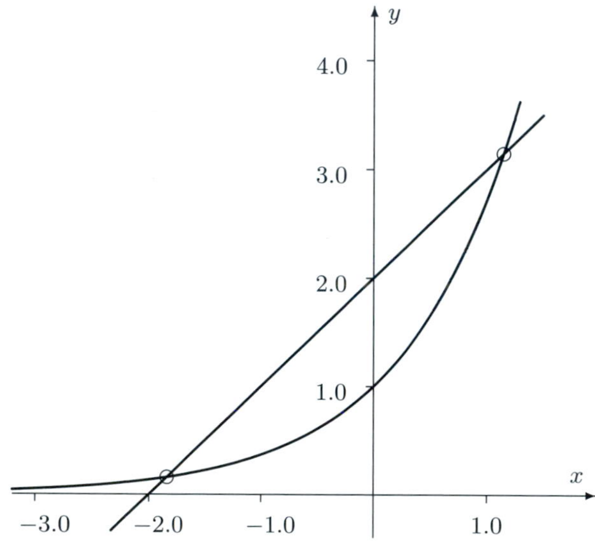

Fig. 1.4. Graphs of $y = \mathrm{e}^x$ and $y = x + 2$.

The equation $f(x) = 0$ may be written in the equivalent form

$$x = \ln(x + 2)\,,$$

which suggests a simple iteration defined by $g(x) = \ln(x + 2)$. We shall show that the positive solution ξ_1 is a stable fixed point of g, while ξ_2 is an unstable fixed point of g.

Clearly, $g'(x) = 1/(x + 2)$, so $0 < g'(\xi_1) < 1$, since ξ_1 is the positive solution. Therefore, by Theorem 1.5, the sequence (x_k) defined by the iteration

$$x_{k+1} = \ln(x_k + 2)\,, \qquad k = 0, 1, 2, \ldots\,, \tag{1.17}$$

will converge to the positive solution, ξ_1, provided that the starting value x_0 is sufficiently close to it.[1] As $0 < g'(\xi_1) < 1/3$, the asymptotic rate of convergence of (x_k) to ξ_1 is certainly greater than $\log_{10} 3$.

On the other hand, $g'(\xi_2) > 1$ since $-2 < \xi_2 < -1$, so the sequence (x_k) defined by (1.17) cannot converge to the solution ξ_2. It is not difficult to prove that for $x_0 > \xi_2$ the sequence (x_k) converges to ξ_1 while if $x_0 < \xi_2$ the sequence will decrease monotonically until $x_k \le -2$ for some k, and then the iteration breaks down as $g(x_k)$ becomes undefined.

The equation $f(x) = 0$ may also be written in the form $x = \mathrm{e}^x - 2$, suggesting the sequence (x_k) defined by the iteration

$$x_{k+1} = \mathrm{e}^{x_k} - 2\,, \qquad k = 0, 1, 2, \dots .$$

In this case $g(x) = \mathrm{e}^x - 2$ and $g'(x) = \mathrm{e}^x$. Hence $g'(\xi_1) > 1$, $g'(\xi_2) < \mathrm{e}^{-1}$, showing that the sequence (x_k) may converge to ξ_2, but cannot converge to ξ_1. It is quite straightforward to show that the sequence converges to ξ_2 for any $x_0 < \xi_1$, but diverges to $+\infty$ when $x_0 > \xi_1$.

As a third alternative, consider rewriting the equation $f(x) = 0$ as $x = g(x)$ where the function g is defined by $g(x) = x(\mathrm{e}^x - x)/2$; the fixed points of the associated iteration $x_{k+1} = g(x_k)$ are the solutions ξ_1 and ξ_2 of $f(x) = 0$, and also the point 0. For this iteration neither of the fixed points, ξ_1 or ξ_2, is stable, and the sequence (x_k) either converges to 0 or diverges to $\pm\infty$.

Evidently the given equation may be written in many different forms, leading to iterations with different properties. ◇

1.4 Relaxation and Newton's method

In the previous section we saw how various ingenious devices lead to iterations which may or may not converge to the desired solutions of a given equation $f(x) = 0$. We would obviously benefit from a more generally applicable iterative method which would, except possibly in special cases, produce a sequence (x_k) that always converges to a required solution. One way of constructing such a sequence is by relaxation.

[1] In fact, by applying the Contraction Mapping Theorem on an arbitrary bounded closed interval $[0, M]$ where $M > \xi_1$, we conclude that the sequence (x_k) defined by the iteration (1.17) will converge to ξ_1 from any positive starting value x_0.

Definition 1.5 *Suppose that f is a real-valued function, defined and continuous in a neighbourhood of a real number ξ.* **Relaxation** *uses the sequence (x_k) defined by*

$$x_{k+1} = x_k - \lambda f(x_k)\,, \qquad k = 0, 1, 2, \ldots\,, \tag{1.18}$$

where $\lambda \neq 0$ is a fixed real number whose choice will be made clear below, and x_0 is a given starting value near ξ.

If the sequence (x_k) defined by (1.18) converges to ξ, then ξ is a solution of the equation $f(x) = 0$, as we assume that f is continuous.

It is clear from (1.18) that relaxation is a simple iteration of the form $x_{k+1} = g(x_k)$, $k = 0, 1, 2, \ldots$, with $g(x) = x - \lambda f(x)$. Suppose now, further, that f is differentiable in a neighbourhood of ξ. It then follows that $g'(x) = 1 - \lambda f'(x)$ for all x in this neighbourhood; hence, if $f(\xi) = 0$ and $f'(\xi) \neq 0$, the sequence (x_k) defined by the iteration $x_{k+1} = g(x_k)$, $k = 0, 1, 2, \ldots$, will converge to ξ if we choose λ to have the same sign as $f'(\xi)$, to be not too large, and take x_0 sufficiently close to ξ. This idea is made more precise in the next theorem.

Theorem 1.7 *Suppose that f is a real-valued function, defined and continuous in a neighbourhood of a real number ξ, and let $f(\xi) = 0$. Suppose further that f' is defined and continuous in some neighbourhood of ξ, and let $f'(\xi) \neq 0$. Then, there exist positive real numbers λ and δ such that the sequence (x_k) defined by the relaxation iteration (1.18) converges to ξ for any x_0 in the interval $[\xi - \delta, \xi + \delta]$.*

Proof Suppose that $f'(\xi) = \alpha$, and that α is positive. If $f'(\xi)$ is negative, the proof is similar, with appropriate changes of sign. Since f' is continuous in some neighbourhood of ξ, we can find a positive real number δ such that $f'(x) \geq \frac{1}{2}\alpha$ in the interval $[\xi - \delta, \xi + \delta]$. Let M be an upper bound for $f'(x)$ in this interval. Hence $M \geq \frac{1}{2}\alpha$. In order to fix the value of the real number λ, we begin by noting that, for any $\lambda > 0$,

$$1 - \lambda M \leq 1 - \lambda f'(x) \leq 1 - \tfrac{1}{2}\lambda\alpha\,, \qquad x \in [\xi - \delta, \xi + \delta]\,.$$

We now choose λ so that these extreme values are equal and opposite, *i.e.*, $1 - \lambda M = -\vartheta$ and $1 - \frac{1}{2}\lambda\alpha = \vartheta$ for a suitable nonnegative real number ϑ. There is a unique value of ϑ for which this holds; it is given by the formula

$$\vartheta = \frac{2M - \alpha}{2M + \alpha}\,,$$

corresponding to

$$\lambda = \frac{4}{2M + \alpha}\,.$$

On defining $g(x) = x - \lambda f(x)$, we then deduce that

$$|g'(x)| \le \vartheta < 1\,, \qquad x \in [\xi - \delta, \xi + \delta]\,. \tag{1.19}$$

Thus we can apply Theorem 1.5 to conclude that the sequence (x_k) defined by the relaxation iteration (1.18) converges to ξ, provided that x_0 is in the interval $[\xi - \delta, \xi + \delta]$. The asymptotic rate of convergence of the relaxation iteration (1.18) to ξ is at least $-\log_{10}\vartheta$. □

We can now extend the idea of relaxation by allowing λ to be a continuous function of x in a neighbourhood of ξ rather than just a constant. This suggests an iteration

$$x_{k+1} = x_k - \lambda(x_k) f(x_k)\,, \qquad k = 0, 1, 2, \ldots\,,$$

corresponding to a simple iteration with $g(x) = x - \lambda(x)f(x)$. If the sequence (x_k) converges, the limit ξ will be a solution of $f(x) = 0$, except possibly when $\lambda(\xi) = 0$. Moreover, as we have seen, the asymptotic rate of convergence is determined by $g'(\xi)$. Since $f(\xi) = 0$, it follows that $g'(\xi) = 1 - \lambda(\xi)f'(\xi)$, and (1.19) suggests using a function λ which makes $1 - \lambda(\xi)f'(\xi)$ small. The obvious choice is $\lambda(x) = 1/f'(x)$, and leads us to Newton's method.[1]

Definition 1.6 *Newton's method for the solution of $f(x) = 0$ is defined by*

$$x_{k+1} = x_k - \frac{f(x_k)}{f'(x_k)}\,, \qquad k = 0, 1, 2, \ldots\,, \tag{1.20}$$

with prescribed starting value x_0. We implicitly assume in the defining formula (1.20) that $f'(x_k) \neq 0$ for all $k \geq 0$.

[1] Isaac Newton was born on 4 January 1643 in Woolsthorpe, Lincolnshire, England and died on 31 March 1727 in London, England. According to the calendar used in England at the time, Newton was born on Christmas day 1642, and died on 21 March 1727: the Gregorian calendar was not adopted in England until 1752. Newton made revolutionary advances in mathematics, physics, astronomy and optics; his contributions to the foundations of calculus were marred by priority disputes with Leibniz. Newton was appointed to the Lucasian chair at Cambridge at the age of 27. In 1705, two years after becoming president of the Royal Society (a position to which he was re-elected each year until his death), Newton was knighted by Queen Anne; he was the first scientist to be honoured in this way. Newton's *Philosophiae naturalis principia mathematica* is one of the most important scientific books ever written.

Newton's method is a simple iteration with $g(x) = x - f(x)/f'(x)$. Its geometric interpretation is illustrated in Figure 1.5: the tangent to the curve $y = f(x)$ at the point $(x_k, f(x_k))$ is the line with the equation $y - f(x_k) = f'(x_k)(x - x_k)$; it meets the x-axis at the point $(x_{k+1}, 0)$.

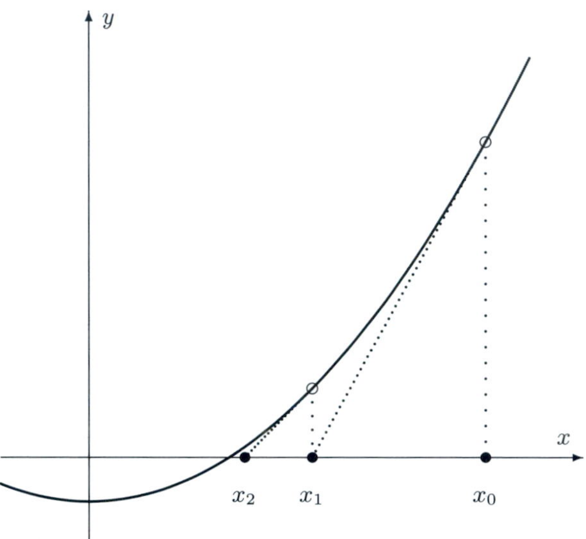

Fig. 1.5. Newton's method.

We could apply Theorem 1.5 to prove the convergence of this iteration, but since generally it converges much faster than ordinary relaxation it is better to apply a special form of proof. First, however, we give a formal definition of quadratic convergence.

Definition 1.7 *Suppose that* $\xi = \lim_{k\to\infty} x_k$. *We say that the sequence* (x_k) *converges to* ξ **with at least order** $q > 1$, *if there exist a sequence* (ε_k) *of positive real numbers converging to* 0, *and* $\mu > 0$, *such that*

$$|x_k - \xi| \le \varepsilon_k\,, \quad k = 0, 1, 2, \ldots, \qquad \textit{and} \qquad \lim_{k\to\infty} \frac{\varepsilon_{k+1}}{\varepsilon_k^q} = \mu\,. \tag{1.21}$$

If (1.21) holds with $\varepsilon_k = |x_k - \xi|$ *for* $k = 0, 1, 2, \ldots$, *then the sequence* (x_k) *is said to converge to* ξ **with order** q. *In particular, if* $q = 2$, *then we say that the sequence* (x_k) *converges to* ξ **quadratically**.

We note that unlike the definition of linear convergence where μ was required to belong to the interval $(0, 1)$, all we demand here is that $\mu > 0$. The reason is simple: when $q > 1$, (1.21) implies suitably rapid decay of the sequence (ε_k) irrespective of the size of μ.

Example 1.8 *Let $c > 1$ and $q > 1$. The sequence (x_k) defined by $x_k = c^{-q^k}$, $k = 0, 1, 2, \ldots$, converges to 0 with order q.*

Theorem 1.8 (Convergence of Newton's method) *Suppose that f is a continuous real-valued function with continuous second derivative f'', defined on the closed interval $I_\delta = [\xi - \delta, \xi + \delta]$, $\delta > 0$, such that $f(\xi) = 0$ and $f''(\xi) \neq 0$. Suppose further that there exists a positive constant A such that*

$$\frac{|f''(x)|}{|f'(y)|} \leq A \qquad \forall\, x, y \in I_\delta\,.$$

If $|\xi - x_0| \leq h$, where h is the smaller of δ and $1/A$, then the sequence (x_k) defined by Newton's method (1.20) converges quadratically to ξ.

Proof Suppose that $|\xi - x_k| \leq h = \min\{\delta, 1/A\}$, so that $x_k \in I_\delta$. Then, by Taylor's Theorem (Theorem A.4), expanding about the point $x_k \in I_\delta$,

$$0 = f(\xi) = f(x_k) + (\xi - x_k) f'(x_k) + \frac{(\xi - x_k)^2}{2} f''(\eta_k)\,, \tag{1.22}$$

for some η_k between ξ and x_k, and therefore in the interval I_δ. Recalling (1.20), this shows that

$$\xi - x_{k+1} = -\frac{(\xi - x_k)^2 f''(\eta_k)}{2f'(x_k)}\,. \tag{1.23}$$

Since $|\xi - x_k| \leq \frac{1}{A}$, we have $|\xi - x_{k+1}| \leq \frac{1}{2}|\xi - x_k|$. As we are given that $|\xi - x_0| \leq h$ it follows by induction that $|\xi - x_k| \leq 2^{-k} h$ for all $k \geq 0$; hence (x_k) converges to ξ as $k \to \infty$.

Now, η_k lies between ξ and x_k, and therefore (η_k) also converges to ξ as $k \to \infty$. Since f' and f'' are continuous on I_δ, it follows from (1.23) that

$$\lim_{k \to \infty} \frac{|x_{k+1} - \xi|}{|x_k - \xi|^2} = \left| \frac{f''(\xi)}{2f'(\xi)} \right|\,, \tag{1.24}$$

which, according to Definition 1.7, implies quadratic convergence of the sequence (x_k) to ξ with $\mu = |f''(\xi)/2f'(\xi)|$, $\mu \in (0, A/2]$. □

The conditions of the theorem implicitly require that $f'(\xi) \neq 0$, for otherwise the quantity $f''(x)/f'(y)$ could not be bounded in a neighbourhood of ξ. (See Exercises 6 and 7 for what happens when $f'(\xi) = 0$.)

One can show that if $f''(\xi) = 0$ and we assume that $f(x)$ has a continuous third derivative, and require certain quantities to be bounded, then the convergence is *cubic* (*i.e.*, convergence with order $q = 3$).

It is possible to demonstrate that Newton's method converges over a wider interval, if we assume something about the signs of the derivatives.

Theorem 1.9 *Suppose that the function f satisfies the conditions of Theorem 1.8 and also that there exists a real number X, $X > \xi$, such that in the interval $J = [\xi, X]$ both f' and f'' are positive. Then, the sequence (x_k) defined by Newton's method (1.20) converges quadratically to ξ from any starting value x_0 in J.*

Proof It follows from (1.23) that if $x_k \in J$, then $x_{k+1} > \xi$. Moreover, since $f'(x) > 0$ on J, f is monotonic increasing on J. As $f(\xi) = 0$, it then follows that $f(x) > 0$ for $\xi < x \leq X$. Hence, $\xi < x_{k+1} < x_k$, $k \geq 0$. Since the sequence (x_k) is bounded and monotonic decreasing, it is convergent; let $\eta = \lim_{k\to\infty} x_k$. Clearly, $\eta \in J$. Further, passing to the limit $k \to \infty$ in (1.20) we have that $f(\eta) = 0$. However, ξ is the only solution of $f(x) = 0$ in J, so $\eta = \xi$, and the sequence converges to ξ.

Having shown that the sequence (x_k) converges, the fact that it converges quadratically follows as in the proof of Theorem 1.8. □

We remark that the same result holds for other possible signs of f' and f'' in a suitable interval J. (See Exercise 8.) The interval J does not have to be bounded; considering, for instance, $f(x) = \mathrm{e}^x - x - 2$ from Example 1.7, it is clear that $f'(x)$ and $f''(x)$ are both positive in the unbounded interval $(0, \infty)$, and the Newton iteration converges to the positive solution of the equation $f(x) = 0$ from any positive starting value x_0.

Note that the definition of quadratic convergence only refers to the behaviour of the sequence for sufficiently large k. In the same example we find that the convergence of the Newton iteration from a large positive value of x_0 is initially very slow. (See Exercise 3.) The possibility of this early behaviour is often emphasised by saying that the convergence of Newton's method is *ultimately* quadratic.

1.5 The secant method

So far we have considered iterations which can be written in the form $x_{k+1} = g(x_k)$, $k \geq 0$, so that the new value is expressed in terms of the old one. It is also possible to define an iteration of the form $x_{k+1} = g(x_k, x_{k-1})$, $k \geq 1$, where the new value is expressed in terms of two previous values. In particular, we shall consider two applications of this idea, leading to the secant method and the method of bisection, respectively.

Remark 1.3 *We note in passing that one can consider more general iterative methods of the form*

$$x_{k+1} = g(x_k, x_{k-1}, \ldots, x_{k-\ell}), \qquad k = \ell, \ell+1, \ldots,$$

with $\ell \geq 1$ fixed; here, we shall confine ourselves to the simplest case when $\ell = 1$ as this is already sufficiently illuminating.

Using Newton's method to solve a nonlinear equation $f(x) = 0$ requires explicit knowledge of the first derivative f' of the function f. Unfortunately, in many practical situations f' is not explicitly available or it can only be obtained at high computational cost. In such cases, the value $f'(x_k)$ in (1.20) can be approximated by a difference quotient; that is,

$$f'(x_k) \approx \frac{f(x_k) - f(x_{k-1})}{x_k - x_{k-1}}.$$

Replacing $f'(x_k)$ in (1.20) by this difference quotient leads us to the following definition.

Definition 1.8 *The* **secant method** *is defined by*

$$x_{k+1} = x_k - f(x_k)\left(\frac{x_k - x_{k-1}}{f(x_k) - f(x_{k-1})}\right), \qquad k = 1, 2, 3, \ldots, \qquad (1.25)$$

where x_0 and x_1 are given starting values. It is implicitly assumed here that $f(x_k) - f(x_{k-1}) \neq 0$ for all $k \geq 1$.

The method is illustrated in Figure 1.6. The new iterate x_{k+1} is obtained from x_{k-1} and x_k by drawing the chord joining the points $P(x_{k-1}, f(x_{k-1}))$ and $Q(x_k, f(x_k))$, and using as x_{k+1} the point at which this chord intersects the x-axis. If x_{k-1} and x_k are close together and f

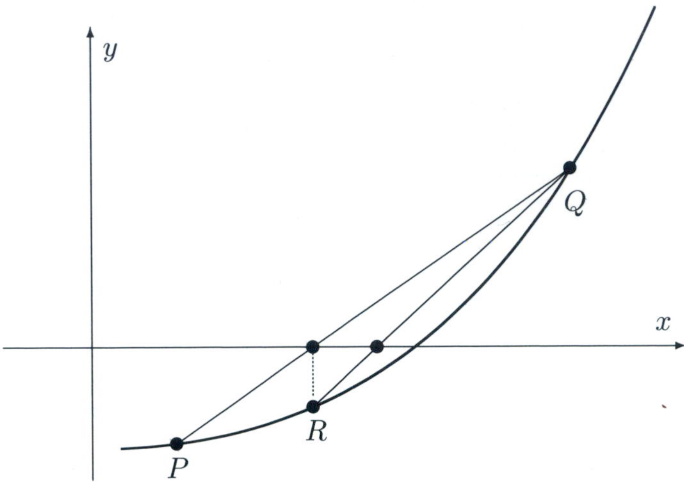

Fig. 1.6. Secant method.

is differentiable, x_{k+1} is approximately the same as the value supplied by Newton's method, which uses the tangent at the point Q.

Theorem 1.10 *Suppose that f is a real-valued function, defined and continuously differentiable on an interval $I = [\xi - h, \xi + h]$, $h > 0$, with centre point ξ. Suppose further that $f(\xi) = 0$, $f'(\xi) \neq 0$. Then, the sequence (x_k) defined by the secant method (1.25) converges at least linearly to ξ provided that x_0 and x_1 are sufficiently close to ξ.*

Proof Since $f'(\xi) \neq 0$, we may suppose that $f'(\xi) = \alpha > 0$; only minor changes are needed in the proof when $f'(\xi)$ is negative. Since f' is continuous on I, corresponding to any $\varepsilon > 0$ we can choose an interval $I_\delta = [\xi - \delta, \xi + \delta]$, with $0 < \delta \leq h$, such that

$$|f'(x) - \alpha| < \varepsilon \,, \qquad x \in I_\delta \,. \tag{1.26}$$

Choosing $\varepsilon = \frac{1}{4}\alpha$ we see that

$$0 < \tfrac{3}{4}\alpha < f'(x) < \tfrac{5}{4}\alpha \,, \qquad x \in I_\delta \,. \tag{1.27}$$

From (1.25) and using the Mean Value Theorem (Theorem A.3) together with the fact that $f(\xi) = 0$, we obtain

$$\xi - x_{k+1} = \xi - x_k + \frac{(x_k - \xi) f'(\vartheta_k)}{f'(\varphi_k)} \,, \tag{1.28}$$

Table 1.3. *Comparison of the secant method and Newton's method for the solution of* $e^x - x - 2 = 0$.

	Secant method	Newton's method
0	1.000000	1.000000
1	3.000000	1.163953
2	1.036665	1.146421
3	1.064489	1.146193
4	1.153299	1.146193
5	1.145745	
6	1.146191	
7	1.146193	

where ϑ_k is between x_k and ξ, and φ_k lies between x_k and x_{k-1}. Hence, if $x_{k-1} \in I_\delta$ and $x_k \in I_\delta$, then also $\vartheta_k \in I_\delta$ and $\varphi_k \in I_\delta$. Therefore,

$$|\xi - x_{k+1}| \leq |\xi - x_k| \left| 1 - \frac{5\alpha/4}{3\alpha/4} \right| = \tfrac{2}{3}|\xi - x_k|\,. \tag{1.29}$$

Thus, $x_{k+1} \in I_\delta$ and the sequence (x_k) converges to ξ at least linearly, with rate at least $\log_{10}(3/2)$, provided that $x_0 \in I_\delta$ and $x_1 \in I_\delta$. □

In fact, it can be shown that

$$\lim_{k\to\infty} \frac{|x_{k+1} - \xi|}{|x_k - \xi|^q} = \mu \tag{1.30}$$

where μ is a positive constant and $q = \frac{1}{2}(1 + \sqrt{5}) \approx 1.6$, so that the convergence of the sequence (x_k) to ξ is faster than linear, but not as fast as quadratic. (See Exercise 10.)

This is illustrated in Table 1.3, which compares two iterative methods for the solution of $f(x) = 0$ with $f\colon x \mapsto e^x - x - 2$; the first is the secant method, starting from $x_0 = 1$, $x_1 = 3$, while the second is Newton's method starting from $x_0 = 1$.

This experiment shows the faster convergence of Newton's method, but it must be remembered that each iteration of Newton's method requires the calculation of both $f(x_k)$ and $f'(x_k)$, while each iteration of the secant method requires the calculation of $f(x_k)$ only (as $f(x_{k-1})$ has already been computed). In our examples the computations are quite trivial, but in a practical situation the calculation of each value of $f(x_k)$ and $f'(x_k)$ may demand a substantial amount of work, and then each iteration of Newton's method is likely to involve at least twice as much work as one iteration of the secant method.

1.6 The bisection method

Suppose that f is a real-valued function defined and continuous on a bounded closed interval $[a, b]$ of the real line and such that $f(\xi) = 0$ for some $\xi \in (a, b)$. A very simple iterative method for the solution of the nonlinear equation $f(x) = 0$ can be constructed by beginning with an interval $[a_0, b_0]$ which is known to contain the required solution ξ (*e.g.*, one may choose $[a_0, b_0]$ as the interval $[a, b]$ itself, with $a_0 = a$ and $b_0 = b$), and successively halving its size.

More precisely, we proceed as follows. Let $k \geq 0$, and suppose that it is known that $f(a_k)$ and $f(b_k)$ have opposite signs; we then conclude from Theorem 1.1 that the interval (a_k, b_k) contains a solution of $f(x) = 0$. Consider the midpoint c_k of the interval (a_k, b_k) defined by

$$c_k = \tfrac{1}{2}(a_k + b_k)\,, \tag{1.31}$$

and evaluate $f(c_k)$. If $f(c_k)$ is zero, then we have located a solution ξ of $f(x) = 0$, and the iteration stops. Else, we define the new interval (a_{k+1}, b_{k+1}) by

$$(a_{k+1}, b_{k+1}) = \begin{cases} (a_k, c_k) & \text{if } f(c_k)f(b_k) > 0\,, \\ (c_k, b_k) & \text{if } f(c_k)f(b_k) < 0\,, \end{cases} \tag{1.32}$$

and repeat this procedure.

This may at first seem to be a very crude method, but it has some important advantages. The analysis of convergence is trivial; the size of the interval containing ξ is halved at each iteration, so the sequence (c_k) defined by the bisection method converges linearly, with rate $\rho = \log_{10} 2$. Even Newton's method may often converge more slowly than this in the early stages, when the starting value is far from the desired solution. Moreover, the convergence analysis assumes only that the function f is continuous, and requires no bounds on the derivatives, nor even their existence.[1] Once we can find an interval $[a_0, b_0]$ such that $f(a_0)$ and $f(b_0)$ have opposite signs, we can guarantee convergence to a solution, and that after k iterations the solution ξ will lie in an interval of length

[1] Consider, for example, solving the equation $f(x) = 0$, where the function f is defined by (1.2). Even though f is not differentiable at the point $x = 1.05$, the bisection method is applicable. It has to be noted, however, that for functions of this kind it is not always easy to find an interval $[a_0, b_0]$ in which f changes sign.

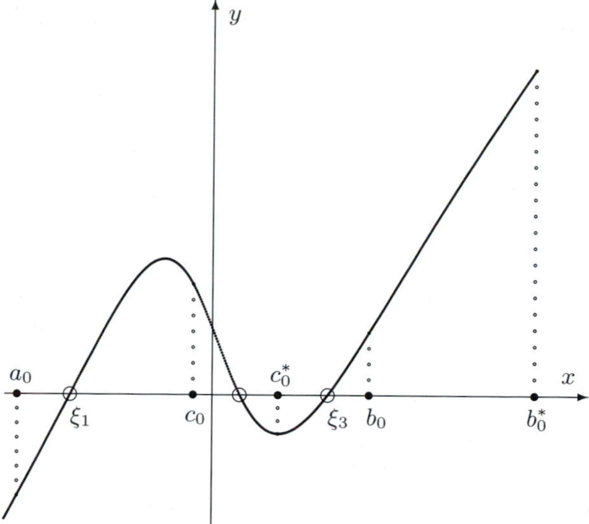

Fig. 1.7. Bisection; from the initial interval $[a_0, b_0]$ the next interval is $[a_0, c_0]$, but starting from $[a_0, b_0^*]$ the next interval is $[c_0^*, b_0^*]$.

$(b_0 - a_0)/2^k$. The bisection method is therefore very robust, though Newton's method will always win once the current iterate is sufficiently close to ξ.

If the initial interval $[a_0, b_0]$ contains more than one solution, the limit of the bisection method will depend on the positions of these solutions. Figure 1.7 illustrates a possible situation, where $[a_0, b_0]$ contains three solutions. Since $f(c_0)$ has the same sign as $f(b_0)$ the second interval is $[a_0, c_0]$, and the sequence (c_k) of midpoints defined by (1.31) converges to the solution ξ_1. If however the initial interval is $[a_0, b_0^*]$ the sequence of midpoints converges to the solution ξ_3.

1.7 Global behaviour

We have already seen how an iteration will often converge to a limit if the starting value is sufficiently close to that limit. The behaviour of the iteration, when started from an arbitrary starting value, can be very complicated. In this section we shall consider two examples. No theorems will be stated: our aim is simply to illustrate various kinds of behaviour.

First consider the simple iteration defined by

$$x_{k+1} = g(x_k)\,, \quad k = 0, 1, 2, \ldots\,, \qquad \text{where } g(x) = a\,x(1-x)\,, \quad (1.33)$$

which is often known as the **logistic equation**. We require the constant a to lie in the range $0 < a \leq 4$, for then if the starting value x_0 is in the interval $[0, 1]$, then all members of the sequence (x_k) also lie in $[0, 1]$. The function g has two fixed points: $x = 0$ and $x = 1 - 1/a$. The fixed point at 0 is stable if $0 < a < 1$, and the fixed point at $1 - 1/a$ is stable if $1 < a < 3$. The behaviour of the iteration for these values of a is what might be expected from this information, but for larger values of the parameter a the behaviour of the sequence (x_k) becomes increasingly complicated.

For example, when $a = 3.4$ there is no stable fixed point, and from any starting point the sequence eventually oscillates between two values, which are 0.45 and 0.84 to two decimal digits. These are the two stable fixed points of the double iteration

$$x_{k+1} = g^*(x_k)\,, \qquad g^*(x) = g(g(x)) = a^2x(1-x)[1-ax(1-x)]\,. \quad (1.34)$$

When $3 < a < 1 + \sqrt{6}$, the fixed points of g^* are the two fixed points of g, that is 0 and $1 - 1/a$, and also

$$\frac{1}{2}\left(1 + \frac{1}{a} \pm \frac{1}{a}\left[a^2 - 2a - 3\right]^{1/2}\right). \quad (1.35)$$

This behaviour is known as a stable two-cycle (see Exercise 12).

When $a > 1 + \sqrt{6}$ all the fixed points of g^* are unstable. For example, when $a = 3.5$ all sequences (x_k) defined by (1.33) tend to a stable 4-cycle, taking successive values 0.50, 0.87, 0.38 and 0.83.

For larger values of the parameter a the sequences become chaotic. For example, when $a = 3.99$ there are no stable fixed points or limit-cycles, and the members of any sequence appear random. In fact it can be shown that for such values of a the members of the sequence are *dense* in a subinterval of $[0, 1]$: there exist real numbers α and β, $\alpha < \beta$, such that any subinterval of (α, β), however small, contains an infinite subsequence of (x_k). For the value $a = 3.99$ the maximal interval (α, β) is $(0.00995, 0.99750)$ to five decimal digits. Starting from $x_0 = 0.75$ we find that the interval $(0.70, 0.71)$, for example, contains the subsequence

$$x_{16}, x_{164}, x_{454}, x_{801}, x_{812}, \ldots\,. \quad (1.36)$$

The sequence does not show any apparent regular behaviour. The calculation is extremely sensitive: if we replace x_0 by $x_0 + \delta x_0$, and write

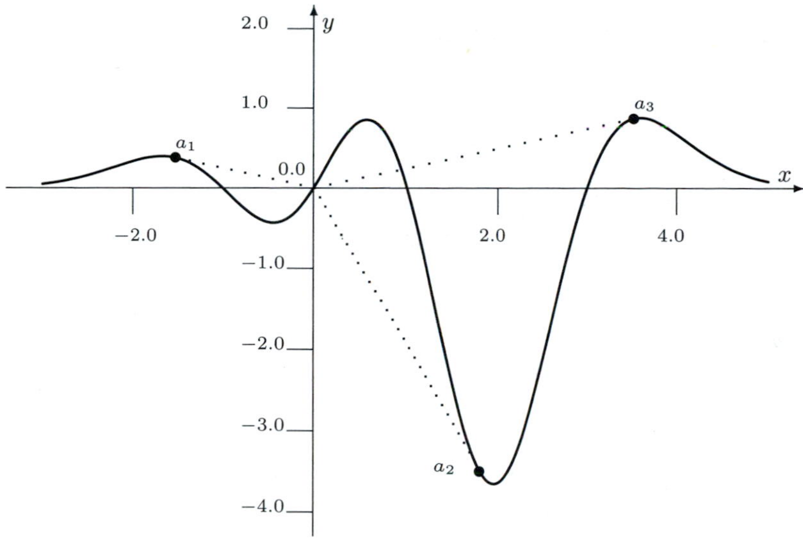

Fig. 1.8. Global behaviour of Newton's method.

$x_k + \delta x_k$ for the resulting perturbed value of x_k, it is easy to see that

$$\delta x_{k+1} = a(1 - 2x_k)\delta x_k\,,$$

provided that the changes δx_k are so small that $a(\delta x_k)^2$ can be ignored. With $x_0 = 0.75$ as above we find from the same calculation that $\delta x_{812}/\delta x_0$ is about 10^{231}, so that to determine x_{812} with reasonable accuracy it is necessary to carry through the whole calculation using 250 decimal digits.

Our second example, of more practical importance, is of Newton's method applied to a function f with several zeros. The example is

$$f(x) = x(x^2 - 1)(x - 3)\exp(-\tfrac{1}{2}(x - 1)^2)\,; \qquad (1.37)$$

the graph of the function is shown in Figure 1.8. The function has zeros at -1, 0, 1 and 3. The sequence generated by the Newton iteration will converge to one of these solutions if the starting value is fairly close to it. Moreover, the geometric interpretation of the iteration shows that if the starting point is sufficiently large in absolute value the iteration diverges rapidly to ∞; the iteration behaves as if the function had a zero at infinity, and the sequence can be loosely described as 'converging to ∞'. With this interpretation some numerical experimentation soon shows that from any starting value Newton's method eventually converges to a solution, which might be $\pm\infty$. However, it is certainly *not* true that the sequence converges to the solution closest to the starting point; indeed, if this were true, no sequence could converge to ∞. It is easy to see why the behaviour is much more complicated than this.

The Newton iteration converges to the solution at 0 from any point in the interval $(-0.327, 0.445)$. As we see from Figure 1.8, the iteration will converge exactly to 0 in one iteration if we start from the x-coordinate of any of the points a_1, a_2 and a_3; at each of these three points the tangent to the curve passes through the origin. Since f is continuous, this means that there is an open interval surrounding each of these points from which the Newton iteration will converge to 0. The maximal such intervals are $(-1.555, -1.487)$, $(1.735, 1.817)$ and $(3.514, 3.529)$ to three decimal digits. In the same way, there are several points at which the tangent to the curve passes through the point $(A_1, 0)$, where A_1 is the x-coordinate of the point a_1. Starting from one of these points, the Newton iteration will evidently converge exactly to the solution at 0 in two steps; surrounding each of these points there is an open interval from which the iteration will converge to 0.

Now suppose we define the sets S_m, $m = -1, 0, 1, 3, \infty, -\infty$, where S_m consists of those points from which the Newton iteration converges to the zero at m. Then, an extension of the above argument shows that each of the sets S_m is the union of an infinite number of disjoint open intervals. The remarkable property of these sets is that, if ξ is a boundary point of one of the sets S_m, then it is also a boundary point of all the other sets as well. This means that any neighbourhood of such a point ξ, however small, contains an infinite number of members of each of the sets S_m. For example, we have seen that the iteration starting from any point in the interval $(-0.327, 0.445)$ converges to 0. We find that the end of this interval lies between 0.4457855 and 0.4457860; Table 1.4 shows the limits of various Newton iterations starting from points near this boundary. Each of these points is, of course, itself surrounded by an open interval which gives the same limit.

1.8 Notes

Theorem 1.2 is a special case of Brouwer's Fixed Point Theorem. Luitzen Egbertus Jan Brouwer (1881–1966) was professor of set theory, function theory and axiomatics at the University of Amsterdam, and made major contributions to topology. Brouwer was a mathematical genius with

Table 1.4. *Limit of Newton's method near a boundary point.*

x_0	Limit
0.4457840	0
0.4457845	0
0.4457850	0
0.4457855	0
0.4457860	1
0.4457865	$-\infty$
0.4457870	-1
0.4457875	-1
0.4457880	$-\infty$
0.4457885	$-\infty$
0.4457890	$+\infty$
0.4457895	3
0.4457900	1

strong mystical and philosophical leanings. For an historical overview of Brouwer's life and work we refer to the recent book of Dirk Van Dalen, *Mystic, Geometer, and Intuitionist. The Life of L.E.J. Brouwer: the Dawning Revolution*, Clarendon Press, Oxford, 1999.

The Contraction Mapping Theorem, as stated here, is a simplified version of Banach's fixed point theorem. Stefan Banach[1] founded modern functional analysis and made outstanding contributions to the theory of topological vector spaces, measure theory, integration, the theory of sets, and orthogonal series. For an inspiring account of Banach's life and times, see R. Kaluza, *Through the Eyes of a Reporter: the Life of Stefan Banach*, Birkhäuser, Boston, MA, 1996.

In our definitions of linear convergence and convergence with order q, we followed Definitions 2.1 and 2.2 in Chapter 4 of

- WALTER GAUTSCHI, *Numerical Analysis: an Introduction*, Birkhäuser, Boston, MA, 1997.

Exciting surveys of the history of Newton's method are available in T. Ypma, Historical development of the Newton–Raphson method, *SIAM Rev.* **37**, 531–551, 1995, H. Goldstine, *History of Numerical Analysis from the Sixteenth through the Nineteenth Century*, Springer, New York, 1977; and in Chapter 6 of Jean-Luc Chabert (Editor), *A History of Algorithms from the Pebble to the Microchip*, Springer, New York, 1999. As

[1] 30 March 1892, Kraków, Austria–Hungary (now in Poland) – 31 August 1945, Lvov, Ukraine, USSR (now independent).

is noted in these sources, Newton's *De analysi per aequationes numero terminorum infinitas*, probably dating from mid-1669, is sometimes regarded as the historical source of the method, despite the fact that, surprisingly, there is no trace in this tract of the familiar recurrence relation $x_{k+1} = x_k - f(x_k)/f'(x_k)$ bearing Newton's name, nor is there a mention of the idea of derivative. Instead, the paper contains an example of a cubic polynomial whose roots are found by purely algebraic and rather complicated substitutions. In 1690, Joseph Raphson (1648–1715) in the Preface to his *Analysis aequationum universalis* describes his version of Newton's method as 'not only, I believe, not of the same origin, but also, certainly, not with the same development' as Newton's method. Further improvements to the method, and its form as we know it today, were given by Thomas Simpson in his *Essays in Mathematicks* (1740). Simpson presents it as 'a new method for the solution of equations' using the 'method of fluxions', *i.e.*, derivatives. It is argued in Ypma's article that Simpson's contributions to this subject have been underestimated, and 'it would seem that the Newton–Raphson–Simpson method is a designation more nearly representing facts of history of this method which lurks inside millions of modern computer programs and is printed with Newton's name attached in so many textbooks'.

The convergence analysis of Newton's method was initiated in the first half of the twentieth century by L.V. Kantorovich.[1] More recently, Smale,[2] Dedieu and Shub,[3] and others have provided significant insight into the properties of Newton's method. A full discussion of the global behaviour of the logistic equation (1.33), and other examples, will be found in P.G. Drazin, *Nonlinear Systems*, Cambridge University Press, Cambridge, 1992, particularly Chapters 1 and 3.

The secant method is also due to Newton (cf. Section 3 of Ypma's paper cited above), and is found in a collection of unpublished notes termed 'Newton's Waste Book' written around 1665.

In this chapter, we have been concerned with the iterative solution of equations for a real-valued function of a single real variable. In Chapter 4, we shall discuss the iterative solution of nonlinear systems of equations

[1] L.V. Kantorovich, Functional analysis and applied mathematics, *Uspekhi Mat. Nauk* **3**, 89–185, 1948; English transl., Rep. 1509, National Bureau of Standards, Washington, DC, 1952.

[2] Steve Smale, Newton's method estimates from data at one point, in *The Merging of Disciplines: New Directions in Pure, Applied and Computational Mathematics*, R. Ewing, K. Gross, C. Martin, Eds., Springer, New York, 185–196, 1986.

[3] Jean-Pierre Dedieu and Michael Shub, Multihomogeneous Newton methods, *Math. Comput.* **69** (231), 1071–1098, 2000.

of the form $\boldsymbol{f}(\boldsymbol{x}) = \mathbf{0}$ where $\boldsymbol{f}\colon \mathbb{R}^n \to \mathbb{R}^n$. There, corresponding to the case of $n = 2$, we shall say more about the solution of equations of the form $f(z) = 0$ where f is a complex-valued function of a single complex variable z.

This chapter has been confined to generally applicable iterative methods for the solution of a single nonlinear equation of the form $f(x) = 0$ for a real-valued function f of a single real variable. In particular, we have not discussed specialised methods for the solution of polynomial equations or the various techniques for locating the roots of polynomials in the complex plane and on the real line (by Budan and Fourier, Descartes, Hurwitz, Lobachevskii, Newton, Schur and others), although in Chapter 5 we shall briefly touch on one such polynomial root-finding method due to Sturm.[1] For a historical survey of the solution of polynomial equations and a review of recent advances in this field, we refer to the article of Victor Pan, Solving a polynomial equation: some history and recent progress, *SIAM Rev.* **39**, 187–220, 1997.

Exercises

1.1 The iteration defined by $x_{k+1} = \frac{1}{2}(x_k^2 + c)$, where $0 < c < 1$, has two fixed points ξ_1, ξ_2, where $0 < \xi_1 < 1 < \xi_2$. Show that

$$x_{k+1} - \xi_1 = \tfrac{1}{2}(x_k + \xi_1)(x_k - \xi_1)\,, \qquad k = 0, 1, 2, \ldots\,,$$

and deduce that $\lim_{k\to\infty} x_k = \xi_1$ if $0 \le x_0 < \xi_2$. How does the iteration behave for other values of x_0?

1.2 Define the function g by $g(0) = 0$, $g(x) = -x \sin^2(1/x)$ for $0 < x \le 1$. Show that g is continuous, and that 0 is the only fixed point of g in the interval $[0, 1]$. By considering the iteration $x_{n+1} = g(x_n)$, $n = 0, 1, 2, \ldots$, starting, first from $x_0 = 1/(k\pi)$, and then from $x_0 = 2/((2k+1)\pi)$, where k is an integer, show that according to Definition 1.3 the critical point is neither stable nor unstable.

1.3 Newton's method is applied to the solution of

$$\mathrm{e}^x - x - 2 = 0\,.$$

[1] For further details in this direction, we refer to M.A. Jenkins and J.F. Traub, A three-stage algorithm for real polynomials using quadratic iterations, *SIAM J. Numer. Anal.* **7**, 545–566, 1970, A.S. Householder, *The Numerical Treatment of a Single Nonlinear Equation,* McGraw–Hill, New York, 1970, and A. Ralston and P. Rabinowitz, *A First Course in Numerical Analysis*, Second Edition, McGraw–Hill, New York, 1978.

Show that if the starting value is positive, the iteration converges to the positive solution, and if the starting value is negative it converges to the negative solution. Obtain approximate expressions for x_1 if (i) $x_0 = 100$ and (ii) $x_0 = -100$, and describe the subsequent behaviour of the iteration. About how many iterations would be required to obtain the solution to six decimal digits in these two cases?

1.4 Consider the iteration

$$x_{k+1} = x_k - \frac{[f(x_k)]^2}{f(x_k + f(x_k)) - f(x_k)}, \qquad k = 0, 1, 2, \dots,$$

for the solution of $f(x) = 0$. Explain the connection with Newton's method, and show that (x_k) converges quadratically if x_0 is sufficiently close to the solution. Apply this method to the same example as in Example 1.7, $f(x) = \mathrm{e}^x - x - 2$, and verify quadratic convergence beginning from $x_0 = 1$. Experiment with calculations beginning from $x_0 = 10$ and from $x_0 = -10$, and account for their behaviour.

1.5 It is sometimes said that Newton's method converges quadratically, and therefore in the successive approximations to the solution the number of correct digits doubles each time. Explain why this is not generally correct. Suppose that $f''(x)$ is defined and continuous in a neighbourhood of ξ and that x_k agrees with the solution ξ to m decimal digits; give an estimate of the number of correct decimal digits in x_{k+1}.

Illustrate your estimate by using Newton's method to determine the positive zero of $f(x) = \mathrm{e}^x - x - 1.000000005$, which is close to 0.0001; use $x_0 = 0.0005$.

1.6 Suppose that $f(\xi) = f'(\xi) = 0$, $f''(\xi) \neq 0$, so that f has a double root at ξ, and that f'' is defined and continuous in a neighbourhood of ξ. If (x_k) is a sequence obtained by Newton's method, show that

$$\xi - x_{k+1} = -\tfrac{1}{2}\frac{(\xi - x_k)^2 f''(\eta_k)}{f'(x_k)} = \tfrac{1}{2}(\xi - x_k)\frac{f''(\eta_k)}{f''(\chi_k)},$$

where η_k and χ_k both lie between ξ and x_k. Suppose, further, that $0 < m < |f''(x)| < M$ for all x in the interval $[\xi - \delta, \xi + \delta]$ for some $\delta > 0$, where $M < 2m$; show that if x_0 lies in this interval the iteration converges to ξ, and that convergence is linear, with rate $\log_{10} 2$. Verify this conclusion by finding the solution of $e^x = 1 + x$, beginning from $x_0 = 1$.

1.7 Extend the result of the previous exercise to a case where f has a triple root at ξ, so that $f(\xi) = f'(\xi) = f''(\xi) = 0$, $f'''(\xi) \neq 0$.

1.8 Suppose that the function f has a continuous second derivative, that $f(\xi) = 0$, and that in the interval $[X, \xi]$, with $X < \xi$, $f'(x) > 0$ and $f''(x) < 0$. Show that the Newton iteration, starting from any x_0 in $[X, \xi]$, converges to ξ.

1.9 The secant method is used to determine solutions of the equation $x^2 - 1 = 0$. Starting from $x_0 = 1 + \varepsilon$, $x_1 = -1 + \varepsilon$, show that $x_2 = \frac{1}{2}\varepsilon + \mathcal{O}(\varepsilon^2)$, and determine x_3, x_4 and x_5, neglecting terms of order $\mathcal{O}(\varepsilon^2)$. Explain why, at least for sufficiently small values of ε, the sequence (x_k) converges to the solution -1.

Repeat the calculation with x_0 and x_1 interchanged, so that $x_0 = -1 + \varepsilon$ and $x_1 = 1 + \varepsilon$, and show that the sequence now converges to the solution 1.

1.10 Write the secant iteration in the form

$$x_{k+1} = \frac{x_k\, f(x_{k-1}) - x_{k-1}\, f(x_k)}{f(x_{k-1}) - f(x_k)}, \qquad k = 1, 2, 3, \ldots .$$

Supposing that f has a continuous second derivative in a neighbourhood of the solution ξ of $f(x) = 0$, and that $f'(\xi) > 0$ and $f''(\xi) > 0$, define

$$\varphi(x_k, x_{k-1}) = \frac{x_{k+1} - \xi}{(x_k - \xi)(x_{k-1} - \xi)},$$

where x_{k+1} has been expressed in terms of x_k and x_{k-1}. Find an expression for

$$\psi(x_{k-1}) = \lim_{x_k \to \xi} \varphi(x_k, x_{k-1}),$$

and then determine $\lim_{x_{k-1} \to \xi} \psi(x_{k-1})$. Deduce that

$$\lim_{x_k, x_{k-1} \to \xi} \varphi(x_k, x_{k-1}) = f''(\xi)/2f'(\xi).$$

Now assume that

$$\lim_{k \to \infty} \frac{|x_{k+1} - \xi|}{|x_k - \xi|^q} = A.$$

Show that $q - 1 - 1/q = 0$, and hence that $q = \frac{1}{2}(1+\sqrt{5})$. Deduce finally that

$$\lim_{k\to\infty} \frac{|x_{k+1} - \xi|}{|x_k - \xi|^q} = \left(\frac{f''(\xi)}{2f'(\xi)}\right)^{q/(1+q)} .$$

1.11 A variant of the secant method defines two sequences (u_k) and (v_k) such that all the values $f(u_k)$, $k = 0, 1, 2, \ldots$, have one sign, and all the values $f(v_k)$, $k = 0, 1, 2, \ldots$, have the opposite sign. From the numbers u_k and v_k the secant formula is used to define

$$w_k = \frac{u_k f(v_k) - v_k f(u_k)}{f(v_k) - f(u_k)}, \qquad k = 0, 1, 2, \ldots ;$$

we define $u_{k+1} = w_k$, $v_{k+1} = v_k$ if $f(w_k)$ has the same sign as $f(u_k)$, and otherwise $u_{k+1} = u_k$, $v_{k+1} = w_k$. Suppose that f'' is defined and continuous on the interval $[u_0, v_0]$, and that, for some K, f'' has constant sign in $[u_K, v_K]$. Explain, graphically or otherwise, why either $u_k = u_K$ for all $k \geq K$, or $v_k = v_K$ for all $k \geq K$. Deduce that the method converges linearly, and determine the asymptotic rate of convergence; explain clearly what you mean by convergence of this method. What advantages, if any, do you think this method has compared with the secant method of Definition 1.8?

1.12 A *two-cycle* of the iteration defined by the function g is a pair of distinct numbers a, b such that $b = g(a)$ and $a = g(b)$. Use the fact that a and b are fixed points of the iteration defined by the function $h(x) = g(g(x))$ to give a definition of *stability* for a two-cycle. Show that if $|g'(a)\, g'(b)| < 1$, then the two-cycle is stable, and that if $|g'(a)\, g'(b)| > 1$ the two-cycle is not stable.

Show that if a, b is a two-cycle for Newton's method for the function f, and if $|f(a)f(b)f''(a)f''(b)| < [f'(a)f'(b)]^2$, then the two-cycle is stable.

Show that Newton's method for the solution of $f(x) = 0$ with

$$f\colon x \mapsto x(x^2 - 1)$$

has a two-cycle of the form $a, -a$, and find the value of a; is this two-cycle stable?

2

Solution of systems of linear equations

2.1 Introduction

In Chapter 1 we considered the solution of a single equation of the form $f(x) = 0$ where f is a real-valued function defined and continuous on a closed interval of the real line. The simplest example of this kind is the linear equation $ax = b$ where a and b are given real numbers, with $a \neq 0$, whose solution is

$$x = a^{-1}b\,, \tag{2.1}$$

trivially. Of course, we could have expressed the solution as $x = b/a$ as in Chapter 1, but, as you will see in a moment, writing $x = a^{-1}b$ is much more revealing in the present context. In this chapter we shall consider a different generalisation of this elementary problem:

> Let A be an $n \times n$ matrix with a_{ij} as its entry in row i and column j
> and $\boldsymbol{b}$ a given column vector of size n with jth entry b_j;
> find a column vector $\boldsymbol{x}$ of size n such that $A\boldsymbol{x} = \boldsymbol{b}$.

Denoting by x_i the ith entry of the vector $\boldsymbol{x}$, we can also write $A\boldsymbol{x} = \boldsymbol{b}$ in the following expanded form:

$$\left.\begin{array}{l} a_{11}x_1 + a_{12}x_2 + \cdots + a_{1n}x_n = b_1\,, \\ a_{21}x_1 + a_{22}x_2 + \cdots + a_{2n}x_n = b_2\,, \\ \cdots\cdots\cdots\cdots\cdots\cdots\cdots\cdots\cdots\cdots \\ a_{n1}x_1 + a_{n2}x_2 + \cdots + a_{nn}x_n = b_n\,. \end{array}\right\} \tag{2.2}$$

Recall that in order to ensure that for real numbers a and b the single linear equation $ax = b$ has a unique solution, we need to assume that $a \neq 0$. In the case of the simultaneous system (2.2) of n linear equations in n unknowns we shall have to make an analogous assumption on the matrix A.

To do so, we introduce the following definition.

Definition 2.1 *The set of all* $m \times n$ *matrices with real entries is denoted by* $\mathbb{R}^{m\times n}$. *A matrix of size* $n \times n$ *will be called a square matrix of order* n, *or simply a matrix of* **order** n. *The* **determinant** *of a square matrix* $A \in \mathbb{R}^{n\times n}$ *is the real number* $\det(A)$ *defined as follows:*

$$\det(A) = \sum_{\text{perm}} \text{sign}(\nu_1, \nu_2, \ldots, \nu_n) a_{1\nu_1} a_{2\nu_2} \ldots a_{n\nu_n} \, .$$

The summation is over all $n!$ *permutations* $(\nu_1, \nu_2, \ldots, \nu_n)$ *of the integers* $1, 2, \ldots, n$, *and* $\text{sign}(\nu_1, \nu_2, \ldots, \nu_n) = +1$ *or* -1 *depending on whether the* n*-tuple* $(\nu_1, \nu_2, \ldots, \nu_n)$ *is an even or odd permutation of* $(1, 2, \ldots, n)$, *respectively. An even (odd) permutation is obtained by an even (odd) number of exchanges of two adjacent elements in the array* $(1, 2, \ldots, n)$. *A matrix* $A \in \mathbb{R}^{n\times n}$ *is said to be* **nonsingular** *when its determinant* $\det(A)$ *is nonzero.*

The **inverse matrix** A^{-1} of a nonsingular matrix $A \in \mathbb{R}^{n\times n}$ is defined as the element of $\mathbb{R}^{n\times n}$ such that $A^{-1}A = AA^{-1} = I$, where I is the $n \times n$ identity matrix

$$I = \begin{pmatrix} 1 & 0 & \ldots & 0 \\ 0 & 1 & \ldots & 0 \\ \ldots & \ldots & \ldots & \ldots \\ 0 & 0 & \ldots & 1 \end{pmatrix} . \tag{2.3}$$

In order to find an explicit expression for A^{-1} in terms of the elements of the matrix A, we recall from linear algebra that, for each $i = 1, 2, \ldots, n$,

$$a_{i1}A_{k1} + a_{i2}A_{k2} + \cdots + a_{in}A_{kn} = \begin{cases} \det(A) & \text{if } i = k\,, \\ 0 & \text{if } i \neq k\,, \end{cases} \tag{2.4}$$

where $A_{ij} = (-1)^{i+j}\text{Cof}(a_{ij})$ and $\text{Cof}(a_{ij})$, called the **cofactor** of a_{ij}, is the determinant of the $(n-1) \times (n-1)$ matrix obtained by erasing from $A \in \mathbb{R}^{n\times n}$ row i and column j. Then, it is a trivial matter to show using (2.4) that A^{-1} has the form

$$A^{-1} = \frac{1}{\det(\text{A})} \begin{pmatrix} A_{11} & A_{21} & \ldots & A_{n1} \\ A_{12} & A_{22} & \ldots & A_{n2} \\ \ldots & \ldots & \ldots & \ldots \\ A_{1n} & A_{2n} & \ldots & A_{nn} \end{pmatrix} . \tag{2.5}$$

Having found an explicit formula for the matrix A^{-1}, we now multiply both sides of the equation $A\boldsymbol{x} = \boldsymbol{b}$ on the left by A^{-1} to deduce that

$A^{-1}(A\boldsymbol{x}) = A^{-1}\boldsymbol{b}$; finally, since $A^{-1}(A\boldsymbol{x}) = (A^{-1}A)\boldsymbol{x} = I\boldsymbol{x} = \boldsymbol{x}$, it follows that

$$\boldsymbol{x} = A^{-1}\boldsymbol{b}\,, \tag{2.6}$$

where the inverse A^{-1} of the nonsingular matrix A is given in terms of the entries of A by (2.5).[1]

An alternative approach to the solution of the linear system $A\boldsymbol{x} = \boldsymbol{b}$, called Cramer's rule, proceeds by expressing the ith entry of $\boldsymbol{x}$ as

$$x_i = D_i/D\,, \qquad i = 1, 2, \ldots, n\,,$$

where $D = \det(A)$, and D_i is the $n \times n$ determinant obtained by replacing the ith column of D by the entries of $\boldsymbol{b}$. Evidently, we must require that A is nonsingular, *i.e.*, that $D = \det(A) \neq 0$. Thus, all we need to do to solve $A\boldsymbol{x} = \boldsymbol{b}$ is to evaluate the $n + 1$ determinants $D, D_1, \ldots, D_n$, each of them $n \times n$, and check that $D = \det(A)$ is nonzero; the final calculation of the elements $x_i, i = 1, 2, \ldots, n$, is then trivial.[2]

The purpose of our next example is to illustrate the application of Cramer's rule.

Example 2.1 *Suppose that we wish to solve the system of linear equations*

$$\begin{aligned} x_1 + x_2 + x_3 &= 6\,, \\ 2x_1 + 4x_2 + 2x_3 &= 16\,, \\ -x_1 + 5x_2 - 4x_3 &= -3\,. \end{aligned}$$

The solution of such a small system can easily be found in terms of determinants, by Cramer's rule. This gives

$$x_1 = D_1/D\,, \quad x_2 = D_2/D\,, \quad x_3 = D_3/D\,,$$

[1] By the way, on comparing (2.6) with (2.1) you will notice that (2.1) is a special case of (2.6) when $n = 1$.

[2] Gabriel Cramer (31 July 1704, Geneva, Switzerland – 4 January 1752, Bagnols-sur-Cèze, France). In the 1730s Colin Maclaurin (February 1698, Kilmodan, Cowal, Argyllshire, Scotland – 14 June 1746, Edinburgh, Scotland) wrote his *Treatise of Algebra* which was not published until 1748, two years after his death. It contained the first published results on determinants proving Cramer's rule for 2×2 and 3×3 systems and indicating how the 4×4 case would work. Cramer gave the general rule for $n \times n$ systems without proof in the Appendix to his paper 'Introduction to the analysis of algebraic curves' (1750), motivated by the desire to find the equation of a plane curve passing through a number of given points.

where

$$D = \begin{vmatrix} 1 & 1 & 1 \\ 2 & 4 & 2 \\ -1 & 5 & -4 \end{vmatrix}, \qquad D_1 = \begin{vmatrix} 6 & 1 & 1 \\ 16 & 4 & 2 \\ -3 & 5 & -4 \end{vmatrix},$$

with similar expressions for D_2 and D_3. To obtain the solution we therefore need to evaluate four determinants. ◇

Now you may think that since, for $A \in \mathbb{R}^{n\times n}$ nonsingular, we have expressed the solution to $A\boldsymbol{x} = \boldsymbol{b}$ in the 'closed form'

$$\boldsymbol{x} = A^{-1}\boldsymbol{b}$$

and have even found a formula for A^{-1} in terms of the coefficients of A, or may simply compute the entries of $\boldsymbol{x}$ directly using Cramer's rule, the story about the simultaneous set of linear equations (2.2) has reached its happy ending. We are sorry to disappoint you: a disturbing tale is about to unfold.

Imagine the following example: let $n = 100$, say, and suppose that you have been given all 10000 entries of a 100×100 matrix A, together with the entries of a 100-component column vector $\boldsymbol{b}$. To avoid trivialities, let us suppose that none of the entries of A or $\boldsymbol{b}$ is equal to 0. Question: *Does the linear system $A\boldsymbol{x} = \boldsymbol{b}$ have a solution? If it does, how would you find, say, the 53rd entry of the solution vector $\boldsymbol{x}$?* Of course, you could calculate the determinant of A and check whether it is equal to zero; if not, you could then calculate the determinant D_{53} obtained by replacing the 53rd column of A by the vector $\boldsymbol{b}$, and the required result, by Cramer's rule, is then the ratio of these two determinants. How much time do you think you would need to accomplish this task? An hour? A day? A month?

I imagine that you do not have a large enough sheet of paper in front of you to write down this 100×100 matrix. Let us therefore start with a somewhat simpler setting. Assume that n is any integer, $n \geq 2$, and denote by d_n the number of arithmetic operations that are required to calculate $\det(A)$ for $A \in \mathbb{R}^{n\times n}$. For example, for a 2×2 matrix,

$$\det(A) = a_{11}a_{22} - a_{12}a_{21}\,;$$

this evaluation requires 3 arithmetic operations – 2 multiplications and 1 subtraction – giving $d_2 = 3$. In general, we can calculate $\det(A)$ by expanding it in the elements of its first row. This requires multiplying each of the n elements in the first row of A by a subdeterminant of size

$n-1$ (a total of $n(d_{n-1}+1)$ operations) and summing the n resulting numbers (another $n-1$ operations). Thus,

$$d_n = n(d_{n-1}+1) + n - 1\,, \quad n \geq 3\,, \qquad d_2 = 3\,. \tag{2.7}$$

Let us write $d_n = c_n n!$ and substitute this into (2.7) to obtain

$$c_n = c_{n-1} + 2\frac{1}{(n-1)!} - \frac{1}{n!}\,, \quad n \geq 3\,, \qquad c_2 = \frac{3}{2}\,. \tag{2.8}$$

Now, summing (2.8) from $n=3$ to k for $k \geq 3$ yields, on letting $0! = 1$,

$$c_k = \sum_{n=0}^{k-1} \frac{1}{n!} - \frac{1}{k!}.$$

As $\sum_{n=0}^{\infty}(1/n!) = \mathrm{e}$, it follows that

$$\lim_{k\to\infty} c_k = \mathrm{e}\,.$$

Thus,[1] $d_n \sim \mathrm{e}\,n!$ as $n \to \infty$. In order to compute the solution of a system of n simultaneous linear equations by Cramer's rule we need to evaluate $n+1$ determinants, each of size $n \times n$, so the total number of operations required is about $(n+1)d_n \sim \mathrm{e}\,(n+1)!$ as $n \to \infty$.

For $n = 100$, this means approximately $101!\,\mathrm{e} \approx 2.56\times 10^{160}$ arithmetic operations.[2] Today's fastest parallel computers are capable of teraflop speeds, *i.e.*, 10^{12} floating point operations per second; therefore, the computing time for our solution would be around $2.56 \times 10^{160}/10^{12} = 2.56 \times 10^{148}$ seconds, or a staggering 8.11×10^{140} years. According to the prevailing theoretical position, the Universe began in a violent explosion, the Big Bang, about $12.5(\pm 3) \times 10^{9}$ years ago. So please put that large sheet of paper away quickly! We need to discover a more efficient approach.

Incidentally, you might notice that in the expansion of all the determinants involved in Cramer's rule all the smaller subdeterminants occur many times over, so the number of operations involved can be reduced by avoiding such repetitions. However, a more careful analysis shows

[1] For two sequences (a_n) and (b_n), we shall write $a_n \sim b_n$ if $\lim_{n\to\infty}(a_n/b_n) = 1$.

[2] While on the subject of calculating factorials of large integers, let us mention **Stirling's formula** which states that $n! \sim \sqrt{2\pi}n^{n+1/2}\mathrm{e}^{-n}$ as $n \to \infty$ (J. Stirling, *Methodus differentialis*, 1730). Stirling's approximation can be made more precise as the double inequality

$$\sqrt{2\pi}n^{n+1/2}\mathrm{e}^{-n+1/(12n+1)} < n! < \sqrt{2\pi}n^{n+1/2}\mathrm{e}^{-n+1/(12n)}$$

(H. Robbins, A remark on Stirling's formula *Amer. Math. Monthly* **62**, 26–29, 1955).

that we cannot by this means reduce the total by more than a factor of about n, which hardly affects our conclusion.

Our other approach to solving $A\boldsymbol{x} = \boldsymbol{b}$, based on computing A^{-1} from (2.5) and writing $\boldsymbol{x} = A^{-1}\boldsymbol{b}$, is equally inefficient: in order to compute the inverse of an $n \times n$ matrix A using determinants, one has to calculate the determinant of A as well as n^2 determinants of size $n-1$ each of which then has to be divided by $\det(A)$, requiring a total of approximately

$$\mathrm{e}\,n! + n^2\mathrm{e}\,(n-1)! + n^2 \sim \mathrm{e}\,(n+1)!$$

arithmetic operations, just the same as before.

The aim of this chapter is to develop alternative methods for the solution of the system of linear equations $A\boldsymbol{x} = \boldsymbol{b}$. We begin by considering a classical technique, Gaussian elimination.[1] We shall then explore its relationship to the factorisation $A = LU$ of the matrix A where L is lower triangular and U is upper triangular. It will be seen that by using the Gaussian elimination the number of arithmetic operations required to solve the linear system $A\boldsymbol{x} = \boldsymbol{b}$ with an $n \times n$ matrix A is approximately $\frac{2}{3}n^3$ – a dramatic reduction from the $\mathcal{O}(\mathrm{e}\,(n+1)!)$ operation count associated with matrix inversion using determinants.[2]

We conclude the chapter with a discussion of another classical idea attributed to Gauss:[3] the least squares method for the solution of the system of linear equations $A\boldsymbol{x} = \boldsymbol{b}$ where $A \in \mathbb{R}^{m\times n}$, $\boldsymbol{x}$ is the column vector of unknowns of size n and $\boldsymbol{b}$ a given column vector of size m.

2.2 Gaussian elimination

The technique for solving systems of linear algebraic equations that we shall describe in this section was developed by Carl Friedrich Gauss and was first published in his *Theoria motus corporum coelestium in sectionibus conicis solem ambientium* (1809), a major two-volume treatise on the motion of celestial bodies. Gauss was concerned with the study of

[1] Carl Friedrich Gauss (30 April 1777, Brunswick, Duchy of Brunswick, Holy Roman Empire (now Germany) – 23 February 1855, Göttingen, Hanover, Germany) made outstanding contributions to mathematics, physics and astronomy. He gave the first proof, in 1799, of the Fundamental Theorem of Algebra. Gauss worked in differential geometry, number theory, algebra and non-Euclidean geometry.

[2] Note, for example, that $\frac{2}{3}100^3 \approx 0.67 \times 10^6 \ll 101!\,\mathrm{e} \approx 2.56 \times 10^{160}$. On a computer that performs 10^{12} floating operations a second a calculation requiring 10^6 operations *via* Gaussian elimination would take 10^{-6} seconds, as opposed to the 8.11×10^{140} years using Cramer's rule or formula (2.5).

[3] See, however, the bibliographical notes at the end of the chapter about the priority dispute between Legendre and Gauss.

the asteroid Pallas, and derived a set of six linear equations with six unknowns, also giving a systematic method for its solution.

The method proceeds by successively eliminating the elements below the diagonal of the matrix of the linear system until the matrix becomes triangular, when the solution of the system is very easy. This technique is now known under the name **Gaussian elimination**.[1]

Before we embark on the general description of Gaussian elimination, let us illustrate its basic steps through a simple example; this is the same as Example 2.1 above, written out again for convenience.

Example 2.2 *Consider the system of linear equations*

$$\begin{aligned} x_1 + \ x_2 + \ x_3 &= \ 6\,, \\ 2x_1 + 4x_2 + 2x_3 &= 16\,, \\ -x_1 + 5x_2 - 4x_3 &= -3\,. \end{aligned}$$

It is convenient to rewrite this in the form $A\boldsymbol{x} = \boldsymbol{b}$ where $A \in \mathbb{R}^{3\times 3}$ and $\boldsymbol{x}$ and $\boldsymbol{b}$ are column vectors of size 3; thus,

$$\begin{pmatrix} 1 & 1 & 1 \\ 2 & 4 & 2 \\ -1 & 5 & -4 \end{pmatrix} \begin{pmatrix} x_1 \\ x_2 \\ x_3 \end{pmatrix} = \begin{pmatrix} 6 \\ 16 \\ -3 \end{pmatrix}. \tag{2.9}$$

We begin by adding the first row, multiplied by -2, to the second row, and adding the first row to the third row, giving the new system

$$\begin{pmatrix} 1 & 1 & 1 \\ \mathit{0} & 2 & 0 \\ \mathit{0} & 6 & -3 \end{pmatrix} \begin{pmatrix} x_1 \\ x_2 \\ x_3 \end{pmatrix} = \begin{pmatrix} 6 \\ 4 \\ 3 \end{pmatrix}. \tag{2.10}$$

The newly created 0 entries in the first column have been typeset in italics. Now adding the new second row, multiplied by -3, to the third row, we find

$$\begin{pmatrix} 1 & 1 & 1 \\ \mathit{0} & 2 & 0 \\ \mathit{0} & \mathit{0} & -3 \end{pmatrix} \begin{pmatrix} x_1 \\ x_2 \\ x_3 \end{pmatrix} = \begin{pmatrix} 6 \\ 4 \\ -9 \end{pmatrix}, \tag{2.11}$$

[1] The idea of this elimination process was already known to the Chinese two thousand years ago. The book *Jiu zhang suan shu* (English translation, by K. Shen *et al.*: *The Nine Chapters on the Mathematical Art*, Oxford University Press, 1999) contained an example of the elimination for a system of five equations with five unknowns. This book was very influential in the history of Chinese mathematics, and is the earliest specialised mathematical work in China that survived to the present day. Although it is unclear when its mathematical content was produced, it is estimated that the book was assembled during the Han dynasty in the first century AD.

which can easily be solved for the unknowns in the reverse order, beginning with $x_3 = 3$. ◇

Each of these successive row operations can be expressed as a multiplication on the left of the matrix $A \in \mathbb{R}^{n\times n}$, $n \geq 2$ (in our example $n = 3$), of the system of linear equations by a transformation matrix. Writing $E^{(rs)}$ for the $n \times n$ matrix whose only nonzero element is $e_{rs} = 1$, we see that the product

$$(I + \mu_{rs}E^{(rs)})A \tag{2.12}$$

is the same as the original matrix A, except that the elements of row s, multiplied by a real number μ_{rs}, have been added to the corresponding elements of row r. Here I denotes the $n \times n$ identity matrix defined by (2.3). In the elimination process we always add a multiple of an earlier row to a later row in the matrix, so that $1 \leq s < r \leq n$ in (2.12); the transformation matrix $I + \mu_{rs}E^{(rs)}$ is therefore lower triangular in the following sense.

Definition 2.2 *Let n be an integer, $n \geq 2$. The matrix $L \in \mathbb{R}^{n\times n}$ is said to be* **lower triangular** *if $l_{ij} = 0$ for every i and j with $1 \leq i < j \leq n$. The matrix $L \in \mathbb{R}^{n\times n}$ is called* **unit lower triangular** *if it is lower triangular, and also the diagonal elements are all equal to unity, that is $l_{ii} = 1$ for $i = 1, 2, \ldots, n$.*

Thus the matrix $I + \mu_{rs}E^{(rs)} \in \mathbb{R}^{n\times n}$ appearing in (2.12) is unit lower triangular if $1 \leq s < r \leq n$, and the above elimination process can be expressed by multiplying A on the left successively by the unit lower triangular matrices $I + \mu_{rs}E^{(rs)}$ for $r = s+1, \ldots, n$ and $s = 1, \ldots, n-1$, with $\mu_{rs} \in \mathbb{R}$; there are $\frac{1}{2}n(n-1)$ of these matrices, one for each element of A below the diagonal (since there are n elements on the diagonal and, therefore, $1 + \cdots + (n-1) = \frac{1}{2}(n^2 - n)$ elements below the diagonal). The next theorem lists the technical tools which are required for proving that the resulting product is a lower triangular matrix.

Theorem 2.1 *The following statements hold for any integer $n \geq 2$:*

(i) *the product of two lower triangular matrices of order n is lower triangular of order n;*

(ii) *the product of two unit lower triangular matrices of order n is unit lower triangular of order n;*

(iii) *a lower triangular matrix is nonsingular if, and only if, all the diagonal elements are nonzero; in particular, a unit lower triangular matrix is nonsingular;*

(iv) *the inverse of a nonsingular lower triangular matrix of order n is lower triangular of order n;*

(v) *the inverse of a unit lower triangular matrix of order n is unit lower triangular of order n.*

Proof The proofs of parts (i), (ii), (iii) and (v) are very straightforward, and are left as an exercise.

Part (iv) is proved by induction; it is easily verified for a nonsingular lower triangular matrix of order 2, using (2.5). Let $n > 2$, suppose that (iv) is true for all nonsingular lower triangular matrices of order k, with $2 \le k < n$, and let L be a nonsingular lower triangular matrix of order $k+1$. Both L and its inverse L^{-1} can be partitioned by their last row and column:

$$L = \begin{pmatrix} L_1 & \mathbf{0} \\ \boldsymbol{r}^{\mathrm{T}} & \alpha \end{pmatrix}, \qquad L^{-1} = \begin{pmatrix} X & \boldsymbol{y} \\ \boldsymbol{z}^{\mathrm{T}} & \beta \end{pmatrix},$$

where L_1 is a nonsingular lower triangular matrix of order k and $X \in \mathbb{R}^{k\times k}$; α and β are real numbers and $\boldsymbol{r}$, $\boldsymbol{z}$ and $\boldsymbol{y}$ are column vectors of size k. Since the product LL^{-1} is the identity matrix of order $k+1$, we have

$$L_1 X = I_k\,, \quad L_1\boldsymbol{y} = \mathbf{0}\,, \quad \boldsymbol{r}^{\mathrm{T}} X + \alpha\boldsymbol{z}^{\mathrm{T}} = \mathbf{0}^{\mathrm{T}}\,, \quad \boldsymbol{r}^{\mathrm{T}}\boldsymbol{y} + \alpha\beta = 1\,;$$

here I_k signifies the identity matrix of order k. Thus $X = L_1^{-1}$, which is lower triangular of order k by the inductive hypothesis, and $\boldsymbol{y} = \mathbf{0}$ given that L_1 is nonsingular; the remaining two equations determine $\boldsymbol{z}$ and β on noting that $\alpha \neq 0$ (given that L is nonsingular). This shows that L^{-1} is lower triangular of order $k+1$, and the inductive step is complete; consequently, (iv) is true for any $n \ge 2$. □

We shall also require the concept of upper triangular matrix.

Definition 2.3 *Let n be an integer, $n \ge 2$. The matrix $U \in \mathbb{R}^{n\times n}$ is said to be* **upper triangular** *if $u_{ij} = 0$ for every i and j with $1 \le j < i \le n$.*

We note that results analogous to those in the preceding theorem concerning lower triangular matrices are also valid for upper triangular matrices (replacing the words 'lower triangular' by 'upper triangular' throughout).

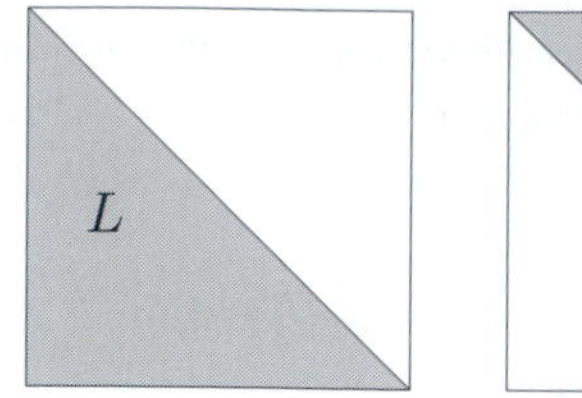

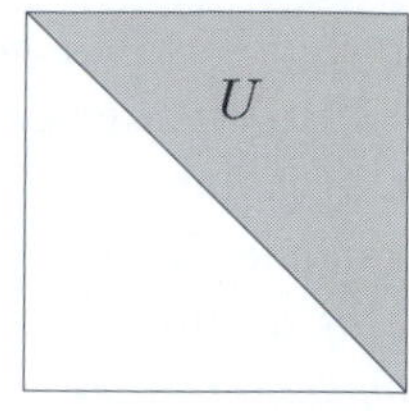

Fig. 2.1. LU factorisation of $A \in \mathbb{R}^{n\times n}$: $A = LU$. The matrix $L \in \mathbb{R}^{n\times n}$ is unit lower triangular and $U \in \mathbb{R}^{n\times n}$ is upper triangular.

The elimination process for $A \in \mathbb{R}^{n\times n}$ may now be written as follows:

$$L_{(N)}L_{(N-1)}\dots L_{(1)}A = U\,, \qquad N = \tfrac{1}{2}n(n-1)\,, \tag{2.13}$$

where $U \in \mathbb{R}^{n\times n}$ is an upper triangular matrix and each of the matrices $L_{(j)} \in \mathbb{R}^{n\times n}$, $j = 1,\dots,N$, is unit lower triangular of order n and has the form $I + \mu_{rs}E^{(rs)}$ with $1 \le s < r \le n$, where I is the identity matrix of order n. That is,

$$L_{(1)} = I+\mu_{21}E^{(21)},\ L_{(2)} = I+\mu_{31}E^{(31)},\ \dots,\ L_{(N)} = I+\mu_{n\,n-1}E^{(n\,n-1)}.$$

It is easy to see that $E^{(rs)}E^{(rs)} = \delta_{rs}E^{(rs)}$, where

$$\delta_{rs} = \begin{cases} 1 & \text{for } r = s\,, \\ 0 & \text{for } r \ne s \end{cases}$$

is known as the **Kronecker delta.**[1] Thus, for $1 \le s < r \le n$, the inverse of the matrix $I + \mu_{rs}E^{(rs)}$ is the lower triangular matrix $I - \mu_{rs}E^{(rs)}$, which corresponds to the subtraction of row s, multiplied by μ_{rs}, from row r. Hence

$$A = L_{(1)}^{-1}\dots L_{(N)}^{-1}U = LU\,, \tag{2.14}$$

where L, as the product of a finite number of unit lower triangular matrices of order n, is itself unit lower triangular of order n by Theorem 2.1(ii); see Figure 2.1.

2.3 LU factorisation

Having seen that the Gaussian elimination process gives rise to the factorisation $A = LU$ of the matrix $A \in \mathbb{R}^{n\times n}$, $n \ge 2$, where L is unit

[1] Leopold Kronecker (7 December 1823, Liegnitz, Prussia, Germany (now Legnica, Poland) – 29 December 1891, Berlin, Germany) made significant contributions to the theory of elliptic functions, the theory of ideals and the algebra of quadratic forms.

lower triangular and U is upper triangular, we shall now show how to calculate the elements of L and U directly. Equating the elements of A and LU we conclude that

$$a_{ij} = \sum_{k=1}^{n} l_{ik} u_{kj}\,, \qquad 1 \le i, j \le n\,. \tag{2.15}$$

Recalling that L and U are lower and upper triangular respectively, we see that, in fact, the range of k in this sum extends only up to $\min\{i, j\}$, the smaller of the numbers i and j. Taking the two cases separately gives

$$a_{ij} = \sum_{k=1}^{j} l_{ik} u_{kj}\,, \qquad 1 \le j < i \le n\,, \tag{2.16}$$

$$a_{ij} = \sum_{k=1}^{i} l_{ik} u_{kj}\,, \qquad 1 \le i \le j \le n\,. \tag{2.17}$$

Rearranging these equations, and using the fact that $l_{ii} = 1$ for all $i = 1, 2, \ldots, n$, we find that

$$l_{ij} = \frac{1}{u_{jj}}\left\{a_{ij} - \sum_{k=1}^{j-1} l_{ik} u_{kj}\right\}, \qquad i = 2, \ldots, n\,,\; j = 1, \ldots, i-1\,, \tag{2.18}$$

$$u_{ij} = a_{ij} - \sum_{k=1}^{i-1} l_{ik} u_{kj}\,, \qquad i = 1, \ldots, n\,,\; j = i, \ldots, n\,, \tag{2.19}$$

with the convention that sums over empty index sets are equal to zero. Thus, the elements of U in the first row are $u_{1j} = a_{1j}$, $j = 1, 2, \ldots, n$, and the elements of L in the first column are $l_{11} = 1$ and $l_{i1} = a_{i1}/u_{11}$, $i = 2, \ldots, n$.

The equations (2.18) and (2.19) can now be used for the calculation of the elements l_{ij} and u_{ij}. For each value of i, starting with $i = 2$, we calculate first l_{ij}, for $j = 1, \ldots, i-1$ in order, and then the values of u_{ij}, for $j = i, \ldots, n$, again in increasing order. We then move on to the same calculation for $i + 1$, and so on until $i = n$. In the calculation of l_{ij} we need the values of u_{kj}, $1 \le k \le j < i - 1$, from previous rows, and we also need the values of l_{ik}, $1 \le k \le j - 1$, in the same row but in previous columns; a similar argument applies to the calculation of u_{ij}. When carried out in this order, all the values required at each step have already been calculated.

Of course, we must ensure that the calculation does not fail because of division by zero; this requires that none of the u_{jj}, $j = 1, \ldots, n-1$, in the formula (2.18) is zero. To investigate this possibility we use the properties of certain submatrices of A.

Definition 2.4 *Suppose that $A \in \mathbb{R}^{n\times n}$ with $n \geq 2$, and let $1 \leq k \leq n$. The* **leading principal submatrix** *of order k of A is defined as the matrix $A^{(k)} \in \mathbb{R}^{k\times k}$ whose element in row i and column j is equal to the element of the matrix A in row i and column j for $1 \leq i, j \leq k$.*

Armed with this definition, we can now formulate the main result of this section. It provides a sufficient condition for ensuring that the algorithm (2.18), (2.19) for calculating the entries of the matrices L and U in the LU factorisation $A = LU$ of a matrix $A \in \mathbb{R}^{n\times n}$ does not break down due to division by zero in (2.18).

Theorem 2.2 *Let $n \geq 2$, and suppose that $A \in \mathbb{R}^{n\times n}$ is such that every leading principal submatrix $A^{(k)} \in \mathbb{R}^{k\times k}$ of A of order k, with $1 \leq k < n$, is nonsingular. (Note that A itself is not required to be nonsingular.) Then, A can be factorised in the form $A = LU$, where $L \in \mathbb{R}^{n\times n}$ is unit lower triangular and $U \in \mathbb{R}^{n\times n}$ is upper triangular.*

Proof The proof is by induction on the order n. Let us begin by verifying the statement of the theorem for $n = 2$. We intend to show that any 2×2 matrix

$$A = \begin{pmatrix} a & b \\ c & d \end{pmatrix},$$

with $a \neq 0$, is equal to the product of a unit lower triangular matrix L of order 2 and an upper triangular matrix U of order 2; that is, we wish to establish the existence of

$$L = \begin{pmatrix} 1 & 0 \\ m & 1 \end{pmatrix}, \qquad U = \begin{pmatrix} u & v \\ 0 & \eta \end{pmatrix},$$

such that $LU = A$, where m, u, v and η are four real numbers, to be determined. Equating the product LU with A, we deduce that

$$u = a\,, \quad v = b\,, \quad mu = c\,, \quad mv + \eta = d\,.$$

Since $a \neq 0$ by hypothesis, the first of these equalities implies that $u \neq 0$ also; hence $m = c/u$, $v = b$, and $\eta = d - mv$. Thus we have shown the existence of the required matrices L and U in $\mathbb{R}^{2\times 2}$ and completed the proof for $n = 2$.

Now, suppose that the statement of the theorem has already been verified for matrices of order k, $2 \le k < n$; suppose that $A \in \mathbb{R}^{(k+1)\times(k+1)}$ and all leading principal submatrices of A of order k and less are nonsingular. We mimic the proof in the case of $n = 2$ by partitioning A into blocks by the last row and column:

$$A = \begin{pmatrix} A^{(k)} & \boldsymbol{b} \\ \boldsymbol{c}^{\mathrm{T}} & d \end{pmatrix},$$

where $A^{(k)} \in \mathbb{R}^{k\times k}$ is a nonsingular matrix (all of whose leading principal submatrices are themselves nonsingular), $\boldsymbol{b}$, $\boldsymbol{c}$ are column vectors of size k, and d is a real number. According to our inductive hypothesis, there exist a unit lower triangular matrix $L^{(k)}$ of order k and an upper triangular matrix $U^{(k)}$ of order k such that $A^{(k)} = L^{(k)}U^{(k)}$. Thus we shall seek the desired unit lower triangular matrix L of order $k+1$ and the upper triangular matrix U of order $k+1$ in the form

$$L = \begin{pmatrix} L^{(k)} & \mathbf{0} \\ \boldsymbol{m}^{\mathrm{T}} & 1 \end{pmatrix} \quad \text{and} \quad U = \begin{pmatrix} U^{(k)} & \boldsymbol{v} \\ \mathbf{0}^{\mathrm{T}} & \eta \end{pmatrix}$$

where $\boldsymbol{m}$ and $\boldsymbol{v}$ are column vectors of size k and η is a real number, to be determined from the requirement that the product LU be equal to the matrix A. On equating LU with A, we obtain

$$L^{(k)}U^{(k)} = A^{(k)}\,, \quad L^{(k)}\boldsymbol{v} = \boldsymbol{b}\,, \quad \boldsymbol{m}^{\mathrm{T}}U^{(k)} = \boldsymbol{c}^{\mathrm{T}}\,, \quad \boldsymbol{m}^{\mathrm{T}}\boldsymbol{v} + \eta = d\,.$$

The first of these four equalities provides no new information. However, we can use the remaining three to determine the column vectors $\boldsymbol{v}$ and $\boldsymbol{m}$ and the real number η. Since $L^{(k)}$ is unit lower triangular, its determinant is equal to 1; therefore $L^{(k)}$ is nonsingular. This means that the second equation uniquely determines the unknown column vector $\boldsymbol{v}$. Further, since $A^{(k)} = L^{(k)}U^{(k)}$, we conclude that

$$\det(A^{(k)}) = \det(L^{(k)}U^{(k)}) = \det(L^{(k)})\det(U^{(k)}) = \det(U^{(k)})\,;$$

given that $\det(A^{(k)}) \neq 0$ by the inductive hypothesis, this implies that $\det(U^{(k)}) \neq 0$ also, and therefore the third equation uniquely determines $\boldsymbol{m}$. Having found $\boldsymbol{v}$ and $\boldsymbol{m}$, the fourth equation yields $\eta = d - \boldsymbol{m}^{\mathrm{T}}\boldsymbol{v}$. Thus we have shown the existence of the desired matrices L and U of order $k+1$, and the inductive step is complete.[1] □

[1] In the last paragraph we made use of the **Binet–Cauchy Theorem** which states that for three matrices A, B, C in $\mathbb{R}^{k\times k}$ with $A = BC$, we have $\det(A) = \det(B)\det(C)$. This result was proved in 1812 independently by Augustin-Louis Cauchy (1789–1857) and Jacques Philippe Marie Binet (1786–1856).

2.4 Pivoting

The aim of this section is to show that even if the matrix A does not satisfy the conditions of Theorem 2.2, by permuting rows and columns it can be transformed into a new matrix $\tilde{A}$ of the same size so that $\tilde{A}$ admits an LU factorisation.

Example 2.3 *Consider, for example, the system obtained from (2.9) by replacing the coefficient of x_1 in the first equation by zero. Then, the leading element in the matrix A is zero, the computation fails at the first step, and the LU factorisation of A does not exist. However if we interchange the first two equations we obtain a new matrix $\tilde{A}$ which is the same as A but with the first two rows interchanged,*

$$\tilde{A} = \begin{pmatrix} 2 & 4 & 2 \\ 0 & 1 & 1 \\ -1 & 5 & -4 \end{pmatrix}. \tag{2.20}$$

Since the leading principal submatrices of order 1 and 2 of $\tilde{A}$ are non-singular, by Theorem 2.2 the matrix $\tilde{A}$ now has the required LU factorisation, which is easily computed.

A computation which fails when an element is exactly zero is also likely to run into difficulties when that element is nonzero but of very small absolute value; the problem stems from the presence of rounding errors. The basic operation in the elimination process consists of multiplying the elements of one row of the matrix by a scalar μ_{rs}, and adding to the elements of another row. The multiplication operation will always introduce a rounding error, so the elements which are multiplied by μ_{rs} will already contain a rounding error from operations with earlier rows of the matrix; these errors will therefore themselves be multiplied by μ_{rs} before adding to the new row. The errors will be magnified if $|\mu_{rs}| > 1$, and will be greatly magnified if $|\mu_{rs}| \gg 1$.

The accumulation of rounding errors alluded to in the previous paragraph can be alleviated by permuting the rows of the matrix. Thus, at each stage of the elimination process we interchange two rows, if necessary, so that the largest element in the current column lies on the diagonal. This process is known as **pivoting**. Clearly, when pivoting is performed none of the multipliers μ_{rs} have absolute value greater than unity. The process is easily formalised by introducing permutation matrices. This leads us to our next definition.

Definition 2.5 *Suppose that $n \geq 2$. A matrix $P \in \mathbb{R}^{n\times n}$ in which every element is either 0 or 1, and whose every row and every column contain exactly one nonzero element, is called a* **permutation matrix**.

Example 2.4 *Here are three of the possible 3! permutation matrices in $\mathbb{R}^{3\times 3}$:*

$$\begin{pmatrix} 1 & 0 & 0 \\ 0 & 1 & 0 \\ 0 & 0 & 1 \end{pmatrix}, \quad \begin{pmatrix} 0 & 1 & 0 \\ 1 & 0 & 0 \\ 0 & 0 & 1 \end{pmatrix}, \quad \begin{pmatrix} 0 & 0 & 1 \\ 1 & 0 & 0 \\ 0 & 1 & 0 \end{pmatrix}.$$

The proof of our next result is elementary and is left to the reader.

Lemma 2.1 *Let $n \geq 2$ and suppose that $P \in \mathbb{R}^{n\times n}$ is a permutation matrix. Then, the following statements hold:*

(i) *given that I is the identity matrix of order n, the matrix P can be obtained from I by permuting rows;*
(ii) *if $Q \in \mathbb{R}^{n\times n}$ is another permutation matrix, then the products PQ and QP are also permutation matrices;*
(iii) *let $P^{(rs)} \in \mathbb{R}^{n\times n}$ denote the* **interchange matrix**, *obtained from the identity matrix $I \in \mathbb{R}^{n\times n}$ by interchanging rows r and s; any interchange matrix is a permutation matrix; moreover, any permutation matrix of order n can be written as a product of interchange matrices of order n;*
(iv) *the determinant of a permutation matrix $P \in \mathbb{R}^{n\times n}$ is equal to 1 or -1, depending on whether P is obtained from the identity matrix of order n by an even or odd number of permutations of rows, respectively; in particular, a permutation matrix is nonsingular.*

Now we are ready to prove the next theorem.

Theorem 2.3 *Let $n \geq 2$ and $A \in \mathbb{R}^{n\times n}$. There exist a permutation matrix P, a unit lower triangular matrix L, and an upper triangular matrix U, all three in $\mathbb{R}^{n\times n}$, such that*

$$PA = LU\,. \tag{2.21}$$

Proof The proof is by induction on the order n. Let $n = 2$ and consider the matrix

$$A = \begin{pmatrix} a & b \\ c & d \end{pmatrix}.$$

If $a \neq 0$, the proof follows from Theorem 2.2 with P taken as the 2×2 identity matrix. If $a = 0$ but $c \neq 0$, we take

$$P = \begin{pmatrix} 0 & 1 \\ 1 & 0 \end{pmatrix}$$

and write

$$PA = \begin{pmatrix} c & d \\ 0 & b \end{pmatrix} = \begin{pmatrix} 1 & 0 \\ 0 & 1 \end{pmatrix} \begin{pmatrix} c & d \\ 0 & b \end{pmatrix} \equiv LU\,.$$

If $a = 0$ and $c = 0$, the result trivially follows by writing

$$\begin{pmatrix} 0 & b \\ 0 & d \end{pmatrix} = \begin{pmatrix} 1 & 0 \\ 0 & 1 \end{pmatrix} \begin{pmatrix} 0 & b \\ 0 & d \end{pmatrix} \equiv LU$$

and taking P as the 2×2 identity matrix. That completes the proof for $n = 2$.

Now, suppose that $A \in \mathbb{R}^{(k+1)\times(k+1)}$ and assume that the theorem holds for every matrix of order k with $2 \leq k < n$. We begin by locating the element in the first column of A which has the largest absolute value, or any one of them if there is more than one such element, and interchange rows if required; if the largest element is in row r we interchange rows 1 and r. We then partition the new matrix according to the first row and column, writing

$$P^{(1r)}A = \begin{pmatrix} \alpha & \boldsymbol{w}^{\mathrm{T}} \\ \boldsymbol{p} & B \end{pmatrix} = \begin{pmatrix} 1 & \boldsymbol{0}^{\mathrm{T}} \\ \boldsymbol{m} & I \end{pmatrix} \begin{pmatrix} \alpha & \boldsymbol{v}^{\mathrm{T}} \\ \boldsymbol{0} & C \end{pmatrix} \tag{2.22}$$

where α is the element of largest absolute value in the first column, $B, C \in \mathbb{R}^{k\times k}$, and $\boldsymbol{p}$, $\boldsymbol{w}$, $\boldsymbol{m}$ and $\boldsymbol{v}$ are column vectors of size k, with $\boldsymbol{m}$, $\boldsymbol{v}$ and C to be determined. Writing out the product we find that

$$\left.\begin{aligned} \boldsymbol{v}^{\mathrm{T}} &= \boldsymbol{w}^{\mathrm{T}}\,, \\ \alpha\boldsymbol{m} &= \boldsymbol{p}\,, \\ C &= B - \boldsymbol{m}\boldsymbol{v}^{\mathrm{T}}\,. \end{aligned}\right\} \tag{2.23}$$

If $\alpha = 0$, then the first column of A consists entirely of zeros ($\boldsymbol{p} = \boldsymbol{0}$); in this case we can evidently choose $\boldsymbol{m} = \boldsymbol{0}$, $\boldsymbol{v} = \boldsymbol{w}$ and $C = B$. Suppose now that $\alpha \neq 0$; then $\boldsymbol{m} = (1/\alpha)\boldsymbol{p}$, so that all the elements of $\boldsymbol{m}$ have absolute value less than or equal to unity, since α is the largest in absolute value element in the first column. By the inductive hypothesis we can write

$$P^*C = L^*U^*\,, \tag{2.24}$$

where P^*, L^*, $U^* \in \mathbb{R}^{k\times k}$, P^* is a permutation matrix, L^* is unit lower triangular, and U^* is upper triangular. Hence, by (2.22) and (2.23),

$$P^{(1r)}A = \begin{pmatrix} 1 & \mathbf{0}^{\mathrm{T}} \\ \mathbf{0} & P^* \end{pmatrix} \begin{pmatrix} 1 & \mathbf{0}^{\mathrm{T}} \\ P^*\boldsymbol{m} & L^* \end{pmatrix} \begin{pmatrix} \alpha & \boldsymbol{v}^{\mathrm{T}} \\ \mathbf{0} & U^* \end{pmatrix} \tag{2.25}$$

since $P^*P^* = I$. Now, defining the permutation matrix P by

$$P = \begin{pmatrix} 1 & \mathbf{0}^{\mathrm{T}} \\ \mathbf{0} & P^* \end{pmatrix} P^{(1r)}, \tag{2.26}$$

we obtain

$$PA = \begin{pmatrix} 1 & \mathbf{0}^{\mathrm{T}} \\ P^*\boldsymbol{m} & L^* \end{pmatrix} \begin{pmatrix} \alpha & \boldsymbol{v}^{\mathrm{T}} \\ \mathbf{0} & U^* \end{pmatrix}, \tag{2.27}$$

which is the required factorisation of $A \in \mathbb{R}^{(k+1)\times(k+1)}$. This completes the inductive step. The theorem therefore holds for every matrix of order $n \geq 2$. □

The proof of this theorem also contains an algorithm for constructing the permutation matrix P, and the matrices L and U. The permutation matrix is conveniently described by specifying the sequence of interchanges: given the $n-1$ integers $p_1, p_2, \ldots, p_{n-1}$, the matrix P is the product of the permutation matrices which interchange rows 1 and p_1, 2 and p_2, and so on.

2.5 Solution of systems of equations

Consider the linear system $A\boldsymbol{x} = \boldsymbol{b}$ where $A \in \mathbb{R}^{n\times n}$ and $\boldsymbol{x}$ and $\boldsymbol{b}$ are column vectors of size n. According to Theorem 2.3 there exist a permutation matrix $P \in \mathbb{R}^{n\times n}$, a unit lower triangular matrix $L \in \mathbb{R}^{n\times n}$ and an upper triangular matrix $U \in \mathbb{R}^{n\times n}$ such that $PA = LU$. Having obtained the LU factorisation of the matrix PA, the solution of the system of linear equations $A\boldsymbol{x} = \boldsymbol{b}$ is straightforward: multiplying both sides of $A\boldsymbol{x} = \boldsymbol{b}$ on the left by the permutation matrix P, we obtain that

$$PA\boldsymbol{x} = P\boldsymbol{b}; \tag{2.28}$$

equivalently, $LU\boldsymbol{x} = P\boldsymbol{b}$. On defining $\boldsymbol{y} = U\boldsymbol{x}$ we can rewrite (2.28) as the following coupled set of linear equations:

$$L\boldsymbol{y} = P\boldsymbol{b}, \qquad U\boldsymbol{x} = \boldsymbol{y}. \tag{2.29}$$

Assuming that the matrix P and the LU factorisation of PA are already known, there are three stages to the calculation of $\boldsymbol{x}$:

`Step 1.`First we apply the sequence of permutations to the vector $\boldsymbol{b}$, to produce $P\boldsymbol{b}$;
`Step 2.[Forward substitution]` We then solve the lower triangular system $L\boldsymbol{y} = P\boldsymbol{b}$, calculating the elements in the order $y_1,\ y_2, \dots, y_n$;
`Step 3.[Backsubstitution]` Finally the required solution $\boldsymbol{x}$ is obtained from the upper triangular system $U\boldsymbol{x} = \boldsymbol{y}$, calculating the elements of $\boldsymbol{x}$ in the reverse order, $x_n,\ x_{n-1}, \dots, x_1$.

`Step 3` will break down if any of the diagonal elements of U are zero, but if this happens the matrix A is singular.

The next section is devoted to assessing the amount of computational work for this algorithm.

2.6 Computational work

In this section we shall show that the work involved in factorising an $n \times n$ matrix in the form $A = LU$ is proportional to n^3. An estimate of the amount of computational work of this kind is important in deciding in advance how long a calculation would take for a very large matrix, and is also useful in comparing different methods for the solution of a given problem. For example, in the next chapter we shall derive a method for solving a system of equations with a symmetric positive definite matrix; that method requires only half the amount of work involved in the standard LU factorisation algorithm which takes no account of symmetry.

Accurate estimates of the time taken by a computation are very complicated and require some detailed knowledge of the computer being used. The estimates which we shall give are simple but crude; they are normally good enough for the types of comparisons we have just mentioned.

We see from (2.18) that the calculation of l_{ij} requires $j-1$ multiplications, $j-2$ additions, 1 subtraction and 1 division, a total of $2j-1$ operations. In the same way, (2.19) shows that the calculation of u_{ij} requires $2i-2$ operations.[1] Recalling that, for any integer $k \geq 2$,

$$1 + \cdots + k = \tfrac{1}{2}k(k+1) \quad \text{and} \quad 1^2 + \cdots + k^2 = \tfrac{1}{6}k(k+1)(2k+1)\,,$$

we then deduce that the total number of operations involved in the LU

[1] We do not count the row interchanges in the number of 'operations'.

factorisation is

$$\sum_{i=2}^{n}\sum_{j=1}^{i-1}(2j-1)+\sum_{i=1}^{n}\sum_{j=i}^{n}2(i-1)=\tfrac{1}{6}n(n-1)(4n+1)\,.$$

It is enough to say that the number of multiplications required is about $\frac{2}{3}n^3-\frac{1}{2}n^2$, for moderately large values of n.

Having constructed the factorisation we can now count the number of operations required to compute the vectors $\boldsymbol{y}$ and $\boldsymbol{x}$ in (2.29). Given the vector $P\boldsymbol{b}$, the elements of $\boldsymbol{y}$ are obtained from

$$y_1=(P\boldsymbol{b})_1\,,\qquad y_i=(P\boldsymbol{b})_i-\sum_{j=1}^{i-1}l_{ij}y_j\,,\quad i=2,3,\ldots,n\,,\tag{2.30}$$

which requires $2i-2$ operations. Summing over i this gives a total of $n(n-1)$. The calculation of the elements of $\boldsymbol{x}$ is similar:

$$x_i=\frac{1}{u_{ii}}\left(y_i-\sum_{j=i+1}^{n}u_{ij}x_j\right),\qquad i=1,2,\ldots,n\,.\tag{2.31}$$

This requires $2(n-i)+1$ operations, giving a total of n^2.

The total number of operations involved in the solution of the system of equations is therefore approximately $\frac{2}{3}n^3-\frac{1}{2}n^2$ for the factorisation, followed by $n(n-1)+n^2=2n^2-n$ for the solution of the two triangular systems, that is, approximately $\frac{2}{3}n^3+\frac{3}{2}n^2$, ignoring terms of size $\mathcal{O}(n)$.

We often need to solve a number of systems of this kind, all with different right-hand sides, but with the same matrix. We then need only factorise the matrix once, and the total number of multiplications required for k right-hand sides becomes approximately $\frac{2}{3}n^3+\left(2k-\frac{1}{2}\right)n^2$. When k is fairly large it might appear that it would be more efficient to form the inverse matrix A^{-1}, and then multiply each right-hand side by the inverse; but we shall show that it is not so.

To form the inverse matrix we first factorise the matrix A, and then solve n systems, with the right-hand sides being the vectors which constitute the columns of the identity matrix. Because these right-hand sides have a special form, there is the possibility of saving some work; some careful counting shows that the total can be reduced from $\frac{2}{3}n^3+2n^3=\frac{8}{3}n^3$ to an approximate total of $2n^3$ operations. It is easy to see that multiplying a vector by the inverse matrix requires $n(2n-1)$ operations; hence the whole computation of first constructing the inverse matrix, and then multiplying each right-hand side by the inverse, requires a total of $2n^3+2kn^2$ multiplications (ignoring terms of size $\mathcal{O}(n)$). This is always greater than the previous value $\frac{2}{3}n^3 + \left(2k - \frac{1}{2}\right) n^2$, whether k is small or large. The most efficient way of solving this problem is to construct and save the L and U factors of A, rather than to form the inverse of A.

2.7 Norms and condition numbers

The analysis of the effects of rounding error on solutions of systems of linear equations requires an appropriate measure. This is provided by the concept of **norm** defined below. In order to motivate the axioms of norm stated in Definition 2.6, we note that the set $\mathbb{R}$ of real numbers is a linear space, and that the **absolute value** function

$$v \in \mathbb{R} \mapsto |v| = \begin{cases} v & \text{if } v \geq 0, \\ -v & \text{if } v < 0 \end{cases}$$

has the following properties:

- $|v| \geq 0$ for any $v \in \mathbb{R}$, and $|v| = 0$ if, and only if, $v = 0$;
- $|\lambda v| = |\lambda|\,|v|$ for all $\lambda \in \mathbb{R}$ and all $v \in \mathbb{R}$;
- $|u + v| \leq |u| + |v|$ for all u and v in $\mathbb{R}$.

The absolute value $|v|$ of a real number v measures the distance between v and 0 (the zero element of the linear space $\mathbb{R}$). Our next definition aims to generalise this idea to an arbitrary linear space $\mathcal{V}$ over the field $\mathbb{R}$ of real numbers: even though the discussion in the present chapter is confined to finite-dimensional linear spaces of vectors ($\mathcal{V} = \mathbb{R}^n$) and square matrices ($\mathcal{V} = \mathbb{R}^{n\times n}$), norms over other linear spaces, including infinite-dimensional function spaces, will appear elsewhere in the text (see Chapters 8, 9, 11 and 14).

Definition 2.6 *Suppose that $\mathcal{V}$ is a linear space over the field $\mathbb{R}$ of real numbers. The nonnegative real-valued function $\|\cdot\|$ is said to be a* **norm** *on the space $\mathcal{V}$ provided that it satisfies the following axioms:*

❶ *$\|v\| = 0$ if, and only if, $v = 0$ in $\mathcal{V}$;*
❷ *$\|\lambda v\| = |\lambda|\,\|v\|$ for all $\lambda \in \mathbb{R}$ and all v in $\mathcal{V}$;*
❸ *$\|u+v\| \leq \|u\|+\|v\|$ for all u and v in $\mathcal{V}$ (the triangle inequality).*

A linear space $\mathcal{V}$, equipped with a norm, is called a **normed linear space***.*

Remark 2.1 *If $\mathcal{V}$ is a linear space over the field $\mathbb{C}$ of complex numbers, then $\mathbb{R}$ in the second axiom of Definition 2.6 should be replaced by $\mathbb{C}$, with $|\lambda|$ signifying the modulus of $\lambda \in \mathbb{C}$.*

Any norm on the linear space $\mathcal{V} = \mathbb{R}^n$ will be called a **vector norm**. Three vector norms are in common use in numerical linear algebra: the 1-norm $\|\cdot\|_1$, the 2-norm (or Euclidean norm) $\|\cdot\|_2$, and the ∞-norm $\|\cdot\|_\infty$; these are defined below.

Definition 2.7 *The* **1-norm** *of the vector* $\boldsymbol{v} = (v_1, \dots, v_n)^{\mathrm{T}} \in \mathbb{R}^n$ *is defined by*

$$\|\boldsymbol{v}\|_1 = \sum_{i=1}^{n} |v_i| \,. \tag{2.32}$$

Definition 2.8 *The* **2-norm** *of the vector* $\boldsymbol{v} = (v_1, \dots, v_n)^{\mathrm{T}} \in \mathbb{R}^n$ *is defined by* $\|\boldsymbol{v}\|_2 = \left(\boldsymbol{v}^{\mathrm{T}}\boldsymbol{v}\right)^{1/2}$. *In other words,*

$$\|\boldsymbol{v}\|_2 = \left\{\sum_{i=1}^{n} |v_i|^2\right\}^{1/2} . \tag{2.33}$$

Definition 2.9 *The* **∞-norm** *of the vector* $\boldsymbol{v} = (v_1, \dots, v_n)^{\mathrm{T}} \in \mathbb{R}^n$ *is defined by*

$$\|\boldsymbol{v}\|_\infty = \max_{i=1}^{n} |v_i| \,. \tag{2.34}$$

When $n = 1$, each of these norms collapses to the absolute value, $|\cdot|$, the simplest example of a norm on $\mathcal{V} = \mathbb{R}$.

It is easy to show that $\|\cdot\|_1$ and $\|\cdot\|_\infty$ obey all axioms of a norm. For the 2-norm the first two axioms are still trivial to verify; to show that the triangle inequality is satisfied by the 2-norm requires use of the Cauchy[1]–Schwarz[2] inequality.

Lemma 2.2 (Cauchy–Schwarz inequality)

$$\left|\sum_{i=1}^{n} u_i v_i\right| \le \|\boldsymbol{u}\|_2 \|\boldsymbol{v}\|_2 \qquad \forall\, \boldsymbol{u},\, \boldsymbol{v} \in \mathbb{R}^n \,. \tag{2.35}$$

[1] Augustin-Louis Cauchy (21 August 1789, Paris, France – 23 May 1857, Sceaux (near Paris), France) made very significant contributions to algebra and number theory. He was one of the founders of modern mathematical analysis, the theory of complex functions, and the mathematics of elasticity theory.

[2] Karl Herman Amandus Schwarz (25 January 1843, Hermsdorf, Silesia, Germany (now in Poland) – 30 November 1921, Berlin, Germany) succeeded Karl Weierstrass as Professor of Mathematics at Berlin in 1892. Outside mathematics he acted as captain of the local Voluntary Fire Brigade, and helped the station-master at the local railway station by closing the doors of the trains.

Proof The proof of this inequality is rather simple: for any $\boldsymbol{u}$ and $\boldsymbol{v}$ in $\mathbb{R}^n$, and all $\lambda \in \mathbb{R}$,

$$\begin{aligned} 0 &\leq \|\lambda \boldsymbol{u} + \boldsymbol{v}\|_2^2 = \sum_{i=1}^n (\lambda u_i + v_i)^2 \\ &= \lambda^2 \sum_{i=1}^n |u_i|^2 + 2\lambda \sum_{i=1}^n u_i v_i + \sum_{i=1}^n |v_i|^2 . \end{aligned} \tag{2.36}$$

Hence, the expression on the right is a nonnegative quadratic polynomial in $\lambda \in \mathbb{R}$, of the form $A\lambda^2 + B\lambda + C$; therefore, the associated discriminant,

$$B^2 - 4AC = \left(2 \sum_{i=1}^n u_i v_i\right)^2 - 4 \left(\sum_{i=1}^n |u_i|^2\right) \left(\sum_{i=1}^n |v_i|^2\right),$$

is nonpositive. This implies (2.35) on recalling Definition 2.8. □

The triangle inequality for the 2-norm is now deduced as follows: letting $\lambda = 1$ in (2.36) and using (2.35), it follows that

$$\begin{aligned} \|\boldsymbol{u} + \boldsymbol{v}\|_2^2 &= \|\boldsymbol{u}\|_2^2 + 2\sum_{i=1}^n u_i v_i + \|\boldsymbol{v}\|_2^2 \\ &\leq \|\boldsymbol{u}\|_2^2 + 2\|\boldsymbol{u}\|_2 \|\boldsymbol{v}\|_2 + \|\boldsymbol{v}\|_2^2 \\ &= (\|\boldsymbol{u}\|_2 + \|\boldsymbol{v}\|_2)^2 , \end{aligned}$$

which yields the triangle inequality in the 2-norm on taking square roots. Hence $\|\cdot\|_2$ satisfies all three axioms of norm.

The 1-norm and the 2-norm on $\mathbb{R}^n$ are special cases of the p-norm, defined on $\mathbb{R}^n$, for $p \geq 1$, by

$$\|\boldsymbol{v}\|_p = \left\{ \sum_{i=1}^n |v_i|^p \right\}^{1/p} . \tag{2.37}$$

The first two axioms of norm are trivial to verify for $\|\cdot\|_p$; however, showing the triangle inequality is less straightforward (except for $p = 1$, and for $p = 2$, as we have already seen before); we shall now sketch the proof of this for $p > 1$. The starting point is the following result, known as Young's inequality.[1]

[1] William Henry Young (20 October 1863, London, England – 7 July 1942, Lausanne, Switzerland) studied mathematics at Peterhouse, Cambridge. His most important contributions were to the calculus of functions of several variables. Young was elected Fellow of the Royal Society in 1907; he was president of the London Mathematical Society (1922–1924) and president of the International Union of Mathematicians (1929–1936).

Theorem 2.4 (Young's inequality) *Let $p, q > 1$, $(1/p) + (1/q) = 1$. Then, for any two nonnegative real numbers a and b,*

$$ab \leq \frac{a^p}{p} + \frac{b^q}{q}.$$

Proof If either $a = 0$ or $b = 0$ the inequality holds trivially. Let us therefore suppose that $a > 0$ and $b > 0$. We recall that a function $x \in \mathbb{R} \mapsto f(x) \in \mathbb{R}$ is said to be **convex** if

$$f(\theta x + (1-\theta)y) \leq \theta f(x) + (1-\theta) f(y)$$

for all $\theta \in [0,1]$, and all x and y in $\mathbb{R}$; *i.e.*, for any x and y in $\mathbb{R}$ the graph of the function f between the points $(x, f(x))$ and $(y, f(y))$ lies below the chord that connects these two points. Note that the function $x \mapsto \mathrm{e}^x$ is convex. Therefore, with $\theta = 1/p$ and $1 - \theta = 1/q$, we get that

$$ab = \mathrm{e}^{\ln a + \ln b} = \mathrm{e}^{(1/p)\ln a^p + (1/q)\ln b^q} \leq \frac{1}{p}\mathrm{e}^{\ln a^p} + \frac{1}{q}\mathrm{e}^{\ln a^q} = \frac{a^p}{p} + \frac{b^q}{q},$$

and the proof is complete. (When $p = q = 2$ the proof is trivial: as $(a-b)^2 \geq 0$ also $2ab \leq a^2 + b^2$, and hence the required result.) □

The next step is to establish Hölder's inequality;[1] it is a generalisation of the Cauchy–Schwarz inequality.

Theorem 2.5 (Hölder's inequality) *Let $p, q > 1$, $(1/p) + (1/q) = 1$. Then, for any $\boldsymbol{u} \in \mathbb{R}^n$ and $\boldsymbol{v} \in \mathbb{R}^n$, we have*

$$\left|\sum_{i=1}^{n} u_i v_i\right| \leq \|\boldsymbol{u}\|_p \|\boldsymbol{v}\|_q .$$

Proof If either $\boldsymbol{u} = \boldsymbol{0}$ or $\boldsymbol{v} = \boldsymbol{0}$ the inequality holds trivially. Let us therefore suppose that $\boldsymbol{u} \neq \boldsymbol{0}$ and $\boldsymbol{v} \neq \boldsymbol{0}$, and consider the vectors $\tilde{\boldsymbol{u}}$ and $\tilde{\boldsymbol{v}}$ in $\mathbb{R}^n$ with components $\tilde{u}_i = u_i/\|\boldsymbol{u}\|_p$ and $\tilde{v}_i = v_i/\|\boldsymbol{v}\|_q$, respectively, $i = 1, 2, \ldots, n$. By Young's inequality,

$$\left|\sum_{i=1}^{n} \tilde{u}_i \tilde{v}_i\right| \leq \sum_{i=1}^{n} |\tilde{u}_i \tilde{v}_i| \leq \frac{1}{p}\sum_{i=1}^{n} |\tilde{u}_i|^p + \frac{1}{q}\sum_{i=1}^{n} |\tilde{v}_i|^q = \frac{1}{p} + \frac{1}{q} = 1 .$$

Inserting the defining expressions for $\tilde{u}_i$ and $\tilde{v}_i$ into the left-most expression in this chain, the result follows. □

[1] Otto Ludwig Hölder (22 December 1859, Stuttgart, Germany – 29 August 1937, Leipzig, Germany) contributed to group theory; we owe him the concepts of factor group, and inner and outer automorphisms. Hölder discovered the inequality now named after him in 1884 while working on the convergence of Fourier series.

The triangle inequality in the p-norm is referred to as Minkowski's inequality.[1]

Theorem 2.6 (Minkowski's inequality) *Let* $1 \le p \le \infty$ *and* $\boldsymbol{u}, \boldsymbol{v} \in \mathbb{R}^n$*. Then,*

$$\|\boldsymbol{u} + \boldsymbol{v}\|_p \le \|\boldsymbol{u}\|_p + \|\boldsymbol{v}\|_p .$$

Proof As we noted earlier, the proof of this inequality for $p = 1$ and $p = \infty$ is easy. Let us therefore focus on the case $1 < p < \infty$. In the nontrivial case of $\boldsymbol{u} \neq \mathbf{0}$ and $\boldsymbol{v} \neq \mathbf{0}$, Hölder's inequality yields

$$\begin{aligned}
\|\boldsymbol{u} + \boldsymbol{v}\|_p^p &= \sum_{i=1}^n |u_i + v_i|^p \le \sum_{i=1}^n |u_i + v_i|^{p-1} \left(|u_i| + |v_i|\right) \\
&\le \left(\sum_{i=1}^n |u_i + v_i|^p\right)^{\frac{p-1}{p}} \left(\left(\sum_{i=1}^n |u_i|^p\right)^{\frac{1}{p}} + \left(\sum_{i=1}^n |v_i|^p\right)^{\frac{1}{p}}\right) \\
&= \|\boldsymbol{u} + \boldsymbol{v}\|_p^{p-1} \left(\|\boldsymbol{u}\|_p + \|\boldsymbol{v}\|_p\right) ,
\end{aligned}$$

and hence the desired result on dividing through by $\|\boldsymbol{u} + \boldsymbol{v}\|_p^{p-1}$. □

Remark 2.2 *For a nonzero element* $\boldsymbol{u}$ *in* $\mathbb{R}^n$*, let* $\tilde{\boldsymbol{u}} = (\|\boldsymbol{u}\|_\infty)^{-1}\boldsymbol{u}$*. Clearly,* $1 \le \|\tilde{\boldsymbol{u}}\|_p \le n^{1/p}$*, and hence* $\lim_{p\to\infty} \|\tilde{\boldsymbol{u}}\|_p = 1$*. Therefore,*

$$\|\boldsymbol{u}\|_\infty = \lim_{p\to\infty} \|\boldsymbol{u}\|_p , \qquad \boldsymbol{u} \in \mathbb{R}^n .$$

This identity justifies our use of the notation $\|\cdot\|_\infty$ *for the maximum norm, defined by* $\|\boldsymbol{u}\|_\infty = \max_{i=1}^n |u_i|$*, and our terminology:* ∞-**norm**.

Remark 2.3 *We note here that* $\|\cdot\|_p$*,* $1 \le p \le \infty$*, is also a norm on the linear space* $\mathbb{C}^n$ *of* n*-component vectors with complex entries, over the field* $\mathbb{C}$ *of complex numbers, provided that* $|v_i|$ *in the definitions (2.37), (2.34) of* $\|\cdot\|_p$ *is interpreted as the modulus of the complex number* v_i*.*

In order to highlight the difference between $\|\cdot\|_1$, $\|\cdot\|_2$ and $\|\cdot\|_\infty$, in Figure 2.2 we plot the 'unit spheres' (or 'unit circles', in the case of $n = 2$) corresponding to these three norms on $\mathcal{V} = \mathbb{R}^2$. We recall that

[1] Hermann Minkowski (22 June 1864, Alexotas, Russia (now Kaunas, Lithuania) – 12 January 1909, Göttingen, Germany) held a chair at the University of Göttingen, where he was exposed to Hilbert's work on mathematical physics. Minkowski realised that the ideas of Lorentz and Einstein can be best understood in terms of non-Euclidean geometry, with space and time coupled into a four-dimensional continuum. He died at the age of 44 from a ruptured appendix.

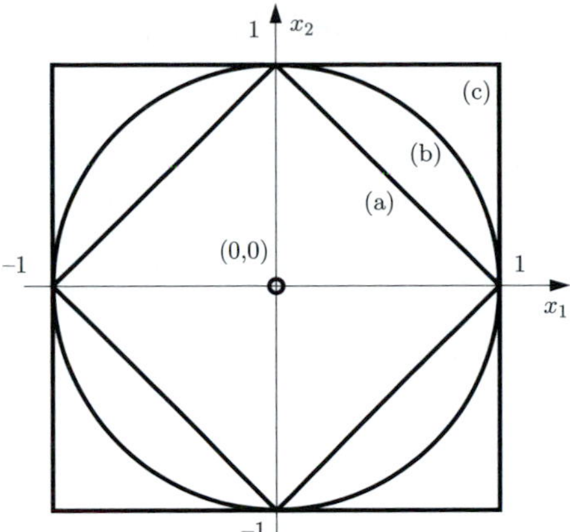

Fig. 2.2. 'Unit circles' in the linear space $\mathcal{V} = \mathbb{R}^2$ with respect to three vector norms: (a) the 1-norm; (b) the 2-norm; (c) the ∞-norm.

the unit sphere in a normed linear space $\mathcal{V}$, with norm $\|\cdot\|$, is defined as the set $\{\boldsymbol{v} \in \mathcal{V}: \|\boldsymbol{v}\| = 1\}$. It can be seen from Figure 2.2 that

$$\{\boldsymbol{v} \in \mathbb{R}^2: \|\boldsymbol{v}\|_1 \leq 1\} \subset \{\boldsymbol{v} \in \mathbb{R}^2: \|\boldsymbol{v}\|_2 \leq 1\} \subset \{\boldsymbol{v} \in \mathbb{R}^2: \|\boldsymbol{v}\|_\infty \leq 1\}.$$

We leave it to the reader as an exercise to show that analogous inclusions hold in $\mathbb{R}^n$ for any $n \geq 1$. (See Exercise 8.)

The unit sphere in a normed linear space $\mathcal{V}$ with norm $\|\cdot\|$ is the boundary of the closed unit ball $\bar{B}_1(\mathbf{0})$ centred at $\mathbf{0}$ defined by

$$\bar{B}_1(\mathbf{0}) = \{\boldsymbol{v} \in \mathcal{V}: \|\boldsymbol{v}\| \leq 1\}.$$

Analogously, the open unit ball centred at $\mathbf{0}$ is defined by

$$B_1(\mathbf{0}) = \{\boldsymbol{v} \in \mathcal{V}: \|\boldsymbol{v}\| < 1\}.$$

More generally, for $\varepsilon > 0$ and $\boldsymbol{\xi} \in \mathcal{V}$,

$$\bar{B}_\varepsilon(\boldsymbol{\xi}) = \{\boldsymbol{v} \in \mathcal{V}: \|\boldsymbol{v} - \boldsymbol{\xi}\| \leq \varepsilon\}$$

is the **closed ball** of radius ε centred at $\boldsymbol{\xi}$; analogously,

$$B_\varepsilon(\boldsymbol{\xi}) = \{\boldsymbol{v} \in \mathcal{V}: \|\boldsymbol{v} - \boldsymbol{\xi}\| < \varepsilon\}$$

is the **open ball** of radius ε centred at $\boldsymbol{\xi}$.

Any norm on the linear space $\mathbb{R}^{n\times n}$ of $n \times n$ matrices with real entries will be referred to as a **matrix norm**. In particular, we shall now consider matrix norms which are induced by vector norms in a sense that will be made precise in the next definition.

Definition 2.10 *Given any norm $\|\cdot\|$ on the space $\mathbb{R}^n$ of n-dimensional vectors with real entries, the* **subordinate matrix norm** *on the space $\mathbb{R}^{n\times n}$ of $n \times n$ matrices with real entries is defined by*

$$\|A\| = \max_{\boldsymbol{v}\in\mathbb{R}^n_*} \frac{\|A\boldsymbol{v}\|}{\|\boldsymbol{v}\|}. \tag{2.38}$$

In (2.38) we used $\mathbb{R}^n_*$ to denote $\mathbb{R}^n \setminus \{\mathbf{0}\}$, where, for sets A and B, $A \setminus B = \{x \in A\colon x \notin B\}$.

Remark 2.4 *Let $\mathbb{C}^{n\times n}$ denote the linear space of $n \times n$ matrices with complex entries over the field $\mathbb{C}$ of complex numbers. Given any norm $\|\cdot\|$ on the linear space $\mathbb{C}^n$, the* **subordinate matrix norm** *on $\mathbb{C}^{n\times n}$ is defined by*

$$\|A\| = \max_{\boldsymbol{v}\in\mathbb{C}^n_*} \frac{\|A\boldsymbol{v}\,\|}{\|\boldsymbol{v}\,\|},$$

where $\mathbb{C}^n_* = \mathbb{C}^n \setminus \{\mathbf{0}\}$.

It is easy to show that a subordinate matrix norm satisfies the axioms of norm listed in Definition 2.6; the details are left as an exercise. Definition 2.10 implies that, for $A \in \mathbb{R}^{n\times n}$,

$$\|A\boldsymbol{v}\,\| \le \|A\|\,\|\boldsymbol{v}\,\|, \qquad \text{for all } \boldsymbol{v} \in \mathbb{R}^n .$$

In a relation like this any vector norm may be used, but of course it is necessary to use the same norm throughout. It follows from Definition 2.10 that, in any subordinate matrix norm $\|\cdot\|$ on $\mathbb{R}^{n\times n}$,

$$\|I\| = 1$$

where I is the $n \times n$ identity matrix.

Given any vector $\boldsymbol{v}$ in $\mathbb{R}^n$, it is a trivial matter to evaluate each of the three norms $\|\boldsymbol{v}\,\|_1$, $\|\boldsymbol{v}\,\|_2$, $\|\boldsymbol{v}\,\|_\infty$; however, it is not yet obvious how one can calculate the corresponding subordinate matrix norm of a given matrix A in $\mathbb{R}^{n\times n}$. Definition 2.10 is unhelpful in this respect: calculating $\|A\|$ *via* (2.38) would involve the unpleasant task of maximising the function $\boldsymbol{v} \mapsto \|A\boldsymbol{v}\,\|/\|\boldsymbol{v}\,\|$ over $\mathbb{R}^n_*$ (or, equivalently, maximising $\boldsymbol{w} \mapsto \|A\boldsymbol{w}\|$ over the unit sphere $\{\boldsymbol{w} \in \mathbb{R}^n\colon \|\boldsymbol{w}\| = 1\}$). This difficulty is resolved by the following three theorems.

Theorem 2.7 *The matrix norm subordinate to the vector norm* $\|\cdot\|_\infty$ *can be expressed, for an* $n \times n$ *matrix* $A = (a_{ij})_{1\le i,j\le n} \in \mathbb{R}^{n\times n}$*, as*

$$\|A\|_\infty = \max_{i=1}^{n} \sum_{j=1}^{n} |a_{ij}|\,. \tag{2.39}$$

This result is often loosely expressed by saying that the ∞*-norm of a matrix is its largest row-sum.*

Proof Given an arbitrary vector $\boldsymbol{v}$ in $\mathbb{R}^n_*$, write $K = \|\boldsymbol{v}\|_\infty$, so that $|v_j| \le K$ for $j = 1, 2, \ldots, n$. Then,

$$|(A\boldsymbol{v})_i| = \left|\sum_{j=1}^{n} a_{ij} v_j\right| \le \sum_{j=1}^{n} |a_{ij}|\,|v_j| \le K \sum_{j=1}^{n} |a_{ij}|\,, \qquad i = 1, 2, \ldots, n\,.$$

Now we define

$$C = \max_{i=1}^{n} \sum_{j=1}^{n} |a_{ij}| \tag{2.40}$$

and note that

$$\frac{\|A\boldsymbol{v}\|_\infty}{\|\boldsymbol{v}\|_\infty} = \frac{\max_{i=1}^{n} |(A\boldsymbol{v})_i|}{\|\boldsymbol{v}\|_\infty} = \frac{\max_{i=1}^{n} |(A\boldsymbol{v})_i|}{K} \le C \qquad \forall\, \boldsymbol{v} \in \mathbb{R}^n_*\,.$$

Hence, $\|A\|_\infty \le C$.

Next we show that $\|A\|_\infty \ge C$. To do so, we take $\boldsymbol{v}$ to be a vector in $\mathbb{R}^n_*$ each of whose entries is ± 1, with the choice of sign to be made clear below. In the definition of C, equation (2.40), let m be the value of i for which the maximum is attained, or any one of the values if there is more than one. Then, in the vector $\boldsymbol{v}$ we give the element v_j the same sign as that of a_{mj}; if a_{mj} happens to be zero, the choice of the sign of v_j is irrelevant. With this definition of $\boldsymbol{v}$ we see at once that

$$\|A\boldsymbol{v}\|_\infty = \max_{i=1}^{n} \left|\sum_{j=1}^{n} a_{ij} v_j\right| \ge \left|\sum_{j=1}^{n} a_{mj} v_j\right| = \sum_{j=1}^{n} |a_{mj}|\,|v_j| = \sum_{j=1}^{n} |a_{mj}| = C\,.$$

As $\|\boldsymbol{v}\|_\infty = 1$, it follows that

$$\|A\boldsymbol{v}\|_\infty \ge C\|\boldsymbol{v}\|_\infty\,,$$

which means that $\|A\|_\infty \ge C$. Hence $\|A\|_\infty = C$, as required. $\square$

Theorem 2.8 *The matrix norm subordinate to the vector norm* $\|\cdot\|_1$ *can be expressed, for an* $n \times n$ *matrix* $A = (a_{ij})_{1 \le i,j \le n} \in \mathbb{R}^{n\times n}$, *as*

$$\|A\|_1 = \max_{j=1}^{n} \sum_{i=1}^{n} |a_{ij}| \,.$$

This is often loosely expressed by saying that the 1-norm of a matrix is its largest column-sum. The proof of this theorem is very similar to that of the previous one, and is left as an exercise (see Exercise 7). Note that Theorems 2.7 and 2.8 mean that the 1-norm of a matrix $A = (a_{ij})_{1\le i,j\le n}$ is the ∞-norm of the transpose $A^{\mathrm{T}} = (a_{ji})_{1\le i,j\le n}$ of the matrix.

Before we state a characterisation of the subordinate matrix 2-norm, we recall the following definition from linear algebra.

Definition 2.11 *Suppose that* $A \in \mathbb{R}^{n\times n}$. *A complex number* λ, *for which the set of linear equations*

$$A\boldsymbol{x} = \lambda \boldsymbol{x}$$

has a nontrivial solution $\boldsymbol{x} \in \mathbb{C}^n_* = \mathbb{C}^n \setminus \{\mathbf{0}\}$, *is called an* **eigenvalue** *of* A*; the associated solution* $\boldsymbol{x} \in \mathbb{C}^n_*$ *is called an* **eigenvector** *of* A *(corresponding to* λ*).*

Now we are ready to state our result.

Theorem 2.9 *Let* $A \in \mathbb{R}^{n\times n}$ *and denote the eigenvalues of the matrix* $B = A^{\mathrm{T}}A$ *by* λ_i, $i = 1, 2, \ldots, n$. *Then,*

$$\|A\|_2 = \max_{i=1}^{n} \lambda_i^{1/2}.$$

Proof Note first that the matrix B is symmetric, *i.e.*, $B = B^{\mathrm{T}}$; therefore all of its eigenvalues are real and the associated eigenvectors belong to $\mathbb{R}^n_*$. (You may wish to prove this: consult the proof of Theorem 3.1, part (ii), for a hint.) Moreover, all eigenvalues of B are nonnegative, since if $\boldsymbol{v} \in \mathbb{R}^n_*$ is an eigenvector of B and λ is the associated eigenvalue λ, then

$$A^{\mathrm{T}} A\boldsymbol{v} = B\boldsymbol{v} = \lambda \boldsymbol{v}$$

and therefore

$$\lambda = \frac{\boldsymbol{v}^{\mathrm{T}} A^{\mathrm{T}} A \boldsymbol{v}}{\boldsymbol{v}^{\mathrm{T}}\boldsymbol{v}} = \frac{\|A\boldsymbol{v}\|_2^2}{\|\boldsymbol{v}\|_2^2} \ge 0\,.$$

Suppose that the vectors $\boldsymbol{w}_i \in \mathbb{R}^n_*$, $i = 1, 2, \ldots, n$, are eigenvectors of B corresponding to the eigenvalues λ_i, $i = 1, 2, \ldots, n$. Since B is symmetric we may assume that the vectors $\boldsymbol{w}_i$ are orthogonal, *i.e.*, $\boldsymbol{w}_i^{\mathrm{T}}\boldsymbol{w}_j = 0$ for $i \neq j$, and we can normalise them so that $\boldsymbol{w}_i^{\mathrm{T}}\boldsymbol{w}_i = 1$ for $i = 1, 2, \ldots, n$. Now choose an arbitrary vector $\boldsymbol{u}$ in $\mathbb{R}^n_*$ and express it as a linear combination of the vectors $\boldsymbol{w}_i$, $i = 1, 2, \ldots, n$:

$$\boldsymbol{u} = c_1\boldsymbol{w}_1 + \cdots + c_n\boldsymbol{w}_n \,.$$

Then,

$$B\boldsymbol{u} = c_1\lambda_1\boldsymbol{w}_1 + \cdots + c_n\lambda_n\boldsymbol{w}_n \,.$$

We may assume, without loss of generality, that

$$(0 \leq)\, \lambda_1 \leq \lambda_2 \leq \cdots \leq \lambda_n \,.$$

Using the orthonormality of the vectors $\boldsymbol{w}_i$, $i = 1, 2, \ldots, n$, we get that

$$\begin{aligned} \|A\boldsymbol{u}\|_2^2 &= \boldsymbol{u}^{\mathrm{T}}A^{\mathrm{T}}A\boldsymbol{u} = \boldsymbol{u}^{\mathrm{T}}B\boldsymbol{u} \\ &= c_1^2\lambda_1 + \cdots + c_n^2\lambda_n \\ &\leq (c_1^2 + \cdots + c_n^2)\lambda_n \\ &= \lambda_n\|\boldsymbol{u}\|_2^2 \,, \end{aligned} \tag{2.41}$$

for any vector $\boldsymbol{u} \in \mathbb{R}^n_*$. Hence $\|A\|_2^2 \leq \lambda_n$. To prove equality we simply choose $\boldsymbol{u} = \boldsymbol{w}_n$ in (2.41), so that $c_1 = \cdots = c_{n-1} = 0$ and $c_n = 1$. □

The square roots of the (nonnegative) eigenvalues of $A^{\mathrm{T}}A$ are referred to as the **singular values** of A. Thus we have shown that the 2-norm of a matrix A is equal to the largest singular value of A.

If the matrix A is symmetric, then $B = A^{\mathrm{T}}A = A^2$, and the eigenvalues of B are just the squares of the eigenvalues of A. In this special case the 2-norm of A is the largest of the absolute values of its eigenvalues.

Theorem 2.10 *Given that* $\|\cdot\|$ *is a subordinate matrix norm on* $\mathbb{R}^{n\times n}$,

$$\|AB\| \leq \|A\|\,\|B\|$$

for any two matrices A *and* B *in* $\mathbb{R}^{n\times n}$.

Proof From the definition of subordinate matrix norm,

$$\|AB\| \;=\; \max_{\boldsymbol{v}\in\mathbb{R}^n_*} \frac{\|AB\boldsymbol{v}\,\|}{\|\boldsymbol{v}\,\|}\,.$$

As

$$\|AB\boldsymbol{v}\,\| \leq \|A\|\,\|B\boldsymbol{v}\,\|$$

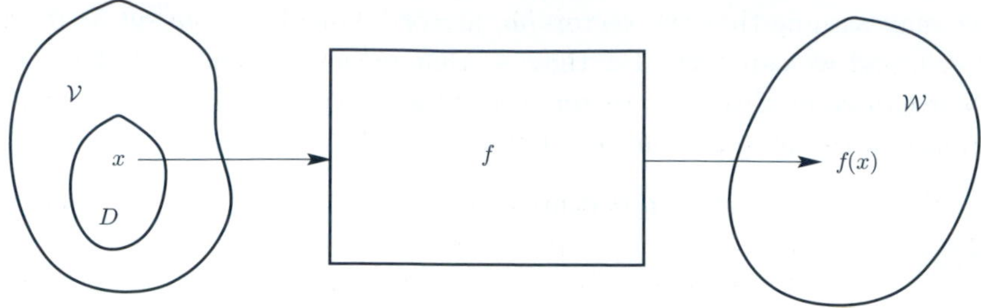

Fig. 2.3. 'Input' $x \in D \subset \mathcal{V}$ and 'output' $f(x) \in \mathcal{W}$ for a mapping $f\colon \mathcal{V} \to \mathcal{W}$.

for all $\boldsymbol{v} \in \mathbb{R}^n_*$, we have

$$\begin{aligned} \|AB\| &\leq \max_{\boldsymbol{v}\in\mathbb{R}^n_*} \frac{\|A\|\,\|B\boldsymbol{v}\,\|}{\|\boldsymbol{v}\,\|} \\ &= \|A\| \max_{\boldsymbol{v}\in\mathbb{R}^n_*} \frac{\|B\boldsymbol{v}\,\|}{\|\boldsymbol{v}\,\|} \\ &= \|A\|\,\|B\|\,, \end{aligned}$$

and hence the desired result. □

Now we are ready to embark on the study of sensitivity to perturbations in the problem of matrix inversion. In order to motivate the concept of *condition number* of a matrix which will play a key role in the analysis, we begin with a discussion of 'conditioning' in a slightly more general context.

Consider a mapping f from a subset D of a normed linear space $\mathcal{V}$ with norm $\|\cdot\|_{\mathcal{V}}$ into another normed linear space $\mathcal{W}$ with norm $\|\cdot\|_{\mathcal{W}}$, depicted in Figure 2.3, where $x \in D \subset \mathcal{V}$ is regarded as the 'input' for f and $f(x) \in \mathcal{W}$ is the 'output'. We shall be concerned with the sensitivity of the output to perturbations in the input; therefore, as a measure of sensitivity, we define the **absolute condition number** of f by

$$\mathrm{Cond}(f) = \sup_{\substack{x,y\in D\subset\mathcal{V}\\ x\neq y}} \frac{\|f(y) - f(x)\|_{\mathcal{W}}}{\|y - x\|_{\mathcal{V}}}\,. \tag{2.42}$$

If $\mathrm{Cond}(f) = +\infty$ or if $1 \ll \mathrm{Cond}(f) < +\infty$, we say that the mapping f is **ill-conditioned**.

Example 2.5 *Consider the function $f\colon x \in D \mapsto \sqrt{x}$, where D is a closed subinterval of $[0, \infty)$. Clearly, if $D = [1, 2]$, then* $\mathrm{Cond}(f) = 1/2$, *while if* $D = [0,1]$, *then* $\text{Cond}(f) = +\infty$. *Indeed, in the latter case, perturbing* $x = 0$ *to* $x = \varepsilon^2$, $0 < \varepsilon \ll 1$, *leads to a perturbation of the function value* $f(0) = 0$ *to* $f(\varepsilon^2) = \varepsilon = \frac{1}{\varepsilon}\varepsilon^2$*: a magnification by a factor* $\frac{1}{\varepsilon} \gg 1$ *in comparison with the size of the perturbation in* x.

When $\|f(y)-f(x)\|_{\mathcal{W}}/\|y-x\|_{\mathcal{V}}$ exhibits large variation as (x,y) ranges through $D \times D$, it is more helpful to consider a finer, local measure of conditioning, the **absolute local condition number**, at $x \in D \subset \mathcal{V}$, of the function f, defined by

$$\text{Cond}_x(f) = \sup_{\substack{\delta x \in \mathcal{V}\setminus\{0\} \\ x+\delta x \in D}} \frac{\|f(x+\delta x) - f(x)\|_{\mathcal{W}}}{\|\delta x\|_{\mathcal{V}}}. \tag{2.43}$$

Example 2.6 *Let us consider the function* $f\colon x \in D \mapsto \sqrt{x}$, *defined on the interval* $D = (0,\infty)$. *The absolute local condition number of* f *at* $x \in D$ *is* $\text{Cond}_x(f) = 1/(2\sqrt{x})$. *Clearly,* $\lim_{x\to 0+} \text{Cond}_x(f) = +\infty$, $\lim_{x\to+\infty} \text{Cond}_x(f) = 0$.

Although the definitions (2.42) and (2.43) seem intuitive, they are not always satisfactory from the practical point of view since they depend on the magnitudes of $f(x)$ and x. A more convenient definition of conditioning is arrived at by rescaling (2.43) by the norms of $f(x)$ and x. This leads us to the notion of **relative local condition number**

$$\text{cond}_x(f) = \sup_{\substack{\delta x \in \mathcal{V}\setminus\{0\} \\ x+\delta x \in D}} \frac{\|f(x+\delta x) - f(x)\|_{\mathcal{W}}/\|f(x)\|_{\mathcal{W}}}{\|\delta x\|_{\mathcal{V}}/\|x\|_{\mathcal{V}}},$$

where it is implicitly assumed that $x \in \mathcal{V}\setminus\{0\}$ and $f(x) \in \mathcal{W}\setminus\{0\}$. The next example highlights the difference between the absolute local condition number and the relative local condition number of f.

Example 2.7 *Let us consider the function* $f\colon x \in D \mapsto \sqrt{x}$, *defined on the interval* $D = (0,\infty)$. *Recall from the preceding example that the absolute local condition number of* f *at* $x \in D$ *approaches* $+\infty$ *as* x *tends to zero. In contrast with this, the relative local condition number of* f *is* $\text{cond}_x(f) = 1/2$ *for all* $x \in D$.

You may also wish to ponder the following, seemingly paradoxical, observation: $\lim_{\varepsilon\to 0} \text{cond}_\varepsilon(\sin) = 1$ *and* $\lim_{\varepsilon\to 0} \text{cond}_{\pi-\varepsilon}(\sin) = \infty$, *even though* $\sin 0 = \sin\pi = 0$ *and* $\text{Cond}_0(\sin) = \text{Cond}_\pi(\sin) = 1$.

Since the present section is concerned with the solution of the linear system $A\boldsymbol{x} = \boldsymbol{b}$, where $A \in \mathbb{R}^{n\times n}$ is nonsingular and $\boldsymbol{b} \in \mathbb{R}^n$, let us consider the relative local condition number of the mapping

$$A^{-1}\cdot : \boldsymbol{b} \in \mathbb{R}^n \mapsto A^{-1}\boldsymbol{b} \in \mathbb{R}^n$$

at $\boldsymbol{b} \in \mathbb{R}^n_* = \mathbb{R}^n \setminus \{\mathbf{0}\}$. We suppose that $\mathbb{R}^n$ has been equipped with a vector norm $\|\cdot\|$ and, since there is no danger of confusion, we denote the associated subordinate matrix norm by $\|\cdot\|$ also. Noting that $A^{-1}\cdot$ is defined on the whole of $\mathbb{R}^n$, it follows that $D = \mathcal{V} = \mathbb{R}^n$, $\mathcal{W} = \mathbb{R}^n$ and we deduce that

$$\begin{aligned} \mathrm{cond}_{\boldsymbol{b}}(A^{-1}\cdot) &= \sup_{\boldsymbol{\delta b} \in \mathbb{R}^n_*} \frac{\|A^{-1}(\boldsymbol{b}+\boldsymbol{\delta b}) - A^{-1}\boldsymbol{b}\| \,/\, \|A^{-1}\boldsymbol{b}\|}{\|\boldsymbol{\delta b}\| \,/\, \|\boldsymbol{b}\|} \\ &= \|A^{-1}\| \frac{\|\boldsymbol{b}\|}{\|A^{-1}\boldsymbol{b}\|}\,. \end{aligned}$$

Since $\|\boldsymbol{b}\| = \|A(A^{-1}\boldsymbol{b})\| \le \|A\|\,\|A^{-1}\boldsymbol{b}\|$, we conclude that

$$\mathrm{cond}_{\boldsymbol{b}}(A^{-1}\cdot) \le \|A^{-1}\|\,\|A\|\,. \tag{2.44}$$

If now, instead, we consider the mapping

$$A\cdot : \boldsymbol{x} \in \mathbb{R}^n \mapsto A\boldsymbol{x} \in \mathbb{R}^n,$$

an identical argument shows that, for $\boldsymbol{x} \in \mathbb{R}^n_*$,

$$\mathrm{cond}_{\boldsymbol{x}}(A\cdot) \le \|A\|\,\|A^{-1}\|\,. \tag{2.45}$$

The inequalities (2.44) and (2.45) indicate that the number $\|A^{-1}\|\,\|A\| = \|A\|\,\|A^{-1}\|$ plays a relevant role in the analysis of sensitivity to perturbations in numerical linear algebra; therefore we adopt the following definition.

Definition 2.12 *The* **condition number** *of a nonsingular matrix A is defined by*

$$\kappa(A) = \|A\|\,\|A^{-1}\|\,.$$

Clearly, $\kappa(A^{-1}) = \kappa(A)$. Further, since $AA^{-1} = I$, it follows from Theorem 2.10 that $\kappa(A) \ge 1$ for every matrix A. If $\kappa(A) \gg 1$, the matrix is said to be **ill-conditioned**. Evidently the condition number of a matrix is unaffected by scaling all its elements by multiplying by a nonzero constant.[1]

[1] We note in passing that, more generally, the condition number of a matrix $A \in \mathbb{R}^{m\times n}$ is defined by $\kappa(A) = \|A\|\,\|A^{+}\|$ where A^{+} is the Moore–Penrose generalised inverse of A. In the special case when $m = n$ and A is nonsingular, $A^{+} = A^{-1}$. For further details in this direction, we refer to the Notes at the end of the chapter. Here, the norm $\|\cdot\|$ on $\mathbb{R}^{m\times n}$ is defined as in (2.38). Theorems 2.7 and 2.8 are

There is a condition number for each norm; for example, if we use the 2-norm, then $\kappa_2(A) = \|A\|_2 \, \|A^{-1}\|_2$, and so on. Indeed, the size of the condition number of a matrix $A \in \mathbb{R}^{n\times n}$ is strongly dependent on the choice of the norm in $\mathbb{R}^n$. In order to illustrate the last point, let us consider the unit lower triangular matrix $A \in \mathbb{R}^{n\times n}$ defined by

$$A = \begin{pmatrix} 1 & 0 & 0 & 0 & \dots & 0 \\ 1 & 1 & 0 & 0 & \dots & 0 \\ 1 & 0 & 1 & 0 & \dots & 0 \\ 1 & 0 & 0 & 1 & \dots & 0 \\ \dots & \dots & \dots & \dots & \dots & \dots \\ 1 & 0 & 0 & 0 & \dots & 1 \end{pmatrix}, \tag{2.46}$$

and note that its inverse is

$$A^{-1} = \begin{pmatrix} 1 & 0 & 0 & 0 & \dots & 0 \\ -1 & 1 & 0 & 0 & \dots & 0 \\ -1 & 0 & 1 & 0 & \dots & 0 \\ -1 & 0 & 0 & 1 & \dots & 0 \\ \dots & \dots & \dots & \dots & \dots & \dots \\ -1 & 0 & 0 & 0 & \dots & 1 \end{pmatrix}.$$

Since

$$\|A\|_1 = n \qquad \text{and} \qquad \|A^{-1}\|_1 = n\,,$$

it follows that $\kappa_1(A) = n^2$. On the other hand,

$$\|A\|_\infty = 2 \qquad \text{and} \qquad \|A^{-1}\|_\infty = 2\,.$$

so that $\kappa_\infty(A) = 4 \ll n^2 = \kappa_1(A)$ when $n \gg 1$. (A question for the curious: how does the condition number $\kappa_2(A)$ of the matrix A in (2.46) depend on the size n of A? See Exercise 11.)

It is left as an exercise to show that for a nonsingular symmetric matrix A (*i.e.*, when $A^{\mathrm{T}} = A$), the 2-norm condition number $\kappa_2(A)$ is the ratio of the largest of the absolute values of the eigenvalues of A to the smallest of the absolute values of the eigenvalues (see Exercise 9).

easily extended to show that, for $A \in \mathbb{R}^{m\times n}$,

$$\|A\|_\infty = \max_{i=1}^{m} \sum_{j=1}^{n} |a_{ij}| \quad \text{and} \quad \|A\|_1 = \max_{j=1}^{n} \sum_{i=1}^{m} |a_{ij}|\,.$$

The 2-norm of A, $\|A\|_2$, is equal to the largest singular value of A, *i.e.*, the square root of the largest eigenvalue of the matrix $A^{\mathrm{T}}A \in \mathbb{R}^{n\times n}$, just as in Theorem 2.9.

We can now assess the sensitivity of the solution of the system $A\boldsymbol{x} = \boldsymbol{b}$ to changes in the right-hand side vector $\boldsymbol{b}$.

Theorem 2.11 *Suppose that $A \in \mathbb{R}^{n\times n}$ is a nonsingular matrix, $\boldsymbol{b} \in \mathbb{R}^n_*$, $A\boldsymbol{x} = \boldsymbol{b}$ and $A(\boldsymbol{x} + \boldsymbol{\delta x}) = \boldsymbol{b} + \boldsymbol{\delta b}$, with $\boldsymbol{\delta x}$, $\boldsymbol{\delta b} \in \mathbb{R}^n$. Then, $\boldsymbol{x} \in \mathbb{R}^n_*$ and*

$$\frac{\|\boldsymbol{\delta x}\|}{\|\boldsymbol{x}\|} \leq \kappa(A)\frac{\|\boldsymbol{\delta b}\|}{\|\boldsymbol{b}\|}\,.$$

Proof Evidently,

$$\boldsymbol{b} = A\boldsymbol{x} \quad \text{and} \quad \boldsymbol{\delta x} = A^{-1}(\boldsymbol{b} + \boldsymbol{\delta b}) - \boldsymbol{x} = A^{-1}\boldsymbol{\delta b}\,.$$

As $\boldsymbol{b} \neq \boldsymbol{0}$ by hypothesis, the first of these two equalities implies that $\boldsymbol{x} \neq \boldsymbol{0}$. Further,

$$\|\boldsymbol{b}\| \leq \|A\|\,\|\boldsymbol{x}\| \quad \text{and} \quad \|\boldsymbol{\delta x}\| \leq \|A^{-1}\|\,\|\boldsymbol{\delta b}\|\,.$$

The result follows immediately by multiplying these inequalities. □

Owing to the effect of rounding errors during the calculation, the numerical solution of $A\boldsymbol{x} = \boldsymbol{b}$ will not be exact. The numerical solution may be written $\boldsymbol{x}+\boldsymbol{\delta x}$, and we shall usually find that this vector satisfies the equation $A(\boldsymbol{x}+\boldsymbol{\delta x}) = \boldsymbol{b}+\boldsymbol{\delta b}$, where the elements of $\boldsymbol{\delta b}$ are very small. If the matrix A has a large condition number, however, the elements of $\boldsymbol{\delta x}$ may not be so small. An example of this will be presented in the next section.

2.8 Hilbert matrix

We consider the Hilbert matrix[1] H_n of order n, whose elements are

$$h_{ij} = \frac{1}{i+j-1}\,, \qquad i,j = 1,2,\ldots,n\,.$$

This matrix is symmetric and positive definite (*i.e.*, $H_n^{\mathrm{T}} = H_n$, and $\boldsymbol{x}^{\mathrm{T}} H_n \boldsymbol{x} > 0$ for all $\boldsymbol{x} \in \mathbb{R}^n_*$), and therefore all of its eigenvalues are real and positive (cf. Theorem 3.1, part (ii)). However, H_n becomes very nearly singular as n increases. Table 2.1 shows the largest and smallest eigenvalues, and the 2-norm condition number $\kappa_2(H_n)$ of H_n, for various values of n.

[1] David Hilbert (23 January 1862, Königsberg, Prussia (now Kaliningrad, Russia) – 14 February 1943, Göttingen, Germany) was the most prominent member of the Göttingen school of mathematics. He made significant contributions to many areas of the subject, including algebra, geometry, number theory, calculus of variations, functional analysis, integral equations, and the foundations of mathematics.

Table 2.1. *Eigenvalues and condition number of the Hilbert matrix H_n.*

n	$\lambda_{\max}$	$\lambda_{\min}$	$\kappa_2(H_n)$
5	1.6	3.3×10^{-6}	4.8×10^{5}
10	1.8	1.1×10^{-13}	1.6×10^{13}
15	1.8	3.0×10^{-21}	6.1×10^{20}
20	1.9	7.8×10^{-29}	2.5×10^{28}
25	2.0	1.9×10^{-36}	1.0×10^{36}

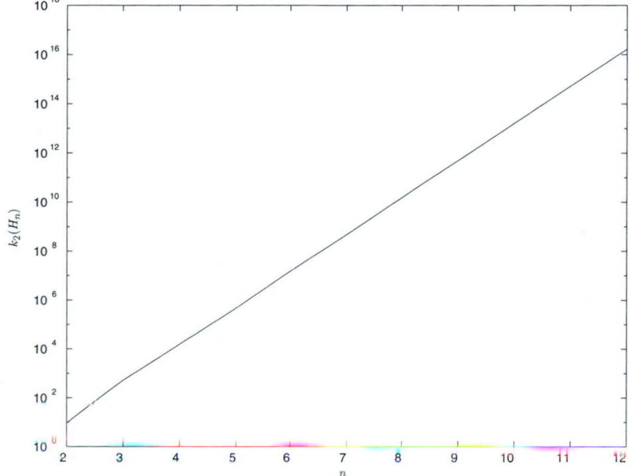

Fig. 2.4. Condition number $\kappa_2(H_n)$ of the Hilbert matrix H_n of size $n = 2, 3, \ldots, 12$ in the 2-norm, against n, in a semilogarithmic-scale plot.

Figure 2.4 depicts the logarithm of the condition number $\kappa_2(H_n)$ in the 2-norm of the Hilbert matrix H_n against its order, n; the straight line in our semilogarithmic-scale plot indicates that $\kappa_2(H_n)$, as a function of n, exhibits exponential growth. Indeed, it can be shown that

$$\kappa_2(H_n) \sim \frac{\left(\sqrt{2}+1\right)^{4n+4}}{2^{15/4}\sqrt{\pi n}} \qquad \text{as } n \to \infty \,.$$

We now define the vector $\boldsymbol{b}$ with elements $b_i = \sum_{j=1}^{n}(j/(i+j-1))$, $i = 1, 2, \ldots, n$, chosen so that the solution of $A\boldsymbol{x} = \boldsymbol{b}$, with $A = H_n$, is the vector $\boldsymbol{x}$ with elements $x_i = i$, $i = 1, 2, \ldots, n$. We obtain a numerical solution of the system, using the method described in Section

2.5 to give the calculated vector $\boldsymbol{x}+\boldsymbol{\delta x}$, and then compute the residual $\boldsymbol{\delta b}$ from $A(\boldsymbol{x}+\boldsymbol{\delta x}) = \boldsymbol{b}+\boldsymbol{\delta b}$. The calculation uses arithmetic operations correct to 15 decimal digits, which is roughly the accuracy used by many computer systems. The results are listed in Table 2.2.

Table 2.2. *Rounding errors in the solution of $H_n\boldsymbol{x} = \boldsymbol{b}$, where H_n is the Hilbert matrix of order n and $\boldsymbol{x} = (1, 2, \ldots, n)^{\mathrm{T}}$.*

n	$\lVert\boldsymbol{\delta b}\rVert_2/\lVert\boldsymbol{b}\rVert_2$	$\lVert\boldsymbol{\delta x}\rVert_2/\lVert\boldsymbol{x}\rVert_2$
5	1.2×10^{-15}	8.5×10^{-11}
10	1.7×10^{-15}	1.3×10^{-3}
15	2.8×10^{-15}	4.1
20	6.3×10^{-15}	8.7
25	1.9×10^{-13}	5.5×10^{2}

The relative size of the residual is, in nearly every case, about the size of the basic rounding error, 10^{-15}. The resulting errors in $\boldsymbol{x}$ are smaller than the bound given by Theorem 2.11, as might be expected, since that bound corresponds to the worst possible case. At any rate, for the Hilbert matrix of order greater than 14 the error is larger than the calculated solution itself, which renders the calculated solution meaningless. For matrices of this kind the condition number and the bound given by Theorem 2.11 are so large that they have little practical relevance, though they do indicate that, due to sensitivity to rounding errors, the numerical calculations are of unreliable accuracy.

The Hilbert matrix is, of course, a rather extreme example of an ill-conditioned matrix. However, we shall meet it in an important problem in Section 9.3 concerning the least squares approximation of a function by polynomials, where we shall see how a reformulation of the problem using an orthonormal basis avoids the disastrous loss of accuracy that would otherwise occur. In the next section, we introduce the idea of least squares approximation in the context of linear algebra and consider the solution of the resulting system of linear equations using the QR algorithm; this, too, relies on the notion of (ortho)normalisation.

2.9 Least squares method

Up to now, we have been dealing with systems of linear equations of the form $A\boldsymbol{x} = \boldsymbol{b}$ where $A \in \mathbb{R}^{n\times n}$. However, it is frequently the case in practical problems (typically, in problems of data-fitting) that the matrix A is not square but rectangular, and we have to solve a linear system of equations $A\boldsymbol{x} = \boldsymbol{b}$ with $A \in \mathbb{R}^{m\times n}$, $\boldsymbol{b} \in \mathbb{R}^m$, with $m > n$; since there are more equations than unknowns, in general such a system will have no solution. Consider, for example, the linear system (with $m = 3$, $n = 2$)

$$\begin{pmatrix} 3 & 1 \\ 1 & 1 \\ 4 & 2 \end{pmatrix} \begin{pmatrix} x_1 \\ x_2 \end{pmatrix} = \begin{pmatrix} 1 \\ 0 \\ 2 \end{pmatrix} ;$$

by adding the first two of the three equations and comparing the result with the third, it is easily seen that there is no solution. If, on the other hand, $m < n$, then the situation is reversed and there may be an infinite number of solutions. Consider, for example, the linear system (with $m = 1$, $n = 2$)

$$(3 \;\; 1) \begin{pmatrix} x_1 \\ x_2 \end{pmatrix} = 1 ;$$

any vector $\boldsymbol{x} = (\mu, 1 - 3\mu)^{\mathrm{T}}$, with $\mu \in \mathbb{R}$, is a solution to this system.

Suppose that $m \geq n$; we may then need to find a vector $\boldsymbol{x} \in \mathbb{R}^n$ which satisfies $A\boldsymbol{x} - \boldsymbol{b} \approx \mathbf{0}$ in $\mathbb{R}^m$ as nearly as possible in some sense. This suggests that we define the residual vector $\boldsymbol{r} = A\boldsymbol{x} - \boldsymbol{b}$ and require to minimise a certain norm of $\boldsymbol{r}$ in $\mathbb{R}^m$. From the practical point of view, it is particularly convenient to minimise the residual vector $\boldsymbol{r}$ in the 2-norm on $\mathbb{R}^m$; this leads to the **least squares** problem:

$$\underset{\boldsymbol{x} \in \mathbb{R}^n}{\text{Minimise}} \quad \|A\boldsymbol{x} - \boldsymbol{b}\|_2 .$$

This is clearly equivalent to minimising the square of the norm; so, on noting that

$$\|A\boldsymbol{x} - \boldsymbol{b}\|_2^2 = (A\boldsymbol{x} - \boldsymbol{b})^{\mathrm{T}}(A\boldsymbol{x} - \boldsymbol{b}) ,$$

the problem may be restated as

$$\underset{\boldsymbol{x} \in \mathbb{R}^n}{\text{Minimise}} \quad (A\boldsymbol{x} - \boldsymbol{b})^{\mathrm{T}}(A\boldsymbol{x} - \boldsymbol{b}) .$$

Since

$$(A\boldsymbol{x} - \boldsymbol{b})^{\mathrm{T}}(A\boldsymbol{x} - \boldsymbol{b}) = \boldsymbol{x}^{\mathrm{T}}A^{\mathrm{T}}A\boldsymbol{x} - 2\boldsymbol{x}^{\mathrm{T}}A^{\mathrm{T}}\boldsymbol{b} + \boldsymbol{b}^{\mathrm{T}}\boldsymbol{b} ,$$

the quantity to be minimised is a nonnegative quadratic function of the n components of the vector $\boldsymbol{x}$; the minimum therefore exists, and may be found by equating to zero the partial derivatives with respect to the components. This leads to the system of equations

$$B\boldsymbol{x} = A^{\mathrm{T}}\boldsymbol{b}, \qquad \text{where } B = A^{\mathrm{T}}A.$$

The matrix B is symmetric, and if A has full rank, n, then B is nonsingular; it is called the **normal** matrix, and the system $B\boldsymbol{x} = A^{\mathrm{T}}\boldsymbol{b}$ is called the system of **normal equations**.

The normal equations have important theoretical properties, but do not lead to a satisfactory numerical algorithm, except for fairly small problems. The difficulty is that in a practical least squares problem the matrix A is likely to be quite ill-conditioned, and $B = A^{\mathrm{T}}A$ will then be extremely ill-conditioned. For example, if

$$A = \begin{pmatrix} \varepsilon & 0 \\ 0 & 1 \end{pmatrix}$$

where $\varepsilon \in (0, 1)$, then $\kappa_2(A) = \varepsilon^{-1} > 1$, while

$$\kappa_2(B) = \kappa_2(A^{\mathrm{T}}A) = \varepsilon^{-2} = \varepsilon^{-1}\kappa_2(A) \gg \kappa_2(A)$$

when $0 < \varepsilon \ll 1$. If possible, one should avoid using a method which leads to such a dramatic deterioration of the condition number.

There are various alternative techniques which avoid the direct construction of the normal matrix $A^{\mathrm{T}}A$, and so do not lead to this extreme ill-conditioning. Here we shall describe just one algorithm, which begins by factorising the matrix A, but using an orthogonal matrix rather than the lower triangular factor as in Section 2.3.

Theorem 2.12 *Suppose that $A \in \mathbb{R}^{m\times n}$ where $m \geq n$. Then, A can be written in the form*

$$A = \hat{Q}\hat{R},$$

where $\hat{R}$ is an upper triangular $n \times n$ matrix, and $\hat{Q}$ is an $m \times n$ matrix which satisfies

$$\hat{Q}^{\mathrm{T}}\hat{Q} = I_n, \tag{2.47}$$

where I_n is the $n \times n$ identity matrix; see Figure 2.5. If $\operatorname{rank}(A) = n$, *then $\hat{R}$ is nonsingular.*

Proof We use induction on n, the number of columns in A. The theorem clearly holds when $n = 1$ so that A has only one column. Indeed, writing $\mathbf{c}$ for this column vector and assuming that $\mathbf{c} \neq \mathbf{0}$, the matrix $\hat{Q}$ has just

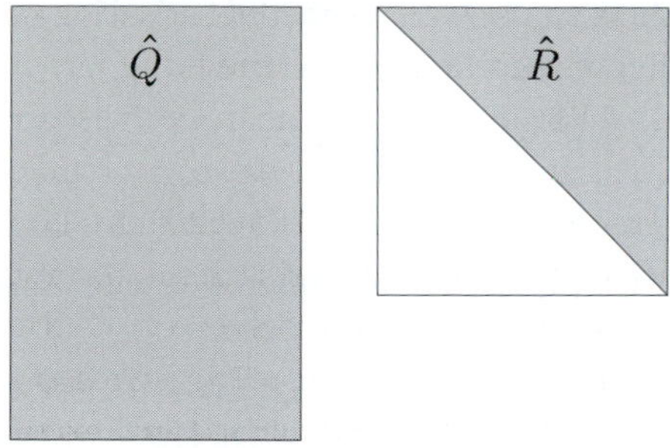

Fig. 2.5. QR factorisation of $A \in \mathbb{R}^{m\times n}$, $m \geq n$: $A = \hat{Q}\hat{R}$, $\hat{Q} \in \mathbb{R}^{m\times n}$, $\hat{Q}^{\mathrm{T}}\hat{Q} = I_n$, and the matrix $\hat{R} \in \mathbb{R}^{n\times n}$ is upper triangular.

one column, the vector $\boldsymbol{c}/\|\boldsymbol{c}\|_2$, and $\hat{R}$ has a single element, $\|\boldsymbol{c}\|_2$. In the special case where $\boldsymbol{c}$ is the zero vector we can choose $\hat{R}$ to have the single element 0, and $\hat{Q}$ to have a single column which can be an arbitrary vector in $\mathbb{R}^m$ whose 2-norm is equal to 1.

Suppose that the theorem is true when $n = k$, where $1 \leq k < m$. Consider a matrix A which has m rows and $k+1$ columns, partitioned as

$$A = (A_k \;\; \boldsymbol{a})\,,$$

where $\boldsymbol{a} \in \mathbb{R}^m$ is a column vector and A_k has k columns. To obtain the desired factorisation $\hat{Q}\hat{R}$ of A we seek $\hat{Q} = (\hat{Q}_k \;\; \boldsymbol{q})$ and

$$\hat{R} = \begin{pmatrix} \hat{R}_k & \boldsymbol{r} \\ \mathbf{0}^{\mathrm{T}} & \alpha \end{pmatrix}$$

such that

$$A = (A_k \;\; \boldsymbol{a}) = (\hat{Q}_k \;\; \boldsymbol{q}) \begin{pmatrix} \hat{R}_k & \boldsymbol{r} \\ \mathbf{0}^{\mathrm{T}} & \alpha \end{pmatrix}.$$

Multiplying this out and requiring that $\hat{Q}^{\mathrm{T}}\hat{Q} = I_{k+1}$, the identity matrix of order $k+1$, we conclude that

$$A_k = \hat{Q}_k\hat{R}_k\,, \tag{2.48}$$

$$\boldsymbol{a} = \hat{Q}_k\boldsymbol{r} + \boldsymbol{q}\alpha\,, \tag{2.49}$$

$$\hat{Q}_k^{\mathrm{T}}\hat{Q}_k = I_k\,, \tag{2.50}$$

$$\boldsymbol{q}^{\mathrm{T}}\hat{Q}_k = \mathbf{0}^{\mathrm{T}}\,, \tag{2.51}$$

$$\boldsymbol{q}^{\mathrm{T}}\boldsymbol{q} = 1\,. \tag{2.52}$$

These equations show that $\hat{Q}_k\hat{R}_k$ is the factorisation of A_k, which exists by the inductive hypothesis, and then lead to

$$\begin{aligned} \boldsymbol{r} &= \hat{Q}_k^{\mathrm{T}}\boldsymbol{a}\,, \\ \boldsymbol{q} &= (1/\alpha)(\boldsymbol{a} - \hat{Q}_k\hat{Q}_k^{\mathrm{T}}\boldsymbol{a})\,, \end{aligned}$$

where $\alpha = \|\boldsymbol{a} - \hat{Q}_k\hat{Q}_k^{\mathrm{T}}\boldsymbol{a}\|_2$. The number α is the constant required to ensure that the vector $\boldsymbol{q}$ is normalised.

The construction fails when $\boldsymbol{a} - \hat{Q}_k\hat{Q}_k^{\mathrm{T}}\boldsymbol{a} = \mathbf{0}$, for then the vector $\boldsymbol{q}$ cannot be normalised. In this case we choose $\boldsymbol{q}$ to be any normalised vector in $\mathbb{R}^m$ which is orthogonal in $\mathbb{R}^m$ to all the columns of $\hat{Q}_k$, for then $\boldsymbol{q}^{\mathrm{T}}\hat{Q}_k = \mathbf{0}^{\mathrm{T}}$ as required. The condition at the beginning of the proof, that $k < m$, is required by the fact that when $k = m$ the matrix $\hat{Q}_m$ is a square orthogonal matrix, and there is no vector $\boldsymbol{q}$ in $\mathbb{R}^m \setminus \{\mathbf{0}\}$ such that $\boldsymbol{q}^{\mathrm{T}}\hat{Q}_m = \mathbf{0}^{\mathrm{T}}$.

With these definitions of $\boldsymbol{q}, \boldsymbol{r}, \alpha, \hat{Q}_k$ and $\hat{R}_k$ we have constructed the required factors of A, showing that the theorem is true when $n = k+1$. Since it holds when $n = 1$ the induction is complete.

Now, for the final part, suppose that $\mathrm{rank}(A) = n$. If $\hat{R}$ were singular, there would exist a nonzero vector $\boldsymbol{p} \in \mathbb{R}^n$ such that $\hat{R}\boldsymbol{p} = \mathbf{0}$; then, $A\boldsymbol{p} = \hat{Q}\hat{R}\boldsymbol{p} = \mathbf{0}$, and hence $\mathrm{rank}(A) < n$, contradicting our hypothesis that $\mathrm{rank}(A) = n$. Therefore, if $\mathrm{rank}(A) = n$, then $\hat{R}$ is nonsingular. □

The matrix factorisation whose existence is asserted in Theorem 2.12 is called the **QR factorisation**. Here, we shall present its use in the solution of least squares problems. In Chapter 5 we shall revisit the idea in a different context which concerns the numerical solution of eigenvalue problems.

Theorem 2.13 *Suppose that $A \in \mathbb{R}^{m\times n}$, with $m \geq n$ and $\mathrm{rank}(A) = n$, and let $\mathbf{b} \in \mathbb{R}^m$. Then, there exists a unique least squares solution of the system of equations $A\boldsymbol{x} = \mathbf{b}$: a vector $\boldsymbol{x}$ in $\mathbb{R}^n$ which minimises the function $\boldsymbol{y} \mapsto \|A\boldsymbol{y} - \mathbf{b}\|_2$ over all $\boldsymbol{y}$ in $\mathbb{R}^n$. The vector $\boldsymbol{x}$ can be obtained by finding the factors $\hat{Q}$ and $\hat{R}$ of A defined in Theorem 2.12, and then solving the nonsingular upper triangular system $\hat{R}\boldsymbol{x} = \hat{Q}^{\mathrm{T}}\mathbf{b}$.*

Proof The matrix $\hat{Q}$ has m rows and n columns, with $m \geq n$, and it satisfies

$$\hat{Q}^{\mathrm{T}}\hat{Q} = I_n\,.$$

We shall suppose that $m > n$, the case $m = n$ being a trivial special case with

$$\boldsymbol{x} = A^{-1}\boldsymbol{b} = (\hat{Q}\hat{R})^{-1}\boldsymbol{b} = \hat{R}^{-1}\hat{Q}^{-1}\boldsymbol{b} = \hat{R}^{-1}\hat{Q}^{\mathrm{T}}\boldsymbol{b}\,,$$

and hence $\hat{R}\boldsymbol{x} = \hat{Q}^{\mathrm{T}}\boldsymbol{b}$, as required.

For $m > n$ now, the vector $\boldsymbol{b} \in \mathbb{R}^m$ can be written as the sum of two vectors:

$$\boldsymbol{b} = \boldsymbol{b}_q + \boldsymbol{b}_r\,,$$

where $\boldsymbol{b}_q$ is in the linear space spanned by the n columns of the matrix $\hat{Q}$, and $\boldsymbol{b}_r$ is in the orthogonal complement of this space in $\mathbb{R}^m$. The vector $\boldsymbol{b}_q$ is a linear combination of the columns of $\hat{Q}$, and $\boldsymbol{b}_r$ is orthogonal to every column of $\hat{Q}$; *i.e.*, there exists $\boldsymbol{c} \in \mathbb{R}^n$ such that

$$\boldsymbol{b} = \boldsymbol{b}_q + \boldsymbol{b}_r\,, \quad \boldsymbol{b}_q = \hat{Q}\boldsymbol{c}\,, \quad \hat{Q}^{\mathrm{T}}\boldsymbol{b}_r = \boldsymbol{0}\,. \tag{2.53}$$

Now, suppose that $\boldsymbol{x}$ is the solution of $\hat{R}\boldsymbol{x} = \hat{Q}^{\mathrm{T}}\boldsymbol{b}$, and that $\boldsymbol{y}$ is any vector in $\mathbb{R}^n$. Then,

$$\begin{aligned}
A\boldsymbol{y} - \boldsymbol{b} &= \hat{Q}\hat{R}\boldsymbol{y} - \boldsymbol{b} \\
&= \hat{Q}\hat{R}(\boldsymbol{y} - \boldsymbol{x}) + \hat{Q}\hat{R}\boldsymbol{x} - \boldsymbol{b} \\
&= \hat{Q}\hat{R}(\boldsymbol{y} - \boldsymbol{x}) + \hat{Q}\hat{Q}^{\mathrm{T}}\boldsymbol{b} - \boldsymbol{b} \\
&= \hat{Q}\hat{R}(\boldsymbol{y} - \boldsymbol{x}) + \hat{Q}\hat{Q}^{\mathrm{T}}\boldsymbol{b}_q - \boldsymbol{b}_q + \hat{Q}\hat{Q}^{\mathrm{T}}\boldsymbol{b}_r - \boldsymbol{b}_r \\
&= \hat{Q}\hat{R}(\boldsymbol{y} - \boldsymbol{x}) + \hat{Q}\hat{Q}^{\mathrm{T}}\hat{Q}\boldsymbol{c} - \boldsymbol{b}_q - \boldsymbol{b}_r \\
&= \hat{Q}\hat{R}(\boldsymbol{y} - \boldsymbol{x}) - \boldsymbol{b}_r\,,
\end{aligned}$$

where we have used (2.53) repeatedly; in particular, the last equality follows by noting that $\hat{Q}^{\mathrm{T}}\hat{Q} = I_n$. Hence

$$\begin{aligned}
\|A\boldsymbol{y} - \boldsymbol{b}\|_2^2 &= (\boldsymbol{y} - \boldsymbol{x})^{\mathrm{T}}\hat{R}^{\mathrm{T}}\hat{Q}^{\mathrm{T}}\hat{Q}\hat{R}(\boldsymbol{y} - \boldsymbol{x}) + \boldsymbol{b}_r^{\mathrm{T}}\boldsymbol{b}_r - 2(\boldsymbol{y} - \boldsymbol{x})^{\mathrm{T}}\hat{R}^{\mathrm{T}}\hat{Q}^{\mathrm{T}}\boldsymbol{b}_r \\
&= \|\hat{R}(\boldsymbol{y} - \boldsymbol{x})\|_2^2 + \|\boldsymbol{b}_r\|^2 \\
&\geq \|\boldsymbol{b}_r\|^2
\end{aligned}$$

since $\hat{Q}^{\mathrm{T}}\boldsymbol{b}_r = \boldsymbol{0}$. Thus $\|A\boldsymbol{y} - \boldsymbol{b}\|_2$ is smallest when $\hat{R}(\boldsymbol{y} - \boldsymbol{x}) = \boldsymbol{0}$, which implies that $\boldsymbol{y} = \boldsymbol{x}$, since the matrix $\hat{R}$ is nonsingular. Hence $\boldsymbol{x}$, defined as the solution of $\hat{R}\boldsymbol{x} = \hat{Q}^{\mathrm{T}}\boldsymbol{b}$, is the required least squares solution. □

2.10 Notes

There are many good books on the subject of numerical linear algebra which cover the topics discussed in this chapter in much greater detail, and address questions which we have not touched on here. Without any attempt to be exhaustive, we single out four texts from the vast literature. The first two books on the list below are well-known monographs on the subject, while the last two are excellent textbooks.

- G.H. Golub and C.F. Van Loan, *Matrix Computations,* Third Edition, Johns Hopkins University Press, Baltimore, 1996.
- N.J. Higham, *Accuracy and Stability of Numerical Algorithms,* SIAM, Philadelphia, 1996.
- L.N. Trefethen and D. Bau, III, *Numerical Linear Algebra,* SIAM, Philadelphia, 1997.
- P.G. Ciarlet, *Introduction to Numerical Linear Algebra and Optimisation,* Cambridge University Press, Cambridge, 1989.

As we have already noted in Section 2.2, the invention of the elimination technique is attributed to Gauss who published the method in his *Theoria motus* (1809), although the idea was already known to the Chinese two thousand years ago. Gauss himself was concerned with positive definite systems. The method was extended to linear systems with general matrices by Jacobi.[1] The interpretation of Gaussian elimination as matrix factorisation is due to P.S. Dwyer: A matrix presentation of least squares and correlation theory with matrix justification of improved methods of solutions, *Ann. Math. Stat.* **15**, 82–89, 1944.

The sensitivity of Gaussian elimination to rounding errors was studied by Wilkinson[2] in Error analysis of direct methods of matrix inversion, *J. Assoc. Comput. Math.* **8**, 281–330, 1961. The idea of pivoting was used as early as 1947 by von Neumann[3] and Goldstein.[4] The concept of the condition number of a matrix was introduced by Turing[5] in Rounding-off errors in matrix processes, *Quart. J. Mech. Appl. Math.* **1**, 287–308, 1948. Our treatment of condition numbers follows the textbook of Trefethen and Bau, cited above.

[1] Carl Gustav Jacob Jacobi (10 December 1804, Potsdam, Prussia, Holy Roman Empire (now Germany) – 18 February 1851, Berlin, Germany) had made important contributions to the theory of elliptic functions and differential equations. The English translation, by G.W. Stuart, of Jacobi's German original article is available from the Internet on `ftp://thales.cs.umd.edu/pub/biographical/xhist.html`

[2] James Hardy Wilkinson (27 September 1919, Strood, Kent, England – 5 October 1986, London, England).

[3] John von Neumann (28 December 1903, Budapest, Austria–Hungary (now in Hungary) – 8 February 1957, Washington DC, USA).

[4] Sydney Goldstein (3 December 1903, Hull, England – 22 January 1989, Belmont, Massachusetts, USA).

[5] Alan Mathison Turing (23 June 1912, London, England – 7 June 1954, Wilmslow, Cheshire, England).

Normed linear spaces play a key role in functional analysis (see, for example, K. Yosida, *Functional Analysis*, Third Edition, Springer, Berlin, 1971, page 30). Here, we have concentrated on finite-dimensional normed linear spaces over the field of real numbers.

The relevance of norms in numerical linear algebra was highlighted by Householder[1] in his book *The Theory of Matrices in Numerical Analysis,* Blaisdell, New York, 1964.

The idea of least squares fitting is due to Gauss, who invented the method in the 1790s. However, it was the French mathematician Legendre[2] who first published the method in 1806 in a book on determining the orbits of comets. Legendre's method involved a number of observations taken at equal intervals and he assumed that the comet followed a parabolic path, so he ended up with more equations than there were unknowns. Legendre then applied his methods to the data known for two comets. In an Appendix to the book Legendre described the least squares method of fitting a curve to the data available. Gauss published his version of the least squares method in 1809 and, although acknowledging that it had already appeared in Legendre's book, Gauss nevertheless claimed priority for himself. This greatly hurt Legendre, leading to one of the infamous priority disputes in the history of mathematics. A recent exhaustive monograph on numerical algorithms for least squares problems is due to Å. Björk: *Numerical Methods for Least Squares Problems,* SIAM, Philadelphia, 1996.

The version of the QR factorisation considered here is the *reduced version*, following the terminology in Chapter 7 of Trefethen and Bau. In the *full version* of the QR factorisation for a matrix $A \in \mathbb{R}^{m\times n}$, we have $A = QR$, where $Q \in \mathbb{R}^{m\times m}$, $R \in \mathbb{R}^{m\times n}$ (cf. Chapter 5).

In a footnote to Definition 2.12 we mentioned the Moore–Penrose generalised inverse A^+ of a matrix $A \in \mathbb{R}^{m\times n}$. A^+ can be defined through the singular value decomposition of A (cf. L.N. Trefethen and D. Bau, III: *Numerical Linear Algebra*, SIAM, Philadelphia, 1997). Recall that the singular values of A are the square roots of the (nonnegative) eigenvalues of the matrix $A^{\mathrm{T}}A$.

[1] Alton Scott Householder (5 May 1904, Rockford, Illinois, USA – 4 July 1993, Malibu, California, USA) was one of the pioneers of numerical linear algebra. Householder's obituary by G.W. Stuart, published in *SIAM News*, is available from `http://www.inf.ethz.ch/research/wr/conferences/householder/stewart.html`

[2] Adrien-Marie Legendre (18 September 1752, Paris, France – 10 January 1833, Paris, France).

Theorem 2.14 (Singular value decomposition) *Let $A \in \mathbb{R}^{m\times n}$; then, there exist $U \in \mathbb{R}^{m\times n}$, $\Sigma \in \mathbb{R}^{n\times n}$ and $V \in \mathbb{R}^{n\times n}$ such that*

$$A = U\Sigma V^{\mathrm{T}},$$

where Σ is a diagonal matrix whose diagonal entries, σ_{ii}, $i = 1, 2, \ldots, n$, are the singular values of A, $U^{\mathrm{T}}U = I_n$ and $V^{\mathrm{T}}V = I_n$, with I_n denoting the $n \times n$ identity matrix.

The Moore–Penrose generalised inverse of the diagonal matrix $\Sigma \in \mathbb{R}^{n\times n}$ is defined as the diagonal matrix $\Sigma^+ \in \mathbb{R}^{n\times n}$ whose diagonal entries are

$$\sigma_{ii}^+ = \begin{cases} \sigma_{ii}^{-1} & \text{if } \sigma_{ii} \neq 0, \\ 0 & \text{if } \sigma_{ii} = 0. \end{cases}$$

The generalised inverse $A^+ \in \mathbb{R}^{n\times m}$ of a matrix $A \in \mathbb{R}^{m\times n}$ with singular value decomposition $A = U\Sigma V^{\mathrm{T}}$ is defined by

$$A^+ = V\Sigma^+U^{\mathrm{T}}.$$

In the special case when $m = n$ and $A \in \mathbb{R}^{n\times n}$ is nonsingular, the n singular values of A are all nonzero and therefore $\Sigma^+ = \Sigma^{-1}$. Hence, also, $A^+ = A^{-1}$, which then justifies the use of the terminology 'generalised inverse' for the matrix A^+ defined above.

Exercises

2.1 Let $n \geq 2$. Given the matrix $A = (a_{ij}) \in \mathbb{R}^{n\times n}$, the permutation matrix $Q \in \mathbb{R}^{n\times n}$ reverses the order of the rows of A, so that $(QA)_{i,j} = a_{n+1-i,j}$. If $L \in \mathbb{R}^{n\times n}$ is a lower triangular matrix, what is the structure of the matrix QLQ?

Show how to factorise $A \in \mathbb{R}^{n\times n}$ in the form $A = UL$, where $U \in \mathbb{R}^{n\times n}$ is unit upper triangular and $L \in \mathbb{R}^{n\times n}$ is lower triangular. What conditions on A will ensure that the factorisation exists? Give an example of a square matrix A which cannot be factorised in this way.

2.2 Let $n \geq 2$. Consider a matrix $A \in \mathbb{R}^{n\times n}$ whose every leading principal submatrix of order less than n is nonsingular. Show that A can be factored in the form $A = LDU$, where $L \in \mathbb{R}^{n\times n}$ is unit lower triangular, $D \in \mathbb{R}^{n\times n}$ is diagonal and $U \in \mathbb{R}^{n\times n}$ is unit upper triangular.

If the factorisation $A = LU$ is known, where L is unit lower triangular and U is upper triangular, show how to find the factors of the transpose A^{T}.

2.3 Let $n \geq 2$ and suppose that the matrix $A \in \mathbb{R}^{n \times n}$ is nonsingular. Show by induction, as in Theorem 2.3, that there are a permutation matrix $P \in \mathbb{R}^{n \times n}$, a lower triangular matrix $L \in \mathbb{R}^{n \times n}$, and a unit upper triangular matrix $U \in \mathbb{R}^{n \times n}$ such that $PA = LU$.

By finding a suitable 2×2 matrix A, or otherwise, show that this may not be true if A is singular.

2.4 The lower triangular matrix $L \in \mathbb{R}^{n \times n}$, $n \geq 2$, is nonsingular, and the vector $\boldsymbol{b} \in \mathbb{R}^n$ is such that $b_i = 0$, $i = 1, 2, \ldots, k$, with $1 \leq k \leq n$. The vector $\boldsymbol{y} \in \mathbb{R}^n$ is the solution of $L\boldsymbol{y} = \boldsymbol{b}$. Show, by partitioning L, that $y_j = 0$, $j = 1, 2, \ldots, k$. Hence give an alternative proof of Theorem 2.1(iv), that the inverse of a nonsingular lower triangular matrix is itself lower triangular.

2.5 Given a matrix $A \in \mathbb{R}^{n \times n}$, define the matrix $B \in \mathbb{R}^{n \times 2n}$ in which the first n columns are the columns of A, and the last n columns are the columns of the identity matrix I_n. Consider the following computational scheme. Treat the rows of the matrix B in order, so that $j = 1, 2, \ldots, n$. Multiply every element in row j by the reciprocal of the diagonal element, $1/b_{jj}$; then, replace every element b_{ik} which is not in row j, so that $i \neq j$, by $b_{ik} - b_{ij}b_{jk}$.

Show that the result is equivalent to multiplying B on the left by a sequence of matrices. Explain why, at the end of the computation, the first n columns of B are the columns of the identity matrix I_n, and the last n columns are the columns of the inverse matrix A^{-1}. Give a condition on the matrix A which will ensure that the computation does not break down.

Show that the process as described requires approximately $2n^3$ multiplications, but that, if the multiplications in which one of the factors is zero are not counted, the total is approximately n^3.

2.6 Use the method of Exercise 5 to find the inverse of the matrix

$$A = \begin{pmatrix} 2 & 4 & 2 \\ 1 & 0 & 3 \\ 3 & 1 & 2 \end{pmatrix}.$$

2.7 Suppose that for a matrix $A \in \mathbb{R}^{n\times n}$,

$$\sum_{i=1}^{n} |a_{ij}| \leq C\,, \quad j = 1, 2, \ldots, n\,.$$

Show that, for any vector $\boldsymbol{x} \in \mathbb{R}^n$,

$$\sum_{i=1}^{n} |(A\boldsymbol{x})_i| \leq C\|\boldsymbol{x}\|_1\,.$$

Find a nonzero vector $\boldsymbol{x}$ for which equality can be achieved, and deduce that

$$\|A\|_1 = \max_{j=1}^{n} \sum_{i=1}^{n} |a_{ij}|\,.$$

2.8 (i) Show that, for any vector $\boldsymbol{v} = (v_1, \ldots, v_n)^{\mathrm{T}} \in \mathbb{R}^n$,

$$\|\boldsymbol{v}\|_\infty \leq \|\boldsymbol{v}\|_2 \text{ and } \|\boldsymbol{v}\|_2^2 \leq \|\boldsymbol{v}\|_1\,\|\boldsymbol{v}\|_\infty\,.$$

In each case give an example of a nonzero vector $\boldsymbol{v}$ for which equality is attained. Deduce that $\|\boldsymbol{v}\|_\infty \leq \|\boldsymbol{v}\|_2 \leq \|\boldsymbol{v}\|_1$. Show also that $\|\boldsymbol{v}\|_2 \leq \sqrt{n}\,\|\boldsymbol{v}\|_\infty$.
(ii) Show that, for any matrix $A \in \mathbb{R}^{m\times n}$,

$$\|A\|_\infty \leq \sqrt{n}\,\|A\|_2 \text{ and } \|A\|_2 \leq \sqrt{m}\,\|A\|_\infty\,.$$

In each case give an example of a matrix A for which equality is attained. (See the footnote following Definition 2.12 for the meaning of $\|A\|_1$, $\|A\|_2$ and $\|A\|_\infty$ when $A \in \mathbb{R}^{m\times n}$.)

2.9 Prove that, for any nonsingular matrix $A \in \mathbb{R}^{n\times n}$,

$$\kappa_2(A) = \left(\frac{\lambda_n}{\lambda_1}\right)^{1/2}\,,$$

where λ_1 is the smallest and λ_n is the largest eigenvalue of the matrix $A^{\mathrm{T}}A$.

Show that the condition number $\kappa_2(Q)$ of an orthogonal matrix Q is equal to 1. Conversely, if $\kappa_2(A) = 1$ for the matrix A, show that all the eigenvalues of $A^{\mathrm{T}}A$ are equal; deduce that A is a scalar multiple of an orthogonal matrix.

2.10 Let $A \in \mathbb{R}^{n\times n}$. Show that if λ is an eigenvalue of $A^{\mathrm{T}}A$, then

$$0 \leq \lambda \leq \|A^{\mathrm{T}}\|\,\|A\|\,,$$

provided that the same subordinate matrix norm is used for

both A and A^{T}. Hence show that, for any nonsingular $n \times n$ matrix A,

$$\kappa_2(A) \le \{\kappa_1(A)\,\kappa_\infty(A)\}^{1/2} .$$

2.11 For the matrix defined by (2.46) write down the matrix $A^{\mathrm{T}}A$. Show that any vector $\boldsymbol{x} \neq \mathbf{0}$ is an eigenvector of $A^{\mathrm{T}}A$ with eigenvalue $\lambda = 1$, provided that $x_1 = 0$ and $x_2 + \cdots + x_n = 0$. Show also that there are two eigenvectors with $x_2 = \cdots = x_n$ and find the corresponding eigenvalues. Deduce that

$$\kappa_2(A) = \tfrac{1}{2}(n+1)\left(1 + \sqrt{1 - \tfrac{4}{(n+1)^2}}\right) .$$

2.12 Let $B \in \mathbb{R}^{n\times n}$ and denote by I the identity matrix of order n. Show that if the matrix $I - B$ is singular, then there exists a nonzero vector $\boldsymbol{x} \in \mathbb{R}^n$ such that $(I-B)\boldsymbol{x} = \mathbf{0}$; deduce that $\|B\| \ge 1$, and hence that, if $\|A\| < 1$, then the matrix $I - A$ is nonsingular.

Now suppose that $A \in \mathbb{R}^{n\times n}$ with $\|A\| < 1$. Show that

$$(I-A)^{-1} = I + A(I-A)^{-1} ,$$

and hence that

$$\|(I-A)^{-1}\| \le 1 + \|A\|\,\|(I-A)^{-1}\| .$$

Deduce that

$$\|(I-A)^{-1}\| \le \frac{1}{1 - \|A\|} .$$

2.13 Let $A \in \mathbb{R}^{n\times n}$ be a nonsingular matrix and $\boldsymbol{b} \in \mathbb{R}^n_*$. Suppose that $A\boldsymbol{x} = \boldsymbol{b}$ and $(A+\delta A)(\boldsymbol{x}+\boldsymbol{\delta x}) = \boldsymbol{b}$, and that $\|A^{-1}\,\delta A\| < 1$. Use the result of Exercise 12 to show that

$$\frac{\|\boldsymbol{\delta x}\|}{\|\boldsymbol{x}\|} \le \frac{\|A^{-1}\,\delta A\|}{1 - \|A^{-1}\,\delta A\|} .$$

2.14 Suppose that $A \in \mathbb{R}^{n\times n}$ is a nonsingular matrix, and $\boldsymbol{b} \in \mathbb{R}^n_*$. Given that $A\boldsymbol{x} = \boldsymbol{b}$ and $A(\boldsymbol{x}+\boldsymbol{\delta x}) = \boldsymbol{b} + \boldsymbol{\delta b}$, Theorem 2.11 states that

$$\frac{\|\boldsymbol{\delta x}\|}{\|\boldsymbol{x}\|} \le \kappa(A)\frac{\|\boldsymbol{\delta b}\|}{\|\boldsymbol{b}\|} .$$

By considering the eigenvectors of $A^{\mathrm{T}}A$, show how to find vectors $\boldsymbol{b}$ and $\boldsymbol{\delta b}$ for which equality is attained, when using the 2-norm.

2.15 Find the QR factorisation of the matrix

$$A = \begin{pmatrix} 9 & -6 \\ 12 & -8 \\ 0 & 20 \end{pmatrix},$$

and hence find the least squares solution of the system of linear equations

$$\begin{aligned} 9x - 6y &= 300\,, \\ 12x - 8y &= 600\,, \\ 20y &= 900\,. \end{aligned}$$

3

Special matrices

3.1 Introduction

In this chapter we show how one can modify the elimination method for the solution of $A\boldsymbol{x} = \boldsymbol{b}$ when the matrix A has certain special properties. In particular when $A \in \mathbb{R}^{n\times n}$ is symmetric and positive definite the amount of computational work can be halved. For matrices with a band structure, having nonzero elements only in positions close to the diagonal, the efficiency can be improved even more dramatically.

3.2 Symmetric positive definite matrices

Definition 3.1 *The matrix $A = (a_{ij}) \in \mathbb{R}^{n\times n}$ is said to be* **symmetric** *if $a_{ij} = a_{ji}$ for all i and j in the set $\{1, 2, \ldots, n\}$; i.e., if $A = A^{\mathrm{T}}$. The set of all symmetric matrices $A \in \mathbb{R}^{n\times n}$ will be denoted by $\mathbb{R}^{n\times n}_{\mathrm{sym}}$. A matrix $A \in \mathbb{R}^{n\times n}$ is called* **positive definite** *if*

$$\boldsymbol{x}^{\mathrm{T}} A\boldsymbol{x} > 0$$

for every vector $\boldsymbol{x} \in \mathbb{R}^n_* = \mathbb{R}^n \setminus \{\mathbf{0}\}$.

Example 3.1 *Consider the matrix $A \in \mathbb{R}^{2\times 2}$,*

$$A = \begin{pmatrix} a & b \\ c & d \end{pmatrix}$$

and a vector $\boldsymbol{x} = (x_1, x_2)^{\mathrm{T}} \in \mathbb{R}^2_* = \mathbb{R}^2 \setminus \{\mathbf{0}\}$.

Clearly, $\boldsymbol{x}^{\mathrm{T}} A\boldsymbol{x} = ax_1^2 + (b + c)x_1x_2 + dx_2^2$. The quadratic form on the right-hand side is positive for all real numbers x_1, x_2 such that

$\boldsymbol{x} = (x_1, x_2)^{\mathrm{T}} \neq (0,0)^{\mathrm{T}} = \mathbf{0}$ if, and only if,

$$a > 0\,, \quad d > 0 \text{ and } (b+c)^2 < 4ad\,.$$

We see that if $A \in \mathbb{R}^{2\times 2}$ is positive definite, then the diagonal elements of A are positive. Further, noting that the third inequality can be rewritten as

$$(b-c)^2 < 4(ad - bc) = 4\,\det(A)\,,$$

we deduce that the determinant of a positive definite matrix $A \in \mathbb{R}^{2\times 2}$ is positive. This, of course, is still true in the special case when $A \in \mathbb{R}^{2\times 2}_{\mathrm{sym}}$, *i.e.*, when $b = c$. ◇

The next theorem extends the observations of the last example to any symmetric positive definite matrix $A \in \mathbb{R}^{n\times n}$.

Theorem 3.1 *Suppose that $n \geq 2$ and $A = (a_{ij}) \in \mathbb{R}^{n\times n}_{\mathrm{sym}}$ is positive definite; then:*

(i) *all the diagonal elements of A are positive, that is, $a_{ii} > 0$, for $i = 1, 2, \ldots, n$;*
(ii) *all the eigenvalues of A are real and positive, and the eigenvectors of A belong to $\mathbb{R}^n_*$;*
(iii) *the determinant of A is positive;*
(iv) *every submatrix B of A obtained by deleting any set of rows and the corresponding set of columns from A is symmetric and positive definite; in particular, every leading principal submatrix is positive definite;*
(v) *$a_{ij}^2 < a_{ii}a_{jj}$ for all i and j in $\{1, 2, \ldots, n\}$ such that $i \neq j$;*
(vi) *the element of A with largest absolute value lies on the diagonal;*
(vii) *if α is the largest of the diagonal elements of A, then*

$$|a_{ij}| \leq \alpha \qquad \forall\, i, j \in \{1, 2, \ldots, n\}\,.$$

Proof (i) Consider the vector $\boldsymbol{x} \in \mathbb{R}^n$ with only one nonzero element, in position $i \in \{1, 2, \ldots, n\}$. Since A is positive definite and $\boldsymbol{x} \in \mathbb{R}^n_*$, it follows that $x_i a_{ii} x_i = \boldsymbol{x}^{\mathrm{T}} A \boldsymbol{x} > 0$, and therefore $a_{ii} > 0$.

(ii) Suppose that $\lambda \in \mathbb{C}$ is an eigenvalue of A and let $\boldsymbol{x} \in \mathbb{C}^n_* = \mathbb{C}^n \backslash \{\mathbf{0}\}$ denote the associated eigenvector. Further, let $\bar{\boldsymbol{x}}$ denote the vector in $\mathbb{C}^n_*$ whose ith element is the complex conjugate of the ith element of

$\boldsymbol{x}$, $i = 1, 2, \ldots, n$. As $A\boldsymbol{x} = \lambda\boldsymbol{x}$, it follows that $\bar{\boldsymbol{x}}^{\mathrm{T}}A\boldsymbol{x} = \lambda(\bar{\boldsymbol{x}}^{\mathrm{T}}\boldsymbol{x})$, and therefore, using the symmetry of A,

$$\boldsymbol{x}^{\mathrm{T}}A\bar{\boldsymbol{x}} = \boldsymbol{x}^{\mathrm{T}}A^{\mathrm{T}}\bar{\boldsymbol{x}} = (\bar{\boldsymbol{x}}^{\mathrm{T}}A\boldsymbol{x})^{\mathrm{T}} = (\lambda(\bar{\boldsymbol{x}}^{\mathrm{T}}\boldsymbol{x}))^{\mathrm{T}} = \lambda(\boldsymbol{x}^{\mathrm{T}}\bar{\boldsymbol{x}})\,.$$

Complex conjugation then yields $\bar{\boldsymbol{x}}^{\mathrm{T}}A\boldsymbol{x} = \bar{\lambda}(\bar{\boldsymbol{x}}^{\mathrm{T}}\boldsymbol{x})$, and hence $\lambda(\bar{\boldsymbol{x}}^{\mathrm{T}}\boldsymbol{x}) = \bar{\lambda}(\bar{\boldsymbol{x}}^{\mathrm{T}}\boldsymbol{x})$. As $\boldsymbol{x} \neq \boldsymbol{0}$, it follows that $\lambda = \bar{\lambda}$; i.e., λ is a real number.

The fact that the eigenvector associated with λ has real elements follows by noting that all elements of the singular matrix $A - \lambda I$ are real numbers. Therefore, the column vectors of $A - \lambda I$ are linearly dependent in $\mathbb{R}^n$. Hence there exist n real numbers $x_1, \ldots, x_n$ such that $(A - \lambda I)\boldsymbol{x} = \boldsymbol{0}$, where $\boldsymbol{x} = (x_1, \ldots, x_n)^{\mathrm{T}}$.

Finally, as $A\boldsymbol{x} = \lambda\boldsymbol{x}$ with $\lambda \in \mathbb{R}$ and $\boldsymbol{x} \in \mathbb{R}^n_*$, we have that $\boldsymbol{x}^{\mathrm{T}}A\boldsymbol{x} = \lambda\boldsymbol{x}^{\mathrm{T}}\boldsymbol{x}$. Since $\lambda = \boldsymbol{x}^{\mathrm{T}}A\boldsymbol{x}/\boldsymbol{x}^{\mathrm{T}}\boldsymbol{x}$ and A is positive definite, λ is the ratio of two positive real numbers and therefore also real and positive.

(iii) This follows from the fact that the determinant of A is equal to the product of its eigenvalues, and the previous result. Indeed, since A is symmetric, there exist an orthogonal matrix X and a diagonal matrix Λ, whose diagonal elements are the eigenvalues λ_i, $i = 1, 2, \ldots, n$, of A, such that $A = X^{\mathrm{T}}\Lambda X = X^{-1}\Lambda X$. By the Binet–Cauchy Theorem (see Chapter 2, end of Section 2.3),

$$\begin{aligned}\det(A) &= \det(X^{-1})\det(\Lambda)\det(X) \\ &= \frac{1}{\det(X)}\det(\Lambda)\det(X) \\ &= \det(\Lambda) = \lambda_1 \ldots \lambda_n > 0\,.\end{aligned}$$

(iv) Consider the vector $\boldsymbol{x} \in \mathbb{R}^n_*$ with zeros in the positions corresponding to the rows which have been deleted. Then,

$$\boldsymbol{x}^{\mathrm{T}}A\boldsymbol{x} = \boldsymbol{y}^{\mathrm{T}}B\boldsymbol{y}$$

where B is the submatrix of A containing the rows and columns which remain after deletion, and $\boldsymbol{y}$ is the vector consisting of the elements of $\boldsymbol{x}$ which were not deleted. Since the expression on the left is positive, the same is true of the expression on the right, for all vectors $\boldsymbol{y}$ except the zero vector. Therefore B is positive definite.

(v) By the previous result the 2×2 submatrix consisting of rows and columns r and s of A is positive definite, and its determinant is therefore positive.

(vi) This follows from the previous result, since it shows that $|a_{ij}|$ cannot exceed the greater of a_{ii} and a_{jj}.

(vii) This follows at once from the previous result. □

The converses of two of these results are also true:

(i) If all the eigenvalues of the symmetric matrix $A \in \mathbb{R}^{n\times n}$ are positive, then A is positive definite;
(ii) If the determinant of each leading principal submatrix of a matrix $A \in \mathbb{R}^{n\times n}$ is positive, then A is positive definite.

The proof of the second result is involved and will not be given here;[1] see, however, Example 3.1 for the case of $n = 2$. The proof of the first statement, on the other hand, is quite simple and proceeds as follows.

Since $A \in \mathbb{R}^{n\times n}$ is symmetric, it has a complete set of orthonormal eigenvectors $\boldsymbol{v}_1, \dots, \boldsymbol{v}_n$ in $\mathbb{R}^n_*$, and the corresponding eigenvalues $\lambda_1, \dots, \lambda_n$ are all real. Given any vector $\boldsymbol{x} \in \mathbb{R}^n_*$, it can be expressed as

$$\boldsymbol{x} = \sum_{i=1}^{n} \alpha_i \boldsymbol{v}_i$$

where $\alpha_i \in \mathbb{R}$, $i = 1, 2, \dots, n$, and $\alpha_1^2 + \cdots + \alpha_n^2 = \boldsymbol{x}^{\mathrm{T}}\boldsymbol{x} > 0$. Since $A\boldsymbol{v}_i = \lambda_i \boldsymbol{v}_i$, $i = 1, 2, \dots, n$, it follows that

$$A\boldsymbol{x} = \sum_{i=1}^{n} \alpha_i \lambda_i \boldsymbol{v}_i \,.$$

As $\boldsymbol{v}_j^{\mathrm{T}}\boldsymbol{v}_i = 0$ for $i \neq j$ and $\boldsymbol{v}_i^{\mathrm{T}}\boldsymbol{v}_i = 1$, we deduce that

$$\begin{aligned} \boldsymbol{x}^{\mathrm{T}} A \boldsymbol{x} &= \sum_{i=1}^{n} \lambda_i \alpha_i^2 \\ &\geq \left(\min_{i=1}^{n} \lambda_i\right) \sum_{i=1}^{n} \alpha_i^2 > 0 \,, \end{aligned}$$

since $\min_{i=1}^{n} \lambda_i > 0$; therefore A is positive definite.

For a symmetric positive definite matrix A we can now obtain an LU factorisation $A = LU$ in which $U = L^{\mathrm{T}}$.

Theorem 3.2 *Suppose that $n \geq 2$ and $A \in \mathbb{R}^{n\times n}_{\mathrm{sym}}$ is a positive definite matrix; then, there exists a lower triangular matrix $L \in \mathbb{R}^{n\times n}$ such that*

$$A = LL^{\mathrm{T}} \,.$$

This is known as the **Cholesky factorisation**[2] *of A.*

[1] For more details, see R.A. Horn and C.R. Johnson, *Matrix Analysis*, Cambridge University Press, 1992, Theorem 7.2.5.

[2] 'André-Louis Cholesky (1875–1918) was a French military officer involved in geodesy and surveying in Crete and North Africa just before World War I. He

Proof Since A is symmetric and positive definite, all the leading principal submatrices of A are positive definite, and hence by Theorem 2.2 the usual LU factorisation exists, with

$$A = L^{(1)}U^{(1)},$$

$L^{(1)} \in \mathbb{R}^{n\times n}$ a unit lower triangular and $U^{(1)} \in \mathbb{R}^{n\times n}$ an upper triangular matrix. In this factorisation the product of the leading principal submatrices of $L^{(1)}$ and $U^{(1)}$ of order k is the leading principal submatrix of A of order k, $1 \le k \le n$. Since the determinant of this submatrix is positive and all the diagonal elements of $L^{(1)}$ are unity, it follows that

$$u_{11}^{(1)}u_{22}^{(1)}\dots u_{kk}^{(1)} > 0\,, \qquad k = 1,2,\dots,n\,.$$

Thus all the diagonal elements of $U^{(1)}$ are positive. If we now define D to be the diagonal matrix with elements $d_{ii} = \sqrt{u_{ii}^{(1)}}$, $i = 1,2,\dots,n$, we can write

$$A = L^{(1)}U^{(1)} = (L^{(1)}D)(D^{-1}U^{(1)}) = LU\,,$$

where now $l_{ii} = u_{ii} = \sqrt{u_{ii}^{(1)}}$. The symmetry of the matrix A shows that

$$LU = A = A^{\mathrm{T}} = U^{\mathrm{T}}L^{\mathrm{T}}\,,$$

so that

$$U(L^{\mathrm{T}})^{-1} = L^{-1}U^{\mathrm{T}}\,.$$

In this equality the left-hand side is upper triangular, and the right-hand side is lower triangular, and hence both sides must be diagonal. Therefore, $U = D^*L^{\mathrm{T}}$, where D^* is a diagonal matrix; but U and L^{T} have the same diagonal elements, so $D^* = I$ and $U = L^{\mathrm{T}}$.

The same argument shows that L and L^{T} are unique, except for the arbitrary choice of the signs of the square roots in the definition of the diagonal matrix D. If we make the natural choice, taking all the square roots to be positive, then the diagonal elements of L are positive, and the factorisation is unique. □

developed the method now named after him to compute solutions to the normal equations for some least squares data fitting problems arising in geodesy. His work was posthumously published on his behalf in 1924 by a fellow officer, Benoit, in the *Bulletin Géodésique*.' – Cleve Moler, *NA-Digest*, February 18, 1990, Volume 90, Issue 07, `http://www.netlib.org/na-digest-html/90/v90n07.html`

In practice we construct the elements of L directly, rather than forming $L^{(1)}$ and $U^{(1)}$ first. This is done in a similar way to the LU factorisation. Suppose that $i \le j$; we then require that

$$a_{ij} = \sum_{k=1}^{i} l_{ik} l_{jk}\,, \qquad 1 \le i \le j \le n\,. \tag{3.1}$$

Note that we have used the fact that $(L^{\mathrm{T}})_{kj} = l_{jk}$; the sum only extends up to $k = i$ since L is lower triangular. The same equation will also hold for $i > j$, since A is symmetric. For $i = j$, equation (3.1) gives

$$l_{11} = a_{11}^{1/2}\,, \qquad l_{ii} = \left\{ a_{ii} - \sum_{k=1}^{i-1} l_{ik}^2 \right\}^{1/2}, \qquad 1 < i \le n\,. \tag{3.2}$$

As A is a positive definite matrix, $a_{11} > 0$ and therefore l_{11} is a positive real number. Further, as we have seen in the proof of the preceding theorem, $l_{ii} > 0$, $i = 2, 3, \ldots, n$. We find similarly that

$$l_{ji} = \frac{1}{l_{ii}} \left\{ a_{ij} - \sum_{k=1}^{i-1} l_{ik} l_{jk} \right\}, \qquad 1 \le i < j \le n\,. \tag{3.3}$$

These equations now enable us to calculate the elements of L in succession. For each $i \in \{1, 2, \ldots, n-1\}$, we first calculate l_{ii} from (3.2), and then calculate $l_{i+1\,i},\ \ l_{i+2\,i},\ \ldots, l_{ni}$ from (3.3). Finally, we compute l_{nn} using (3.2).

As, by hypothesis, the matrix $A \in \mathbb{R}^{n \times n}_{\mathrm{sym}}$ is positive definite, the required factorisation exists, so we can be sure that the divisor l_{ii} in (3.3), and the expression in the curly brackets in (3.2) whose square root is taken, will be positive. Thus, (3.2) implies that

$$l_{11}^2 = a_{11}\,, \qquad \max_{k=1}^{i-1} l_{ik}^2 < a_{ii}\,, \qquad i = 2, 3, \ldots, n\,.$$

The elements of the factor L cannot therefore grow very large, and no pivoting is necessary.

The evaluation of l_{ii} from (3.2) requires $i-1$ multiplications, $i-1$ subtractions and one square root operation, a total of $2i-1$ operations. The calculation of each l_{ji} from (3.3) also requires $2i-1$ operations. The total number of operations required to construct L is therefore

$$\sum_{i=1}^{n} \sum_{j=i}^{n} (2i-1) = \sum_{i=1}^{n} (2i-1)(1+n-i) = \tfrac{1}{6} n(n+1)(2n+1)\,.$$

For large n the number of operations required is approximately $\frac{1}{3}n^3$, which, as might be expected, is half the number given in Section 2.6 for the LU factorisation of a nonsymmetric matrix.

3.3 Tridiagonal and band matrices

As we shall see in the final chapters, in the numerical solution of boundary value problems for second-order differential equations one encounters a particular kind of matrix whose elements are mostly zeros, except for those along its main diagonal and the two adjacent diagonals. Matrices of this kind are referred to as tridiagonal. In order to motivate the definition of tridiagonal matrix stated in Definition 3.2 below, we begin with an example which is discussed in more detail in Chapter 13.

Example 3.2 *Consider the two-point boundary value problem*

$$\begin{aligned} -\frac{\mathrm{d}^2 y}{\mathrm{d}x^2} + r(x)y &= f(x)\,, \qquad x \in (0,1)\,, \\ y(0) &= 0\,, \quad y(1) = 0\,, \end{aligned}$$

where r and f are continuous functions of x defined on the interval $[0,1]$.

The numerical solution of the boundary value problem proceeds by selecting an integer $n \geq 4$, choosing a step size $h = 1/n$, and subdividing the interval $[0,1]$ by the points $x_k = kh$, $k = 0, 1, \ldots, n$. The numerical approximation to $y(x_k)$, the value of the analytical solution y at the point $x = x_k$, is denoted by Y_k. The values Y_k are obtained by solving the set of linear equations

$$-\frac{Y_{k+1} - 2Y_k + Y_{k-1}}{h^2} + r(x_k)Y_k = f(x_k)$$

for $k = 1, 2, \ldots, n-1$, together with the boundary conditions

$$Y_0 = 0\,, \quad Y_n = 0\,.$$

Equivalently,

$$\begin{aligned} a_k Y_{k-1} + c_k Y_k + b_k Y_{k+1} &= d_k\,, \qquad k = 1, 2, \ldots, n-1\,, \\ Y_0 &= 0\,, \quad Y_n = 0\,, \end{aligned}$$

where

$$a_k = b_k = -1/h^2, \quad c_k = 2/h^2 + r(x_k), \quad d_k = f(x_k),$$

for $k = 1, 2, \ldots, n-1$.

Clearly, for $1 < k < n-1$, the kth equation in the linear system above involves only three of the $n-1$ unknowns: Y_{k-1}, Y_k and Y_{k+1}. ◇

The example motivates the following definition of a tridiagonal (or triple diagonal) matrix.

Definition 3.2 *Suppose that $n \geq 3$. A matrix $T = (t_{ij}) \in \mathbb{R}^{n\times n}$ is said to be* **tridiagonal** *if it has nonzero elements only on the main diagonal and the two adjacent diagonals; i.e.,*

$$t_{ij} = 0 \quad \text{if} \quad |i-j| > 1, \qquad i, j \in \{1, 2, \ldots, n\}.$$

Such matrices are also sometimes called **triple diagonal**.

It is easy to see that in the LU factorisation process of a tridiagonal matrix $T \in \mathbb{R}^{n\times n}$, without row interchanges, the unit lower triangular matrix $L \in \mathbb{R}^{n\times n}$ and the upper triangular matrix $U \in \mathbb{R}^{n\times n}$ each have only two elements in each row. Writing T in the compact notation

$$T = \begin{pmatrix} b_1 & c_1 & & & & \\ a_2 & b_2 & c_2 & & & \\ & a_3 & b_3 & c_3 & & \\ \cdots & \cdots & \cdots & \cdots & \cdots & \cdots \\ \cdots & \cdots & \cdots & \cdots & \cdots & \cdots \\ & & & & a_n & b_n \end{pmatrix}, \tag{3.4}$$

the factorisation may be written $T = LU$ where

$$L = \begin{pmatrix} 1 & & & & \\ l_2 & 1 & & & \\ & l_3 & 1 & & \\ \cdots & \cdots & \cdots & \cdots & \cdots \\ & & & l_n & 1 \end{pmatrix} \tag{3.5}$$

and

$$U = \begin{pmatrix} u_1 & v_1 & & & \\ & u_2 & v_2 & & \\ & & u_3 & v_3 & \\ \cdots & \cdots & \cdots & \cdots & \cdots \\ & & & & u_n \end{pmatrix}, \tag{3.6}$$

with the convention that the missing elements in these matrices are all equal to zero. It is often convenient to define $a_1 = 0$ and $c_n = 0$. Multiplying L and U shows that $v_j = c_j$, and that the elements l_j and u_j can be calculated from

$$l_j = a_j/u_{j-1}\,, \quad u_j = b_j - l_j c_{j-1}\,, \qquad j = 2, 3, \ldots, n\,, \tag{3.7}$$

starting from $u_1 = b_1$.

Let us suppose that our aim is to solve the system of linear equations $T\boldsymbol{x} = \boldsymbol{r}$, where the matrix $T \in \mathbb{R}^{n\times n}$ is tridiagonal and nonsingular, and $\boldsymbol{r} \in \mathbb{R}^n$. Having calculated the elements of the matrices L and U in the LU factorisation $T = LU$ using (3.7), the forward and backsubstitution are then also very simple. Letting $\boldsymbol{y} = U\boldsymbol{x}$, the equation $L\boldsymbol{y} = \boldsymbol{r}$ gives

$$y_1 = r_1\,, \tag{3.8}$$

$$y_j = r_j - l_j y_{j-1}\,, \quad j = 2, 3, \ldots, n\,, \tag{3.9}$$

and finally from $U\boldsymbol{x} = \boldsymbol{y}$ we get

$$x_n = y_n/u_n\,, \tag{3.10}$$

$$x_j = (y_j - v_j x_{j+1})/u_j\,, \qquad j = n-1, n-2, \ldots, 1\,. \tag{3.11}$$

The LU factorisation of a tridiagonal matrix requires approximately $3n$ operations. The forward and backsubstitution together involve approximately $5n$ operations. Thus, the whole solution process requires approximately $8n$ operations. The total amount of work is therefore far less than for a full matrix, being of order n for large n, compared with $\frac{2}{3}n^3$ for a full matrix. The method we have described is a minor variation on what is often known as the *Thomas algorithm*.[1]

So far we have assumed that pivoting was not necessary; clearly any interchange of rows will destroy the tridiagonal structure of T. However, it is easy to see that the only interchanges required will be between two adjacent rows.

Theorem 3.3 *Suppose that $n \geq 3$ and $T \in \mathbb{R}^{n\times n}$ is a tridiagonal matrix; then, there exists a permutation matrix $P \in \mathbb{R}^{n\times n}$ such that*

$$PA = L^{(1)}U^{(1)} \tag{3.12}$$

[1] After Llewellyn H. Thomas, a distinguished physicist, who in the 1950s held positions at Columbia University and at IBM's Watson Research Laboratory. He is probably best known in connection with the Thomas–Fermi electron gas model. The terminology 'Thomas algorithm' comes from David Young. Thomas, L.H., *Elliptic Problems in Linear Difference Equations over a Network*, Watson Sci. Comput. Lab. Rept, Columbia University, New York, 1949. See *NA-Digest* V.96, 09, `http://www.netlib.org/cgi-bin/mfs/02/96/v96n09.html`

where $L^{(1)} \in \mathbb{R}^{n\times n}$ is unit lower triangular with at most two nonzero elements in each row, and $U^{(1)} \in \mathbb{R}^{n\times n}$ is upper triangular with at most three nonzero elements in each row.

The proof of this theorem is left as an exercise (see Exercise 6). It shows that the effect of pivoting is at worst to lead to an additional superdiagonal in the upper triangular factor.

In an important class of problems it is also easy to show that pivoting is unnecessary. We have shown this to be true for a symmetric positive definite matrix, and we can now show that it is also true for a tridiagonal matrix which is strictly diagonally dominant.

Definition 3.3 *A matrix $A \in \mathbb{R}^{n\times n}$ is said to be* **diagonally dominant** *if*

$$|a_{ii}| \geq \sum_{\substack{j=1\\ j\neq i}}^{n} |a_{ij}|\,, \qquad i = 1, 2, \ldots, n\,;$$

A is said to be **strictly diagonally dominant** *if strict inequality holds for each i.*

Theorem 3.4 *Suppose that $n \geq 3$, $T \in \mathbb{R}^{n\times n}$ is tridiagonal, as in (3.4), and*

$$|b_j| > |a_j| + |c_j|\,, \qquad j = 1, 2, \ldots, n \tag{3.13}$$

(with the convention $a_1 = 0$, $c_n = 0$); then T is nonsingular, and it can be written, without pivoting, in the form $T = LU$ where $L \in \mathbb{R}^{n\times n}$ is unit lower triangular and $U \in \mathbb{R}^{n\times n}$ is upper triangular. The condition (3.13) ensures that the matrix T is strictly diagonally dominant.

Proof We first show by induction that $|u_j| > |c_j|$ for all $j = 1, 2, \ldots, n$. This inequality trivially holds for $j = 1$ since

$$|u_1| = |b_1| > |a_1| + |c_1| = |c_1|\,.$$

Now let $j \in \{2, \ldots, n\}$ and adopt the inductive hypothesis:

$$\texttt{Hyp}_{j-1}: \qquad |u_{k-1}| > |c_{k-1}| \qquad \forall\, k \in \{2, \ldots, j\}\,.$$

(As we have already seen, $\texttt{Hyp}_1$ is true.) Then, from (3.7) we see that

$$\begin{aligned} |u_j| &\geq \left|\, |b_j| - |a_j| \left|\frac{c_{j-1}}{u_{j-1}}\right| \,\right| \\ &\geq \big|\, |b_j| - |a_j| \,\big| > |c_j| \end{aligned} \tag{3.14}$$

by the condition of strict diagonal dominance (3.13), which then shows that $\texttt{Hyp}_j$ holds. That completes the inductive step.

We have thus proved that $|u_j| > |c_j|$ for all $j = 1, 2, \ldots, n$. In particular, we deduce that $u_j \neq 0$ for all $j \in \{1, 2, \ldots, n\}$; hence the LU factorisation $T = LU$ defined by (3.7) exists. Further,

$$\det(T) = \det(L)\det(U) = \det(U) = u_1 u_2 \ldots u_n \neq 0\,,$$

so T is nonsingular.

The formulae (3.7) and the inequalities $|u_j| > |c_j|$, $j = 1, 2, \ldots, n$, now imply that

$$\begin{aligned} |u_j| &\leq |b_j| + |l_j|\,|c_{j-1}| \\ &= |b_j| + |a_j c_{j-1}|/|u_{j-1}| \\ &\leq |b_j| + |a_j|\,, \qquad j = 1, 2, \ldots, n\,, \end{aligned} \tag{3.15}$$

so the elements u_j cannot grow large, and rounding errors are kept under control without pivoting. □

It is easy to see that the same result holds under the weaker assumption that the matrix is diagonally dominant, but not necessarily strictly diagonally dominant, provided that we also require that all the elements c_j, $j = 1, 2, \ldots, n-1$, are nonzero (see Exercise 5).

Note also that the matrix constructed in Example 3.2 satisfies this condition, provided that the function r is nonnegative; this often holds in practical boundary value problems.

If the matrix $T \in \mathbb{R}^{n \times n}$ is symmetric and positive definite, as well as tridiagonal, it can be factorised in the form $T = LL^{\mathrm{T}}$, where $L \in \mathbb{R}^{n \times n}$ is lower triangular with nonzero elements only on and immediately below the diagonal. If we use the notation $d_i = l_{ii}$, $e_i = l_{i\,i-1}$ we easily find from (3.2) and (3.3) that the elements can be calculated in succession from the following formulae:

$$\begin{aligned} d_1 &= b_1^{1/2}\,, \\ e_i &= c_{i-1}/d_{i-1}\,, \quad d_i = \left(b_i - e_i^2\right)^{1/2}\,, \qquad i = 2, 3, \ldots, n\,. \end{aligned}$$

This calculation involves about $4n$ operations. Including also the work required by the forward and backsubstitution stages, the complete solution of $T\boldsymbol{x} = \boldsymbol{b}$ will be found to involve about $10n$ operations. For the tridiagonal matrix the Cholesky factorisation method thus requires more work for the complete solution than the Thomas algorithm; in this case there is no particular advantage in exploiting the symmetry of the matrix in this way.

Fig. 3.1. The asterisks indicate the 36 nonzero elements in this 10×10 Band(1,2) matrix.

More generally, a system of equations may often involve a matrix of band type.

Definition 3.4 $B \in \mathbb{R}^{n\times n}$ *is a* **band matrix** *if there exist nonnegative integers* $p < n$ *and* $q < n$ *such that* $b_{ij} = 0$ *for all* $i, j \in \{1, 2, \ldots, n\}$ *such that* $p < i - j$ *or* $q < j - i$. *The band is of width* $p + q + 1$, *with* p *elements to the left of the diagonal and* q *elements to the right of the diagonal, in each row. Such a matrix is said to be* Band(p, q).

Thus, for example, a tridiagonal matrix is Band(1,1), and an $n \times n$ lower triangular matrix is Band($n - 1$,0).

An example of a Band(1,2) matrix $A \in \mathbb{R}^{10\times 10}$ is shown in Figure 3.1, where each nonzero element in the matrix is identified by an asterisk. In addition to its main diagonal, the matrix has nonzero elements on its lower subdiagonal and two of its superdiagonals.

It is easy to see that, provided that no interchanges are necessary, such a band matrix can be written in the form $B = LU$, where L is Band(p,0) and U is Band(0,q) (see Exercise 7). It is also fairly simple to count the operations required in this calculation; the result is approximately proportional to $np(p + 2q)$ when n is moderately large. The most common situation has $q = p$, and then the number of operations is approximately proportional to np^2. As in the tridiagonal case, this is much smaller than n^3 when p and q are fairly small compared with n.

3.4 Monotone matrices

If a positive real number a is increased by $\varepsilon > 0$ to $a + \varepsilon$, then its reciprocal a^{-1} decreases to $(a + \varepsilon)^{-1}$. It is not usually true, however, that if we increase some or all of the elements of a nonsingular matrix $A \in \mathbb{R}^{n\times n}$, then the elements of the inverse $A^{-1} \in \mathbb{R}^{n\times n}$ will decrease. This useful property holds for the class of monotone matrices defined below.

The discussion in this section is not related to Gaussian elimination and LU factorisation, but it is of relevance in the iterative solution of systems of linear equations with monotone matrices which arise in the course of numerical approximation of boundary value problems for certain ordinary and partial differential equations.

Definition 3.5 *The nonsingular matrix $A \in \mathbb{R}^{n\times n}$ is said to be* **monotone** *if all the elements of the inverse A^{-1} are nonnegative.*

Example 3.3 *Suppose that a and d are positive real numbers, and b and c are nonnegative real numbers such that $ad > bc$. Then,*

$$A = \begin{pmatrix} a & -b \\ -c & d \end{pmatrix}$$

is a monotone matrix. This is easily seen by considering the inverse of the matrix A,

$$A^{-1} = \frac{1}{ad - bc} \begin{pmatrix} d & b \\ c & a \end{pmatrix},$$

and noting that all elements of A^{-1} are nonnegative.

Next we introduce the concept of ordering in $\mathbb{R}^n$ and $\mathbb{R}^{n\times n}$.

Definition 3.6 *For vectors $\boldsymbol{x}$ and $\boldsymbol{y}$ in $\mathbb{R}^n$ we use the notation*

$$\boldsymbol{x} \succeq \boldsymbol{y}$$

to mean that

$$x_i \geq y_i\,, \qquad i = 1, 2, \ldots, n\,.$$

In the same way, for matrices A and B in $\mathbb{R}^{n\times n}$ we write

$$A \succeq B$$

to mean that

$$a_{ij} \geq b_{ij}\,, \qquad i, j = 1, 2, \ldots, n\,.$$

The sign $\succeq$ is read 'succeeds or is equal to' or, simply, 'is greater than or equal to'.

Note that, given two arbitrary matrices A and B in $\mathbb{R}^{n\times n}$, in general none of $A \succeq B$, $A = B$ and $B \succeq A$ will be true. Therefore the relation $\succeq$ is a partial, rather than a total, ordering on $\mathbb{R}^{n\times n}$; the same is true of the ordering $\succeq$ on $\mathbb{R}^n$.

Theorem 3.5 *(i) Suppose that the nonsingular matrix $A \in \mathbb{R}^{n\times n}$ is monotone, $\mathbf{b}, \mathbf{c} \in \mathbb{R}^n$, and the vectors $\boldsymbol{x}$ and $\boldsymbol{y}$ in $\mathbb{R}^n$ are the solutions of*

$$A\boldsymbol{x} = \boldsymbol{b}\,, \quad A\boldsymbol{y} = \boldsymbol{c}\,,$$

respectively. If $\mathbf{b} \succeq \mathbf{c}$, then $\boldsymbol{x} \succeq \boldsymbol{y}$.

(ii) Suppose that A and B are nonsingular matrices in $\mathbb{R}^{n\times n}$ and that both are monotone. If $A \succeq B$, then $B^{-1} \succeq A^{-1}$.

Proof (i) Since the elements of A^{-1} are nonnegative and

$$\boldsymbol{x} - \boldsymbol{y} = A^{-1}(\boldsymbol{b} - \boldsymbol{c})\,,$$

the result follows from the fact that all elements of the vector $A^{-1}(\boldsymbol{b}-\boldsymbol{c})$ appearing on the right-hand side of this equality are nonnegative.

(ii) Since $A \succeq B$ and all the elements of B^{-1} are nonnegative, it follows that

$$B^{-1}A \succeq B^{-1}B = I\,.$$

In the same way, since all the elements of A^{-1} are nonnegative, it follows that

$$B^{-1} = B^{-1}A\,A^{-1} \succeq A^{-1},$$

as required. □

The following theorem will be useful in Chapter 13.

Theorem 3.6 *Suppose that $n \geq 3$ and $T \in \mathbb{R}^{n\times n}$ is a tridiagonal matrix of the form (3.4) with the properties*

$$a_i < 0\,, \quad i = 2, 3, \ldots, n\,, \qquad c_i < 0\,, \quad i = 1, 2, \ldots, n-1\,,$$

and

$$a_i + b_i + c_i \geq 0\,, \qquad i = 1, 2, \ldots, n\,,$$

where we have followed the convention that $a_1 = 0$, $c_n = 0$; then, the matrix T is monotone.

Proof Let $k \in \{1, 2, \ldots, n\}$. Column k of the inverse T^{-1} is the solution of the linear system $T\boldsymbol{c}^{(k)} = \boldsymbol{e}^{(k)}$, where $\boldsymbol{e}^{(k)}$ is column k of the identity matrix of size n, having a single nonzero element, 1, in row k. By applying the Thomas algorithm to this linear system, it is easy to deduce by induction from (3.7) that $l_j \le 0$, $u_j \ge 0$ and $v_j \le 0$ for all j; the argument is very similar to the proof of Theorem 3.4. It then follows from (3.8) and (3.9) that, in the notation of the Thomas algorithm, the vectors $\boldsymbol{y}$ and $\boldsymbol{x}$ have nonnegative elements. Hence column k of the inverse T^{-1} has nonnegative elements. Since the same is true for each $k \in \{1, 2, \ldots, n\}$, it follows that T is monotone. □

3.5 Notes

Symmetric systems of linear algebraic equations arise in the numerical solution of self-adjoint boundary value problems for differential equations with real-valued coefficients.

For further details on the Cholesky factorisation, the reader may consult any of the books listed in the Notes at the end of Chapter 2, particularly Chapter 10 of N.J. Higham, *Accuracy and Stability of Numerical Algorithms*, SIAM, Philadelphia, 1996.

Classical iterative methods for the solution of systems of linear equations with monotone matrices are discussed, for example, in

- RICHARD S. VARGA, *Matrix Iterative Analysis*, Prentice–Hall, Englewood Cliffs, NJ, 1962.

A more recent reference on iterative algorithms for linear systems is

- OWE AXELSON, *Iterative Solution Methods*, Cambridge University Press, Cambridge, 1996.

In particular, Chapter 6 of Axelson's book considers the relevance of monotone matrices in the context of iterative solution of systems of linear equations.

Theorem 3.6 is a slight variation on the following general result.

Theorem 3.7 *A sufficient condition for $A \in \mathbb{R}^{n\times n}$ to be a monotone matrix is that A is an* ***M-matrix****, that is, (a) $a_{ij} \le 0$ for all $i, j \in \{1, 2, \ldots, n\}$ such that $i \ne j$, and (b) there exists a vector $\boldsymbol{g} \in \mathbb{R}^n$ with positive elements such that all elements of $A\boldsymbol{g} \in \mathbb{R}^n$ are positive.*

Exercises

3.1 Find the Cholesky factorisation of the matrix

$$A = \begin{pmatrix} 4 & 6 & 2 \\ 6 & 10 & 3 \\ 2 & 3 & 5 \end{pmatrix}.$$

3.2 Use the method of Cholesky factorisation to solve the system of equations

$$\begin{aligned} x_1 - 2x_2 + 2x_3 &= 4, \\ -2x_1 + 5x_2 - 3x_3 &= -7, \\ 2x_1 - 3x_2 + 6x_3 &= 10. \end{aligned}$$

3.3 Let $n \geq 3$. The $n \times n$ tridiagonal matrix T has the diagonal elements

$$T_{ii} = 2, \quad i = 1, 2, \dots, n,$$

and the off-diagonal elements

$$T_{i\,i+1} = T_{i+1\,i} = -1, \quad i = 1, 2, \dots, n-1.$$

In the factorisation $T = LU$, where $L \in \mathbb{R}^{n \times n}$ is unit lower triangular and $U \in \mathbb{R}^{n \times n}$ is upper triangular, show that

$$L_{i+1\,i} = -i/(i+1), \qquad i = 1, 2, \dots, n,$$

and find expressions for the elements of U. What is the determinant of T?

3.4 Let $n \geq 3$ and $1 \leq k \leq n$. Define the vector $\boldsymbol{v}^{(k)} \in \mathbb{R}^n$ with elements given by

$$v_i^{(k)} = \begin{cases} i(n+1-k), & i = 1, \dots, k, \\ k(n+1-i), & i = k+1, \dots, n. \end{cases}$$

Evaluate M_{kj}, the inner product of the vector $\boldsymbol{v}^{(k)}$ with column j of the matrix T defined in Exercise 3. (The inner product $\langle \boldsymbol{v}, \boldsymbol{w} \rangle$ of two vectors $\boldsymbol{v}$ and $\boldsymbol{w}$ in $\mathbb{R}^n$ is defined as the real number $\boldsymbol{v}^{\mathrm{T}}\boldsymbol{w}$.) Hence give expressions for the elements of the inverse matrix T^{-1}, and verify that this inverse is symmetric. Find the ∞-norm of the inverse, $\|T^{-1}\|_\infty$, and show that the condition number of T is

$$\kappa_\infty(T) = \frac{1}{2}(n+1)^2, \quad n \text{ odd}.$$

What is the condition number $\kappa_\infty(T)$ when n is even?

3.5 Given that $n \geq 3$, in the notation of Theorem 3.4 suppose that

$$|b_j| \geq |a_j| + |c_j|, \quad j = 1, 2, \ldots, n,$$

and

$$|c_j| > 0, \quad j = 1, 2, \ldots, n-1,$$

with the convention that $a_1 = 0$ and $c_n = 0$. Show that the factorisation $T = LU$ exists without pivoting, and can be constructed by the Thomas algorithm. Show further that T is non-singular. Give an example of a matrix T which satisfies these conditions, except that $c_k = 0$ for some $k \in \{1, 2, \ldots, n-1\}$ and such that T is singular and cannot be written in the form $T = LU$ without pivoting.

3.6 Let $n \geq 3$ and suppose that the matrix $T \in \mathbb{R}^{n \times n}$ is tridiagonal. Show that there exists a permutation matrix $P \in \mathbb{R}^{n \times n}$ such that

$$PA = L^{(1)}U^{(1)}$$

where $L^{(1)} \in \mathbb{R}^{n \times n}$ is unit lower triangular with at most two nonzero elements in each row, and $U^{(1)} \in \mathbb{R}^{n \times n}$ is upper triangular with at most three nonzero elements in each row.

3.7 Suppose that the matrix B is Band(p,q), and that there exists a factorisation $B = LU$ without row interchanges. Show that L is Band(p,0) and U is Band(0,q).

3.8 Suppose that $n \geq 4$, that the matrix $A \in \mathbb{R}^{n \times n}$ is Band(3,3), and has the LU factorisation $A = LU$, so that $L \in \mathbb{R}^{n \times n}$ is Band(3,0) and $U \in \mathbb{R}^{n \times n}$ is Band(0,3). Suppose also that $a_{i+2,i} = 0$, $a_{i,i+2} = 0$ for $i = 1, 2, \ldots, n-2$. By considering u_{24} and l_{42}, or otherwise, show that in general the elements $l_{i+2,i}$ and $u_{i,i+2}$ are not zero.

4

Simultaneous nonlinear equations

4.1 Introduction

In Chapter 1 we discussed iterative methods for the solution of a single nonlinear equation of the form $f(x) = 0$ where f is a continuous real-valued function of a single real variable. In Chapters 2 and 3, on the other hand, we were concerned with direct (as opposed to iterative) methods for systems of linear equations. The purpose of the present chapter is to extend the techniques developed in Chapter 1 to systems of simultaneous nonlinear equations for functions of several real variables. We shall concentrate on two methods: the generalisation of simple iteration, usually referred to as simultaneous iteration, and Newton's method.

Given that $\boldsymbol{x} = (x_1, \dots, x_n)^{\mathrm{T}} \in \mathbb{R}^n$, as in Chapters 2 and 3 we denote by $\|\boldsymbol{x}\|_\infty$ the ∞-norm of $\boldsymbol{x}$ defined by

$$\|\boldsymbol{x}\|_\infty = \max_{i=1}^{n} |x_i| \,.$$

Throughout the chapter, $\mathbb{R}^n$ will be thought of as a linear space equipped with the ∞-norm; with only minor alterations all of our results can be restated in the p-norm with $p \in [1, \infty)$ on replacing $\|\cdot\|_\infty$ by $\|\cdot\|_p$ throughout. We begin with some basic definitions which involve the concept of *open ball* defined in Section 2.7.

Let $\boldsymbol{\xi} \in \mathbb{R}^n$; the open ball in $\mathbb{R}^n$ (with respect to the ∞-norm) of radius $\varepsilon > 0$ and centre $\boldsymbol{\xi}$ is defined as the set

$$B_\varepsilon(\boldsymbol{\xi}) = \{\boldsymbol{x} \in \mathbb{R}^n\colon \|\boldsymbol{x} - \boldsymbol{\xi}\|_\infty < \varepsilon\} \,.$$

A set $D \subset \mathbb{R}^n$ is said to be an **open set** in $\mathbb{R}^n$ if for every $\boldsymbol{\xi} \in D$ there exists $\varepsilon = \varepsilon(\boldsymbol{\xi}) > 0$ such that $B_\varepsilon(\boldsymbol{\xi}) \subset D$ (see Figure 4.1). For example, any open ball in $\mathbb{R}^n$ is an open set in $\mathbb{R}^n$. Given $\boldsymbol{\xi} \in \mathbb{R}^n$, any open set

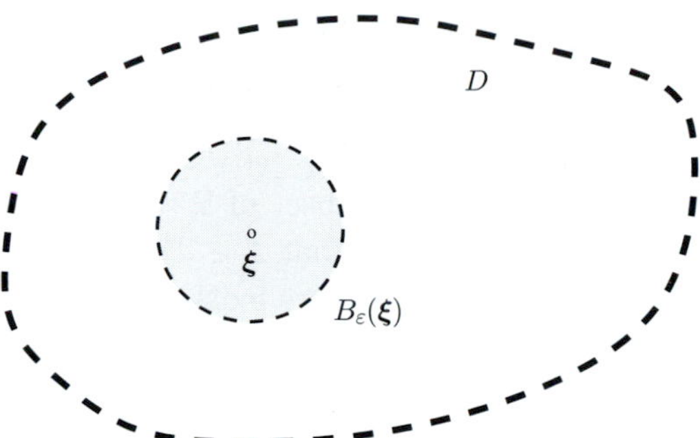

Fig. 4.1. Open set D: for each $\boldsymbol{\xi} \in D$ there exists $\varepsilon = \varepsilon(\boldsymbol{\xi})$ such that the open ball $B_\varepsilon(\boldsymbol{\xi})$ of radius ε and centre $\boldsymbol{\xi}$ is contained in D.

$N(\boldsymbol{\xi}) \subset \mathbb{R}^n$ containing $\boldsymbol{\xi}$ will be called a **neighbourhood** of $\boldsymbol{\xi}$; thus, any open set in $\mathbb{R}^n$ is a neighbourhood of each of its elements.

A set $D \subset \mathbb{R}^n$ is said to be a **closed set** in $\mathbb{R}^n$ if its complement $\mathbb{R}^n \setminus D$ is an open set in $\mathbb{R}^n$. For example, the closed ball of radius $\varepsilon > 0$ and centre $\boldsymbol{\xi}$, defined by

$$\bar{B}_\varepsilon(\boldsymbol{\xi}) = \{\boldsymbol{x} \in \mathbb{R}^n \colon \|\boldsymbol{x} - \boldsymbol{\xi}\|_\infty \leq \varepsilon\},$$

is a closed set in $\mathbb{R}^n$.

A sequence $(\boldsymbol{x}^{(k)}) \subset \mathbb{R}^n$ is called a **Cauchy sequence** in $\mathbb{R}^n$ if for any $\varepsilon > 0$ there exists a positive integer $k_0 = k_0(\varepsilon)$ such that

$$\|\boldsymbol{x}^{(k)} - \boldsymbol{x}^{(m)}\|_\infty < \varepsilon \qquad \forall\, k, m \geq k_0 .$$

We shall make use of the fact that $\mathbb{R}^n$ is **complete**: that is, if $(\boldsymbol{x}^{(k)})$ is a Cauchy sequence in $\mathbb{R}^n$, then there exists $\boldsymbol{\xi}$ in $\mathbb{R}^n$ such that $(\boldsymbol{x}^{(k)})$ converges to $\boldsymbol{\xi}$; *i.e.*,

$$\lim_{k\to\infty} \|\boldsymbol{x}^{(k)} - \boldsymbol{\xi}\|_\infty = 0 . \tag{4.1}$$

For the sake of brevity, we shall write $\lim_{k\to\infty} \boldsymbol{x}^{(k)} = \boldsymbol{\xi}$ instead of (4.1).

Lemma 4.1 *Suppose that D is a nonempty closed subset of $\mathbb{R}^n$ and $(\boldsymbol{x}^{(k)}) \subset D$ is a Cauchy sequence in $\mathbb{R}^n$. Then, $\lim_{k\to\infty} \boldsymbol{x}^{(k)} = \boldsymbol{\xi}$ exists and $\boldsymbol{\xi} \in D$.*

Proof As $(\boldsymbol{x}^{(k)})$ is a Cauchy sequence in $\mathbb{R}^n$, there exists $\boldsymbol{\xi} \in \mathbb{R}^n$ such that $\lim_{k\to\infty} \boldsymbol{x}^{(k)} = \boldsymbol{\xi}$. It remains to prove that $\boldsymbol{\xi} \in D$. Suppose, otherwise, that $\boldsymbol{\xi}$ belongs to the open set $\mathbb{R}^n \setminus D$. Then, there exists

$\varepsilon > 0$ such that $B_\varepsilon(\boldsymbol{\xi}) \subset \mathbb{R}^n \setminus D$. As $(\boldsymbol{x}^{(k)}) \subset D$, no member of the sequence $(\boldsymbol{x}^{(k)})$ can enter $B_\varepsilon(\boldsymbol{\xi})$. This, however, contradicts the fact that $(\boldsymbol{x}^{(k)})$ converges to $\boldsymbol{\xi}$. The contradiction implies that $\boldsymbol{\xi} \in D$. □

Suppose that D is a nonempty subset of $\mathbb{R}^n$ and $\boldsymbol{f}$: $D(\subset \mathbb{R}^n) \to \mathbb{R}^n$ is a function defined on D. Given that $\boldsymbol{\xi} \in D$, we shall say that $\boldsymbol{f}$ is continuous at $\boldsymbol{\xi}$ if for every $\varepsilon > 0$ there exists $\delta = \delta(\varepsilon) > 0$ such that, for every $\boldsymbol{x} \in B_\delta(\boldsymbol{\xi}) \cap D$,

$$\|\boldsymbol{f}(\boldsymbol{x}) - \boldsymbol{f}(\boldsymbol{\xi})\|_\infty < \varepsilon .$$

When a function $\boldsymbol{f}$, defined on the set D, is continuous at each point of D, it is said to be a **continuous function** on D.

Lemma 4.2 *Let D be a nonempty subset of $\mathbb{R}^n$ and $\boldsymbol{f}$: $D(\subset \mathbb{R}^n) \to \mathbb{R}^n$ a function, defined and continuous on D. If $(\boldsymbol{x}^{(k)}) \subset D$ converges in $\mathbb{R}^n$ to $\boldsymbol{\xi} \in D$, then $\lim_{k\to\infty} \boldsymbol{f}(\boldsymbol{x}^{(k)}) = \boldsymbol{f}(\boldsymbol{\xi})$.*

Proof Due to the continuity of $\boldsymbol{f}$ at $\boldsymbol{\xi} \in D$, given $\varepsilon > 0$, there exists $\delta = \delta(\varepsilon) > 0$ such that if $\|\boldsymbol{x} - \boldsymbol{\xi}\|_\infty < \delta$ for some $\boldsymbol{x} \in D$, then

$$\|\boldsymbol{f}(\boldsymbol{x}) - \boldsymbol{f}(\boldsymbol{\xi})\|_\infty < \varepsilon . \tag{4.2}$$

Further, as $(\boldsymbol{x}^{(k)})$ converges to $\boldsymbol{\xi}$, there exists $k_0 = k_0(\delta) = k_0(\delta(\varepsilon))$ such that

$$\|\boldsymbol{x}^{(k)} - \boldsymbol{\xi}\|_\infty < \delta \qquad \forall\, k \geq k_0 .$$

Hence, taking $\boldsymbol{x} = \boldsymbol{x}^{(k)}$ in (4.2), we deduce that for each $\varepsilon > 0$ there exists k_0 such that

$$\|\boldsymbol{f}(\boldsymbol{x}^{(k)}) - \boldsymbol{f}(\boldsymbol{\xi})\|_\infty < \varepsilon \qquad \forall\, k \geq k_0 ,$$

which means that $\lim_{k\to\infty} \boldsymbol{f}(\boldsymbol{x}^{(k)}) = \boldsymbol{f}(\boldsymbol{\xi})$. □

After this brief preparation, we are ready to embark on the development of numerical algorithms for the solution of systems of simultaneous nonlinear equations.

4.2 Simultaneous iteration

Let D be a nonempty closed subset of $\mathbb{R}^n$ and $\boldsymbol{f}$: $D(\subset \mathbb{R}^n) \to \mathbb{R}^n$ a continuous function defined on D. We shall be concerned with the problem of finding $\boldsymbol{\xi} \in D$ such that $\boldsymbol{f}(\boldsymbol{\xi}) = \mathbf{0}$. If such $\boldsymbol{\xi}$ exists, it is

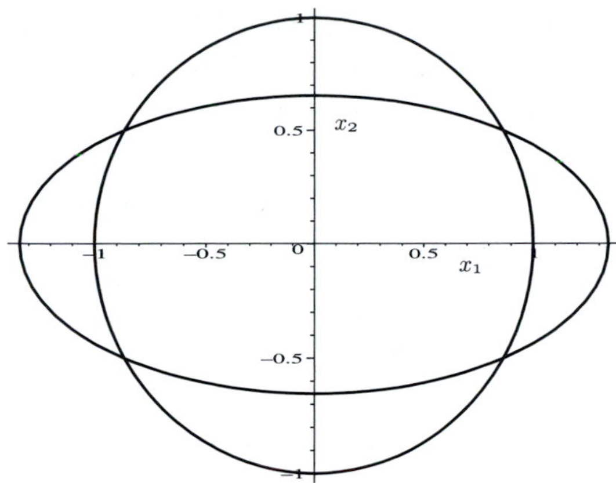

Fig. 4.2. Graphs of the curves $x_1^2 + x_2^2 - 1 = 0$ and $5x_1^2 + 21x_2^2 - 9 = 0$.

called a **solution** to the equation $\boldsymbol{f}(\boldsymbol{x}) = \boldsymbol{0}$ (in D). When written in componentwise form, $\boldsymbol{f}(\boldsymbol{x}) = \boldsymbol{0}$ becomes

$$f_i(x_1, \ldots, x_n) = 0\,, \quad i = 1, \ldots, n\,,$$

a system of n simultaneous nonlinear equations for n unknowns, where $f_1, \ldots, f_n$ are the components of $\boldsymbol{f}$.

Example 4.1 *Consider the system of two simultaneous nonlinear equations in two unknowns, x_1 and x_2, defined by*

$$\begin{aligned} x_1^2 + x_2^2 - 1 &= 0\,, \\ 5x_1^2 + 21x_2^2 - 9 &= 0\,. \end{aligned}$$

Here $\boldsymbol{x} = (x_1, x_2)^{\mathrm{T}}$ and $\boldsymbol{f} = (f_1, f_2)^{\mathrm{T}}$ with

$$\begin{aligned} f_1(x_1, x_2) &= x_1^2 + x_2^2 - 1\,, \\ f_2(x_1, x_2) &= 5x_1^2 + 21x_2^2 - 9\,. \end{aligned}$$

The equation $\boldsymbol{f}(\boldsymbol{x}) = \boldsymbol{0}$ has four solutions:

$$\begin{aligned} \boldsymbol{\xi}_1 &= (-\sqrt{3}/2, 1/2)^{\mathrm{T}}\,, & \boldsymbol{\xi}_2 &= (\sqrt{3}/2, 1/2)^{\mathrm{T}}\,, \\ \boldsymbol{\xi}_3 &= (-\sqrt{3}/2, -1/2)^{\mathrm{T}}\,, & \boldsymbol{\xi}_4 &= (\sqrt{3}/2, -1/2)^{\mathrm{T}}\,. \end{aligned}$$

The curves $f_1(x_1, x_2) = 0$ and $f_2(x_1, x_2) = 0$ are depicted in Figure 4.2. The four solutions correspond to the four points of intersection of the two curves in the figure.

Example 4.2 *Let us suppose that* $A \in \mathbb{R}^{n\times n}$ *and* $\boldsymbol{b} \in \mathbb{R}^n$. *On letting* $\boldsymbol{f}(\boldsymbol{x}) = \boldsymbol{b} - A\boldsymbol{x}$ *we deduce that the problem of solving the system of simultaneous linear equations considered in Chapters 2 and 3 can be restated in the form: find* $\boldsymbol{x} \in \mathbb{R}^n$ *such that* $\boldsymbol{f}(\boldsymbol{x}) = \mathbf{0}$.

Let us assume that we have transformed the equation $\boldsymbol{f}(\boldsymbol{x}) = \mathbf{0}$ into an equivalent form $\boldsymbol{g}(\boldsymbol{x}) = \boldsymbol{x}$, where $\boldsymbol{g}\colon \mathbb{R}^n \to \mathbb{R}^n$ is a continuous function, defined on the closed subset $D \subset \mathbb{R}^n$, such that $\boldsymbol{g}(D) \subset D$. For example, one can choose $\boldsymbol{g}(\boldsymbol{x}) = \boldsymbol{x} - \alpha\boldsymbol{f}(\boldsymbol{x})$, with $\alpha \in \mathbb{R}$ a suitable parameter. By 'equivalent' we mean that $\boldsymbol{\xi} \in D$ satisfies $\boldsymbol{f}(\boldsymbol{\xi}) = \mathbf{0}$ if, and only if, $\boldsymbol{g}(\boldsymbol{\xi}) = \boldsymbol{\xi}$. Any $\boldsymbol{\xi} \in D$ such that $\boldsymbol{g}(\boldsymbol{\xi}) = \boldsymbol{\xi}$ is called a **fixed point** of the function $\boldsymbol{g}$ in D. Thus the problem of finding a solution $\boldsymbol{\xi} \in D$ to the equation $\boldsymbol{f}(\boldsymbol{x}) = \mathbf{0}$ has been converted into one of finding a fixed point in D of the function $\boldsymbol{g}$. We embark on the latter task by considering the natural extension to $\mathbb{R}^n$ of the simple iteration discussed in Section 1.2 for the solution of the scalar nonlinear equation $g(x) = x$.

Definition 4.1 *Suppose that* $\boldsymbol{g}\colon \mathbb{R}^n \to \mathbb{R}^n$ *is a function, defined and continuous on a closed subset* D *of* $\mathbb{R}^n$, *such that* $\boldsymbol{g}(D) \subset D$. *Given that* $\boldsymbol{x}^{(0)} \in D$, *the recursion defined by*

$$\boxed{\boldsymbol{x}^{(k+1)} = \boldsymbol{g}(\boldsymbol{x}^{(k)}), \qquad k = 0, 1, 2, \dots ,} \tag{4.3}$$

is called a **simultaneous iteration**. *For* $n = 1$ *the recursion (4.3) is just the simple iteration considered in (1.3).*

Note that here we use the superscript k as the sequence index; following the convention adopted in Chapters 2 and 3, we reserve subscripts for labelling the entries of vectors. Thus $x_i^{(k)}$ is entry i of the vector $\boldsymbol{x}^{(k)}$, the kth member of the sequence $(\boldsymbol{x}^{(k)})$. The motivation behind the definition of the simultaneous iteration (4.3) is, of course, our hope that, under suitable conditions on $\boldsymbol{g}$ and D, the sequence $(\boldsymbol{x}^{(k)})$ will converge to a fixed point $\boldsymbol{\xi}$ of $\boldsymbol{g}$.

Two remarks are in order at this point. First, it is easy to show that if a sequence of vectors $(\boldsymbol{x}^{(k)})$ converges in $\mathbb{R}^n$ to $\boldsymbol{\xi}$ in the norm $\|\cdot\|_\infty$, then it also converges to this same limit in the norm $\|\cdot\|_p$ for any $p \in [1, \infty)$. To see this, note that

$$\|\boldsymbol{w}\|_\infty \le \|\boldsymbol{w}\|_p \le n^{1/p}\|\boldsymbol{w}\|_\infty \qquad \forall\, \boldsymbol{w} \in \mathbb{R}^n\,, \tag{4.4}$$

for $1 \le p < \infty$, and take $\boldsymbol{w} = \boldsymbol{x}^{(k)} - \boldsymbol{\xi}$ to deduce that, as $k \to \infty$, convergence in the ∞-norm implies convergence in the p-norm for any $p \in [1, \infty)$, and *vice versa*. Thus, in this sense, the choice of norm on $\mathbb{R}^n$ is irrelevant. Second, the assumption that D is a closed set is crucial in our discussion. If D is not closed, $\boldsymbol{g}\colon D \to D$ need not have a fixed point in D, even if $\boldsymbol{x}^{(k)} \in D$ for all $k \ge 0$ and $(\boldsymbol{x}^{(k)})$ converges in $\mathbb{R}^n$. We verify this claim through a simple example.

Example 4.3 *Suppose that D is the open unit disc in $\mathbb{R}^2$ in the ∞-norm, which is just the square defined by $-1 < x_1 < 1$, $-1 < x_2 < 1$. Consider the simultaneous iteration defined by (4.3), where $\boldsymbol{x}^{(0)} = \boldsymbol{0} \in D$, and*

$$\boldsymbol{g}(\boldsymbol{x}) = \tfrac{1}{2}(\boldsymbol{x} + \boldsymbol{u}), \quad \boldsymbol{u} = (1,1)^{\mathrm{T}}.$$

If $\|\boldsymbol{x}\|_\infty < 1$ it is easy to see that $\|\boldsymbol{g}(\boldsymbol{x})\|_\infty < 1$; hence, starting the iteration $\boldsymbol{x}^{(k+1)} = \boldsymbol{g}(\boldsymbol{x}^{(k)})$ from $\boldsymbol{x}^{(0)} = \boldsymbol{0}$, it follows that $\boldsymbol{x}^{(k)} \in D$ for all $k \ge 0$. The definition of $\boldsymbol{g}$ implies at once that

$$\boldsymbol{x}^{(k+1)} - \boldsymbol{u} = \tfrac{1}{2}(\boldsymbol{x}^{(k)} - \boldsymbol{u}),$$

and therefore

$$\|\boldsymbol{x}^{(k+1)} - \boldsymbol{u}\|_\infty = \tfrac{1}{2}\|\boldsymbol{x}^{(k)} - \boldsymbol{u}\|_\infty = \cdots = \left(\tfrac{1}{2}\right)^{k+1}\|\boldsymbol{x}^{(0)} - \boldsymbol{u}\|_\infty = \left(\tfrac{1}{2}\right)^{k+1},$$

from which it is obvious that the sequence $(\boldsymbol{x}^{(k)})$ converges in $\mathbb{R}^2$ to the limit $\boldsymbol{u}$. However, $\boldsymbol{u} \notin D$, since $\boldsymbol{u}$ lies on the unit circle in the ∞-norm that represents the boundary of the open set D. ◇

Up to now we have been assuming that the function $\boldsymbol{g}\colon \mathbb{R}^n \to \mathbb{R}^n$ is defined and continuous on a closed subset D of $\mathbb{R}^n$. In order to ensure that $\boldsymbol{g}$ has a (unique) fixed point in D, we strengthen our hypotheses on the function $\boldsymbol{g}$.

Definition 4.2 *Suppose that $\boldsymbol{g}\colon \mathbb{R}^n \to \mathbb{R}^n$ is defined on a closed subset D of $\mathbb{R}^n$. If there exists a positive constant L such that,*

$$\|\boldsymbol{g}(\boldsymbol{x}) - \boldsymbol{g}(\boldsymbol{y})\|_\infty \le L\,\|\boldsymbol{x} - \boldsymbol{y}\|_\infty \tag{4.5}$$

for all $\boldsymbol{x}$ and $\boldsymbol{y}$ in D, then we say that $\boldsymbol{g}$ satisfies a **Lipschitz condition** *on D in the ∞-norm. The number L is called a* **Lipschitz constant** *for $\boldsymbol{g}$ in the ∞-norm. In particular, if $L \in (0,1)$, then $\boldsymbol{g}$ is said to be a* **contraction** *on D in the ∞-norm.*

Any function $\boldsymbol{g}$ that satisfies a Lipschitz condition on a set D is continuous on D. For let $\boldsymbol{x}_0 \in D$ and $\varepsilon > 0$; then, on defining $\delta = \varepsilon/L$, we deduce from (4.5) that if $\|\boldsymbol{x} - \boldsymbol{x}_0\|_\infty < \delta$ for some $\boldsymbol{x} \in D$, then

$$\|\boldsymbol{g}(\boldsymbol{x}) - \boldsymbol{g}(\boldsymbol{x}_0)\|_\infty \leq L\,\|\boldsymbol{x} - \boldsymbol{x}_0\|_\infty < \varepsilon\,.$$

It follows from (4.4) that if $\boldsymbol{g}$ satisfies a Lipschitz condition on D in the ∞-norm then it also does so in the p-norm for any $p \in [1, \infty)$, and *vice versa.* However, in general, the size of the constant L will depend on the choice of norm. Specifically, if $\boldsymbol{g}$ is a contraction on a set D in the ∞-norm (*i.e.*, (4.5) holds with $L < 1$), then $\boldsymbol{g}$ *need not* be a contraction in the p-norm, unless $L < n^{-1/p}$. (See Exercise 1.) Conversely, if $\boldsymbol{g}$ is a contraction on D in the p-norm for some $p \in [1, \infty)$, it does *not* follow that $\boldsymbol{g}$ is a contraction on D in the ∞-norm.

For example, suppose that $\boldsymbol{g}\colon \mathbb{R}^2 \to \mathbb{R}^2$ is the linear function defined by $\boldsymbol{g}(\boldsymbol{x}) = A\boldsymbol{x}$, where A is the 2×2 matrix

$$A = \begin{pmatrix} 3/4 & 1/3 \\ 0 & 3/4 \end{pmatrix}.$$

This function $\boldsymbol{g}$ satisfies a Lipschitz condition on $\mathbb{R}^2$ in $\|\cdot\|_p$ for any $p \in [1, \infty]$, and if L is a Lipschitz constant for $\boldsymbol{g}$ in the p-norm, then $L \geq \|A\|_p$, in the subordinate matrix norm. It is easy to see that $\|A\|_1 = \|A\|_\infty = 13/12$, and a small calculation gives $\|A\|_2 = 0.935$ to three decimal digits. Hence the function $\boldsymbol{g}$ is a contraction in the 2-norm, but not in the 1- or ∞-norm.

Our next result is a direct generalisation of Theorem 1.3 formulated in Chapter 1.

Theorem 4.1 (**Contraction Mapping Theorem**) *Suppose that D is a closed subset of $\mathbb{R}^n$, $\mathbf{g}\colon \mathbb{R}^n \to \mathbb{R}^n$ is defined on D, and $\mathbf{g}(D) \subset D$. Suppose further that $\mathbf{g}$ is a contraction on D in the ∞-norm. Then, $\mathbf{g}$ has a unique fixed point $\boldsymbol{\xi}$ in D, and the sequence $(\boldsymbol{x}^{(k)})$ defined by (4.3) converges to $\boldsymbol{\xi}$ for any starting value $\boldsymbol{x}^{(0)} \in D$.*

Proof Assuming that $\boldsymbol{g}$ has a fixed point $\boldsymbol{\xi}$ in D, the *uniqueness* of the fixed point is easy to show: for suppose that $\boldsymbol{\eta}$ is also a fixed point of $\boldsymbol{g}$ in D. Then, by (4.5),

$$\|\boldsymbol{\xi} - \boldsymbol{\eta}\|_\infty = \|\boldsymbol{g}(\boldsymbol{\xi}) - \boldsymbol{g}(\boldsymbol{\eta})\|_\infty \leq L\|\boldsymbol{\xi} - \boldsymbol{\eta}\|_\infty\,,$$

i.e., $(1-L)\|\boldsymbol{\xi}-\boldsymbol{\eta}\|_\infty \le 0$. Since $L \in (0,1)$, and $\|\cdot\|_\infty$ is a norm, it follows that $\boldsymbol{\xi}-\boldsymbol{\eta}=\mathbf{0}$, and hence $\boldsymbol{\xi}=\boldsymbol{\eta}$. Consequently, if $\boldsymbol{g}$ has a fixed point in D, then this is the unique fixed point of $\boldsymbol{g}$ in D.

Now, still *assuming* that $\boldsymbol{g}$ possesses a fixed point $\boldsymbol{\xi} \in D$, we shall show that the sequence $(\boldsymbol{x}^{(k)})$ defined by (4.3) converges to $\boldsymbol{\xi}$ for any starting value $\boldsymbol{x}^{(0)} \in D$. By repeating the argument from Chapter 1 which led to (1.10), with the absolute value sign $|\cdot|$ replaced by $\|\cdot\|_\infty$ throughout, we find that

$$\|\boldsymbol{x}^{(k)}-\boldsymbol{\xi}\|_\infty \le L^k \frac{1}{1-L}\|\boldsymbol{x}^{(1)}-\boldsymbol{x}^{(0)}\|_\infty .$$

As $L \in (0,1)$, we deduce that $\lim_{k\to\infty} L^k = 0$, and hence,

$$\lim_{k\to\infty} \|\boldsymbol{x}^{(k)}-\boldsymbol{\xi}\|_\infty = 0 ,$$

showing that the sequence $(\boldsymbol{x}^{(k)})$ defined by (4.3) converges to $\boldsymbol{\xi}$ for any starting value $\boldsymbol{x}^{(0)} \in D$. In particular, if $\varepsilon > 0$, then letting

$$k_0 = k_0(\varepsilon) = \left[\frac{\ln\|\,\boldsymbol{x}^{(1)}-\boldsymbol{x}^{(0)}\,\|_\infty - \ln(\varepsilon(1-L))}{\ln(1/L)}\right] + 1 , \qquad (4.6)$$

we find that

$$L^k \frac{1}{1-L}\|\boldsymbol{x}^{(1)}-\boldsymbol{x}^{(0)}\|_\infty \le \varepsilon$$

for all $k \ge k_0(\varepsilon)$, and therefore

$$\|\boldsymbol{x}^{(k)}-\boldsymbol{\xi}\|_\infty \le \varepsilon , \qquad (4.7)$$

for all $k \ge k_0(\varepsilon)$, as in Chapter 1. A brief comment on the notation: in (4.6), $[x]$ denotes the integer part of the real number x; *i.e.*, $[x]$ is the largest integer such that $[x] \le x$ – just as in Theorem 1.4.

In order to complete the proof of the theorem, it remains to show the *existence* of a fixed point $\boldsymbol{\xi} \in D$ for $\boldsymbol{g}$. In contrast with the proof of existence of a fixed point for a real-valued function of a single real variable presented in Chapter 1, here we cannot rely on the Intermediate Value Theorem (unless, of course, $n = 1$), so we shall develop a different argument. The essence of this will be to show that $(\boldsymbol{x}^{(k)}) \subset D$ is a Cauchy sequence in $\mathbb{R}^n$; for then we can apply Lemmas 4.1 and 4.2 to deduce that the sequence converges to a fixed point $\boldsymbol{\xi}$ of the function $\boldsymbol{g}$.

Let us begin by noting that since $\boldsymbol{g}(D) \subset D$, if $\boldsymbol{x}^{(0)}$ belongs to D, then $\boldsymbol{x}^{(k)} = \boldsymbol{g}(\boldsymbol{x}^{(k-1)}) \in D$ for all $k \ge 1$. Further, since $\boldsymbol{g}$ is a contraction on D in the ∞-norm, we have that

$$\|\boldsymbol{x}^{(k)}-\boldsymbol{x}^{(k-1)}\|_\infty = \|\boldsymbol{g}(\boldsymbol{x}^{(k-1)})-\boldsymbol{g}(\boldsymbol{x}^{(k-2)})\|_\infty \le L\|\boldsymbol{x}^{(k-1)}-\boldsymbol{x}^{(k-2)}\|_\infty$$

for all $k \geq 2$. We then deduce by induction that

$$\|\boldsymbol{x}^{(k)} - \boldsymbol{x}^{(k-1)}\|_\infty \leq L^{k-1}\|\boldsymbol{x}^{(1)} - \boldsymbol{x}^{(0)}\|_\infty , \qquad k \geq 1 . \tag{4.8}$$

Suppose that m and k are positive integers and $m \geq k+1$. Then, by repeated application of the triangle inequality in the ∞-norm and using (4.8), we have that

$$\begin{aligned} \|\boldsymbol{x}^{(m)} - \boldsymbol{x}^{(k)}\|_\infty &= \|(\boldsymbol{x}^{(m)} - \boldsymbol{x}^{(m-1)}) + \cdots + (\boldsymbol{x}^{(k+1)} - \boldsymbol{x}^{(k)})\|_\infty \\ &\leq \|\boldsymbol{x}^{(m)} - \boldsymbol{x}^{(m-1)}\|_\infty + \cdots + \|\boldsymbol{x}^{(k+1)} - \boldsymbol{x}^{(k)}\|_\infty \\ &\leq (L^{m-1} + \cdots + L^k)\|\boldsymbol{x}^{(1)} - \boldsymbol{x}^{(0)}\|_\infty \\ &= L^k(L^{m-k-1} + \cdots + 1)\|\boldsymbol{x}^{(1)} - \boldsymbol{x}^{0)}\|_\infty \\ &\leq L^k \frac{1}{1-L}\|\boldsymbol{x}^{(1)} - \boldsymbol{x}^{(0)}\|_\infty , \end{aligned} \tag{4.9}$$

where, in the transition to the last line, we made use of the fact that the geometric series $1 + L + L^2 + \cdots$, with $L \in (0,1)$, sums to $1/(1-L)$.

As $\lim_{k\to\infty} L^k = 0$, it follows from (4.9) that $(\boldsymbol{x}^{(k)})$ is a Cauchy sequence in $\mathbb{R}^n$; that is, for each $\varepsilon > 0$ there exists $k_0 = k_0(\varepsilon)$ (defined by (4.6) above) such that

$$\|\boldsymbol{x}^{(m)} - \boldsymbol{x}^{(k)}\|_\infty < \varepsilon \qquad \forall\, m, k \geq k_0 = k_0(\varepsilon) . \tag{4.10}$$

Any Cauchy sequence in $\mathbb{R}^n$ is convergent in $\mathbb{R}^n$; consequently, there exists $\boldsymbol{\xi} \in \mathbb{R}^n$ such that $\boldsymbol{\xi} = \lim_{k\to\infty} \boldsymbol{x}^{(k)}$. Further, since $\boldsymbol{g}$ satisfies a Lipschitz condition on D, the discussion in the paragraph following Definition 4.2 shows that $\boldsymbol{g}$ is continuous on D. Hence, by Lemma 4.2,

$$\boldsymbol{\xi} = \lim_{k\to\infty} \boldsymbol{x}^{(k+1)} = \lim_{k\to\infty} \boldsymbol{g}(\boldsymbol{x}^{(k)}) = \boldsymbol{g}\left(\lim_{k\to\infty} \boldsymbol{x}^{(k)}\right) = \boldsymbol{g}(\boldsymbol{\xi}) ,$$

which proves that $\boldsymbol{\xi}$ is a fixed point of $\boldsymbol{g}$.

It remains to show that $\boldsymbol{\xi} \in D$. This follows from Lemma 4.1 since $(\boldsymbol{x}^{(k)}) \subset D$, $\boldsymbol{\xi} = \lim_{k\to\infty} \boldsymbol{x}^{(k)}$ and D is closed. □

As a byproduct of the proof, we deduce from (4.7) that, given a positive tolerance ε, one can compute an approximation $\boldsymbol{x}^{(k)}$ to the unknown solution $\boldsymbol{\xi}$ using (4.3) in no more than $k_0 = k_0(\varepsilon)$ iterations so that the approximation error $\boldsymbol{\xi} - \boldsymbol{x}^{(k)}$, measured in the ∞-norm, is less than ε; the integer $k_0(\varepsilon)$ is defined by (4.6).

The next theorem relates the constant L from the Lipschitz condition (4.5) to the partial derivatives of $\boldsymbol{g}$, giving a more practically useful sufficient condition for convergence.

Definition 4.3 *Let $\boldsymbol{g} = (g_1, \ldots, g_n)^{\mathrm{T}}: \mathbb{R}^n \to \mathbb{R}^n$ be a function defined and continuous in an (open) neighbourhood $N(\boldsymbol{\xi})$ of $\boldsymbol{\xi} \in \mathbb{R}^n$. Suppose further that the first partial derivatives $\frac{\partial g_i}{\partial x_j}$, $j = 1, \ldots, n$, of g_i exist at $\boldsymbol{\xi}$ for $i = 1, \ldots, n$. The* **Jacobian matrix** *$J_g(\boldsymbol{\xi})$ of $\boldsymbol{g}$ at $\boldsymbol{\xi}$ is the $n \times n$ matrix with elements*

$$J_g(\boldsymbol{\xi})_{ij} = \frac{\partial g_i}{\partial x_j}(\boldsymbol{\xi}), \qquad i, j = 1, \ldots, n.$$

Theorem 4.2 *Suppose that $\boldsymbol{g} = (g_1, \ldots, g_n)^{\mathrm{T}}: \mathbb{R}^n \to \mathbb{R}^n$ is defined and continuous on a closed set $D \subset \mathbb{R}^n$. Let $\boldsymbol{\xi} \in D$ be a fixed point of $\boldsymbol{g}$, and suppose that the first partial derivatives $\frac{\partial g_i}{\partial x_j}$, $j = 1, \ldots, n$, of g_i, $i = 1, \ldots, n$, are defined and continuous in some (open) neighbourhood $N(\boldsymbol{\xi}) \subset D$ of $\boldsymbol{\xi}$, with*

$$\|J_g(\boldsymbol{\xi})\|_\infty < 1.$$

Then, there exists $\varepsilon > 0$ such that $\boldsymbol{g}(\bar{B}_\varepsilon(\boldsymbol{\xi})) \subset \bar{B}_\varepsilon(\boldsymbol{\xi})$, and the sequence defined by (4.3) converges to $\boldsymbol{\xi}$ for all $\boldsymbol{x}^{(0)} \in \bar{B}_\varepsilon(\boldsymbol{\xi})$.

Proof The proof is a natural extension of that of Theorem 1.5. We write $K = \|J_g(\boldsymbol{\xi})\|_\infty$. Since the partial derivatives $\frac{\partial g_i}{\partial x_j}$, $i, j = 1, \ldots, n$, are continuous in the neighbourhood $N(\boldsymbol{\xi})$ of $\boldsymbol{\xi}$, we can find a closed ball $\bar{B}_\varepsilon(\boldsymbol{\xi}) \subset N(\boldsymbol{\xi}) \subset D$ of radius ε and centre $\boldsymbol{\xi}$ such that

$$\|J_g(\boldsymbol{z})\|_\infty \le \tfrac{1}{2}(K+1) < 1 \qquad \forall\, \boldsymbol{z} \in \bar{B}_\varepsilon(\boldsymbol{\xi}). \tag{4.11}$$

Now, suppose that $\boldsymbol{x}$ and $\boldsymbol{y}$ are both in $\bar{B}_\varepsilon(\boldsymbol{\xi})$ and, for $i \in \{1, \ldots, n\}$ fixed, define the function $t \mapsto \varphi_i(t)$ of the single variable $t \in [0, 1]$ by

$$\varphi_i(t) = g_i(t\boldsymbol{x} + (1-t)\boldsymbol{y});$$

thus, $\varphi_i(0) = g_i(\boldsymbol{y})$ and $\varphi_i(1) = g_i(\boldsymbol{x})$. The function $t \mapsto \varphi_i(t)$ has a continuous derivative in t on the interval $[0, 1]$; thus, by the Mean Value Theorem (Theorem A.3), there exists $\eta \in (0, 1)$ such that

$$g_i(\boldsymbol{x}) - g_i(\boldsymbol{y}) = \varphi_i(1) - \varphi_i(0) = \varphi_i'(\eta)(1-0) = \varphi_i'(\eta).$$

This means that

$$g_i(\boldsymbol{x}) - g_i(\boldsymbol{y}) = \sum_{j=1}^{n} (x_j - y_j) \frac{\partial g_i}{\partial x_j}(\eta\boldsymbol{x} + (1-\eta)\boldsymbol{y}) \tag{4.12}$$

for $i = 1, \ldots, n$. Now $|x_j - y_j| \le \|\boldsymbol{x} - \boldsymbol{y}\|_\infty$ for all $j \in \{1, \ldots, n\}$, and so (4.12) gives

$$\begin{aligned} |g_i(\boldsymbol{x}) - g_i(\boldsymbol{y})| &\le \|\boldsymbol{x} - \boldsymbol{y}\|_\infty \sum_{j=1}^{n} \left| \frac{\partial g_i}{\partial x_j}(\eta \boldsymbol{x} + (1-\eta)\boldsymbol{y}) \right| \\ &\le \|\boldsymbol{x} - \boldsymbol{y}\|_\infty \, \|J_g(\eta \boldsymbol{x} + (1-\eta)\boldsymbol{y})\|_\infty , \end{aligned}$$

for all $i = 1, \ldots, n$. Consequently, for any $\boldsymbol{x}, \boldsymbol{y} \in \bar{B}_\varepsilon(\boldsymbol{\xi})$,

$$\begin{aligned} \|\boldsymbol{g}(\boldsymbol{x}) - \boldsymbol{g}(\boldsymbol{y})\|_\infty &\le \max_{t \in [0,1]} \|J_g(t\boldsymbol{x} + (1-t)\boldsymbol{y})\|_\infty \|\boldsymbol{x} - \boldsymbol{y}\|_\infty \\ &\le \tfrac{1}{2}(1+K)\|\boldsymbol{x} - \boldsymbol{y}\|_\infty , \qquad (4.13) \end{aligned}$$

due to (4.11), given that $t\boldsymbol{x} + (1-t)\boldsymbol{y} \in \bar{B}_\varepsilon(\boldsymbol{\xi})$ for all $t \in [0,1]$. It follows that $\boldsymbol{g}$ satisfies a Lipschitz condition (4.5), in the ∞-norm, on the closed ball $\bar{B}_\varepsilon(\boldsymbol{\xi})$ with $L = \frac{1}{2}(1+K) < 1$. Furthermore, on selecting $\boldsymbol{y} = \boldsymbol{\xi}$ in (4.13) we get that

$$\|\boldsymbol{g}(\boldsymbol{x}) - \boldsymbol{\xi}\|_\infty = \|\boldsymbol{g}(\boldsymbol{x}) - \boldsymbol{g}(\boldsymbol{\xi})\|_\infty \le \|\boldsymbol{x} - \boldsymbol{\xi}\|_\infty \le \varepsilon$$

for all $\boldsymbol{x} \in \bar{B}_\varepsilon(\boldsymbol{\xi})$. Hence, $\boldsymbol{g}(\bar{B}_\varepsilon(\boldsymbol{\xi})) \subset \bar{B}_\varepsilon(\boldsymbol{\xi})$. The convergence of the iteration (4.3) to $\boldsymbol{\xi}$, for an arbitrary starting value $\boldsymbol{x}^{(0)} \in \bar{B}_\varepsilon(\boldsymbol{\xi})$, now follows from Theorem 4.1. □

We close this section with an example which illustrates the application of the method of simultaneous iteration to the solution of a system of nonlinear equations.

Example 4.4 *Let us consider, as in Example 4.1, the system of two simultaneous nonlinear equations in the unknowns x_1 and x_2, defined by*

$$\begin{aligned} x_1^2 + x_2^2 - 1 &= 0 , \\ 5x_1^2 + 21x_2^2 - 9 &= 0 . \end{aligned}$$

Here $\boldsymbol{x} = (x_1, x_2)^{\mathrm{T}}$ and $\boldsymbol{f} = (f_1, f_2)^{\mathrm{T}}$ with

$$\begin{aligned} f_1(x_1, x_2) &= x_1^2 + x_2^2 - 1 , \\ f_2(x_1, x_2) &= 5x_1^2 + 21x_2^2 - 9 . \end{aligned}$$

Let us suppose that we need to find the solution of the system $\boldsymbol{f}(\boldsymbol{x}) = \mathbf{0}$ in the first quadrant of the (x_1, x_2)-coordinate system.

Of course, the example is a little artificial, since we already know from Example 4.1 that $\boldsymbol{\xi}_2 = (\sqrt{3}/2, 1/2)^{\mathrm{T}}$ is the required solution. In what follows, however, we proceed as if we knew nothing about the location of $\boldsymbol{\xi}_2$. Our aim here is to illustrate the construction of the function $\boldsymbol{g}$ from $\boldsymbol{f}$ and the verification of the hypotheses of Theorem 4.1.

Let us rewrite the two equations as

$$x_1 = \left(1 - x_2^2\right)^{1/2}, \qquad x_2 = \frac{1}{\sqrt{21}}\left(9 - 5x_1^2\right)^{1/2},$$

and define $g_1(x_1, x_2)$ and $g_2(x_1, x_2)$ as the right-hand sides of these, respectively. We consider the simultaneous iteration

$$\boldsymbol{x}^{(k+1)} = \boldsymbol{g}(\boldsymbol{x}^{(k)}), \qquad k = 0, 1, 2, \ldots, \tag{4.14}$$

with suitably chosen $\boldsymbol{x}^{(0)}$ and $\boldsymbol{g} = (g_1, g_2)^{\mathrm{T}}$.

Our first task is to find a closed subset D of $\mathbb{R}^2$ containing the required solution, such that $\boldsymbol{g}$ satisfies the hypotheses of Theorem 4.1 on D. In order to ensure that $\boldsymbol{x} \mapsto \boldsymbol{g}(\boldsymbol{x})$ is real-valued and continuous, and that the partial derivatives of g_1 and g_2 are continuous at $\boldsymbol{x} = (x_1, x_2)^{\mathrm{T}} \in D$, we demand that $|x_2| < 1$ and $|x_1| < 3/(\sqrt{5})$. In fact, since we are looking for a solution in the first quadrant, it is natural to suppose that $x_1 \geq 0$, $x_2 \geq 0$. Hence we let $M = \{\boldsymbol{x} \in \mathbb{R}^2\colon 0 \leq x_1 < 3/\sqrt{5},\ \ 0 \leq x_2 < 1\}$, and we seek D as a suitable closed subset of M.

For $\boldsymbol{x} \in M$, let

$$J_g(\boldsymbol{x}) = \begin{pmatrix} \partial g_1/\partial x_1 & \partial g_1/\partial x_2 \\ \partial g_2/\partial x_1 & \partial g_2/\partial x_2 \end{pmatrix}.$$

Clearly,

$$\frac{\partial g_1}{\partial x_1} = 0, \qquad \frac{\partial g_1}{\partial x_2} = -x_2\left(1 - x_2^2\right)^{-1/2},$$
$$\frac{\partial g_2}{\partial x_1} = -\frac{5}{\sqrt{21}}\, x_1\left(9 - 5x_1^2\right)^{-1/2}, \qquad \frac{\partial g_2}{\partial x_2} = 0,$$

so we conclude that, for any $\boldsymbol{x} \in M$,

$$\|J_g(\boldsymbol{x})\|_\infty = \max\left(x_2\left(1 - x_2^2\right)^{-1/2}, \frac{5}{\sqrt{21}}\, x_1\left(9 - 5x_1^2\right)^{-1/2}\right).$$

In particular, we have $\|J_g(\boldsymbol{x})\|_\infty < 1$ provided that

$$x_2^2 < 1 - x_2^2 \quad \text{and} \quad 25x_1^2 < 21(9 - 5x_1^2),$$

that is, when $x_2^2 < 1/2$ and $x_1^2 < 189/130$. These conditions are clearly satisfied if, for example, $0 \leq x_1 \leq 1$ and $0 \leq x_2 \leq 3/5$. If we now define $D = [0, 1] \times [0, 3/5]$, then, analogously as in (4.13), we have that

$$\|\boldsymbol{g}(\boldsymbol{x}) - \boldsymbol{g}(\boldsymbol{y})\|_\infty \leq \max_{t \in [0,1]} \|J_g(t\boldsymbol{x} + (1 - t)\boldsymbol{y})\|_\infty\, \|\boldsymbol{x} - \boldsymbol{y}\|_\infty$$

for all $\boldsymbol{x}$ and $\boldsymbol{y}$ in D. Therefore, also,

$$\|\boldsymbol{g}(\boldsymbol{x}) - \boldsymbol{g}(\boldsymbol{y})\|_\infty \le L\|\boldsymbol{x} - \boldsymbol{y}\|_\infty$$

with

$$L = \max_{\boldsymbol{z}\in D} \|J_g(\boldsymbol{z})\|_\infty < 1 . \tag{4.15}$$

With our choice of D, (4.15) holds with $L = \max\{0.75, 0.55\} = 0.75 < 1$. Furthermore, it is easy to check that $\boldsymbol{g}(D) \subset D$. Thus we deduce from Theorem 4.1 that $\boldsymbol{g}$ has a unique fixed point in D – we call this fixed point $\boldsymbol{\xi}_2$, for the sake of consistency with the notation in Example 4.1; moreover, the sequence $(\boldsymbol{x}^{(k)})$ defined by (4.14) converges to $\boldsymbol{\xi}_2$.

After all these preparations you are now probably curious to see what the successive iterates look like: Table 4.1 gives a flavour of the behaviour of the sequence $(\boldsymbol{x}^{(k)})$, with the starting value chosen as $\boldsymbol{x}^{(0)} = (0.5, 0.3)^{\mathrm{T}}$. You can see that after 15 iterations the first 5 decimal digits have settled to their correct values.[1]

4.3 Relaxation and Newton's method

We now go on to apply the ideas developed in the previous section to the construction of an iteration which converges to a solution of the equation $\boldsymbol{f}(\boldsymbol{x}) = \boldsymbol{0}$, where $\boldsymbol{f}$: $\mathbb{R}^n \to \mathbb{R}^n$. One way of constructing such a sequence is by relaxation.

Definition 4.4 *The recursion*

$$\boxed{\boldsymbol{x}^{(k+1)} = \boldsymbol{x}^{(k)} - \lambda \boldsymbol{f}(\boldsymbol{x}^{(k)}) , \qquad k = 0, 1, 2, \dots ,} \tag{4.16}$$

where $\boldsymbol{x}^{(0)} \in \mathbb{R}^n$ *is given and where* $\lambda \neq 0$ *is a constant, is called* **simultaneous relaxation**.

Suppose that the sequence $(\boldsymbol{x}^{(k)})$ converges to a limit $\boldsymbol{\xi} \in \mathbb{R}^n$ and $\boldsymbol{f}$ is continuous in a neighbourhood of $\boldsymbol{\xi}$; then, on passing to the limit $k \to \infty$ in (4.16), we deduce that $\boldsymbol{\xi}$ is a solution of the equation $\boldsymbol{f}(\boldsymbol{x}) = \boldsymbol{0}$.

Simultaneous relaxation is evidently a simultaneous iteration defined by taking $\boldsymbol{g}(\boldsymbol{x}) = \boldsymbol{x} - \lambda \boldsymbol{f}(\boldsymbol{x})$.

[1] You may wish to contemplate the following question: *how many iterations should be performed to ensure that all 15 digits have settled to their correct values?* Use inequality (4.6) to get an idea of the (maximum) amount of work involved!

Table 4.1. *The first 15 iterates in the sequence* $\boldsymbol{x}^{(k)} = (x_1^{(k)}, x_2^{(k)})^{\mathrm{T}}$ *defined by (4.14), with starting value* $(0.5, 0.3)^{\mathrm{T}}$. *The exact solution is* $\boldsymbol{\xi}_2 = (\sqrt{3}/2, 1/2)^{\mathrm{T}} = (0.866025403784439, 0.500000000000000)^{\mathrm{T}}$ *to 15 decimal digits.*

k	$x_1^{(k)}$	$x_2^{(k)}$
0	0.500000000000000	0.300000000000000
1	0.953939197667987	0.607492896293956
2	0.794325110362489	0.460331145598201
3	0.887747281827575	0.527583804908580
4	0.849502989281489	0.490845908224662
5	0.871246402792635	0.506703790432366
6	0.862120217116774	0.497835722000956
7	0.867271349636195	0.501604267098156
8	0.865097196405654	0.499485546313646
9	0.866322220091208	0.500382434879534
10	0.865804492286815	0.499877559050176
11	0.866096083560039	0.500091082450647
12	0.865972810920378	0.499970850112656
13	0.866042232825645	0.500021687802653
14	0.866012881963649	0.499993059704778
15	0.866029410728674	0.500005163847862

Theorem 4.3 *Suppose that* $\boldsymbol{f}(\boldsymbol{\xi}) = \mathbf{0}$, *and that all the first partial derivatives of* $\boldsymbol{f} = (f_1, \dots, f_n)^{\mathrm{T}}$ *are defined and continuous in some (open) neighbourhood of* $\boldsymbol{\xi}$, *and satisfy a condition of strict diagonal dominance at* $\boldsymbol{\xi}$; *i.e.,*

$$\frac{\partial f_i}{\partial x_i}(\boldsymbol{\xi}) > \sum_{\substack{j=1 \\ j \neq i}}^{n} \left| \frac{\partial f_i}{\partial x_j}(\boldsymbol{\xi}) \right|, \quad i = 1, 2, \dots, n. \tag{4.17}$$

Then, there exist $\varepsilon > 0$ *and a positive constant* λ *such that the relaxation iteration (4.16) converges to* $\boldsymbol{\xi}$ *for any* $\boldsymbol{x}^{(0)}$ *in the closed ball* $\bar{B}_{\varepsilon}(\boldsymbol{\xi})$ *of radius* ε, *centre* $\boldsymbol{\xi}$.

Proof The elements of the Jacobian matrix $J_g(\boldsymbol{\xi}) = (\gamma_{ij}) \in \mathbb{R}^{n \times n}$ of the function $\boldsymbol{x} \mapsto \boldsymbol{g}(\boldsymbol{x}) = \boldsymbol{x} - \lambda \boldsymbol{f}(\boldsymbol{x})$ at $\boldsymbol{x} = \boldsymbol{\xi}$ are

$$\gamma_{ii}(\boldsymbol{\xi}) = 1 - \lambda \frac{\partial f_i}{\partial x_i}(\boldsymbol{\xi}), \quad \gamma_{ij}(\boldsymbol{\xi}) = -\lambda \frac{\partial f_i}{\partial x_j}(\boldsymbol{\xi}), \quad j \neq i, \quad i, j \in \{1, \dots, n\}.$$

We now define

$$m = \max_{i=1}^{n} \frac{\partial f_i}{\partial x_i}(\boldsymbol{\xi})$$

and then choose $\lambda = 1/m$. Under hypothesis (4.17), $m > 0$ and therefore $\lambda > 0$. This choice of λ ensures that all the diagonal elements $\gamma_{ii}(\boldsymbol{\xi})$, $i = 1, \ldots, n$, of $J_g(\boldsymbol{\xi})$ are nonnegative. Moreover, for any $i \in \{1, \ldots, n\}$,

$$\sum_{j=1}^{n} |\gamma_{ij}(\boldsymbol{\xi})| \quad = \quad 1 - \lambda \frac{\partial f_i}{\partial x_i}(\boldsymbol{\xi}) + \lambda \sum_{\substack{j=1 \\ j \neq i}}^{n} \left| \frac{\partial f_i}{\partial x_j}(\boldsymbol{\xi}) \right| < 1\,,$$

by condition (4.17); consequently, $\|J_g(\boldsymbol{\xi})\|_\infty < 1$. As $\boldsymbol{\xi}$ is a fixed point of $\boldsymbol{g}$, it follows from Theorem 4.2 that there exists $\varepsilon > 0$ such that the iteration (4.16) converges to $\boldsymbol{\xi}$ for all $\boldsymbol{x}^{(0)} \in \bar{B}_\varepsilon(\boldsymbol{\xi})$. □

The condition of strict diagonal dominance will only be satisfied in a small class of problems (although this class does contain some examples of practical importance). More generally it will be necessary to replace the scalar λ by a nonsingular constant matrix Λ, giving a more general relaxation iteration

$$\boldsymbol{x}^{(k+1)} = \boldsymbol{x}^{(k)} - \Lambda \boldsymbol{f}(\boldsymbol{x}^{(k)})\,, \qquad k = 0, 1, 2, \ldots\,.$$

This may be interpreted as trying to solve the new system of equations $\Lambda \boldsymbol{f}(\boldsymbol{x}) = \boldsymbol{0}$. The Jacobian matrix of this system is ΛJ_f, where J_f is the Jacobian matrix of $\boldsymbol{f}$. It is now possible to select the matrix Λ so that $\Lambda J_f(\boldsymbol{\xi})$ has the property of strict diagonal dominance. In principle, this can obviously be done by choosing $\Lambda = [J_f(\boldsymbol{\xi})]^{-1}$, the inverse of the Jacobian matrix of $\boldsymbol{f}$ evaluated at the solution $\boldsymbol{\xi}$. The Jacobian matrix of the new system is then the identity matrix, which clearly satisfies the diagonal dominance condition. However, this choice is not possible in practice, since of course the solution $\boldsymbol{\xi}$ is unknown. If we allow the matrix Λ to be a function of $\boldsymbol{x}$, instead of being constant, the argument above suggests taking

$$\Lambda = [J_f(\boldsymbol{x}^{(k)})]^{-1}\,,$$

leading to Newton's method for a system of equations.

Definition 4.5 *The recursion defined by*

$$\boldsymbol{x}^{(k+1)} = \boldsymbol{x}^{(k)} - [J_f(\boldsymbol{x}^{(k)})]^{-1} \boldsymbol{f}(\boldsymbol{x}^{(k)})\,, \qquad k = 0, 1, 2, \ldots\,, \tag{4.18}$$

where $\boldsymbol{x}^{(0)} \in \mathbb{R}^n$, is called **Newton's method** *(or Newton iteration)* *for the system of equations* $\boldsymbol{f}(\boldsymbol{x}) = \mathbf{0}$. *It is implicitly assumed that the matrix* $J_f(\boldsymbol{x}^{(k)})$ *exists and is nonsingular for each* $k = 0, 1, 2, \ldots$.

The next theorem is concerned with the convergence of Newton's method. As in the scalar case, for a starting value $\boldsymbol{x}^{(0)}$ that is sufficiently close to the solution $\boldsymbol{\xi}$ of $\boldsymbol{f}(\boldsymbol{x}) = \mathbf{0}$, Newton's method converges quadratically. The precise definition of quadratic convergence is given below: it resembles Definition 1.7 of Chapter 1.

Definition 4.6 *Suppose that* $(\boldsymbol{x}^{(k)})$ *is a convergent sequence in* $\mathbb{R}^n$ *and* $\boldsymbol{\xi} = \lim_{k\to\infty} \boldsymbol{x}^{(k)}$. *We say that* $(\boldsymbol{x}^{(k)})$ *converges to* $\boldsymbol{\xi}$ **with at least order** $q > 1$, *if there exist a sequence* (ε_k) *of positive real numbers converging to* 0, *and* $\mu > 0$, *such that*

$$\|\boldsymbol{x}^{(k)} - \boldsymbol{\xi}\|_\infty \le \varepsilon_k\,, \quad k = 0, 1, 2, \ldots\,, \qquad \text{and} \qquad \lim_{k\to\infty} \frac{\varepsilon_{k+1}}{\varepsilon_k^q} = \mu\,. \tag{4.19}$$

If (4.19) holds with $\varepsilon_k = \|\boldsymbol{x}^{(k)} - \boldsymbol{\xi}\|_\infty$, $k = 0, 1, 2, \ldots$, *then the sequence* $(\boldsymbol{x}^{(k)})$ *is said to converge to* $\boldsymbol{\xi}$ **with order** q. *In particular, if* $q = 2$, *then we say that the sequence* $(\boldsymbol{x}^{(k)})$ *converges to* $\boldsymbol{\xi}$ **quadratically**.

Again, due to (4.4), if a sequence $(\boldsymbol{x}^{(k)})$ converges quadratically in the ∞-norm, then it also does so in the p-norm for any $p \in [1, \infty)$, though the constant μ may be different.

Theorem 4.4 *Suppose that* $\boldsymbol{f}(\boldsymbol{\xi}) = \mathbf{0}$, *that in some (open) neighbourhood* $N(\boldsymbol{\xi})$ *of* $\boldsymbol{\xi}$, *where* $\boldsymbol{f}$ *is defined and continuous, all the second-order partial derivatives of* $\boldsymbol{f}$ *are defined and continuous, and that the Jacobian matrix* $J_f(\boldsymbol{\xi})$ *of* $\boldsymbol{f}$ *at the point* $\boldsymbol{\xi}$ *is nonsingular. Then, the sequence* $(\boldsymbol{x}^{(k)})$ *defined by Newton's method (4.18) converges to the solution* $\boldsymbol{\xi}$ *provided that* $\boldsymbol{x}^{(0)}$ *is sufficiently close to* $\boldsymbol{\xi}$; *the convergence of the sequence* $(\boldsymbol{x}^{(k)})$ *to* $\boldsymbol{\xi}$ *is at least quadratic.*

Proof Let us begin by writing Newton's method as a simultaneous iteration $\boldsymbol{x}^{(k+1)} = \boldsymbol{g}(\boldsymbol{x}^{(k)})$, $k = 0, 1, 2, \ldots$, as in (4.3), with $\boldsymbol{x}^{(0)}$ given and

$$\boldsymbol{g}(\boldsymbol{x}) = \boldsymbol{x} - [J_f(\boldsymbol{x})]^{-1} \boldsymbol{f}(\boldsymbol{x})\,.$$

The idea of the proof is to verify that the function $\boldsymbol{g}$ satisfies all the conditions of Theorem 4.2 in a certain closed ball centred at $\boldsymbol{\xi}$, the fixed point of $\boldsymbol{g}$, and thus deduce that the sequence $(\boldsymbol{x}^{(k)})$ converges to $\boldsymbol{\xi}$.

As the function $\boldsymbol{x} \mapsto \det J_f(\boldsymbol{x})$ is continuous in $N(\boldsymbol{\xi})$ and $\det J_f(\boldsymbol{\xi}) \neq 0$, there exists $\varepsilon > 0$ such that $\det J_f(\boldsymbol{x}) \neq 0$ for all $\boldsymbol{x} \in \bar{B}_\varepsilon(\boldsymbol{\xi}) \subset N(\boldsymbol{\xi})$.

Further, as the entries of $[J_f(\boldsymbol{x})]^{-1}$ depend continuously on the entries of $J_f(\boldsymbol{x})$ and since the entries of $J_f(\,\cdot\,)$ are continuous functions of $\boldsymbol{x}$ in $N(\boldsymbol{\xi})$, we deduce that $\boldsymbol{x} \mapsto [J_f(\boldsymbol{x})]^{-1}\boldsymbol{f}(\boldsymbol{x})$ is a continuous function on $\bar{B}_\varepsilon(\boldsymbol{\xi})$; therefore,

$$\boldsymbol{x} \mapsto \boldsymbol{g}(\boldsymbol{x}) = \boldsymbol{x} - [J_f(\boldsymbol{x})]^{-1}\boldsymbol{f}(\boldsymbol{x})$$

is also a continuous function on $\bar{B}_\varepsilon(\boldsymbol{\xi})$. For later reference, we note that $\boldsymbol{x} \mapsto \|[J_f(\boldsymbol{x})]^{-1}\|_\infty$, too, is a continuous function on $\bar{B}_\varepsilon(\boldsymbol{\xi})$, and therefore it is a bounded function on $\bar{B}_\varepsilon(\boldsymbol{\xi})$; we define

$$C = \max_{\boldsymbol{x} \in \bar{B}_\varepsilon(\boldsymbol{\xi})} \|[J_f(\boldsymbol{x})]^{-1}\|_\infty \,.$$

Now, $\boldsymbol{\xi}$ is a fixed point of $\boldsymbol{g}$ and, by the hypotheses of the theorem, the entries of the Jacobian matrix J_g of $\boldsymbol{g}$ are continuous functions of $\boldsymbol{x}$ on $\bar{B}_\varepsilon(\boldsymbol{\xi})$. Furthermore, it is easy to check that all the elements of the Jacobian matrix $J_g(\boldsymbol{x})$ of $\boldsymbol{g}$ vanish at $\boldsymbol{x} = \boldsymbol{\xi}$; see Exercise 6. Hence, $\|J_g(\boldsymbol{\xi})\|_\infty = 0 < 1$, trivially. Thus we have shown that $\boldsymbol{g}\colon \mathbb{R}^n \to \mathbb{R}^n$ satisfies all the conditions of Theorem 4.2 on the closed set $D = \bar{B}_\varepsilon(\boldsymbol{\xi})$, and the convergence of the sequence $(\boldsymbol{x}^{(k)})$ to $\boldsymbol{\xi}$, as $k \to \infty$, follows.

To show that convergence is at least quadratic, we write the iteration in the form

$$J_f(\boldsymbol{x}^{(k)})\,[\boldsymbol{x}^{(k+1)} - \boldsymbol{\xi}] = J_f(\boldsymbol{x}^{(k)})\,[\boldsymbol{x}^{(k)} - \boldsymbol{\xi}] - \boldsymbol{f}(\boldsymbol{x}^{(k)})\,. \tag{4.20}$$

Taylor's Theorem for a function of n variables, Theorem A.7 (including only the first-order terms), implies that, when $\boldsymbol{x}^{(k)} \in \bar{B}_\varepsilon(\boldsymbol{\xi})$,

$$\mathbf{0} = \boldsymbol{f}(\boldsymbol{\xi}) = \boldsymbol{f}(\boldsymbol{x}^{(k)}) + J_f(\boldsymbol{x}^{(k)})[\boldsymbol{\xi} - \boldsymbol{x}^{(k)}] + \mathbf{E}_f\,, \tag{4.21}$$

where

$$\|\mathbf{E}_f\|_\infty \le \tfrac{1}{2}n^2 A_f \|\boldsymbol{\xi} - \boldsymbol{x}^{(k)}\|_\infty^2\,, \tag{4.22}$$

and

$$A_f = \max_{1 \le i,j,l \le n} \max_{\boldsymbol{x} \in \bar{B}_\varepsilon(\boldsymbol{\xi})} \left| \frac{\partial^2 f_i}{\partial x_j \partial x_l}(\boldsymbol{x}) \right|$$

is a bound on all the second-order partial derivatives of $\boldsymbol{f}$ on $\bar{B}_\varepsilon(\boldsymbol{\xi})$. The factor n^2 in (4.22) stems from the fact that, for each $i \in \{1, \dots, n\}$, f_i is a function of n variables and therefore it has n^2 second-order partial derivatives – each bounded by A_f over $\bar{B}_\varepsilon(\boldsymbol{\xi})$. From (4.21) and (4.20) we see that

$$\boldsymbol{x}^{(k+1)} - \boldsymbol{\xi} = [J_f(\boldsymbol{x}^{(k)})]^{-1}\mathbf{E}_f\,,$$

and so

$$\|\boldsymbol{x}^{(k+1)} - \boldsymbol{\xi}\|_\infty \le \tfrac{1}{2} n^2 A_f C \|\boldsymbol{x}^{(k)} - \boldsymbol{\xi}\|_\infty^2 .$$

On writing $M = \frac{1}{2}n^2 A_f C$, we then deduce by induction that

$$\|\boldsymbol{x}^{(k)} - \boldsymbol{\xi}\|_\infty \le \frac{1}{M}\left(M\|\boldsymbol{x}^{(0)} - \boldsymbol{\xi}\|_\infty\right)^{2^k}, \qquad k = 0, 1, 2, \dots .$$

Suppose that $\boldsymbol{x}^{(0)} \in \bar{B}_\varepsilon(\boldsymbol{\xi})$ where $\varepsilon \le \frac{1}{2}\min\{1, 1/M\}$. Then,

$$M\|\boldsymbol{x}^{(0)} - \boldsymbol{\xi}\|_\infty \le \frac{1}{2}, \qquad k = 0, 1, 2, \dots ,$$

and hence

$$\|\boldsymbol{x}^{(k)} - \boldsymbol{\xi}\|_\infty \le \frac{1}{M}\left(\frac{1}{2}\right)^{2^k}$$

This implies that convergence is at least quadratic (on choosing $\varepsilon_k = M^{-1}2^{-2^k}$ and $q = 2$ in Definition 4.6). $\square$

Newton's method is defined in (4.18) by using the inverse of the Jacobian matrix. As we saw in Chapter 2 it is more efficient to avoid inverting a matrix, if possible. In practice the method is therefore implemented by writing (4.18) in the form

$$J_f(\boldsymbol{x}^{(k)})[\boldsymbol{x}^{(k+1)} - \boldsymbol{x}^{(k)}] = -\boldsymbol{f}(\boldsymbol{x}^{(k)}) . \tag{4.23}$$

Given the vector $\boldsymbol{x}^{(k)}$, we calculate $\boldsymbol{f}(\boldsymbol{x}^{(k)})$ and the Jacobian matrix $J_f(\boldsymbol{x}^{(k)}) \in \mathbb{R}^{n\times n}$, and then solve the system of linear equations (4.23) by Gaussian elimination; this gives the increment vector $\boldsymbol{x}^{(k+1)} - \boldsymbol{x}^{(k)}$, which is added to $\boldsymbol{x}^{(k)}$ to obtain the new iterate $\boldsymbol{x}^{(k+1)}$.

Example 4.5 *We close this section with an example which illustrates the application of Newton's method. Consider the simultaneous nonlinear equations*

$$\begin{aligned} f_1(x, y, z) &\equiv x^2 + y^2 + z^2 - 1 = 0 , \\ f_2(x, y, z) &\equiv 2x^2 + y^2 - 4z = 0 , \\ f_3(x, y, z) &\equiv 3x^2 - 4y + z^2 = 0 . \end{aligned}$$

Letting $\boldsymbol{f} = (f_1, f_2, f_3)^{\mathrm{T}}$ *and* $\boldsymbol{x} = (x, y, z)^{\mathrm{T}}$, *the aim of the exercise is to determine the solution to the equation* $\boldsymbol{f}(\boldsymbol{x}) = \boldsymbol{0}$ *contained in the first octant* $\{(x, y, z) \in \mathbb{R}^3\colon x > 0 ,\ y > 0 ,\ z > 0\}$ *in* $\mathbb{R}^3$.

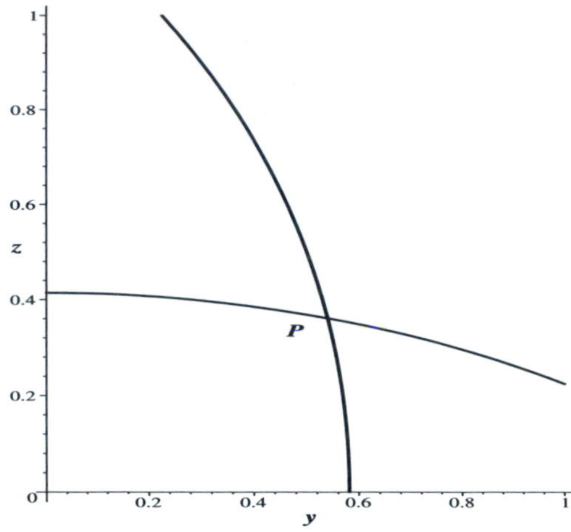

Fig. 4.3. Example 4.5: Projections onto the (y, z)-plane of the intersection-curves of the surfaces $f_1(x, y, z) = 0$ and $f_2(x, y, z) = 0$, and $f_1(x, y, z) = 0$ and $f_3(x, y, z) = 0$. The two curves intersect at the point P whose two coordinates are the y- and z-coordinates of $\boldsymbol{\xi}$, the solution of the system $f_1(x, y, z) = 0$, $f_2(x, y, z) = 0$, $f_3(x, y, z) = 0$.

Note that the Jacobian matrix of $\boldsymbol{f}$ at $\boldsymbol{x} \in \mathbb{R}^3$ is

$$J_f(\boldsymbol{x}) = \begin{pmatrix} 2x & 2y & 2z \\ 4x & 2y & -4 \\ 6x & -4 & 2z \end{pmatrix}.$$

Since the first equation represents a sphere of radius 1 centred at $(0, 0, 0)$, and the second and third equations describe elliptic paraboloids whose axes are aligned with the coordinate semi-axes $(0, 0, z)$, $z \geq 0$, and $(0, y, 0)$, $y \geq 0$, respectively, the point of intersection of the three surfaces belongs to $[0, 1]^3$. Let us denote this point by $\boldsymbol{\xi}$. In order to select a suitable starting value $\boldsymbol{x}^{(0)}$ for the iteration, we observe that the intersection of the first and the second surface is a curve whose projection onto the (y, z)-plane has the equation $y^2 + 2z^2 + 4z = 2$, while the intersection of the first and the third surface is a curve whose projection onto the (y, z)-plane has the equation $3y^2 + 4y + 2z^2 = 3$. The two curves are shown in Figure 4.3; the point P where the curves intersect has the same y- and z-coordinates as $\boldsymbol{\xi}$. The x-coordinate of $\boldsymbol{\xi}$ can be obtained from the first equation in terms of the y- and z-coordinates of P *via* $x = +(1 - y^2 - z^2)^{1/2}$. As the two coordinates of

P are, very roughly, $y \approx 0.5$ and $z \approx 0.5$, it is reasonable to choose as starting value for the Newton iteration the point $\boldsymbol{x}^{(0)} = (0.5, 0.5, 0.5)^{\mathrm{T}}$.

Thus, $\boldsymbol{f}(\boldsymbol{x}^{(0)}) = (-0.25, -1.25, -1.00)^{\mathrm{T}}$ and

$$J_f(\boldsymbol{x}^{(0)}) = \begin{pmatrix} 1 & 1 & 1 \\ 2 & 1 & -4 \\ 3 & -4 & 1 \end{pmatrix}.$$

On solving the system of linear equations

$$J_f(\boldsymbol{x}^{(0)}) \left(\boldsymbol{x}^{(1)} - \boldsymbol{x}^{(0)}\right) = -\boldsymbol{f}(\boldsymbol{x}^{(0)})$$

for $\boldsymbol{x}^{(1)} - \boldsymbol{x}^{(0)}$, we find that $\boldsymbol{x}^{(1)} = (0.875, 0.500, 0.375)^{\mathrm{T}}$. Similarly,

$$\boldsymbol{x}^{(2)} = (0.78981, 0.49662, 0.36993)^{\mathrm{T}},$$
$$\boldsymbol{x}^{(3)} = (0.78521, 0.49662, 0.36992)^{\mathrm{T}}.$$

As $\boldsymbol{f}(\boldsymbol{x}^{(3)}) = 10^{-5}(1, 4, 5)^{\mathrm{T}}$, the vector $\boldsymbol{x}^{(3)}$ can be thought of as a satisfactory approximation to the required solution $\boldsymbol{\xi}$; after rounding to four decimal digits, we have that

$$x = 0.7852, \qquad y = 0.4966, \qquad z = 0.3699.$$

4.4 Global convergence

Much of the discussion of the global convergence of Newton's method for a single equation in Section 1.7 applies, with obvious changes, in the case of several variables. If the system has several solutions, $\boldsymbol{\xi}_1$, $\boldsymbol{\xi}_2$, ..., we can define the corresponding sets $S_1, S_2, \ldots$ in $\mathbb{R}^n$ so that S_j comprises those starting points from which Newton's method converges to $\boldsymbol{\xi}_j$. As before, the sets S_j, $j = 1, 2, \ldots$, have the property that any point on the boundary of one of the sets is also on the boundary of the others. The difference now is that for systems of equations in $\mathbb{R}^n$, $n \geq 2$, these sets can be much more complicated than in the case of a single equation on the real line $\mathbb{R}^1 = \mathbb{R}$.

To illustrate this point for $n = 2$, we return to our earlier example problem, Example 1.7 from Chapter 1, but now extend it to complex variables, so we require to solve $\mathrm{e}^z - z - 2 = 0$ for the complex number $z = x + \imath y$. Separating this equation into real and imaginary parts we obtain a system of two nonlinear equations for the unknowns $x_1 = x$ and $x_2 = y$. The system has the two real solutions which we found in

Chapter 1, and also an infinite number of complex solutions. It is easy to see from the periodic character of $e^{\imath y}$ that the equation has a solution near $w_m = (2m + \frac{1}{2})\imath\pi$, $\imath = \sqrt{-1}$, for integer values of m; a better estimate is given in Exercise 9. It is a good deal more difficult to prove that there are no other solutions.

The behaviour of Newton's method for this problem may be illustrated by showing a picture of the complex plane, with the sets S_j depicted in different colours. In our example we cannot, of course, show more than a small number of the solutions, and cannot use an infinite number of colours. We have therefore coloured the sets with six colours cyclically, so that, for example, the sets $S_1, S_7, S_{13}, \ldots$ have the same colour. The background colour, white, represents the set S_1 of points from which the iteration converges to the real negative root. It includes most of the negative half-plane. Successive pictures in the series from Figure 4.5 to Figure 4.9 show a magnified view of a small region of the previous picture, the region being outlined in black. In Figure 4.4 the black crosses mark the positions of solutions of $f(z) = 0$. The pictures show in a striking way the fractal behaviour of the boundary of a set. Figure 4.9 is very similar to Figure 4.5; the former is a magnified view of a small part of Figure 4.5, with a magnification of about 50000 in each direction. The same sort of behaviour is repeated when the picture is magnified indefinitely.

4.5 Notes

For an introduction to the topology of $\mathbb{R}^n$, including the definitions of open set, closed set, continuity, convergence and Cauchy sequence, the reader is referred to any standard textbook on the subject; see, *e.g.*,

- W. Rudin, *Principles of Mathematical Analysis*, Third Edition, International Series in Pure and Applied Mathematics, McGraw–Hill, New York, Auckland, Düsseldorf, 1976,
- S.A. Douglass, *Introduction to Mathematical Analysis*, Addison–Wesley, Reading, MA, 1996.

Our first remark concerns the Contraction Mapping Theorem, Theorem 4.1, which is a direct generalisation of Theorem 1.3 from Chapter 1. Comparing the proofs of Theorems 1.3 and 4.1, we see that the proof of Theorem 1.3 is much simpler. This is not accidental: in the case of a single equation $x = g(x)$, involving a real-valued function g of a single real variable x, the existence of a fixed point follows directly from

Theorem 1.2, Brouwer's Fixed Point Theorem on a bounded closed interval of the real line. On the other hand, for the simultaneous system of equations $\boldsymbol{x} = \boldsymbol{g}(\boldsymbol{x})$ in $\mathbb{R}^n$ considered in Theorem 4.1 we had to invoke the completeness of $\mathbb{R}^n$ (i.e., the property that every Cauchy sequence in $\mathbb{R}^n$ is a convergent sequence) to show the existence of a fixed point. An alternative, shorter proof of Theorem 4.1 could have been devised by applying Brouwer's Fixed Point Theorem in $\mathbb{R}^n$.

Theorem 4.5 (**Brouwer's Fixed Point Theorem**) *Let us assume that D is a nonempty, closed, bounded and convex subset of $\mathbb{R}^n$. Suppose further that $\boldsymbol{g}\colon \mathbb{R}^n \mapsto \mathbb{R}^n$ is a continuous function defined on D such that $\boldsymbol{g}(D) \subset D$. Then, there exists $\boldsymbol{\xi} \in D$ such that $\boldsymbol{g}(\boldsymbol{\xi}) = \boldsymbol{\xi}$.*

A set $D \subset \mathbb{R}^n$ is said to be convex if, whenever $\boldsymbol{x}$ and $\boldsymbol{y}$ belong to D, also

$$\theta \boldsymbol{x} + (1-\theta)\boldsymbol{y} \in D \qquad \forall \theta \in [0,1] .$$

For example, any nonempty interval of the real line $\mathbb{R}^1 = \mathbb{R}$ is a convex set, as is a nonempty (open or closed) ball in $\mathbb{R}^n$, $n \geq 2$. Unfortunately, when $n \geq 2$ the proof of Theorem 4.5 is nontrivial and is well beyond the scope of this book.[1]

Benoit Mandelbrot (1924–) has been largely responsible for the present interest in fractal geometry and its connections with iterative methods. Mandelbrot highlighted in his book

- B. Mandelbrot, *Fractals: Form, Chance, and Dimension*, W.H. Freeman, San Francisco, 1977,

and, more fully, in

- B. Mandelbrot, *The Fractal Geometry of Nature*, W.H. Freeman, New York, 1983,

the omnipresence of fractals both in mathematics and elsewhere in nature. In relation with the subject of this chapter, we note that the **Mandelbrot set** is a connected set of points in the complex plane defined as follows. Choose a point z_0 in the complex plane, and consider the iteration $z_{n+1} = z_n^2 + z_0$, $n = 0, 1, 2, \ldots$. If the sequence $z_0, z_1, z_2, \ldots$ remains within a distance of 2 from the origin for ever, then the point z_0

[1] For a proof of Theorem 4.5 in the case when D is a closed ball in $\mathbb{R}^n$, see John W. Milnor, *Topology from the Differentiable Viewpoint*, Princeton Landmarks in Mathematics, 1997.

is said to be in the Mandelbrot set. If the sequence diverges from the origin, then the point z_0 is not in the set.

A standard reference for theoretical results concerning the convergence of Newton's method in complete normed linear spaces is

➧ L.V. KANTOROVICH AND G.P. AKILOV, *Functional Analysis*, Second edition, Pergamon Press, Oxford, New York, 1982.

A further significant book in the area of iterative solution of systems of nonlinear equations is the text by

➧ J.M. ORTEGA AND W.C. RHEINBOLDT, *Iterative Solution of Nonlinear Equations in Several Variables*, Reprint of the 1970 original, Classics in Applied Mathematics, 30, SIAM, Philadelphia, 2000.

It gives a comprehensive treatment of the numerical solution of n nonlinear equations in n unknowns, covering asymptotic convergence results for a number of algorithms, including Newton's method, as well as existence theorems for solutions of nonlinear equations based on the use of topological degree theory and Brouwer's Fixed Point Theorem.

Exercises

4.1 Suppose that the function $\boldsymbol{g}$ is a contraction in the ∞-norm, as in (4.5). Use the fact that

$$\|\boldsymbol{g}(\boldsymbol{x}) - \boldsymbol{g}(\boldsymbol{y})\|_p \leq n^{1/p} \|\boldsymbol{g}(\boldsymbol{x}) - \boldsymbol{g}(\boldsymbol{y})\|_\infty$$

to show that $\boldsymbol{g}$ is a contraction in the p-norm if $L < n^{-1/p}$.

4.2 Show that the simultaneous equations $\boldsymbol{f}(x_1, x_2) = \boldsymbol{0}$, where $\boldsymbol{f} = (f_1, f_2)^{\mathrm{T}}$, with

$$f_1(x_1, x_2) = x_1^2 + x_2^2 - 25\,, \qquad f_2(x_1, x_2) = x_1 - 7x_2 - 25\,,$$

have two solutions, one of which is $x_1 = 4$, $x_2 = -3$, and find the other. Show that the function $\boldsymbol{f}$ does not satisfy the conditions of Theorem 4.3 at either of these solutions, but that if the sign of f_2 is changed the conditions are satisfied at one solution, and that if $\boldsymbol{f}$ is replaced by $\boldsymbol{f}^* = (f_2 - f_1, -f_2)^{\mathrm{T}}$, then the conditions are satisfied at the other. In each case, give a value of the relaxation parameter λ which will lead to convergence.

4.3 The complex-valued function $z \mapsto g(z)$ of the complex variable z is holomorphic in a convex region Ω containing the point ζ, at which $g(\zeta) = \zeta$. By applying the Mean Value Theorem (Theorem A.3) to the function φ of the real variable t defined by $\varphi(t) = g((1-t)u + tv)$ show that if u and v lie in Ω, then there is a complex number η in Ω such that

$$g(u) - g(v) = (u - v)g'(\eta)\,.$$

Hence show that if $|g'(\zeta)| < 1$, then the complex iteration defined by $z_{k+1} = g(z_k)$, $k = 0, 1, 2, \ldots$, converges to ζ provided that z_0 is sufficiently close to ζ.

4.4 Suppose that in Exercise 3 the real and imaginary parts of g are u and v, so that $g(x + \imath y) = u(x, y) + \imath v(x, y)$, $\imath = \sqrt{-1}$. Show that the iteration defined by $\boldsymbol{x}^{(k+1)} = \boldsymbol{g}^*(\boldsymbol{x}^{(k)})$, $k = 0, 1, 2, \ldots$, where $\boldsymbol{g}^*(\boldsymbol{x}) = (u(x_1, x_2), v(x_1, x_2))^{\mathrm{T}}$, generates the real and imaginary parts of the sequence defined in Exercise 3. Compare the condition for convergence given in that exercise with the sufficient condition given by Theorem 4.2.

4.5 Verify that the iteration $\boldsymbol{x}^{(k+1)} = \boldsymbol{g}(\boldsymbol{x}^{(k)})$, $k = 0, 1, 2, \ldots$, where $\boldsymbol{g} = (g_1, g_2)^{\mathrm{T}}$ and g_1 and g_2 are functions of two variables defined by

$$g_1(x_1, x_2) = \tfrac{1}{3}(x_1^2 - x_2^2 + 3)\,, \qquad g_2(x_1, x_2) = \tfrac{1}{3}(2x_1x_2 + 1)\,,$$

has the fixed point $\boldsymbol{x} = (1, 1)^{\mathrm{T}}$. Show that the function $\boldsymbol{g}$ does not satisfy the conditions of Theorem 4.3. By applying the results of Exercises 3 and 4 to the complex function g defined by

$$g(z) = \tfrac{1}{3}(z^2 + 3 + \imath)\,, \qquad z \in \mathbb{C}\,, \quad \imath = \sqrt{-1}\,,$$

show that the iteration, nevertheless, converges.

4.6 Suppose that all the second-order partial derivatives of the function $\boldsymbol{f}\colon \mathbb{R}^n \to \mathbb{R}^n$ are defined and continuous in a neighbourhood of the point $\boldsymbol{\xi}$ in $\mathbb{R}^n$, at which $\boldsymbol{f}(\boldsymbol{\xi}) = \boldsymbol{0}$. Assume also that the Jacobian matrix, $J_f(\boldsymbol{x})$, of $\boldsymbol{f}$ is nonsingular at $\boldsymbol{x} = \boldsymbol{\xi}$, and denote its inverse by $K(\boldsymbol{x})$ at all $\boldsymbol{x}$ for which it exists. Defining the Newton iteration by $\boldsymbol{x}^{(k+1)} = \boldsymbol{g}(\boldsymbol{x}^{(k)})$, $k = 0, 1, 2, \ldots$, with $\boldsymbol{x}^{(0)}$ given, where $\boldsymbol{g}(\boldsymbol{x}) = \boldsymbol{x} - K(\boldsymbol{x})\boldsymbol{f}(\boldsymbol{x})$, show that the (i, j)-entry of the Jacobian matrix $J_g(\boldsymbol{x}) \in \mathbb{R}^{n\times n}$ of $\boldsymbol{g}$ is

$$\delta_{ij} - \sum_{r=1}^{k} \frac{\partial K_{ir}}{\partial x_j} f_r - \sum_{r=1}^{k} K_{ir} J_{rj}\,, \qquad i,j = 1,\ldots,n\,,$$

where J_{rj} is the (r,j)-entry of $J_f(\boldsymbol{x})$. Deduce that all the elements of this matrix vanish at the point $\boldsymbol{\xi}$.

4.7 The vector function $\boldsymbol{x} \mapsto \boldsymbol{f}(\boldsymbol{x})$ of two variables is defined by

$$f_1(x_1,x_2) = x_1^2 + x_2^2 - 2\,, \qquad f_2(x_1,x_2) = x_1 - x_2\,.$$

Verify that the equation $\boldsymbol{f}(\boldsymbol{x}) = \mathbf{0}$ has two solutions, $x_1 = x_2 = 1$ and $x_1 = x_2 = -1$. Show that one iteration of Newton's method for the solution of this system gives $\boldsymbol{x}^{(1)} = (x_1^{(1)}, x_2^{(1)})^{\mathrm{T}}$, with

$$x_1^{(1)} = x_2^{(1)} = \frac{\left(x_1^{(0)}\right)^2 + \left(x_2^{(0)}\right)^2 + 2}{2\left(x_1^{(0)} + x_2^{(0)}\right)}\,.$$

Deduce that the iteration converges to $(1,1)^{\mathrm{T}}$ if $x_1^{(0)} + x_2^{(0)}$ is positive, and, if $x_1^{(0)} + x_2^{(0)}$ is negative, the iteration converges to the other solution. Verify that convergence is quadratic.

4.8 Suppose that $\boldsymbol{\xi} = \lim_{k\to\infty} \boldsymbol{x}^{(k)}$ in $\mathbb{R}^n$. Following Definition 1.4, explain what is meant by saying that *the sequence* $(\boldsymbol{x}^{(k)})$ *converges to* $\boldsymbol{\xi}$ *linearly, with asymptotic rate* $-\log_{10}\mu$, where $0 < \mu < 1$.

Given the vector function $\boldsymbol{x} \mapsto \boldsymbol{f}(\boldsymbol{x})$ of two real variables x_1 and x_2 defined by

$$f_1(x_1,x_2) = x_1^2 + x_2^2 - 2\,, \qquad f_2(x_1,x_2) = x_1 + x_2 - 2\,,$$

show that $\boldsymbol{f}(\boldsymbol{\xi}) = \mathbf{0}$ when $\boldsymbol{\xi} = (1,1)^{\mathrm{T}}$. Suppose that $x_1^{(0)} \neq x_2^{(0)}$; show that one iteration of Newton's method for the solution of $\boldsymbol{f}(\boldsymbol{x}) = \mathbf{0}$ with starting value $\boldsymbol{x}^{(0)} = (x_1^{(0)}, x_2^{(0)})^{\mathrm{T}}$ then gives $\boldsymbol{x}^{(1)} = (x_1^{(1)}, x_2^{(1)})^{\mathrm{T}}$ such that $x_1^{(1)} + x_2^{(1)} = 2$. Determine $\boldsymbol{x}^{(1)}$ when

$$x_1^{(0)} = 1 + \alpha\,,\ x_2^{(0)} = 1 - \alpha\,,$$

where $\alpha \neq 0$. Assuming that $x_1^{(0)} \neq x_2^{(0)}$, deduce that Newton's method converges linearly to $(1,1)^{\mathrm{T}}$, with asymptotic rate of convergence $\log_{10} 2$. Why is the convergence not quadratic?

4.9 Suppose that the equation $e^z = z + 2$, $z \in \mathbb{C}$, has a solution

$$z = (2m + \tfrac{1}{2})\imath\pi + \ln[(2m + \tfrac{1}{2})\pi] + \eta\,,$$

where m is a positive integer and $\imath = \sqrt{-1}$. Show that

$$\eta = \ln[1 - \imath(\ln(2m + \tfrac{1}{2})\pi + \eta + 2)/(2m + \tfrac{1}{2}\pi)]$$

and deduce that $\eta = \mathcal{O}(\ln m/m)$ for large m.
(Note that $|\ln(1 + \imath t)| < |t|$ for all $t \in \mathbb{R} \setminus \{0\}$.)

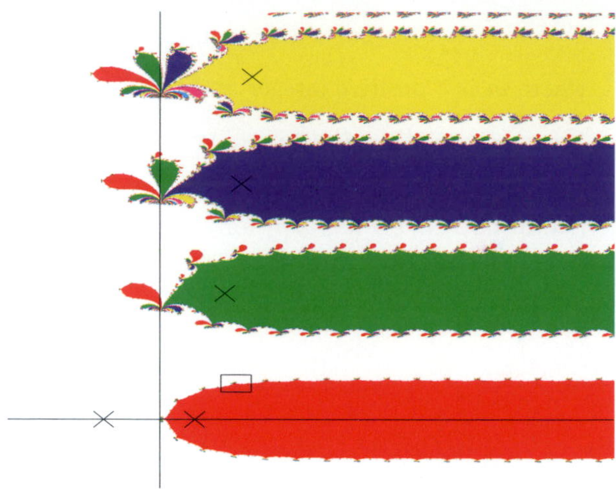

Fig. 4.4. The sets S_k in the region $-5 \leq x \leq 15$, $-4 \leq y \leq 24$ of the complex plane.

Fig. 4.5. The sets S_k in the region $2 \leq x \leq 3$, $1.6 \leq y \leq 2.6$ of the complex plane.

Fig. 4.6. The sets S_k in the region $2.4 \leq x \leq 2.55$, $2.1 \leq y \leq 2.25$ of the complex plane.

Fig. 4.7. The sets S_k in the region $2.4825 \leq x \leq 2.4975$, $2.2075 \leq y \leq 2.2225$ of the complex plane.

Fig. 4.8. The sets S_k in the region $2.4930 \leq x \leq 2.4960$, $2.2100 \leq y \leq 2.2130$ of the complex plane.

Fig. 4.9. The sets S_k in the region $2.493645 \leq x \leq 2.493665$, $2.21073 \leq y \leq 2.21075$ of the complex plane.

5

Eigenvalues and eigenvectors of a symmetric matrix

5.1 Introduction

Eigenvalue problems for symmetric matrices arise in all areas of applied science. The terminology *eigenvalue* comes from the German word *Eigenwert* which means proper or characteristic value. The concept of eigenvalue first appeared in an article on systems of linear differential equations by the French mathematician d'Alembert[1] in the course of studying the motion of a string with masses attached to it at various points.

Let us recall from Chapter 2 the definition of eigenvalue and eigenvector.

Definition 5.1 *Suppose that* $A \in \mathbb{R}^{n\times n}$*. A complex number* λ *for which the set of linear equations*

$$A\boldsymbol{x} = \lambda \boldsymbol{x} \tag{5.1}$$

has a nontrivial solution $\boldsymbol{x} \in \mathbb{C}^n_* = \mathbb{C}^n \setminus \{\mathbf{0}\}$ *is called an* **eigenvalue** *of* A*; the associated solution* $\boldsymbol{x} \in \mathbb{C}^n_*$ *is called an* **eigenvector** *of* A *(corresponding to* λ*).*

[1] Jean le Rond d'Alembert (17 November 1717, Paris, France – 29 October 1783, Paris, France) was abandoned as a newly born child on the steps of the church of St Jean le Rond in Paris and spent his early life in a home for homeless children. d'Alembert was the central mathematical figure among the French Encyclopedists in the period 1751–1772; the Encyclopedia, edited by Jean Diderot, comprised 28 volumes. D'Alembert made a number of significant contributions to the dynamics of rigid bodies, hydrodynamics, aerodynamics, the three-body problem, and the theory of vibrating strings.

In order to motivate the discussion that will follow, we begin with two familiar elementary examples.

In considering the rotation of a rigid body $\Omega \subset \mathbb{R}^3$, the *inertia matrix* is the 3×3 symmetric matrix

$$J = \begin{pmatrix} I_{xx} & -I_{xy} & -I_{xz} \\ -I_{yx} & I_{yy} & -I_{yz} \\ -I_{zx} & -I_{zy} & I_{zz} \end{pmatrix}$$

whose diagonal elements are the moments of inertia about the axes,

$$I_{xx} = \int_\Omega (y^2 + z^2)\,\mathrm{d}\Omega, \;\; I_{yy} = \int_\Omega (z^2 + x^2)\,\mathrm{d}\Omega, \;\; I_{zz} = \int_\Omega (x^2 + y^2)\,\mathrm{d}\Omega,$$

and whose off-diagonal elements are defined by the corresponding products of inertia

$$\begin{aligned} I_{xy} = I_{yx} &= \int_\Omega xy\,\mathrm{d}\Omega\,, \\ I_{yz} = I_{zy} &= \int_\Omega yz\,\mathrm{d}\Omega\,, \\ I_{zx} = I_{xz} &= \int_\Omega zx\,\mathrm{d}\Omega\,. \end{aligned}$$

Then, the eigenvectors of the inertia matrix are the directions of the *principal axes of inertia* of the body, about which free steady rotation is possible, and the eigenvalues are the *principal moments of inertia* about these axes.

A second example, which involves matrices of any order, arises in the solution of systems of linear ordinary differential equations of the form

$$\frac{\mathrm{d}\boldsymbol{x}}{\mathrm{d}t} = A\boldsymbol{x}\,,$$

where $\boldsymbol{x}$ is a vector of n elements, each of which is a function of the independent variable t, and A is an $n \times n$ matrix whose elements are constants. If A were a diagonal matrix, with diagonal elements $a_{ii} = \lambda_i$, $i = 1, 2, \dots, n$, the solution of this system would be straightforward, as each of the equations could be solved separately, giving

$$x_i(t) = x_i(0)\exp(\lambda_i t)\,, \qquad i = 1, 2, \dots, n\,.$$

When A is not a diagonal matrix, suppose that we can find a nonsingular matrix M such that

$$M^{-1}AM = D\,,$$

where D is a diagonal matrix. Then, on letting

$$\boldsymbol{y} = M^{-1}\boldsymbol{x}\,,$$

we easily see that

$$\frac{\mathrm{d}\boldsymbol{y}}{\mathrm{d}t} = M^{-1}AM\boldsymbol{y} = D\boldsymbol{y}\,.$$

The solution of this system of differential equations is straightforward, as we have just seen, and we then find that

$$x_i = (M\boldsymbol{y})_i = \sum_{j=1}^{n} M_{ij}y_j(0)\exp(\lambda_j t)\,,$$

where $\lambda_j = d_{jj}$ is one of the diagonal elements of D. The numbers λ_j, $j = 1, 2, \ldots, n$, are the eigenvalues of the matrix $A \in \mathbb{R}^{n\times n}$, and the columns of M are the eigenvectors of A, so the solution of this system of differential equations requires the calculation of the eigenvalues and eigenvectors of the matrix A.

In systems of differential equations of this kind the matrix A is not necessarily symmetric. In that case, the problem is more difficult; if the eigenvalues of A are not distinct there may not exist a complete set of linearly independent eigenvectors, and then the matrix M will not exist.[1]

In this chapter, we shall develop numerical algorithms for the solution of the algebraic eigenvalue problem (5.1), assuming throughout that $A \in \mathbb{R}^{n\times n}$ is a symmetric matrix. As has been noted above, the analogous problem for a nonsymmetric matrix is more involved, and will not be considered here.[2]

Throughout this chapter, the set of all real-valued symmetric matrices of order n will be denoted by $\mathbb{R}^{n\times n}_{\mathrm{sym}}$; thus, given a matrix $A = (a_{ij})$,

$$A \in \mathbb{R}^{n\times n}_{\mathrm{sym}} \qquad \Leftrightarrow \qquad A \in \mathbb{R}^{n\times n} \quad \& \quad a_{ij} = a_{ji}\,, \quad i, j = 1, 2, \ldots, n\,.$$

We begin with a reminder of some fundamental properties.

[1] Consider, for example,

$$A = \begin{pmatrix} 1 & 2 \\ 0 & 1 \end{pmatrix}.$$

This matrix has one eigenvalue of multiplicity 2, $\lambda_{1/2} = 1$, and only one (linearly independent) eigenvector, $(1, 0)^{\mathrm{T}}$.

[2] The reader is referred to the last four chapters of J.H. Wilkinson's monograph, *The Algebraic Eigenvalue Problem,* The Clarendon Press, Oxford University Press, New York, 1988.

Theorem 5.1 *Suppose that $A \in \mathbb{R}^{n\times n}_{\mathrm{sym}}$; then, the following statements are valid.*

(i) *There exist n linearly independent eigenvectors $\boldsymbol{x}^{(i)} \in \mathbb{R}^n$ and corresponding eigenvalues $\lambda_i \in \mathbb{R}$ such that $A\boldsymbol{x}^{(i)} = \lambda_i \boldsymbol{x}^{(i)}$ for all $i = 1, 2, \ldots, n$.*

(ii) *The function*

$$\lambda \mapsto \det(A - \lambda I) \qquad (5.2)$$

is a polynomial of degree n with leading term $(-1)^n \lambda^n$, called the **characteristic polynomial of** *A. The eigenvalues of A are the zeros of the characteristic polynomial.*

(iii) *If the eigenvalues λ_i and λ_j of A are distinct, then the corresponding eigenvectors $\boldsymbol{x}^{(i)}$ and $\boldsymbol{x}^{(j)}$ are orthogonal in $\mathbb{R}^n$, i.e.,*

$$\boldsymbol{x}^{(i)\mathrm{T}}\boldsymbol{x}^{(j)} = 0 \qquad if\ \lambda_i \neq \lambda_j\,, \qquad i, j \in \{1, 2, \ldots, n\}\,.$$

(iv) *If λ_i is a root of multiplicity m of (5.2), then there is a linear subspace in $\mathbb{R}^n$ of dimension m, spanned by m mutually orthogonal eigenvectors associated with the eigenvalue λ_i.*

(v) *Suppose that each of the eigenvectors $\boldsymbol{x}^{(i)}$ of A is* **normalised**, *in other words, $\boldsymbol{x}^{(i)\mathrm{T}}\boldsymbol{x}^{(i)} = 1$ for $i = 1, 2, \ldots, n$, and let X denote the square matrix whose columns are the normalised (orthogonal) eigenvectors; then, the matrix $\Lambda = X^{\mathrm{T}} A X$ is diagonal, and the diagonal elements of Λ are the eigenvalues of A.*

(vi) *Let $Q \in \mathbb{R}^{n\times n}$ be an orthogonal matrix and define $B \in \mathbb{R}^{n\times n}_{\mathrm{sym}}$ by $B = Q^{\mathrm{T}} A Q$; then, $\det(B - \lambda I) = \det(A - \lambda I)$ for each $\lambda \in \mathbb{R}$. The eigenvalues of B are the same as the eigenvalues of A, and the eigenvectors of B are the vectors $Q^{\mathrm{T}}\boldsymbol{x}^{(i)}$, $i = 1, 2, \ldots, n$.*

(vii) *Any vector $\boldsymbol{v} \in \mathbb{R}^n$ can be expressed as a linear combination of the (ortho)normalised eigenvectors $\boldsymbol{x}^{(i)}$, $i = 1, 2, \ldots, n$, of A, i.e.,*

$$\boldsymbol{v} = \sum_{i=1}^{n} \alpha_i \boldsymbol{x}^{(i)}, \qquad \alpha_i = \boldsymbol{x}^{(i)\mathrm{T}}\boldsymbol{v}\,.$$

(viii) *The trace of A, $\mathrm{Trace}(A) = \sum_{i=1}^{n} a_{ii}$, is equal to the sum of the eigenvalues of A.*

These properties should be familiar; proofs will be found in any standard text on linear algebra.[1]

[1] See, for example, T.S. Blyth and E.F. Robertson, *Basic Linear Algebra*, Springer Undergraduate Mathematics Series, Springer, 1998, A.G. Hamilton, *Linear Algebra*, Cambridge University Press, 1990, or R.A. Horn and C.R. Johnson, *Matrix Analysis*, Cambridge University Press, 1992.

5.2 The characteristic polynomial

Given that $A \in \mathbb{R}^{n\times n}$ and $n \le 4$, it is quite easy to write down the characteristic polynomial $\det(A-\lambda I)$ by expanding the determinant, and then find the roots of this polynomial of degree n in order to determine the eigenvalues of A. If $n > 4$ there is no general closed formula for the roots of a polynomial in terms of its coefficients, and therefore we have to resort to a numerical technique. A further difficulty is that the roots may be very sensitive to small changes in the coefficients of the polynomial, and we find that the effect of rounding errors in the construction of the characteristic polynomial is usually catastrophic.

Example 5.1 *Consider, for example, the diagonal matrix of order* 16 *whose diagonal elements are* $j + \frac{1}{3}$, $j = 1, 2, \ldots, 16$; *the eigenvalues are, of course, just the diagonal elements. Constructing the characteristic polynomial, working with 10 significant digits throughout, gives the result*

$$\lambda^{16} - 141.3333333\lambda^{15} + 9193.333333\lambda^{14} - \cdots .$$

Using a standard numerical algorithm (such as Newton's method) for computing the roots of the polynomial and working with 10 significant digits gives the smallest root as 1.333333331, which is nearly correct to 10 significant digits. The three largest roots, however, are computed as, approximately, $15.5 \pm 1.3\imath$ and 16.7, which are very different from their true values $14.\dot{3}$, $15.\dot{3}$, $16.\dot{3}$, respectively, even though the matrix in this example is of quite modest size, and the eigenvalues are well spaced. Thus we conclude from this example that the numerical method which constructs the characteristic polynomial and finds its roots is completely unsatisfactory for general use, except for matrices of very small size. $\diamond$

The fact that in general the roots of the characteristic polynomial cannot be given in closed form shows that any method must proceed by successive approximation. Although one cannot expect to produce the required eigenvalues exactly in a finite number of steps, we shall see that there exist rapidly convergent iterative methods for computing the eigenvalues and eigenvectors numerically.

5.3 Jacobi's method

This method uses a succession of orthogonal transformations to produce a sequence of matrices which approaches a diagonal matrix in the limit.

Each step in the process involves a matrix representing a plane rotation. We begin with a simple example.

Example 5.2 (The plane rotation matrix in $\mathbb{R}^2$) *Let us suppose that $\varphi \in [-\pi, \pi]$ and consider the matrix $R(\varphi) \in \mathbb{R}^{2\times 2}$ defined by*

$$R(\varphi) = \begin{pmatrix} \cos\varphi & \sin\varphi \\ -\sin\varphi & \cos\varphi \end{pmatrix}.$$

For a vector $\boldsymbol{x} \in \mathbb{R}^2$, $R(\varphi)\boldsymbol{x}$ is the plane rotation of $\boldsymbol{x}$ around the origin by an angle φ (in the clockwise direction when $\varphi > 0$ and in the anticlockwise direction when $\varphi < 0$).

We note in passing that since $\cos(-\varphi) = \cos\varphi$, $\sin(-\varphi) = -\sin\varphi$ and $\cos^2\varphi + \sin^2\varphi = 1$, we have that

$$(R(\varphi))^{\mathrm{T}} = R(-\varphi) \qquad \textit{and} \qquad R(\varphi)\,R(-\varphi) = I\,.$$

Hence $R(\varphi)$ is an orthogonal matrix; i.e.,

$$R(\varphi)R(\varphi)^{\mathrm{T}} = R(\varphi)^{\mathrm{T}}R(\varphi) = I\,,$$

where I is the 2×2 identity matrix.

The next definition extends the notion of plane rotation matrix to $\mathbb{R}^n$.

Definition 5.2 (The plane rotation matrix in $\mathbb{R}^n$) *Suppose that $n \geq 2$, $1 \leq p < q \leq n$ and $\varphi \in [-\pi, \pi]$. We consider the matrix $R^{(pq)}(\varphi) \in \mathbb{R}^{n\times n}$ whose elements are the same as those of the identity matrix $I \in \mathbb{R}^{n\times n}$, except for the four elements*

$$r_{pp} = c\,, \qquad r_{pq} = s\,,$$
$$r_{qp} = -s\,, \qquad r_{qq} = c\,,$$

where $c = \cos\,\varphi$, $s = \sin\,\varphi$.

As in Example 5.2, it is a straightforward matter to show that

$$(R^{(pq)}(\varphi))^{\mathrm{T}} = R^{(pq)}(-\varphi)\,, \qquad R^{(pq)}(\varphi)\,R^{(pq)}(-\varphi) = I\,,$$

and that, therefore,

$$R^{(pq)}(\varphi)(R^{(pq)}(\varphi))^{\mathrm{T}} = (R^{(pq)}(\varphi))^{\mathrm{T}}R^{(pq)}(\varphi) = I\,.$$

Hence $R^{(pq)}(\varphi) \in \mathbb{R}^{n\times n}$ is an orthogonal matrix for any p, q such that $1 \leq p < q \leq n$, and any $\varphi \in [-\pi, \pi]$.

The basic result underlying Jacobi's method is encapsulated in the next theorem.

Theorem 5.2 *Suppose that $A \in \mathbb{R}^{n\times n}_{\mathrm{sym}}$. For each pair of integers (p, q) with $1 \le p < q \le n$, there exists $\varphi \in [-\pi/4, \pi/4]$ such that the (p, q) entry of the symmetric matrix $R^{(pq)}(\varphi)^{\mathrm{T}} A R^{(pq)}(\varphi)$ is equal to 0.*

Proof For the sake of notational simplicity, we shall write R instead of $R^{(pq)}(\varphi)$ throughout the proof, and abbreviate $c = \cos\varphi$ and $s = \sin\varphi$.

Consider the product $A' = AR$. Evidently the only difference between A' and A is in columns p and q; these columns of A' are linear combinations of the same two columns of A:

$$\left.\begin{aligned} a'_{ip} &= a_{ip}c - a_{iq}s \\ a'_{iq} &= a_{ip}s + a_{iq}c \end{aligned}\right\}, \quad i = 1, 2, \dots, n. \tag{5.3}$$

Multiplication of A' by R^{T} on the left gives a similar result, but affects rows p and q, rather than columns p and q. Writing $B = R^{\mathrm{T}} A'$ gives

$$\left.\begin{aligned} b_{pj} &= a'_{pj}c - a'_{qj}s \\ b_{qj} &= a'_{pj}s + a'_{qj}c \end{aligned}\right\}, \quad j = 1, 2, \dots, n. \tag{5.4}$$

Combining these equations shows that $B = R^{\mathrm{T}} A\, R$, where

$$\left.\begin{aligned} b_{pp} &= a_{pp}c^2 - 2a_{pq}sc + a_{qq}s^2, \\ b_{qq} &= a_{pp}s^2 + 2a_{pq}sc + a_{qq}c^2, \\ b_{pq} &= (a_{pp} - a_{qq})sc + a_{pq}(c^2 - s^2) = b_{qp}. \end{aligned}\right\} \tag{5.5}$$

The remaining elements of $B = R^{\mathrm{T}} A\, R$ in columns p and q are given by the expressions

$$\left.\begin{aligned} b_{ip} &= a_{ip}c - a_{iq}s \\ b_{iq} &= a_{ip}s + a_{iq}c \end{aligned}\right\}, \quad i = 1, 2, \dots, n, \quad i \ne p, q.$$

The matrix $B = R^{\mathrm{T}} A\, R$ is evidently symmetric, so the nondiagonal elements of B in rows p and q are also given by the same expressions.

Finally, we note that all the elements of B which do not lie either in row p or q or in column p or q are the same as the corresponding elements of A, that is,

$$b_{ij} = a_{ij}, \qquad \text{if } i \ne p, q \text{ and } j \ne p, q.$$

We see from (5.5) that in order to ensure that b_{pq}, the (p, q)-entry of the matrix $B = R^{\mathrm{T}} A\, R$, is equal to 0, it suffices to choose φ such that

$$\tan 2\varphi = \frac{2a_{pq}}{a_{qq} - a_{pp}}; \tag{5.6}$$

thus we select

$$\varphi = \frac{1}{2} \tan^{-1} \frac{2a_{pq}}{a_{qq} - a_{pp}} \in [-\pi/4, \pi/4] \,. \tag{5.7}$$

To see this, apply the trigonometric identities $c^2 - s^2 = \cos(2\varphi)$ and $sc = \frac{1}{2}\sin(2\varphi)$ to b_{pq} in (5.5), with $b_{pq} = 0$. That completes the proof.[1] □

We can avoid the trigonometric calculations involved in the formula (5.7) for φ by writing $t = s/c$, and seeing that t is required to satisfy

$$(a_{pp} - a_{qq})t + a_{pq}(1 - t^2) = 0 \,. \tag{5.8}$$

If $a_{pq} = 0$, we can ensure that (5.8) holds by selecting $t = 0$ (which corresponds to choosing $\varphi = 0$). If $a_{pq} \neq 0$ and $a_{pp} = a_{qq}$, we put $t = 1$ (corresponding to $\varphi = \pi/4$). Finally, if $a_{pq} \neq 0$ and $a_{pp} \neq a_{qq}$, we solve the quadratic equation (5.8); there will be two distinct real roots, so we define t as the one that is smaller in absolute value. Having selected t, we then use the relation $\sec^2\varphi = 1 + \tan^2\varphi$ to calculate c by $c = 1/(1+t^2)^{1/2}$, and then s from $s = ct$.

Definition 5.3 (The classical Jacobi method) *Let $A \in \mathbb{R}^{n\times n}_{\text{sym}}$ and define $A^{(0)} = A$. Given $k \geq 0$ and $A^{(k)} \in \mathbb{R}^{n\times n}_{\text{sym}}$, the basic step of Jacobi's method computes $A^{(k+1)} \in \mathbb{R}^{n\times n}_{\text{sym}}$ by first locating the largest in absolute value off-diagonal element $(A^{(k)})_{pq} = a^{(k)}_{pq}$ of the matrix $A^{(k)}$, and then setting $A^{(k+1)} = R^{(pq)}(\varphi_k)^{\mathrm{T}} A^{(k)} R^{(pq)}(\varphi_k)$ with φ_k chosen so that $(A^{(k+1)})_{pq} = 0$. This process is then repeated until all the off-diagonal elements are smaller than a given positive tolerance ε.*

In order to show that as $k \to \infty$ the sequence of matrices $(A^{(k)})$ generated by successive steps of the classical Jacobi method converges to a diagonal matrix (whose diagonal entries are the eigenvalues of the original matrix A), we need the following result.

Lemma 5.1 *The sum of squares of the elements of a symmetric matrix is invariant under an orthogonal transformation: that is, if $A \in \mathbb{R}^{n\times n}_{\text{sym}}$*

[1] For future reference, note that a simple calculation based on (5.5) and (5.6) gives

$$b_{ii} - a_{ii} = \begin{cases} 0 & \text{if} \quad i \neq p, q \,, \\ -a_{pq}\tan\varphi & \text{if} \quad i = p \,, \\ a_{pq}\tan\varphi & \text{if} \quad i = q \,. \end{cases}$$

and $B = R^{\mathrm{T}} A\, R$ *where* $R \in \mathbb{R}^{n\times n}$ *is an orthogonal matrix, then*

$$\sum_{i=1}^{n}\sum_{j=1}^{n} b_{ij}^2 = \sum_{i=1}^{n}\sum_{j=1}^{n} a_{ij}^2 \,. \tag{5.9}$$

The quantity

$$\|A\|_F = \left(\sum_{i=1}^{n}\sum_{j=1}^{n} a_{ij}^2\right)^{1/2}$$

is called the **Frobenius norm**[1] of $A \in \mathbb{R}^{n\times n}$. The Frobenius norm of $A \in \mathbb{R}^{n\times n}$ is the 2-norm of A, with A regarded as an element of a linear space of dimension n^2 over the field of real numbers; however, it is *not* a subordinate norm in the sense of Definition 2.10. In particular, the Frobenius norm on $\mathbb{R}^{n\times n}$ is not subordinate to the 2-norm on $\mathbb{R}^n$.

Now, one can express (5.9) equivalently by saying that the Frobenius norm of a symmetric matrix A is invariant under an orthogonal transformation: $\|R^{\mathrm{T}}AR\|_F = \|A\|_F$.

Proof of lemma The sum of squares of the elements of A is the same as the trace of A^2, for

$$\mathrm{Trace}(A^2) = \sum_{i=1}^{n}(A^2)_{ii} = \sum_{i=1}^{n}\sum_{j=1}^{n} a_{ij}a_{ji} = \sum_{i=1}^{n}\sum_{j=1}^{n} a_{ij}^2 \,, \tag{5.10}$$

since A is symmetric. Analogously, as $B = R^{\mathrm{T}}AR$ is symmetric, we have that

$$\mathrm{Trace}(B^2) = \sum_{i=1}^{n}\sum_{j=1}^{n} b_{ij}^2 \,.$$

Thus, it remains to show that $\mathrm{Trace}(B^2) = \mathrm{Trace}(A^2)$. Now,

$$B^2 = (R^{\mathrm{T}} A\, R)(R^{\mathrm{T}} A\, R) = R^{\mathrm{T}} A^2 R \,, \tag{5.11}$$

since R is orthogonal. Hence B^2 is an orthogonal transformation of A^2 which, by virtue of Theorem 5.1 (vi), means that B^2 and A^2 have the same eigenvalues, and therefore the same trace, since the trace is the sum of the eigenvalues (see Theorem 5.1 (viii)). □

[1] Ferdinand Georg Frobenius (26 October 1849, Berlin-Charlottenburg, Prussia, Germany – 3 August 1917, Berlin, Germany), contributed to the theory of analytic functions, representation theory of groups, differential equation theory and the theory of elliptic functions.

Now we are ready to embark on the convergence analysis of the classical Jacobi method.

Theorem 5.3 *Suppose that $A \in \mathbb{R}^{n\times n}_{\mathrm{sym}}$, $n \geq 2$. In the classical Jacobi method the off-diagonal entries in the sequence of matrices $(A^{(k)})$, generated from $A^{(0)} = A$ according to Definition 5.3, converge to 0 in the sense that*

$$\lim_{k\to\infty} \sum_{\substack{i,j=1\\ i\neq j}}^{n} [(A^{(k)})_{ij}]^2 = 0\,. \tag{5.12}$$

Furthermore,

$$\lim_{k\to\infty} \sum_{i=1}^{n} [(A^{(k)})_{ii}]^2 = \mathrm{Trace}(A^2)\,. \tag{5.13}$$

Proof Let a_{pq} be the off-diagonal element of A with largest absolute value, and let $B = (R^{(pq)}(\varphi))^{\mathrm{T}} A\, R^{(pq)}(\varphi)$, where φ is defined by (5.7). Then, letting $c = \cos\varphi$ and $s = \sin\varphi$, we have that

$$\begin{pmatrix} b_{pp} & b_{pq} \\ b_{qp} & b_{qq} \end{pmatrix} = \begin{pmatrix} c & s \\ -s & c \end{pmatrix}^{\mathrm{T}} \begin{pmatrix} a_{pp} & a_{pq} \\ a_{qp} & a_{qq} \end{pmatrix} \begin{pmatrix} c & s \\ -s & c \end{pmatrix},$$

and Lemma 5.1 implies that

$$b_{pp}^2 + 2b_{pq}^2 + b_{qq}^2 = a_{pp}^2 + 2a_{pq}^2 + a_{qq}^2\,.$$

Writing

$$S(A) = \sum_{i,j=1}^{n} a_{ij}^2\,, \quad D(A) = \sum_{i=1}^{n} a_{ii}^2\,, \quad L(A) = \sum_{\substack{i,j=1\\ i\neq j}}^{n} a_{ij}^2\,,$$

it follows that $S(A) = D(A) + L(A)$. Now $S(B) = S(A)$ by Lemma 5.1, and so $D(B) + L(B) = D(A) + L(A)$. The diagonal entries of B are the same as those of A, except the ones in rows p and q, $1 \leq p < q \leq n$. Further, as $b_{pq} = 0$, it follows that $b_{pp}^2 + b_{qq}^2 = a_{pp}^2 + a_{qq}^2 + 2a_{pq}^2$. Therefore,

$$D(B) = D(A) + 2a_{pq}^2\,.$$

Consequently,

$$L(B) = L(A) - 2a_{pq}^2\,.$$

Now a_{pq} is the largest off-diagonal element of A; hence $L(A) \leq N a_{pq}^2$ where $N = n(n-1)$ is the number of off-diagonal elements, and therefore

$$L(B) \leq (1 - 2/N)L(A)\,. \tag{5.14}$$

On writing $A^{(0)} = A$, $A^{(1)} = B$, and generating subsequent members of the sequence $(A^{(k)})$ in a similar manner, as indicated in the algorithm in Definition 5.3, we deduce from (5.14) that

$$0 \le L(A^{(k)}) \le (1 - 2/N)^k L(A)\,, \qquad k = 1, 2, 3, \dots\,, \tag{5.15}$$

where $N \ge 2$. Thus we conclude that $\lim_{k\to\infty} L(A^{(k)}) = 0$.

Now, (5.13) follows from (5.10) and (5.12) on noting that

$$\mathrm{Trace}(A^2) = S(A) = S(A^{(k)}) = D(A^{(k)}) + L(A^{(k)}) \qquad \forall\, k \ge 0\,,$$

and passing to the limit $k \to \infty$: $\mathrm{Trace}(A^2) = \lim_{k\to\infty} D(A^{(k)})$. □

According to Theorem 5.1 (viii) the trace of A^2 is the sum of the eigenvalues of A^2, and the eigenvalues of A^2 are the squares of the eigenvalues of A. Thus, we have shown that the sum of the squares of the diagonal elements in the sequence of matrices $(A^{(k)})$ generated by the classical Jacobi method converges to the sum of the squares of the eigenvalues of A. More work is required to show that for each $i = 1, 2, \dots, n$ the sequence of diagonal elements $(a_{ii}^{(k)})$ converges to an eigenvalue of A as $k \to \infty$. We shall further discuss this question in the final paragraphs of Section 5.4. First, however, we describe another variant of Jacobi's method.

Definition 5.4 (The serial Jacobi method) *This version of Jacobi's method proceeds in a systematic order, using transformations $R^{(pq)}(\varphi)$ to reduce to zero the elements $(1, 2)$, $(1, 3)$, …, $(1, n)$, $(2, 3)$, $(2, 4)$, …, $(2, n)$, …, $(n-1, n)$ in this order. The complete step is then repeated iteratively.*

It is not difficult to prove that this method also converges. Both these variants of the Jacobi method converge quite rapidly; the rate of convergence is in practice much faster than is suggested by (5.15), and in fact it can be shown that convergence is ultimately quadratic.

It is time for an example!

Example 5.3 *Let us consider the 5×5 matrix*

$$A = \begin{pmatrix} 4 & 1 & 2 & 1 & 2 \\ 1 & 3 & 0 & -3 & 4 \\ 2 & 0 & 1 & 2 & 2 \\ 1 & -3 & 2 & 4 & 1 \\ 2 & 4 & 2 & 1 & 1 \end{pmatrix}. \tag{5.16}$$

The values of $D(A^{(k)})$ and $L(A^{(k)})$ after each iteration of the serial Jacobi method, with $A^{(0)} = A$, are shown in Table 5.1. The off-diagonal elements of the third iterate, $A^{(3)}$, are zero to 10 decimal digits. The diagonal elements of $A^{(3)}$, which give the eigenvalues, are

$$8.094,\ 1.690,\ -0.671,\ 7.170,\ -3.282\,.$$

Note that the eigenvalues do not appear in any particular order.

Table 5.1. *Convergence of the serial Jacobi iteration.*

k	$D(A^{(k)})$	$L(A^{(k)})$
0	43.000	88.00000000
1	126.309	4.69087885
2	130.981	0.01948855
3	131.000	0.00000000

This concludes the discussion about the use of Jacobi's method for computing the eigenvalues of a symmetric matrix A. 'Fine,' you might say, 'but how do we determine the *eigenvectors* of A?'

It turns out that by collecting the information accumulated in the course of the Jacobi iteration, it is fairly easy to calculate the eigenvectors of A. We begin by noting that if M is an orthogonal matrix such that $M^{\mathrm{T}}AM = D$, where D is diagonal, then the diagonal elements of D are the eigenvalues of A, and the columns of M are the corresponding eigenvectors of A.

In the course of the Jacobi iteration (be it classical or serial), we have constructed the plane rotations $R^{(p_j q_j)}(\varphi_j)$, $j = 1, 2, \ldots, k$. Thus, an approximation $M^{(k)}$ to the orthogonal matrix M can be obtained by considering the product of these rotation matrices: initially, we put $M^{(0)} = I$ and then we apply the column transformation $R^{(p_j q_j)}(\varphi_j)$ at each step $j = 1, 2, \ldots, k$. This corresponds to multiplying $M^{(j-1)}$ on the right by $R^{(p_j q_j)}(\varphi_j)$ for $j = 1, 2, \ldots, k$, and leads to the orthogonal matrix

$$M^{(k)} = R^{(p_1 q_1)}(\varphi_1) \ldots R^{(p_k q_k)}(\varphi_k)$$

which represents the required approximation to the orthogonal matrix M. The columns of $M^{(k)}$ will be the desired approximate eigenvectors of A corresponding to the approximate eigenvalues which appear along the diagonal of $A^{(k)}$.

The Jacobi method usually converges in a reasonable number of iterations, and is a satisfactory method for small or moderate-sized matrices. However, there are many problems, particularly in the area of numerical solution of partial differential equations, which give rise to very large matrices that are sparse, with most of the elements being zero. A further consideration is that in many practical situations one does not need to compute all the eigenvalues. It is much more common to require a few of the largest eigenvalues and corresponding eigenvectors, or perhaps a few of the smallest. Jacobi's method is not suitable for such problems, as it always produces all the eigenvalues, and will not preserve the sparse structure of a matrix during the course of the iteration. For example, it is easy to see that if Jacobi's method is applied to a symmetric tridiagonal matrix, then at the end of one sweep all but two of the elements of the matrix will in general be nonzero and, although still symmetric, the transformed matrix is no longer tridiagonal. Later on in this chapter we shall consider numerical algorithms for computing selected eigenvalues of a matrix. Thus, as an overture to what will follow, we now outline a 'rough and ready' technique for locating the eigenvalues.

5.4 The Gerschgorin theorems

Gerschgorin's Theorem[1] provides a very simple way of determining a region that contains the eigenvalues of a matrix. It is very general, and does not assume that the matrix is symmetric; in fact we shall allow the elements of a square matrix of order n to be complex and write $A \in \mathbb{C}^{n \times n}$ to express this fact.

Definition 5.5 *Suppose that $n \geq 2$ and $A \in \mathbb{C}^{n \times n}$. The* **Gerschgorin discs** *D_i, $i = 1, 2, \ldots, n$, of the matrix A are defined as the closed circular regions*

$$D_i = \{z \in \mathbb{C}\colon |z - a_{ii}| \leq R_i\} \tag{5.17}$$

in the complex plane, where

$$R_i = \sum_{\substack{j=1 \\ j \neq i}}^{n} |a_{ij}| \tag{5.18}$$

is the radius of D_i.

[1] After S.A. Gerschgorin; see the historical survey of Seiji Fujino and Joachim Fischer, Über S.A. Gerschgorin (1901–1933) [German: About S.A. Gershgorin (1901–1933)], *GAMM Mitt. Ges. Angew. Math. Mech.* **21**, no. 1, 15–19, 1998.

Theorem 5.4 (Gerschgorin's Theorem) *Let $n \geq 2$ and $A \in \mathbb{C}^{n\times n}$. All eigenvalues of the matrix A lie in the region $D = \bigcup_{i=1}^{n} D_i$, where D_i, $i = 1, 2, \ldots, n$, are the Gerschgorin discs of A defined by (5.17), (5.18).*

Proof Suppose that $\lambda \in \mathbb{C}$ and $\boldsymbol{x} \in \mathbb{C}^n \setminus \{\mathbf{0}\}$ are an eigenvalue and the corresponding eigenvector of A, so that

$$\sum_{j=1}^{n} a_{ij}x_j = \lambda x_i\,, \quad i = 1, 2, \ldots, n\,. \tag{5.19}$$

Suppose that x_k, with $k \in \{1, 2, \ldots, n\}$, is the component of $\boldsymbol{x}$ which has largest modulus, or one of those components if more than one have the same modulus. We note in passing that $x_k \neq 0$, given that $\boldsymbol{x} \neq \mathbf{0}$; also,

$$|x_j| \leq |x_k|\,, \quad j = 1, 2, \ldots, n\,. \tag{5.20}$$

This means that

$$\begin{aligned} |\lambda - a_{kk}|\,|x_k| &= |\lambda\, x_k - a_{kk}\, x_k| \\ &= \left|\sum_{j=1}^{n} a_{kj}x_j - a_{kk}x_k\right| \\ &= \left|\sum_{\substack{j=1\\ j\neq k}}^{n} a_{kj}x_j\right| \\ &\leq |x_k| R_k\,, \end{aligned} \tag{5.21}$$

which, on division by $|x_k|$, shows that λ lies in the Gerschgorin disc D_k of radius R_k centred at a_{kk}. Hence, $\lambda \in D = \bigcup_{i=1}^{n} D_i$. □

Theorem 5.5 (Gerschgorin's Second Theorem) *Let $n \geq 2$. Suppose that $1 \leq p \leq n-1$ and that the Gerschgorin discs of the matrix $A \in \mathbb{C}^{n\times n}$ can be divided into two disjoint subsets $D^{(p)}$ and $D^{(q)}$, containing p and $q = n - p$ discs respectively. Then, the union of the discs in $D^{(p)}$ contains p of the eigenvalues, and the union of the discs in $D^{(q)}$ contains $n - p$ eigenvalues. In particular, if one disc is disjoint from all the others, it contains exactly one eigenvalue, and if all the discs are disjoint then each disc contains exactly one eigenvalue.*

Proof We shall use a so-called *homotopy* (or continuation) argument.

For $0 \leq \varepsilon \leq 1$, we consider the matrix $B(\varepsilon) = (b_{ij}(\varepsilon)) \in \mathbb{C}^{n\times n}$, where

$$b_{ij}(\varepsilon) = \begin{cases} a_{ii} & \text{if } i = j\,, \\ \varepsilon a_{ij} & \text{if } i \neq j\,. \end{cases} \tag{5.22}$$

Then, $B(1) = A$, and $B(0)$ is the diagonal matrix whose diagonal elements coincide with those of A. Each of the eigenvalues of $B(0)$ is therefore the centre of one of the Gerschgorin discs of A; thus exactly p of the eigenvalues of $B(0)$ lie in the union of the discs in $D^{(p)}$. Now, the eigenvalues of $B(\varepsilon)$ are the zeros of its characteristic polynomial, which is a polynomial whose coefficients are continuous functions of ε; hence the zeros of this polynomial are also continuous functions of ε. Thus as ε increases from 0 to 1 the eigenvalues of $B(\varepsilon)$ move along continuous paths in the complex plane, and at the same time the radii of the Gerschgorin discs increase from 0 to the radii of the Gerschgorin discs of A. Since p of the eigenvalues lie in the union of the discs in $D^{(p)}$ when $\varepsilon = 0$, and these discs are disjoint from all of the discs in $D(q)$, these p eigenvalues must still lie in the union of the discs in $D^{(p)}$ when $\varepsilon = 1$, and the theorem is proved.

The same proof evidently still applies when the discs can be divided into any number of disjoint subsets. □

Example 5.4 *Consider the matrix*

$$A = \begin{pmatrix} 4.00 & 0.20 & -0.10 & 0.10 \\ 0.20 & -1.00 & -0.10 & 0.05 \\ -0.10 & -0.10 & 3.00 & 0.10 \\ 0.10 & 0.05 & 0.10 & -3.00 \end{pmatrix}. \tag{5.23}$$

Figure 5.1 shows, as solid circles, the Gerschgorin discs for this matrix; for instance, one of the discs has centre at 4.00 *and radius* 0.40*. The discs are clearly disjoint, so that each disc contains one eigenvalue of the matrix. The significance of the dotted circles will be explained in our next example.*

Example 5.5 *Let us consider the matrix A defined by (5.23), and then transform it into $B = KAK^{-1}$, where $K \in \mathbb{R}^{4\times 4}$ is the same as the identity matrix except that $k_{22} = \kappa > 0$.*

This transformation has the effect of multiplying the elements in row 2 by κ, and multiplying the elements in column 2 by $1/\kappa$; the diagonal element a_{22} thus remains unaltered. A small value of κ then means that the second disc of B is smaller than the second disc of A, but the other

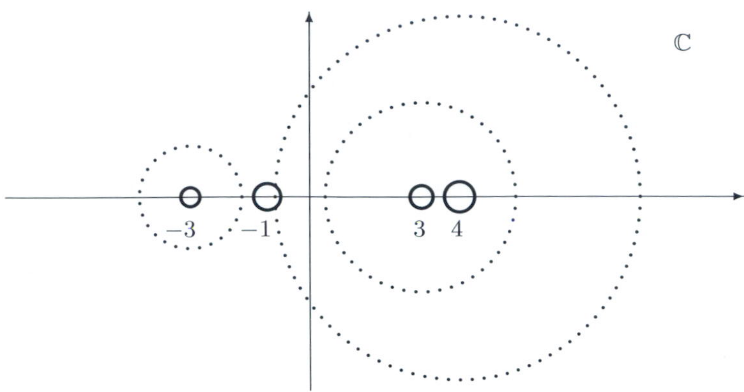

Fig. 5.1. Gerschgorin discs in the complex plane for the matrix A defined in (5.23) (solid circles) and for $B = KAK^{-1}$ (dotted circles). The numbers along the real axis denote the first coordinate of the centre point of each circle (the second coordinate being zero in each case).

discs grow larger. The dotted discs in Figure 5.1 are for the matrix B with $\kappa = 1/23$. For this value the other discs are still just disjoint from the disc centred at -1.00; the disc with centre at 4.00 almost touches the disc with centre at -1.00. The disc with centre -1.00 has radius 0.014, and is too small to be visible in the figure. The eigenvalue in this disc is -1.009 to three decimal digits. The same procedure can be used to reduce the size of each of the discs in turn. ◇

This idea is formalised in the next theorem.

Theorem 5.6 *Let $n \geq 2$, and suppose that in the matrix $A \in \mathbb{C}^{n\times n}$ all the off-diagonal elements are smaller in absolute value than ε, so that $|a_{ij}| < \varepsilon$, for all $i, j \in \{1, 2, \ldots, n\}$ with $i \neq j$. Suppose also that for a particular integer $r \in \{1, 2, \ldots, n\}$ the diagonal element a_{rr} is distant δ from all the other diagonal elements, so that $|a_{rr} - a_{ii}| > \delta$, for all i such that $i \neq r$. Then, provided that*

$$\varepsilon < \frac{\delta}{2(n-1)}, \tag{5.24}$$

there is an eigenvalue λ of A such that

$$|\lambda - a_{rr}| < 2(n-1)\varepsilon^2/\delta\,. \tag{5.25}$$

Proof We apply the **similarity transformation**

$$A \in \mathbb{C}^{n\times n} \mapsto A' = K\,A\,K^{-1} \in \mathbb{C}^{n\times n}\,,$$

where $K \in \mathbb{R}^{n\times n}$ is the same as the identity matrix, except that the diagonal element in row r is chosen to be $k_{rr} = \kappa > 0$. This has the effect of multiplying the off-diagonal elements of row r by κ, and the element in column r of row i, where $i \neq r$, by $1/\kappa$. The Gerschgorin disc from row r then has centre a_{rr} and radius not exceeding $\kappa(n-1)\varepsilon$, and the disc corresponding to row $i \neq r$ has centre a_{ii} and radius not exceeding $(n-2)\varepsilon + \varepsilon/\kappa$.

We now want to reduce the size of disc r by choosing a small value of κ, while keeping it disjoint from the rest. This is easily done by choosing $\kappa = 2\varepsilon/\delta$. The radius of disc r does not exceed $2(n-1)\varepsilon^2/\delta$, and the radius of disc $i \neq r$ does not exceed $(n-2)\varepsilon + \frac{1}{2}\delta$. The sum of these radii therefore satisfies

$$\begin{aligned} R_r + R_i &\leq 2(n-1)\varepsilon^2/\delta + (n-2)\varepsilon + \tfrac{1}{2}\delta \\ &< \varepsilon + (n-2)\varepsilon + \tfrac{1}{2}\delta \\ &< \delta\,, \end{aligned} \tag{5.26}$$

where we have used the given condition (5.24) twice. As the centres a_{rr} and a_{ii} of these discs are distant more than δ from each other, (5.26) shows that the two discs are disjoint, and the required result is proved. □

Theorem 5.6 is sufficient to show that for a matrix satisfying its hypotheses we can find a Gerschgorin disc whose radius is of order ε^2 provided that ε is sufficiently small. It also indicates that the spacing between the diagonal elements is important.

In particular, Theorem 5.6 applies to the matrix $A^{(k)}$ which results after k iterations of the Jacobi method. If at that stage all the off-diagonal elements have magnitude less than ε then there is one eigenvalue in each of the intervals $[a_{ii}^{(k)} - (n-1)\varepsilon, a_{ii}^{(k)} + (n-1)\varepsilon]$, provided that these intervals are disjoint; this follows from Theorem 5.5. If ε is sufficiently small compared with the distances between the diagonal elements of $A^{(k)}$, Theorem 5.6 may be used to give closer bounds on the eigenvalues.

We close this section with some comments on the convergence of the classical Jacobi iteration. According to the Cauchy–Schwarz inequality,

$$\left(\sum_{\substack{i,j=1\\ i\neq j}}^{n} |a_{ij}^{(k)}|\right)^2 \le \sum_{\substack{i,j=1\\ i\neq j}}^{n} 1^2 \sum_{\substack{i,j=1\\ i\neq j}}^{n} |a_{ij}^{(k)}|^2 = n(n-1) \sum_{i,j=1,i\neq j}^{n} |a_{ij}^{(k)}|^2 .$$

Therefore, also,

$$\left(\max_{i=1}^{n} \sum_{\substack{j=1\\ j\neq i}}^{n} |a_{ij}^{(k)}|\right)^2 \le n(n-1) \sum_{\substack{i,j=1\\ i\neq j}}^{n} |a_{ij}^{(k)}|^2 ,$$

so (5.12) implies that

$$\lim_{k\to\infty} \max_{i=1}^{n} \sum_{\substack{j=1\\ j\neq i}}^{n} |a_{ij}^{(k)}| = 0 .$$

In other words, the radii of the Gerschgorin discs for the matrices in the sequence $(A^{(k)})$ converge to 0 as $k \to \infty$. As $A^{(k)}$ and A have identical eigenvalues for all k, it follows from Theorems 5.4 and 5.5 that the set of limiting diagonal entries $\{\lim_{k\to\infty} a_{11}^{(k)}, \dots, \lim_{k\to\infty} a_{nn}^{(k)}\}$ delivered by the Jacobi iteration is equal to the set of eigenvalues of A. This holds irrespective of the spacing between the diagonal entries.

5.5 Householder's method

The general method for finding the eigenvalues of a real symmetric matrix begins by applying an orthogonal transformation to reduce it to a tridiagonal matrix. This can be done in a finite number of steps by using Householder matrices.

Definition 5.6 *Given a vector* $\boldsymbol{v} \in \mathbb{R}^n_*$, *the corresponding* **Householder matrix** $H = H(\boldsymbol{v})$ *of order* n *is defined by*

$$H = I - \frac{2}{\boldsymbol{v}^{\mathrm{T}}\boldsymbol{v}} \boldsymbol{v}\boldsymbol{v}^{\mathrm{T}},$$

where I *is the identity matrix of order* n.

Clearly, for any vector $\boldsymbol{x} \in \mathbb{R}^n$, we have

$$H\boldsymbol{x} = \boldsymbol{x} - 2\frac{\boldsymbol{v}^{\mathrm{T}}\boldsymbol{x}}{\boldsymbol{v}^{\mathrm{T}}\boldsymbol{v}}\boldsymbol{v} ,$$

and hence the vectors $H\boldsymbol{x}$, $\boldsymbol{x}$ and $\boldsymbol{v}$ are coplanar. In particular, if $\boldsymbol{x} \in \mathbb{R}^n$ and $\boldsymbol{v}^{\mathrm{T}}\boldsymbol{x} = \mathbf{0}$ then $H\boldsymbol{x} = \boldsymbol{x}$, and therefore the $(n-1)$-dimensional

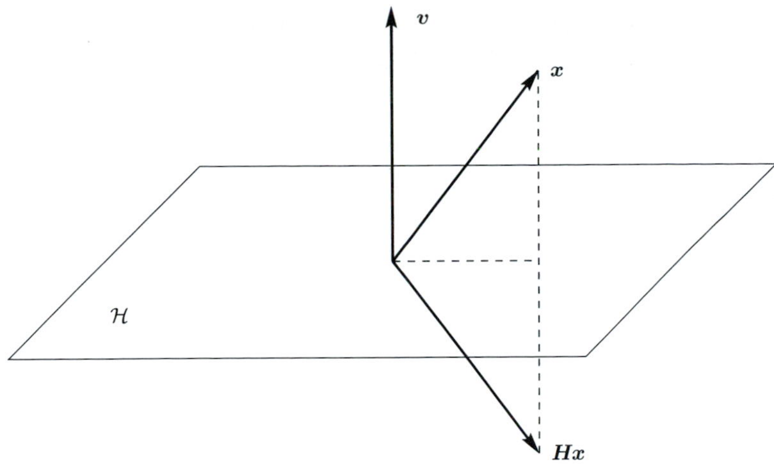

Fig. 5.2. Action of the Householder reflector H: $\boldsymbol{x} \mapsto H\boldsymbol{x}$, corresponding to $\boldsymbol{v} \in \mathbb{R}^n_*$, on a vector $\boldsymbol{x} \in \mathbb{R}^n$. $H\boldsymbol{x}$ is the reflection of $\boldsymbol{x}$ in the hyperplane $\mathcal{H}$ perpendicular to $\boldsymbol{v}$.

hyperplane $\mathcal{H}$ consisting of all vectors $\boldsymbol{x}$ that are perpendicular to $\boldsymbol{v}$ in $\mathbb{R}^n$ is invariant under the mapping $\boldsymbol{x} \mapsto H\boldsymbol{x}$. Finally, for any $\boldsymbol{x} \in \mathbb{R}^n$,

$$\boldsymbol{v}^{\mathrm{T}} H\boldsymbol{x} = -\boldsymbol{v}^{\mathrm{T}}\boldsymbol{x}\,.$$

Hence, if the angle between $\boldsymbol{x}$ and $\boldsymbol{v}$ is denoted by φ, then the angle between $\boldsymbol{v}$ and $H\boldsymbol{x}$ is equal to $\pi+\varphi$. We conclude from these observations that the vector $H\boldsymbol{x}$ is the reflection of $\boldsymbol{x}$ in the hyperplane $\mathcal{H}$. For this reason, the mapping $\boldsymbol{x} \mapsto H\boldsymbol{x}$ is frequently referred to as **Householder reflector**, corresponding to the vector $\boldsymbol{v} \in \mathbb{R}^n_*$ (see Figure 5.2).

Lemma 5.2 *Every Householder matrix is symmetric and orthogonal.*

Proof As $I^{\mathrm{T}} = I$, $(\boldsymbol{v}\boldsymbol{v}^{\mathrm{T}})^{\mathrm{T}} = (\boldsymbol{v}^{\mathrm{T}})^{\mathrm{T}}\boldsymbol{v}^{\mathrm{T}} = \boldsymbol{v}\boldsymbol{v}^{\mathrm{T}}$, and $\boldsymbol{v}^{\mathrm{T}}\boldsymbol{v}$ is a (positive real) number, the symmetry of H follows. The orthogonality of H is a consequence of the identity

$$H^{\mathrm{T}}H = HH^{\mathrm{T}} = H^2 = I - \frac{4}{\boldsymbol{v}^{\mathrm{T}}\boldsymbol{v}}\,\boldsymbol{v}\boldsymbol{v}^{\mathrm{T}} + \frac{4}{(\boldsymbol{v}^{\mathrm{T}}\boldsymbol{v})^2}(\boldsymbol{v}\boldsymbol{v}^{\mathrm{T}})(\boldsymbol{v}\boldsymbol{v}^{\mathrm{T}}) = I\,,$$

since $(\boldsymbol{v}\boldsymbol{v}^{\mathrm{T}})(\boldsymbol{v}\boldsymbol{v}^{\mathrm{T}}) = \boldsymbol{v}(\boldsymbol{v}^{\mathrm{T}}\boldsymbol{v})\boldsymbol{v}^{\mathrm{T}} = (\boldsymbol{v}^{\mathrm{T}}\boldsymbol{v})\boldsymbol{v}\boldsymbol{v}^{\mathrm{T}}$ by the associativity of matrix multiplication. □

Lemma 5.3 *Let $1 \le k < n$ and suppose that H_k is a $k \times k$ Householder matrix. Then, the matrix $H \in \mathbb{R}^{n\times n}$, written in partitioned form as*

$$H = \begin{pmatrix} I_{n-k} & 0 \\ 0^{\mathrm{T}} & H_k \end{pmatrix}$$

where I_{n-k} is the identity matrix of order $n-k$ and 0 is the $(n-k)\times k$ zero matrix, is also a Householder matrix.

The proof of this lemma is straightforward and is left as an exercise. (See Exercise 1.)

Lemma 5.4 *Given any vector $\boldsymbol{x} \in \mathbb{R}^n_*$, there exists a Householder matrix $H \in \mathbb{R}^{n\times n}_{\mathrm{sym}}$ such that all elements of the vector $H\boldsymbol{x}$ are zero, except the first; i.e., $H\boldsymbol{x}$ is a nonzero multiple of $\mathbf{e}_1$, the first column of the identity matrix.*

In geometrical terms this result can be rephrased by saying that for any vector $\boldsymbol{x} \in \mathbb{R}^n_*$ there exists an $(n-1)$-dimensional hyperplane $\mathcal{H}$ passing through the origin in $\mathbb{R}^n$ such that the reflection $H\boldsymbol{x}$ of $\boldsymbol{x}$ in $\mathcal{H}$ is equal to a nonzero multiple of $\mathbf{e}_1$. To find $\mathcal{H}$ it suffices to identify a vector $\boldsymbol{v} \in \mathbb{R}^n_*$ normal to $\mathcal{H}$. Since $\mathcal{H}$ is unaffected by rescaling $\boldsymbol{v}$ (see Definition 5.6), the length of $\boldsymbol{v}$ is immaterial. As noted in the discussion following Definition 5.6, the vectors $H\boldsymbol{x}$, $\boldsymbol{x}$ and $\boldsymbol{v}$ are coplanar. Therefore, we shall seek $\boldsymbol{v} \in \mathbb{R}^n_*$ as a suitable linear combination of $\boldsymbol{x}$ and $\boldsymbol{e}_1$.

Proof of lemma We seek $H = I - [2/(\boldsymbol{v}^{\mathrm{T}}\boldsymbol{v})]\,\boldsymbol{v}\boldsymbol{v}^{\mathrm{T}}$ with $\boldsymbol{v} = \boldsymbol{x} + c\mathbf{e}_1$, where c is a nonzero real number to be determined. Hence,

$$\begin{aligned} \boldsymbol{v}^{\mathrm{T}}\boldsymbol{x} &= \boldsymbol{x}^{\mathrm{T}}\boldsymbol{x} + c\beta\,, \\ \boldsymbol{v}^{\mathrm{T}}\boldsymbol{v} &= \boldsymbol{x}^{\mathrm{T}}\boldsymbol{x} + 2c\beta + c^2\,, \end{aligned}$$

where $\beta = \mathbf{e}_1^{\mathrm{T}}\boldsymbol{x}$ is the first entry of $\boldsymbol{x}$. A simple manipulation then shows that

$$H\boldsymbol{x} = \boldsymbol{x} - \frac{2}{\boldsymbol{v}^{\mathrm{T}}\boldsymbol{v}}\,\boldsymbol{v}(\boldsymbol{v}^{\mathrm{T}}\boldsymbol{x}) = \frac{(c^2 - \boldsymbol{x}^{\mathrm{T}}\boldsymbol{x})\boldsymbol{x} - 2c(\boldsymbol{x}^{\mathrm{T}}\boldsymbol{x} + c\beta)\mathbf{e}_1}{\boldsymbol{x}^{\mathrm{T}}\boldsymbol{x} + 2c\beta + c^2}\,.$$

Thus, $H\boldsymbol{x}$ will be a multiple of $\mathbf{e}_1$ provided that we choose c so that $c^2 = \boldsymbol{x}^{\mathrm{T}}\boldsymbol{x}$. Also, to avoid division by 0, we need to ensure that $\boldsymbol{x}^{\mathrm{T}}\boldsymbol{x} + 2c\beta + c^2 \neq 0$. To do so, note that $c^2 \ge \beta^2$; therefore

$$\boldsymbol{x}^{\mathrm{T}}\boldsymbol{x} + 2c\beta + c^2 \ge (\beta + c)^2 \neq 0\,,$$

provided that $\beta + c \neq 0$, which can be ensured by selecting the appropriate sign for c, that is, by defining

$$c = \begin{cases} (\text{sign } \beta)\sqrt{\boldsymbol{x}^{\mathrm{T}}\boldsymbol{x}} & \text{when } \beta \neq 0\,, \\ \sqrt{\boldsymbol{x}^{\mathrm{T}}\boldsymbol{x}} & \text{when } \beta = 0\,. \end{cases}$$

With this choice of c, we have $H\boldsymbol{x} = -c\,\mathbf{e}_1$, as required. □

We now show how Householder matrices can be used to reduce a given matrix to tridiagonal form.

Theorem 5.7 *Given that $A \in \mathbb{R}^{n\times n}_{\text{sym}}$ and $n \geq 3$, there exists a matrix $Q_n \in \mathbb{R}^{n\times n}_{\text{sym}}$, a product of $n-2$ Householder matrices $H_{(n,k)} \in \mathbb{R}^{n\times n}_{\text{sym}}$, $k = 2, \ldots, n-1$, given by*

$$Q_n = H_{(n,n-1)}H_{(n,n-2)}\ldots H_{(n,2)}$$

such that $Q_n^{\mathrm{T}}AQ_n = T_n$ is tridiagonal; the matrix Q_n is orthogonal.

Proof The proof of the theorem will proceed by induction. Before embarking on this, we make some preparatory observations which highlight the key ideas in the proof.

Consider the matrix $A \in \mathbb{R}^{n\times n}_{\text{sym}}$, partitioned by its first row and column in the form

$$A = \begin{pmatrix} \alpha & \boldsymbol{b}^{\mathrm{T}} \\ \boldsymbol{b} & C \end{pmatrix},$$

where $\alpha \in \mathbb{R}$, $\boldsymbol{b} \in \mathbb{R}^{n-1}$ and $C \in \mathbb{R}^{(n-1)\times(n-1)}_{\text{sym}}$, and define

$$\mathcal{E}^n_1 = \{\boldsymbol{v} \in \mathbb{R}^n \colon \boldsymbol{v} = (\lambda, 0, \ldots, 0)^{\mathrm{T}} \ \text{ for some } \lambda \in \mathbb{R}\}\,.$$

If $\boldsymbol{b}$ happens to belong to $\mathbb{R}^{n-1}_*$, then, by Lemma 5.4, there exists an $(n-1)\times(n-1)$ Householder matrix H_{n-1} such that each element of $H_{n-1}\boldsymbol{b}$, except the first, is equal to 0. If, on the other hand, $\boldsymbol{b} = \mathbf{0}$, then $H_{n-1}\boldsymbol{b} = \mathbf{0}$, trivially. Either way, $H_{n-1}\boldsymbol{b} \in \mathcal{E}^{n-1}_1$.

Let us extend the Householder matrix $H_{n-1} \in \mathbb{R}^{(n-1)\times(n-1)}_{\text{sym}}$, using Lemma 5.3 with $k = n-1$, to a Householder matrix $H_{(n,n-1)} \in \mathbb{R}^{n\times n}_{\text{sym}}$ by defining the $(1,1)$-entry of $H_{(n,n-1)}$ as 1 and choosing the remaining entries in the first row and first column of $H_{(n,n-1)}$ as 0. Then,

$$\begin{aligned} H^{\mathrm{T}}_{(n,n-1)}AH_{(n,n-1)} &= \begin{pmatrix} 1 & \mathbf{0}^{\mathrm{T}} \\ \mathbf{0} & H^{\mathrm{T}}_{n-1} \end{pmatrix}\begin{pmatrix} \alpha & \boldsymbol{b}^{\mathrm{T}} \\ \boldsymbol{b} & C \end{pmatrix}\begin{pmatrix} 1 & \mathbf{0}^{\mathrm{T}} \\ \mathbf{0} & H_{n-1} \end{pmatrix} \\ &= \begin{pmatrix} \alpha & \boldsymbol{d}^{\mathrm{T}} \\ \boldsymbol{d} & D \end{pmatrix}, \end{aligned} \tag{5.27}$$

where

$$\boldsymbol{d} = H_{n-1}^{\mathrm{T}}\boldsymbol{b} = H_{n-1}\boldsymbol{b} \in \mathcal{E}_1^{n-1} \text{ and } D = H_{n-1}^{\mathrm{T}} C\, H_{n-1} \in \mathbb{R}_{\mathrm{sym}}^{(n-1)\times(n-1)}.$$

As $\boldsymbol{d} \in \mathcal{E}_1^{n-1}$, the first row and first column of $H_{(n,n-1)}^{\mathrm{T}} A H_{(n,n-1)}$ are of the desired form. It remains to transform the submatrix D to tridiagonal form. This will be achieved by proceeding inductively.

If $n = 3$, then the 3×3 matrix $H_{(n,n-1)}^{\mathrm{T}} A H_{(n,n-1)}$ is automatically tridiagonal since $\boldsymbol{d} \in \mathcal{E}_1^2$, and we complete the proof by taking $Q_3 = H_{(3,2)}$. We note in passing that if $\mathbf{f} \in \mathcal{E}_1^3$, then

$$Q_3^{\mathrm{T}}\mathbf{f} = H_{(3,2)}^{\mathrm{T}}\mathbf{f} = H_{(3,2)}\mathbf{f} \in \mathcal{E}_1^3,$$

as the $(1,1)$-entry of $H_{(3,2)}$ is 1 and the remaining entries in its first column are all 0.

Let us suppose that $n \geq 4$ and $A \in \mathbb{R}_{\mathrm{sym}}^{n\times n}$. Our inductive hypothesis is that the statement of the theorem has already been established for any real symmetric matrix of order $n-1$, *i.e.*, $D \in \mathbb{R}_{\mathrm{sym}}^{(n-1)\times(n-1)}$ can be transformed into tridiagonal form:

$$Q_{n-1}^{\mathrm{T}} D Q_{n-1} = T_{n-1},$$

where $Q_{n-1} \in \mathbb{R}^{(n-1)\times(n-1)}$ is an orthogonal matrix that is a product of $n-3$ Householder matrices, each of size $(n-1)\times(n-1)$:

$$Q_{n-1} = H_{(n-1,n-2)} \cdots H_{(n-1,2)},$$

and $Q_{n-1}^{\mathrm{T}}\mathbf{f} \in \mathcal{E}_1^{n-1}$ for any vector $\mathbf{f} \in \mathcal{E}_1^{n-1}$. This inductive hypothesis has already been verified above for 3×3 real symmetric matrices.

We now extend each of the $(n-1)\times(n-1)$ matrices $H_{(n-1,k)}$, for $k = 2, \ldots, n-2$, to $n \times n$ Householder matrices $H_{(n,k)}$, $k = 2, \ldots, n-2$, respectively, as in Lemma 5.3, and define

$$Q_n = H_{(n,n-1)} H_{(n,n-2)} \cdots H_{(n,2)}\,.$$

Then, by (5.27),

$$\begin{aligned} Q_n^{\mathrm{T}} A Q_n &= \begin{pmatrix} 1 & \mathbf{0}^{\mathrm{T}} \\ \mathbf{0} & Q_{n-1}^{\mathrm{T}} \end{pmatrix} \begin{pmatrix} \alpha & \boldsymbol{d}^{\mathrm{T}} \\ \boldsymbol{d} & D \end{pmatrix} \begin{pmatrix} 1 & \mathbf{0}^{\mathrm{T}} \\ \mathbf{0} & Q_{n-1} \end{pmatrix} \\ &= \begin{pmatrix} \alpha & \boldsymbol{d}^{\mathrm{T}} Q_{n-1} \\ Q_{n-1}^{\mathrm{T}}\boldsymbol{d} & T_{n-1} \end{pmatrix}. \end{aligned}$$

As $\boldsymbol{d} \in \mathcal{E}_1^{n-1}$, it follows from our inductive hypothesis that $Q_{n-1}^{\mathrm{T}}\boldsymbol{d}$ also belongs to $\mathcal{E}_1^{n-1}$, and therefore the last matrix is tridiagonal. As Q_n is a product of $n-2$ Householder matrices, each of size $n \times n$ and each orthogonal, $Q_n \in \mathbb{R}^{n\times n}$ is itself orthogonal. Moreover, for any $\mathbf{f} \in \mathcal{E}_1^n$ we have $Q_n^{\mathrm{T}}\mathbf{f} \in \mathcal{E}_1^n$, since the $(1,1)$-entry of Q_n is 1 and the remaining entries in the first column of Q_n are 0. This concludes the inductive step, and completes the proof. □

The recursive transformation of a symmetric matrix to tridiagonal form outlined in the proof of Theorem 5.7 is called **Householder's method**. In implementing this method in practice it is important to carry out the transformations efficiently. Counting the arithmetic operations involved is straightforward but tedious, and shows that the complete reduction requires approximately $\frac{1}{3}n^3$ multiplications, for a moderately large value of n.

Example 5.6 *In order to illustrate Householder's method, we return to the matrix A defined in (5.16). The first stage uses the Householder matrix defined by the vector*

$$\boldsymbol{v} = (0.000,\ 4.162,\ 2.000,\ 1.000,\ 2.000)^{\mathrm{T}} . \tag{5.28}$$

The result of the transformation is the matrix

$$\begin{pmatrix} 4.000 & -3.162 & 0.000 & 0.000 & 0.000 \\ -3.162 & 5.300 & 1.232 & -0.332 & 0.284 \\ 0.000 & 1.232 & 1.653 & 3.312 & 0.275 \\ 0.000 & -0.332 & 3.312 & 5.149 & 1.123 \\ 0.000 & 0.284 & 0.275 & 1.123 & -3.102 \end{pmatrix} .$$

The leading element of the matrix is unchanged, and the first row and column have tridiagonal structure.

The second stage uses the Householder matrix with the vector

$$\boldsymbol{v} = (0.000,\ 0.000,\ 2.540,\ -0.332,\ 0.284)^{\mathrm{T}} \tag{5.29}$$

and gives the new matrix

$$\begin{pmatrix} 4.000 & -3.162 & 0.000 & 0.000 & 0.000 \\ -3.162 & 5.300 & -1.308 & 0.000 & 0.000 \\ 0.000 & -1.308 & 0.057 & -2.166 & 0.792 \\ 0.000 & 0.000 & -2.166 & 6.610 & 0.420 \\ 0.000 & 0.000 & 0.792 & 0.420 & -2.967 \end{pmatrix} .$$

This time the leading 2×2 minor is unaltered, and the first two rows and columns have tridiagonal structure.

The final stage uses the Householder matrix with vector

$$\boldsymbol{v} = (0.000,\ 0.000,\ 0.000,\ -4.471,\ 0.792)^{\mathrm{T}} \tag{5.30}$$

and gives the tridiagonal matrix

$$\begin{pmatrix} 4.000 & -3.162 & 0.000 & 0.000 & 0.000 \\ -3.162 & 5.300 & -1.308 & 0.000 & 0.000 \\ 0.000 & -1.308 & 0.057 & 2.306 & 0.000 \\ 0.000 & 0.000 & 2.306 & 5.208 & -3.411 \\ 0.000 & 0.000 & 0.000 & -3.411 & -1.565 \end{pmatrix}. \tag{5.31}$$

The numerical values are quoted here to three decimal digits, for simplicity. ◇

Having shown how to transform a symmetric matrix into tridiagonal form, we can now consider the problem of determining the eigenvalues of a tridiagonal matrix.

5.6 Eigenvalues of a tridiagonal matrix

Before developing a numerical algorithm for calculating the eigenvalues and the eigenvectors of a symmetric tridiagonal matrix, let us spend some time exploring the location of the eigenvalues. The main result of this section is the so-called Sturm sequence property,[1] stated in Theorem 5.9, which enables us to specify the number of eigenvalues of a symmetric tridiagonal matrix which exceed a given real number ϑ. The proof of the Sturm sequence property is based on Cauchy's Interlace Theorem which is of independent interest, and proving the latter is our first task.

To simplify the notation we now write the symmetric tridiagonal matrix in the form

$$T = \begin{pmatrix} a_1 & b_2 & & & & & \\ b_2 & a_2 & b_3 & & & & \\ & b_3 & a_3 & b_4 & & & \\ & & \cdots & \cdots & \cdots & & \\ & & & \cdots & \cdots & \cdots & \\ & & & & \cdots & \cdots & \cdots \\ & & & & b_{n-1} & a_{n-1} & b_n \\ & & & & & b_n & a_n \end{pmatrix}.$$

[1] Jacques Charles François Sturm (22 September 1803, Geneva, Helvetia (now Switzerland) – 18 December 1855, Paris, France). The results discussed here are based on Sturm's paper 'Mémoire sur la résolution des équations numériques', published in *Mémoires présentés par divers savants étrangers à l'Académie royale des sciences, section Sc. math. phys.*, **6**, 273–318, 1835, concerning the number of roots of a polynomial in an interval. In 1826 Sturm made the first accurate determination of the velocity of sound in water working with the Swiss engineer Daniel Colladon. In 1840 Sturm succeeded Poisson in the chair of mechanics in the Faculté des Sciences in Paris.

The determinants of the successive principal minors of a matrix of this form can easily be calculated by recurrence. Defining $p_r(\lambda)$ to be the determinant of the leading principal minor of order r of $T - \lambda I$, we see that

$$\begin{aligned} p_1(\lambda) &= a_1 - \lambda\,, \\ p_2(\lambda) &= (a_2 - \lambda)(a_1 - \lambda) - b_2^2\,. \end{aligned}$$

Expanding $p_r(\lambda)$ in terms of the elements of the last row, and then in terms of the last column, we obtain the relation

$$p_r(\lambda) = (a_r - \lambda)p_{r-1}(\lambda) - b_r^2 p_{r-2}(\lambda)\,, \qquad r = 2, 3, \dots, n\,,$$

with the convention that

$$p_0(\lambda) \equiv 1\,.$$

In the rest of this section we shall assume that all the off-diagonal elements b_i are nonzero. For suppose that $b_k = 0$ for some k in the set $\{2, 3, \dots, n\}$; then, the eigenvalues of the matrix T comprise the eigenvalues of the matrix consisting of the first $k-1$ rows and columns, together with the eigenvalues of the matrix consisting of the last $n-k+1$ rows and columns. These two problems become separated and can be treated independently; if several of the off-diagonal elements are zero, the matrix can be partitioned into a number of smaller matrices which can then be dealt with independently.

Theorem 5.8 (Cauchy's Interlace Theorem) *Let $n \geq 3$. The roots of p_r separate those of p_{r+1}, for $r = 1, 2, \dots, n-1$; i.e., between two consecutive roots of p_{r+1} there is exactly one root of the polynomial p_r, $r = 1, 2, \dots, n-1$.*

Proof The proof is by induction. It is trivial to show that the property holds for $r = 1$: the two roots

$$\tfrac{1}{2}\left[a_1 + a_2 \pm \sqrt{(a_1 - a_2)^2 + 4b_2^2}\right]$$

of p_2 are separated by a_1, the only root of the linear polynomial p_1.

Suppose that the statement is true when $r = i-1$, $2 \leq i \leq n-1$, so that the roots of p_{i-1} separate those of p_i. On denoting by α and β two consecutive roots of p_i, the inductive hypothesis implies that p_{i-1} has exactly one root between α and β, which means that $p_{i-1}(\alpha)$ and

$p_{i-1}(\beta)$ have opposite signs. Now,

$$p_{i+1}(\lambda) = (a_{i+1} - \lambda)p_i(\lambda) - b_{i+1}^2 p_{i-1}(\lambda),$$

so that, as α and β are roots of p_i, it follows that $p_{i+1}(\alpha)$ and $p_{i+1}(\beta)$ also have opposite signs. Hence p_{i+1} has at least one root between α and β. Choosing α and β to be each pair of consecutive roots of p_i in turn we have therefore located $i-1$ roots of p_{i+1}.

Next choose α to be the algebraically smallest root of p_i. It is easy to see that each of the polynomials $p_1, p_2, \ldots, p_n$ tends to $+\infty$ as $\lambda \to -\infty$. By the inductive hypothesis, p_{i-1} has no roots smaller than α, so $p_{i-1}(\alpha)$ is positive; hence from the recurrence relation $p_{i+1}(\alpha)$ is negative, and therefore p_{i+1} must have a root smaller than α. A similar argument shows that p_{i+1} has a root greater than the largest root of p_i, so that we have located all the $i+1$ roots of p_{i+1}. There is exactly one root of p_i between each pair of consecutive roots of p_{i+1}, and the interlacing property follows. □

We have shown in particular that all the roots of each p_r are distinct. Moreover $p_i(\lambda)$ and $p_{i-1}(\lambda)$ cannot both vanish for the same λ, for if this were to happen the recurrence relation would show that this value of λ is a root of p_r for all values of $r \in \{0, 1, \ldots, n\}$; but p_0 evidently never vanishes.

Theorem 5.9 (The Sturm sequence property) *Let us suppose that $\vartheta \in \mathbb{R}$ and consider the sequence $p_i(\vartheta)$, $i = 0, 1, \ldots, n$. The number of agreements in sign between consecutive members of the sequence is the same as the number of eigenvalues of the matrix T which are strictly greater than ϑ.*

Proof Given that $\lambda \in \mathbb{R}$ and $1 \le j \le n$, we write $s_j(\lambda)$ for the number of agreements in sign in the sequence

$$p_0(\lambda), p_1(\lambda), \ldots, p_j(\lambda),$$

and $g_j(\lambda)$ for the number of roots of the polynomial p_j which are strictly greater than λ.

It is trivial to see that $s_1(\vartheta) = g_1(\vartheta)$. The proof now proceeds by induction. Let us suppose that $2 \le k \le n$ and adopt the inductive hypothesis that $s_{k-1}(\vartheta) = g_{k-1}(\vartheta)$; we shall prove that $s_k(\vartheta) = g_k(\vartheta)$.

Under our hypothesis, either $s_k(\vartheta) = s_{k-1}(\vartheta)+1$, if $p_k(\vartheta)$ and $p_{k-1}(\vartheta)$ have the same sign, or $s_k(\vartheta) = s_{k-1}(\vartheta)$ if they have opposite sign. Suppose that ϑ lies in the interval between the two consecutive roots α and

β of p_{k-1}. By Theorem 5.8, there is exactly one root of p_k between α and β; denote this root by φ. As we saw in the proof of the previous theorem $p_k(\lambda)$ is positive when λ is large and negative, and the sign of $p_k(\lambda)$ is determined by the number of roots of p_k which are less than λ. Now, if $\vartheta < \varphi$ then p_k and p_{k-1} have the same number of roots less than ϑ, so that $p_k(\vartheta)$ and $p_{k-1}(\vartheta)$ have the same sign, and $s_k(\vartheta) = s_{k-1}(\vartheta) + 1$. Also if p_k and p_{k-1} have the same number of roots less than ϑ, then p_k must have one more root which is greater than ϑ; this means that $g_k(\vartheta) = g_{k-1}(\vartheta) + 1$. Hence $s_k(\vartheta) = g_k(\vartheta)$. A similar argument shows that $s_k(\vartheta) = g_k(\vartheta)$ in the alternative situation where $\vartheta > \varphi$. It is also a simple matter to modify the argument slightly for the cases where ϑ is less than the smallest root of p_{k-1}, or greater than the largest root of p_{k-1}, and so the inductive step is complete. □

The theorem and proof do not allow for any of the members of the sequence being zero, in which case the sign becomes undefined. A more careful analysis is tedious but not difficult; it shows that the theorem still holds if we adopt the convention that when $p_j(\vartheta)$ is zero it is given the same sign as $p_{j-1}(\vartheta)$. As we have already seen, two consecutive members of the sequence cannot both be zero.

Our next example will illustrate the application of the Sturm sequence property.

Example 5.7 *Determine the second largest eigenvalue of the matrix*

$$A = \begin{pmatrix} 3 & 1 & 0 & 0 \\ 1 & -1 & 2 & 0 \\ 0 & 2 & 1 & 1 \\ 0 & 0 & 1 & 1 \end{pmatrix}. \tag{5.32}$$

If the eigenvalues are λ_j, where $\lambda_1 > \lambda_2 > \lambda_3 > \lambda_4$, we wish to find λ_2. Now, it is easy to see from Theorem 5.4 that all the eigenvalues lie in the interval $[-4, 4]$. We take the midpoint of this interval, and evaluate the Sturm sequence with $\vartheta = 0$, giving

$$p_0(0) = 1\,, \quad p_1(0) = 3\,, \quad p_2(0) = -4\,, \quad p_3(0) = -16\,, \quad p_4(0) = -12\,.$$

In this sequence there are three agreements of sign:

$$(1, 3)\,, \quad (-4, -16) \quad \text{and} \quad (-16, -12)\,.$$

Hence $s_4(0) = 3$, and the matrix has three eigenvalues greater than 0; this means that λ_2 must lie in the right-hand half of the interval

$[-4,4]$, that is, in $[0,4]$. We construct the Sturm sequence for $\vartheta = 2$, the midpoint of the interval, giving

$$p_0(2) = 1\,, \;\; p_1(2) = 1\,, \;\; p_2(2) = -4\,, \;\; p_3(2) = -0\,, \;\; p_4(2) = 4\,.$$

Notice that here $p_3(2)$ is zero, and is given the negative sign to agree with $p_2(2)$. The number of agreements in sign here is two, so two of the eigenvalues are greater than 2, and λ_2 must lie in $[2,4]$, the right-hand half of the interval $[0,4]$. For $\vartheta = 3$ we obtain the sequence

$$1\,, \;\; +0\,, \;\; -1\,, \;\; 2\,, \;\; -3\,,$$

with only one agreement of sign, so this time λ_2 must lie in the left-hand half $[2,3]$ of the interval $[2,4]$, and we repeat the process, taking $\vartheta = \frac{5}{2}$, the midpoint of $[2,3]$. This time the sequence is

$$1\,, \;\; \frac{1}{2}\,, \;\; -\frac{11}{4}\,, \;\; \frac{17}{8}\,, \;\; -\frac{7}{16}\,,$$

with one agreement in sign, showing that $\lambda_2 < 2.5$.

The process of bisection can be repeated as many times as required to locate the eigenvalue to a given accuracy. After 13 stages we find that $\lambda_2 = 2.450$ correct to three decimal digits. ◇

This method is very similar to the usual bisection process for finding a solution of $f(x) = 0$, beginning with an interval $[a,b]$ such that $f(a)$ and $f(b)$ have opposite signs. A great advantage of the Sturm sequence method is that it not only determines the eigenvalue, but also indicates which eigenvalue it is. If we used the Jacobi method of Section 5.3 we would have to determine *all* the eigenvalues, sort them into order, and then choose the second largest eigenvalue as λ_2.

The Sturm sequence method will also determine how many eigenvalues of a matrix lie in a given interval (α, β); all that we need is to construct the Sturm sequences $(p_j(\alpha))_{j=0,1,\dots,n}$ and $(p_j(\beta))_{j=0,1,\dots,n}$; then, the required number of eigenvalues is $s_n(\alpha) - s_n(\beta)$.

It is very important to calculate the sequence $p_j(\vartheta)$ directly from the recurrence relation. For instance, in Example 5.7, with $\vartheta = 2.445$ we obtain

$$\begin{aligned}
p_0(2.445) &= 1\,,\\
p_1(2.445) &= 3 - 2.445 = 0.555\,,\\
p_2(2.445) &= (-1 - 2.445) \times 0.555 - 1 \times 1 = -2.9120\,,\\
p_3(2.445) &= (1 - 2.445) \times -2.9120 - 4 \times 0.555 = 1.9878\,,\\
p_4(2.445) &= (1 - 2.445) \times 1.9878 - 1 \times -2.9120 = 0.0396\,.
\end{aligned}$$

The alternative, to construct explicit forms for the polynomials $p_j(\lambda)$, $j = 0, 1, \ldots, n$, and *then* evaluate $p_j(\vartheta)$ by inserting the value of $\lambda = \vartheta$ into each of the polynomials $p_j(\lambda)$, will lead to the construction of the explicit form of the characteristic polynomial of the matrix, which is $p_n(\lambda)$, and we have already seen that this is affected disastrously by rounding errors. The calculation by direct use of the recurrence relation is perfectly satisfactory.

Example 5.8 *As a second example, we return to the matrix A in (5.16), which has been transformed to the tridiagonal form (5.31), to determine the largest eigenvalue.*

Table 5.2. *Bisection process for the largest eigenvalue. In the table k denotes the iteration number, ϑ_k the kth iterate approximating the unknown eigenvalue λ_1, and $s_4(\vartheta_k)$ signifies the number of sign agreements in the Sturm sequence $p_0(\vartheta_k), \ldots, p_4(\vartheta_k)$.*

k	ϑ_k	$s_4(\vartheta_k)$
1	0.000	3
2	5.463	2
3	8.194	0
4	6.829	2
5	7.511	1
6	7.853	1
7	8.024	1
8	8.109	0
9	8.066	1
10	8.088	1
11	8.098	0
12	8.093	1
13	8.096	0
14	8.094	0
15	8.094	1

Table 5.2 shows the result of the bisection process, using the Sturm sequence. The ∞-norm of the tridiagonal matrix is 10.926, so the process begins with the interval $[-10.926, 10.926]$.[1] The largest eigenvalue

[1] To explain this choice, let us note that if $\lambda \in \mathbb{C}$ is an eigenvalue of $A \in \mathbb{C}^{n\times n}$ and $\boldsymbol{x} \in \mathbb{C}^n \setminus \{\mathbf{0}\}$ is the corresponding eigenvector, then $|\lambda|\,\|\boldsymbol{x}\| = \|\lambda\boldsymbol{x}\| = \|A\boldsymbol{x}\| \le \|A\|\,\|\boldsymbol{x}\|$; *i.e.*, $|\lambda| \le \|A\|$, for any subordinate matrix norm $\|\cdot\|$ and any eigenvalue λ of A.

is 8.094, to three decimal digits, agreeing with the result of Jacobi's method, in Section 5.3. This table also shows how some savings are possible when all the eigenvalues are required. We see from the table that use of $\vartheta = 7.511$ gives 1 agreement in sign, while $\vartheta = 6.829$ gives 2 agreements in sign. The bisection process for the second largest eigenvalue can therefore begin with the interval [6.829, 7.511]. ◇

The method of bisection may appear rather crude, but it has the great advantage of guaranteed success, and is very little affected by rounding errors. Moreover, the amount of work involved is not large. If we have calculated the squares of the off-diagonal entries, b_r^2, of the matrix T in advance, each computation of all members of the sequence requires about $2n$ multiplications. If the bisection process is continued for 40 stages, the eigenvalue will be determined to about nine significant digits, and if we require to calculate m of the eigenvalues to this accuracy, we shall need about $80mn$ multiplications. If m is a good deal smaller than n, the order of the matrix, this is likely to be a great deal smaller than the work involved in the process of reduction to tridiagonal form, which, as we have seen, is about $\frac{1}{3}n^3$ multiplications. In most practical problems it is the initial Householder reduction to tridiagonal form which accounts for most of the computational work.

5.7 The QR algorithm

In this section we discuss briefly the QR algorithm, an alternative method for determining the eigenvalues of a tridiagonal matrix. In principle it could be applied to a full matrix, but it is more efficient to use the Householder method to reduce the matrix to tridiagonal form first. The basis of the method is the QR factorisation of the matrix which we have already encountered in Chapter 2, in the solution of least squares problems. In contrast with Section 2.9, however, where we were concerned with the solution of least squares problems for rectangular matrices $A \in \mathbb{R}^{m\times n}$, here the focus is on eigenvalue problems for symmetric tridiagonal matrices $A \in \mathbb{R}^{n\times n}$; we shall therefore revisit the derivation of the QR factorisation by adopting a slightly different approach from the one proposed in Section 2.9.

5.7.1 The QR factorisation revisited

Suppose that $n \geq 3$ and $A \in \mathbb{R}^{n\times n}$ is a symmetric tridiagonal matrix. We first show how to construct an orthogonal matrix $Q \in \mathbb{R}^{n\times n}$ and an upper triangular matrix $R \in \mathbb{R}^{n\times n}$ such that $A = QR$; the problem is similar to the LU factorisation used in solving systems of linear equations, but here we have an orthogonal matrix Q instead of a lower triangular matrix L.

We construct the matrix Q as a product of plane rotation matrices $R^{(p\,p+1)}(\varphi) \in \mathbb{R}^{n\times n}$ (see Definition 5.2), with a suitably chosen φ. In order to explain what is meant here by 'suitably chosen', we note that in the product

$$B = R^{(p\,p+1)}(\varphi)A \tag{5.33}$$

the element $b_{p+1\,p}$ is easily found to be

$$b_{p+1\,p} = -s\,a_{pp} + c\,a_{p+1\,p}\,,$$

where $s = \sin\varphi$ and $c = \cos\varphi$. We can make $b_{p+1\,p} = 0$ by choosing

$$s = \frac{a_{p+1\,p}}{\rho}\,, \quad c = \frac{a_{pp}}{\rho}\,, \quad \rho = (a_{pp}^2 + a_{p+1\,p}^2)^{1/2}\,. \tag{5.34}$$

We note in passing that

$$\begin{aligned} b_{pp} &= c a_{pp} + s a_{p+1\,p}\,,\\ b_{p\,p+1} &= c a_{p\,p+1} + s a_{p+1\,p+1}\,,\\ b_{p+1\,p+1} &= -s a_{p\,p+1} + c a_{p+1\,p+1}\,. \end{aligned}$$

The remaining elements of B are the same as those of A.

To summarise the important points, upon multiplying the symmetric tridiagonal matrix A on the left by $R^{(p\,p+1)}(\varphi)$, where $c = \cos\varphi$ and $s = \sin\varphi$ in $R^{(p\,p+1)}(\varphi)$ are chosen as indicated in (5.34), we obtain a tridiagonal matrix $B = (b_{ij}) \in \mathbb{R}^{n\times n}$ such that $b_{p+1\,p} = 0$.

After this brief preparation, we embark on the description of the QR factorisation. Let us suppose that we successively multiply A on the left by the $n-1$ plane rotation matrices,

$$Q_1 = R^{(12)}(\varphi_1)\,, \quad Q_2 = R^{(23)}(\varphi_2)\,, \quad \ldots\,, \quad Q_{n-1} = R^{(n-1\,n)}(\varphi_{n-1})\,,$$

with $\varphi_1, \varphi_2, \ldots, \varphi_{n-1}$ selected according to (5.34); more precisely,

$$\begin{aligned} &\text{for } p = 1, 2, \ldots, n-1,\\ &\varphi_p \text{ is chosen so as to set the } (p+1,p)\text{-entry of } Q_p \ldots Q_1 A\\ &\text{to zero}\,. \end{aligned}$$

Given that the elements below the diagonal of the matrix

$$Q_{p-1}\ldots Q_1 A\,, \qquad 2 \le p \le n-1\,,$$

which are already equal to zero, remain zero upon multiplication by the next rotation matrix Q_p in the sequence, we deduce that, after successive multiplications of A on the left by Q_1, $Q_2, \ldots, Q_{n-1}$, the matrix

$$Q_{n-1}\, Q_{n-2} \ldots Q_1\, A = R\,, \tag{5.35}$$

is upper triangular. In fact, since A is tridiagonal, R is tridiagonal and upper triangular; consequently, R is **bidiagonal** in the sense that $R_{ij} = 0$ if $i \neq j, j-1$.

As the matrices $Q_p = R^{(p\,p+1)}(\varphi_p)$, $p = 1, 2, \ldots, n-1$, are orthogonal, and therefore $Q_p^{\mathrm{T}} Q_p = I$, on multiplying (5.35) on the left by $Q_1^{\mathrm{T}}\, Q_2^{\mathrm{T}} \ldots Q_{n-1}^{\mathrm{T}}$, we find that

$$A = Q\, R\,,$$

where

$$Q = Q_1^{\mathrm{T}}\, Q_2^{\mathrm{T}} \ldots Q_{n-1}^{\mathrm{T}}$$

is an orthogonal matrix (as it is a product of orthogonal matrices). The next subsection describes the QR algorithm, based on the QR factorisation, for the numerical solution of the eigenvalue problem (5.1) where the matrix $A \in \mathbb{R}^{n\times n}$ is symmetric and tridiagonal.

5.7.2 The definition of the QR algorithm

Suppose that $A \in \mathbb{R}^{n\times n}$ is symmetric and tridiagonal. The QR algorithm defines a sequence of symmetric tridiagonal matrices $A^{(k)} \in \mathbb{R}^{n\times n}$, $k = 0, 1, 2, \ldots$, starting with $A^{(0)} = A$, as follows.

Suppose that $k \geq 0$. The kth step of the QR algorithm takes the symmetric tridiagonal matrix $A^{(k)}$ and chooses a **shift** $\mu_k \in \mathbb{R}$ (the choice of μ_k will be discussed below), then forming the QR factorisation

$$A^{(k)} - \mu_k\, I = Q^{(k)}\, R^{(k)}\,.$$

We then multiply $Q^{(k)}$ and $R^{(k)}$ in the reverse order, and construct the new matrix $A^{(k+1)}$ defined by

$$A^{(k+1)} = R^{(k)}\, Q^{(k)} + \mu_k\, I\,.$$

Recalling that the matrix $Q^{(k)}$ is orthogonal, it is a simple matter to see that $A^{(k+1)} = Q^{(k)T} A^{(k)} Q^{(k)}$, so that $A^{(k+1)}$ and $A^{(k)}$ have the same eigenvalues. As $A^{(0)} = A$, all matrices in the sequence $(A^{(k)})$ have the same eigenvalues as A itself. It is also easy to show that each of the matrices $A^{(k)}$ is symmetric and tridiagonal. (See Exercise 7.)

The choice of the shift parameter μ_k is very important; if correctly chosen the sequence of matrices $A^{(k)}$ converges very rapidly to a matrix in which one of the off-diagonal elements is zero. If this element is in the first or last row, we have thereby identified one of the eigenvalues; if it is one of the intermediate elements, we can split the matrix into two separate matrices of lower order. In either case we can repeat the iterative process with smaller matrices, until all the eigenvalues are found.

The usual simple choice of the shift parameter in the kth step is

$$\mu_k = a_{nn}^{(k)} ,$$

the last diagonal element of the matrix $A^{(k)}$. In general, after a few steps of the iteration the element at position $(n, n-1)$ will become negligibly small. One of the eigenvalues of the resulting matrix is then the last diagonal element, and we continue the process with the matrix of order $n-1$ obtained by removing the last row and column. There are special circumstances where this choice of shift is unsatisfactory, and other situations where another choice is more efficient, but we shall not discuss the details any further. The proof of the convergence of this method is long and technical; details will be found in the books cited in the Notes at the end of the chapter.

The method does not determine the eigenvalues in any particular order, so if we require only a small number of the largest eigenvalues, for example, the Sturm sequence method is preferable. The usual recommendation is that the QR algorithm should be used on a matrix of order n if more than about $\frac{1}{4}n$ of the eigenvalues are required.

Example 5.9 *We apply the QR algorithm to the tridiagonal matrix (5.31).*

After one step of the iteration the matrix $A^{(1)} = R^{(0)} Q^{(0)} + \mu_0 I$, with $\mu_0 = a_{55}^{(0)} = a_{55}$, is

$$A^{(1)} = \begin{pmatrix} 7.034 & -2.271 & 0 & 0 & 0 \\ -2.271 & 2.707 & -0.744 & 0 & 0 \\ 0 & -0.744 & 5.804 & 3.202 & 0 \\ 0 & 0 & 3.202 & -0.464 & 1.419 \\ 0 & 0 & 0 & 1.419 & -2.082 \end{pmatrix} .$$

In successive iterations $k = 1, 2, 3, 4, 5$, the element $a_{54}^{(k)}$ has the values 1.419, −1.262, 0.965, −0.223, 0.002, and after the next iteration $a_{54}^{(6)}$

vanishes to 10 decimal digits. The element $a_{55}^{(6)}$ is -3.282, which is therefore an eigenvalue.

We then remove the last row and column, and continue the process on the resulting 4×4 matrix. After just one iteration the element at position $(4, 3)$ vanishes to 7 decimal digits, giving the eigenvalue -0.671. We remove the last row and column and continue with the resulting 3×3 matrix. After one iteration of the resulting 3×3 matrix the element at position $(3, 2)$ is 0.0005, and another iteration gives the accurate eigenvalue 1.690. We are now left with a 2×2 matrix, and the calculation of the last two eigenvalues is trivial. The number of iterations required to isolate each eigenvalue reduces as the algorithm reduces the size of the matrix; this sort of behaviour is typical.

The numerical values agree with those obtained by Jacobi's method, and the bisection method. ◇

5.8 Inverse iteration for the eigenvectors

We saw in Section 5.3 that Jacobi's method can also, if required, produce the eigenvectors of the matrix, but the use of Householder's algorithm, in conjunction with the Sturm sequence method or the QR algorithm, only gives the eigenvalues. Suppose that $A \in \mathbb{R}^{n \times n}$ is a symmetric matrix, and assume that we have a good approximation $\vartheta \in \mathbb{R}$ to the required eigenvalue $\lambda \in \mathbb{R}$ of A, and some approximation $\boldsymbol{v}^{(0)} \in \mathbb{R}^n_*$, $\|\boldsymbol{v}^{(0)}\|_2 = 1$, to the associated eigenvector $\boldsymbol{v} \in \mathbb{R}^n_*$, $\|\boldsymbol{v}\|_2 = 1$. It is implicitly assumed that $\vartheta \neq \lambda$ and that ϑ is not an eigenvalue of A, so that the matrix $A - \vartheta I$ is nonsingular. The method of **inverse iteration** defines the sequence of vectors $\boldsymbol{v}^{(k)}$, $k = 0, 1, \ldots$, as follows: given $\boldsymbol{v}^{(k)} \in \mathbb{R}^n_*$, find $\boldsymbol{w}^{(k)} \in \mathbb{R}^n_*$ and then $\boldsymbol{v}^{(k+1)} \in \mathbb{R}^n_*$ from

$$\begin{aligned} (A - \vartheta I)\boldsymbol{w}^{(k)} &= \boldsymbol{v}^{(k)}, \\ \boldsymbol{v}^{(k+1)} &= c_k \boldsymbol{w}^{(k)}, \end{aligned} \tag{5.36}$$

where $c_k = 1/\sqrt{\boldsymbol{w}^{(k)\mathrm{T}}\boldsymbol{w}^{(k)}} = 1/\|\boldsymbol{w}^{(k)}\|_2$. Hence, we conclude that $\|\boldsymbol{v}^{(k)}\|_2 = 1$, $k = 0, 1, 2, \ldots$.

Theorem 5.10 *Suppose that $A \in \mathbb{R}^{n \times n}_{\mathrm{sym}}$. The sequence of vectors $(\boldsymbol{v}^{(k)})$ in $\mathbb{R}^n_*$ defined in the process of inverse iteration (5.36) converges to the normalised eigenvector $\boldsymbol{v} \in \mathbb{R}^n_*$ corresponding to the eigenvalue $\lambda \in \mathbb{R}$ which is closest to $\vartheta \in \mathbb{R}$, provided that λ is a simple eigenvalue and the initial vector $\boldsymbol{v}^{(0)} \in \mathbb{R}^n_*$ is not orthogonal to the vector $\boldsymbol{v}$.*

Proof According to Theorem 5.1 (vii), the vector $\boldsymbol{v}^{(0)}$ can be expressed as a linear combination of the (ortho)normalised eigenvectors $\boldsymbol{x}^{(j)}$ in $\mathbb{R}^n_*$, $j = 1, 2, \dots, n$, of the matrix A in the form

$$\boldsymbol{v}^{(0)} = \sum_{j=1}^{n} \alpha_j \boldsymbol{x}^{(j)}, \qquad \alpha_j = \boldsymbol{v}^{(0)\mathrm{T}} \boldsymbol{x}^{(j)}. \tag{5.37}$$

Let $\lambda_s \in \mathbb{R}$ denote the eigenvalue of A which is closest to $\vartheta \in \mathbb{R}$. We shall prove that the sequence $(\boldsymbol{v}^{(k)})$ converges, as $k \to \infty$, to the eigenvector $\boldsymbol{v} = \boldsymbol{x}^{(s)} \in \mathbb{R}^n_*$ associated with λ_s, provided that $\alpha_s = \boldsymbol{v}^{(0)\mathrm{T}} \boldsymbol{x}^{(s)} \neq 0$. On expanding

$$\boldsymbol{w}^{(0)} = \sum_{j=1}^{n} \beta_j \boldsymbol{x}^{(j)},$$

inserting this expansion into the first line of (5.36) with $k = 0$ and comparing the resulting left-hand side with the expansion (5.37) of $\boldsymbol{v}^{(0)}$ on the right, we find that $(\lambda_j - \vartheta)\beta_j = \alpha_j$. Our hypothesis that $\alpha_s \neq 0$ implies that $\lambda_s \neq \vartheta$. Further, as λ_s is the eigenvalue closest to ϑ, it then follows that $\lambda_j - \vartheta \neq 0$ for all $j \in \{1, 2, \dots, n\}$. Hence,

$$\boldsymbol{v}^{(1)} = c_0 \boldsymbol{w}^{(0)} = c_0 \sum_{j=1}^{n} \frac{\alpha_j}{\lambda_j - \vartheta} \boldsymbol{x}^{(j)}.$$

Repeating this argument for $k = 1, 2, \dots, m-1$ gives

$$\boldsymbol{v}^{(m)} = c_{m-1} \dots c_0 \sum_{j=1}^{n} \frac{\alpha_j}{(\lambda_j - \vartheta)^m} \boldsymbol{x}^{(j)}. \tag{5.38}$$

Now $\boldsymbol{v}^{(m)\mathrm{T}} \boldsymbol{v}^{(m)} = 1$, and therefore,

$$c_{m-1} \dots c_0 = \left[\sum_{j=1}^{n} \frac{\alpha_j^2}{(\lambda_j - \vartheta)^{2m}} \right]^{-1/2}. \tag{5.39}$$

Substituting (5.39) into (5.38), we obtain

$$\boldsymbol{v}^{(m)} = \frac{\sum_{j=1}^{n} \frac{\alpha_j}{(\lambda_j - \vartheta)^m} \boldsymbol{x}^{(j)}}{\left[\sum_{j=1}^{n} \frac{\alpha_j^2}{(\lambda_j - \vartheta)^{2m}} \right]^{1/2}} = \frac{\boldsymbol{x}_s + \sum_{j \neq s} \left(\frac{\alpha_j}{\alpha_s} \right) \left(\frac{\lambda_s - \vartheta}{\lambda_j - \vartheta} \right)^m \boldsymbol{x}^{(j)}}{\left[1 + \sum_{j \neq s} \left(\frac{\alpha_j}{\alpha_s} \right)^2 \left(\frac{\lambda_s - \vartheta}{\lambda_j - \vartheta} \right)^{2m} \right]^{1/2}}.$$

Since

$$\left| \frac{\lambda_s - \vartheta}{\lambda_j - \vartheta} \right| < 1 \qquad \forall j \in \{1, 2, \dots, n\} \setminus \{s\},$$

we find that $\lim_{m \to \infty} \boldsymbol{v}^{(m)} = \boldsymbol{x}_s = \boldsymbol{v}$; that completes the proof. □

If the estimate ϑ is within rounding error of λ_s and the eigenvalues are well spaced, the convergence of the sequence $(\boldsymbol{v}^{(k)})$ will be extremely rapid: usually a couple of iterations will be sufficient.

The proof of Theorem 5.10 breaks down if $\alpha_s = 0$, *i.e.*, when the initial vector $\boldsymbol{v}^{(0)}$ is exactly orthogonal to the required eigenvector. However, this does not mean that the iteration (5.36) will also break down; for the effect of rounding error will almost always introduce a small multiple of the vector $\boldsymbol{x}^{(s)}$ into the expansion of $\boldsymbol{v}^{(0)}$ in terms of the $\boldsymbol{x}^{(j)}$ with $j = 1, 2, \dots, n$, and the required eigenvector will then be obtained in a small number of iterations. This is a useful property of the method, since in practice it is not possible to check whether or not $\boldsymbol{v}^{(0)}$ is orthogonal to $\boldsymbol{v}$, given that the eigenvector $\boldsymbol{v}$ is unknown.

There will also be a problem if there is a multiple eigenvalue, or two eigenvalues are very close together: in the first case $|\lambda_s - \vartheta|/|\lambda_j - \vartheta| = 1$ for some $j \neq s$, and the proof of Theorem 5.10 breaks down; in the second case $|\lambda_s - \vartheta|/|\lambda_j - \vartheta| \approx 1$ for some $j \neq s$, leading to very slow convergence.

The computation of $\boldsymbol{w}^{(k)}$ from (5.36) requires the solution of a system of linear equations whose matrix is $A - \vartheta I$. This matrix will usually be nearly singular – in fact, our objective in choosing ϑ was to make $A - \vartheta I$ exactly singular. In general the solution of such a system is extremely dangerous, because of the effect of rounding errors; in this case, however, the effect of rounding error will be to introduce a multiple of the dominant eigenvector, and this is exactly what is required. An analysis of the effect of rounding errors will confirm this fact, but would take too long here.[1]

There are two ways in which we can implement the inverse iteration process. One obvious possibility would be to use the original matrix $A \in \mathbb{R}^{n\times n}$, as implied in (5.36). An alternative is to replace A in this equation by the tridiagonal matrix $T \in \mathbb{R}^{n\times n}$ supplied by Householder's method. The calculation is then very much quicker, but produces the eigenvector of T; to obtain the corresponding eigenvector of A we must then apply to this vector the sequence of Householder transformations which were used in the original reduction to tridiagonal form. It is easy to show that this is the most efficient method.

[1] For further details, we refer to Sec. 4.3 in B. Parlett, *The Symmetric Eigenvalue Problem*, Prentice–Hall, Englewood Cliffs, NJ, 1980, and Section 7.6.1 in G.H. Golub and C.F. Van Loan, *Matrix Computations*, Third Edition, Johns Hopkins University Press, Baltimore, 1996.

Inverse iteration with the original matrix $A \in \mathbb{R}^{n\times n}$ requires the LU decomposition of A, followed by one or more forward and backsubstitution operations. As we saw in Section 2.6, the LU decomposition requires approximately $\frac{1}{3}n^3$ multiplications. The same process with the tridiagonal matrix T, using the Thomas algorithm, involves only a small multiple of n multiplications.

Having found an eigenvector of the tridiagonal matrix $T \in \mathbb{R}^{n\times n}$, so that

$$T\boldsymbol{v} = \lambda\boldsymbol{v}\,,$$

we use the fact that $Q^{\mathrm{T}}AQ = T$ to write

$$AQ\boldsymbol{v} = \lambda Q\boldsymbol{v}\,,$$

so that the vector $Q\boldsymbol{v}$ is an eigenvector of A. Using Theorem 5.7, this means that the required eigenvector of A is

$$H_{(n,n-1)}\dots H_{(n,2)}\boldsymbol{v}\,,$$

where the matrices $H_{(n,j)} \in \mathbb{R}^{n\times n}$, $j = 2,\dots,n-1$, are Householder matrices. To multiply a vector $\boldsymbol{x}$ by a Householder matrix $H = H(\boldsymbol{u})$ we write

$$H\boldsymbol{x} = (I - \alpha\boldsymbol{u}\boldsymbol{u}^{\mathrm{T}})\boldsymbol{x} = \boldsymbol{x} - \alpha(\boldsymbol{u}^{\mathrm{T}}\boldsymbol{x})\boldsymbol{u}\,.$$

Assuming that $\alpha = 2/(\boldsymbol{u}^{\mathrm{T}}\boldsymbol{u})$ is known, this requires the calculation of the scalar product $\boldsymbol{u}^{\mathrm{T}}\boldsymbol{x}$, and then subtracting a multiple of the vector $\boldsymbol{u}$ from the vector $\boldsymbol{x}$. This evidently involves $2n$ multiplications. Hence the calculation of $Q\boldsymbol{v}$ requires only $2n(n-2)$ multiplications, and the work involved in the whole process is proportional to n^2, instead of n^3. In fact the total is less than $2n(n-2)$, since a more careful count can use the fact that many of the elements in the vector $\boldsymbol{u}$ are known to be zero.

Example 5.10 *Returning to the tridiagonal matrix (5.31), the QR algorithm has given an accurate eigenvalue which is* 8.094 *to three decimal digits. Beginning the inverse iteration (5.36) with a randomly chosen vector* $\boldsymbol{v}^{(0)} \in \mathbb{R}^5_*$*, we find that*

$$\boldsymbol{v}^{(1)} = (-0.0249, -0.0574, -0.3164, 0.4256, 0.8455)^{\mathrm{T}}.$$

Successive iterations make no change in this vector, as might be expected, since the eigenvalue used was accurate to within rounding error.

This is therefore the eigenvector of the tridiagonal matrix (5.31), to within rounding error. To obtain the eigenvector of the original matrix (5.16) we multiply $\boldsymbol{v}^{(1)}$ *in succession by the three Householder matrices defined by the vectors (5.30), (5.29) and (5.28). The result is the eigenvector*

$$\boldsymbol{v} = (-0.0249,\ -0.5952,\ -0.1920,\ -0.2885,\ 0.7246)^{\mathrm{T}}.$$

Using this vector and the accurately calculated eigenvalue, we can check the result, and find that the elements of $A\boldsymbol{v} - \lambda\boldsymbol{v}$ *are of the same order as rounding error.*

5.9 The Rayleigh quotient

In this section we develop a simple technique, based on the concept of Rayleigh quotient,[1] for obtaining an accurate approximation to an eigenvalue of a symmetric matrix when a reasonably accurate approximation to the associated eigenvector is already available.

Definition 5.7 *Given a vector* $\boldsymbol{x} \in \mathbb{R}^n_*$ *and a matrix* $A \in \mathbb{R}^{n\times n}_{\mathrm{sym}}$, *the associated* **Rayleigh quotient** $R(\boldsymbol{x})$ *is defined as the real number*

$$R(\boldsymbol{x}) = \frac{\boldsymbol{x}^{\mathrm{T}} A\, \boldsymbol{x}}{\boldsymbol{x}^{\mathrm{T}}\boldsymbol{x}}\,. \tag{5.40}$$

Clearly, if $\boldsymbol{x} \in \mathbb{R}^n_*$ is an eigenvector corresponding to an eigenvalue $\lambda \in \mathbb{R}$ of a matrix $A \in \mathbb{R}^{n\times n}_{\mathrm{sym}}$, then $R(\boldsymbol{x}) = \lambda$. More generally, if $\boldsymbol{x}$ is any nonzero vector in $\mathbb{R}^n$, then a number of further properties of the Rayleigh quotient are immediate deductions from the expansion of $\boldsymbol{x}$ in terms of the eigenvectors of A.

Theorem 5.11 *Suppose that the matrix* $A \in \mathbb{R}^{n\times n}_{\mathrm{sym}}$ *has the eigenvalues* $\lambda_j \in \mathbb{R}$, $j = 1, 2, \ldots, n$, *and the corresponding normalised eigenvectors* $\boldsymbol{x}^{(j)} \in \mathbb{R}^n_*$, $j = 1, 2, \ldots, n$. *If the vector* $\boldsymbol{x} \in \mathbb{R}^n_*$ *is expressed in terms of*

[1] John William Strutt, Lord Rayleigh (12 November 1842, Langford Grove (near Maldon), Essex, England – 30 June 1919, Terling Place, Witham, Essex, England). In 1879 Rayleigh wrote a paper on travelling waves which set the foundation for the modern theory of solitons. His theory of scattering (1871) was the first correct explanation of why the sky is blue: the intensity of light scattered from small particles is inversely proportional to the fourth power of the wavelength; for this reason, the intensity of the short-wavelength blue component dominates in the scattered light reaching our eyes. From 1879 to 1884 Rayleigh was the second Cavendish Professor of Physics at Cambridge, succeeding Maxwell, and he was awarded the Nobel prize in 1904 for the discovery of the gas argon.

the eigenvectors $\boldsymbol{x}^{(j)}$, $j = 1, 2, \ldots, n$, *as*

$$\boldsymbol{x} = \sum_{j=1}^{n} \alpha_j \, \boldsymbol{x}^{(j)} \,, \tag{5.41}$$

then

$$R(\boldsymbol{x}) = \frac{\sum_{j=1}^{n} \lambda_j \, \alpha_j^2}{\sum_{j=1}^{n} \alpha_j^2} \,. \tag{5.42}$$

On noting that $\boldsymbol{x}^{(i)\mathrm{T}} \boldsymbol{x}^{(j)}$ is equal to 1 when $i = j$ and to 0 otherwise, (5.42) follows trivially by inserting (5.41) into (5.40).

Theorem 5.12 *Let* $A \in \mathbb{R}^{n \times n}_{\mathrm{sym}}$. *For any vector* $\boldsymbol{x} \in \mathbb{R}^n_*$,

$$\lambda_{\min} \le R(\boldsymbol{x}) \le \lambda_{\max} \,, \tag{5.43}$$

where $\lambda_{min} \in \mathbb{R}$ *and* $\lambda_{max} \in \mathbb{R}$ *are respectively the least and greatest of the eigenvalues of* A. *These bounds are attained when* $\boldsymbol{x}$ *is the corresponding eigenvector.*

Proof The inequalities follow immediately from (5.42) by noting that $\lambda_{\min} \le \lambda_j \le \lambda_{\max}$, $j = 1, 2, \ldots, n$. □

Theorem 5.13 *Suppose that* $\boldsymbol{x} \in \mathbb{R}^n_*$ *is a normalised vector, that is,* $\|\boldsymbol{x}\|_2 = 1$. *Assume, further, that* $\boldsymbol{x}^{(k)} \in \mathbb{R}^n_*$ *is the* k*th normalised eigenvector of* $A \in \mathbb{R}^{n \times n}$, *and that*

$$\|\boldsymbol{x} - \boldsymbol{x}^{(k)}\|_2 = \mathcal{O}(\varepsilon)$$

for a small $\varepsilon \in \mathbb{R}$. *Then,*

$$R(\boldsymbol{x}) = \lambda_k + \mathcal{O}(\varepsilon^2) \,.$$

Proof It follows from (5.41) that $\boldsymbol{x}^{\mathrm{T}} \boldsymbol{x}^{(k)} = \alpha_k$, and therefore,

$$\begin{aligned} \|\boldsymbol{x} - \boldsymbol{x}^{(k)}\|_2^2 &= (\boldsymbol{x} - \boldsymbol{x}^{(k)})^{\mathrm{T}} (\boldsymbol{x} - \boldsymbol{x}^{(k)}) \\ &= \|\boldsymbol{x}\|_2^2 - 2\boldsymbol{x}^{\mathrm{T}} \boldsymbol{x}^{(k)} + \|\boldsymbol{x}^{(k)}\|_2^2 \\ &= 2(1 - \alpha_k). \end{aligned}$$

Hence, $\alpha_k = 1 + \mathcal{O}(\varepsilon^2)$. Further,

$$1 = \|\boldsymbol{x}\|_2^2 = \sum_{j=1}^{n} \alpha_j^2$$

$$
\begin{aligned}
&= \alpha_k^2 + \sum_{j\neq k} \alpha_j^2 \\
&= 1 + \mathcal{O}(\varepsilon^2) + \sum_{j\neq k} \alpha_j^2 .
\end{aligned}
$$

Consequently, $\alpha_j = \mathcal{O}(\varepsilon)$ for all $j \neq k$. The result then follows from (5.42) which (with $\sum_{j=1}^{n} \alpha_j^2 = \|\boldsymbol{x}\|_2^2 = 1$) yields that

$$
\begin{aligned}
R(\boldsymbol{x}) &= \lambda_k \alpha_k^2 + \sum_{j\neq k} \lambda_j \alpha_j^2 \\
&= \lambda_k + \mathcal{O}(\varepsilon^2) .
\end{aligned}
$$

□

This important result means that if we have a fairly close approximation $\boldsymbol{x}$ to an eigenvector of A, then the Rayleigh quotient $R(\boldsymbol{x})$ gives very easily a much more accurate approximation to the corresponding eigenvalue.

5.10 Perturbation analysis

It is often necessary to have an estimate of how much the eigenvalues and eigenvectors of a matrix are affected by changes in the elements. Such perturbations may arise, for example, when the matrix elements are obtained by physical measurements which are inexact, or they might result from finite difference approximations of a differential equation, as will be seen in Chapter 13. The last two theorems in this chapter address some of these questions. We begin with the following preliminary result.

Theorem 5.14 *Let $M \in \mathbb{R}^{n\times n}_{\text{sym}}$, with eigenvalues λ_i and corresponding orthonormal eigenvectors $\boldsymbol{v}_i, i = 1, 2, \ldots, n$, and suppose that $\boldsymbol{u} \neq \mathbf{0}$ and $\boldsymbol{w}$ are vectors in $\mathbb{R}^n$ and μ is a real number such that*

$$
(M - \mu I)\boldsymbol{u} = \boldsymbol{w} . \tag{5.44}
$$

Then, at least one eigenvalue λ_j of M satisfies

$$
|\lambda_j - \mu| \leq \|\boldsymbol{w}\|_2 / \|\boldsymbol{u}\|_2 .
$$

Proof If μ is equal to one of the eigenvalues the proof is trivial, so we shall assume that $\mu \neq \lambda_k$, $k = 1, 2, \ldots, n$. We write the vectors $\boldsymbol{u}$ and

$\boldsymbol{w}$ as linear combinations of the eigenvectors of M, so that

$$\boldsymbol{u} = \sum_{k=1}^{n} \alpha_k \boldsymbol{v}_k\,, \qquad \boldsymbol{w} = \sum_{k=1}^{n} \beta_k \boldsymbol{v}_k\,.$$

Substituting in (5.44), we may equate coefficients of the linearly independent vectors $\boldsymbol{v}_k$, $k = 1, 2, \ldots, n$, to deduce that

$$(\lambda_k - \mu)\alpha_k = \beta_k\,, \qquad k = 1, 2, \ldots, n\,.$$

Now suppose that λ_j is the eigenvalue which is closest to μ; this means that

$$|\lambda_j - \mu| \le |\lambda_k - \mu|\,, \qquad k = 1, 2, \ldots, n\,.$$

Since the eigenvectors $\boldsymbol{v}_i$, $i = 1, 2, \ldots, n$, are orthonormal in $\mathbb{R}^n$, we have

$$\sum_{k=1}^{n} \alpha_k^2 = \|\boldsymbol{u}\|_2^2\,, \qquad \sum_{k=1}^{n} \beta_k^2 = \|\boldsymbol{w}\|_2^2\,.$$

Hence

$$\sum_{k=1}^{n} \frac{\beta_k^2}{(\lambda_k - \mu)^2} = \|\boldsymbol{u}\|_2^2\,,$$

which gives

$$\|\boldsymbol{w}\|_2^2 = \sum_{k=1}^{n} \beta_k^2 \ge (\lambda_j - \mu)^2 \sum_{k=1}^{n} \frac{\beta_k^2}{(\lambda_k - \mu)^2} = (\lambda_j - \mu)^2 \|\boldsymbol{u}\|_2^2\,,$$

as required. □

We shall now use this result to show that in the case of a symmetric matrix A, small symmetric perturbations of A lead to small changes in the eigenvalues of A.

Theorem 5.15 (Bauer–Fike Theorem (symmetric case)) *Suppose that $A, E \in \mathbb{R}^{n\times n}_{\text{sym}}$ and $B = A - E$. Assume, further, that the eigenvalues of A are denoted by $\lambda_j, j = 1, 2, \ldots, n$, and μ is an eigenvalue of B. Then, at least one eigenvalue λ_j of A satisfies*

$$|\lambda_j - \mu| \le \|E\|_2\,.$$

Proof This is a straightforward consequence of the previous theorem. Suppose that $\boldsymbol{u}$ is the normalised eigenvector of B corresponding to the eigenvalue μ, so that $B\boldsymbol{u} = \mu\boldsymbol{u}$. Then,

$$(A - \mu I)\boldsymbol{u} = (B + E - \mu I)\boldsymbol{u} = E\boldsymbol{u}\,.$$

It then follows from Theorem 5.14 that there is an eigenvalue λ_j of A such that

$$|\lambda_j - \mu| \le \|E\boldsymbol{u}\|_2 \le \|E\|_2 \, \|\boldsymbol{u}\|_2 = \|E\|_2 \,,$$

as required. □

Example 5.11 *Consider the* 3×3 *Hilbert matrix*

$$A = \begin{pmatrix} 1 & 1/2 & 1/3 \\ 1/2 & 1/3 & 1/4 \\ 1/3 & 1/4 & 1/5 \end{pmatrix}$$

and its perturbation

$$B = \begin{pmatrix} 1.0000 & 0.5000 & 0.3333 \\ 0.5000 & 0.3333 & 0.2500 \\ 0.3333 & 0.2500 & 0.2000 \end{pmatrix}$$

which results by rounding each entry of A *to four decimal digits.*

In this case, $E = A - B$ and $\|E\|_2 = 3.3 \times 10^{-5}$. Let μ be an eigenvalue of B; then, according Theorem 5.15, at least one of the eigenvalues λ_1, λ_2, λ_3 of the matrix A satisfies the inequality

$$|\lambda_j - \mu| \le 3.3 \times 10^{-5} \,. \tag{5.45}$$

Indeed, the true eigenvalues of A and B are, respectively,

$$\lambda_1 = 0.002687338072 \,, \quad \lambda_2 = 0.1223270673 \,, \quad \lambda_3 = 1.408318925 \,,$$

and

$$\mu_1 = 0.002664493933 \,, \quad \mu_2 = 0.1223414532 \,, \quad \mu_3 = 1.408294053 \,.$$

Therefore,

$$\lambda_1 - \mu_1 = 2.29 \times 10^{-5}, \quad \lambda_2 - \mu_2 = -1.44 \times 10^{-5}, \quad \lambda_3 - \mu_3 = 2.49 \times 10^{-5},$$

which is in agreement with (5.45). ◇

5.11 Notes

Theorem 5.15 is a special case of the following general result, known as the Bauer–Fike Theorem.[1]

[1] F.L. Bauer and C.T. Fike, Norms and exclusion theorems, *Num. Math.* **2**, 137–141, 1960.

Theorem 5.16 *Assume that $A \in \mathbb{C}^{n\times n}$ is diagonalisable; i.e., there exists a nonsingular matrix $X \in \mathbb{C}^{n\times n}$ such that $X^{-1}AX = \Lambda$, where Λ is a diagonal matrix whose diagonal entries λ_j, $j = 1, \dots, n$, are the eigenvalues of A. Suppose further that $E \in \mathbb{C}^{n\times n}$, $B = A - E$, and μ is an eigenvalue of B. Then, at least one eigenvalue λ_j of A satisfies*

$$|\lambda_j - \mu| \le \kappa_2(X)\|E\|_2 ,$$

where $\kappa_2(X) = \|X\|_2\,\|X^{-1}\|_2$ is the condition number of the matrix X in the matrix 2-norm $\|\cdot\|_2$ on $\mathbb{C}^{n\times n}$.

In the special case when A, $E \in \mathbb{R}^{n\times n}_{\text{sym}}$, the matrix X can be chosen to be orthogonal; *i.e.*, $X^{-1} = X^{\mathrm{T}}$. Therefore, $\|X\|_2 = \|X^{-1}\|_2 = 1$, and hence $\kappa_2(X) = 1$, in accordance with the inequality stated in Theorem 5.15. Theorems 5.15 and 5.16 estimate how far the eigenvalues of A are perturbed by changes in the elements of A. The question as to how large the changes in the eigenvectors may be is more difficult; it is discussed in detail in

- J.H. Wilkinson, *The Algebraic Eigenvalue Problem,* Clarendon Press, Oxford University Press, New York, 1988.

Chapter 8 of Wilkinson's book outlines the convergence proof of the QR iteration, while the convergence of Jacobi's method is covered in Chapter 5 of that book. For further details, see also Chapter 9 of

- B. Parlett, *The Symmetric Eigenvalue Problem*, Prentice–Hall, Englewood Cliffs, NJ, 1980.

Exercises

5.1 Give a proof of Lemma 5.3.

5.2 Use Householder matrices to transform the matrix

$$A = \begin{pmatrix} 2 & 1 & 2 & 2 \\ 1 & -7 & 6 & 5 \\ 2 & 6 & 2 & -5 \\ 2 & 5 & -5 & 1 \end{pmatrix}$$

to tridiagonal form.

5.3 Use Sturm sequences to show that no eigenvalue of the matrix

$$A = \begin{pmatrix} 3 & 1 & 0 & 0 \\ 1 & 2 & -2 & 0 \\ 0 & -2 & 4 & \alpha \\ 0 & 0 & \alpha & 1 \end{pmatrix}$$

lies in the interval $(0, 1)$ if $5\alpha^2 > 8$, and that exactly one eigenvalue of A lies in this interval if $5\alpha^2 < 8$.

5.4 Given any two nonzero vectors $\boldsymbol{x}$ and $\boldsymbol{y}$ in $\mathbb{R}^n_*$, construct a Householder matrix H such that $H\boldsymbol{x}$ is a scalar multiple of $\boldsymbol{y}$; note that if $H\boldsymbol{x} = c\boldsymbol{y}$, then $c^2 = \boldsymbol{x}^{\mathrm{T}}\boldsymbol{x}/\boldsymbol{y}^{\mathrm{T}}\boldsymbol{y}$. Is the matrix unique?

5.5 Suppose that the matrix $D \in \mathbb{R}^{n\times n}$ is diagonal with distinct diagonal elements $d_{11}, \ldots, d_{nn}$. Let $A \in \mathbb{R}^{n\times n}_{\mathrm{sym}}$, with $|a_{ij}| \leq 1$ for all $i, j \in \{1, 2, \ldots, n\}$, and assume that $\varepsilon \in \mathbb{R}$ is so small that ε^2 can be neglected, and that the matrix $D + \varepsilon A$ has eigenvalue $\lambda + \varepsilon\mu$ and corresponding eigenvector $\mathbf{e} + \varepsilon\mathbf{u}$. Show that $\lambda = d_{jj}$ for some $j \in \{1, 2, \ldots, n\}$ and that $\mu = a_{jj}$. Write down the elements of $\mathbf{e}$, and show that

$$u_i = -\frac{a_{ij}}{d_{ii} - \lambda}, \qquad i \neq j\,.$$

Explain why the requirement that eigenvectors should be normalised implies that $u_j = 0$.

5.6 With the same notation as in Exercise 5, suppose now that $d_{11} = d_{22} = \cdots = d_{kk}$, that $d_{kk}, d_{k+1,k+1}, \ldots, d_{nn}$ are distinct, and that ε^3 can be neglected. Writing the matrices and the eigenvector in partitioned form, so that

$$\begin{pmatrix} d_{11}I_k + \varepsilon A_1 & \varepsilon A_2 \\ \varepsilon A_2^{\mathrm{T}} & D_{n-k} + \varepsilon A_3 \end{pmatrix} \begin{pmatrix} \mathbf{e} + \varepsilon\boldsymbol{u} + \varepsilon^2\boldsymbol{x} \\ \mathbf{f} + \varepsilon\boldsymbol{v} + \varepsilon^2\boldsymbol{y} \end{pmatrix}$$

$$= (\lambda + \varepsilon\mu + \varepsilon^2\nu) \begin{pmatrix} \mathbf{e} + \varepsilon\boldsymbol{u} + \varepsilon^2\boldsymbol{x} \\ \mathbf{f} + \varepsilon\boldsymbol{v} + \varepsilon^2\boldsymbol{y} \end{pmatrix},$$

show that $\lambda = d_{11}$, $\mathbf{f} = \mathbf{0}$, and that μ is an eigenvalue of A_1 with corresponding eigenvector $\mathbf{e}$. Show how $\boldsymbol{v}$ is obtained from the solution of $(D_{n-k} - \lambda I)\boldsymbol{v} = -A_2^{\mathrm{T}}\mathbf{e}$, and that

$$(A_1 - \mu)\boldsymbol{u} = \nu\mathbf{e} - A_2\mathbf{v}\,.$$

Explain how the vector $\boldsymbol{u}$ can be obtained in terms of the eigenvectors and eigenvalues of the matrix A_1, assuming that these eigenvalues are distinct.

5.7 Suppose that $A \in \mathbb{R}^{n\times n}_{\text{sym}}$ is tridiagonal, that $A - \mu I = QR$ and $B = RQ + \mu I$, where $\mu \in \mathbb{R}$, $Q \in \mathbb{R}^{n\times n}$ is a product of plane rotations and $R \in \mathbb{R}^{n\times n}$ is upper triangular and tridiagonal. Show that B can be written as an orthogonal transformation of A, and that B is symmetric. Show also that the only nonzero elements in the matrix B which are below the diagonal lie immediately below the diagonal; deduce that B is tridiagonal.

5.8 Perform one step of the QR algorithm, using the shift $\mu = a_{nn}$, for the matrix

$$A = \begin{pmatrix} 0 & 1 \\ 1 & 0 \end{pmatrix}.$$

Show that the QR algorithm does not converge for this matrix. (This is a special case in which a different shift must be used.)

5.9 Perform one step of the QR algorithm, using the shift $\mu = a_{nn}$, for the matrix

$$A = \begin{pmatrix} 13 & 4 \\ 4 & 10 \end{pmatrix}.$$

5.10 Carry out two steps of inverse iteration for the matrix

$$A = \begin{pmatrix} 2 & 2 \\ 2 & 5 \end{pmatrix},$$

using the eigenvalue estimate $\vartheta = 5$ and the initial vector

$$\boldsymbol{v}^{(0)} = \begin{pmatrix} 1 \\ 1 \end{pmatrix}.$$

Verify that the elements of the vector $\boldsymbol{v}^{(2)}$ agree with those of the true eigenvector with an accuracy of about 5%. Evaluate the Rayleigh quotient using the vector $\boldsymbol{v}^{(2)}$, and verify that the result agrees with the true eigenvalue to about 1 in 3000.

5.11 An eigenvalue and eigenvector of the matrix A may be evaluated by solving the system of nonlinear equations

$$\begin{aligned} (A - \lambda I)\boldsymbol{x} &= \boldsymbol{0}, \\ \boldsymbol{x}^{\mathrm{T}}\boldsymbol{x} &= 1 \end{aligned}$$

for the unknowns λ and $\boldsymbol{x}$. Using Newton's method, starting from estimates $\lambda^{(0)}$ and $\boldsymbol{x}^{(0)}$, show that the next iteration is determined by

$$\begin{aligned} A\,\boldsymbol{\delta x} - \delta\lambda\,\boldsymbol{x}^{(0)} &= -(A - \lambda^{(0)}I)\boldsymbol{x}^{(0)}\,, \\ -\boldsymbol{x}^{(0)\mathrm{T}}\boldsymbol{\delta x} &= \tfrac{1}{2}(\boldsymbol{x}^{(0)\mathrm{T}}\boldsymbol{x}^{(0)} - 1) \end{aligned}$$

and $\boldsymbol{x}^{(1)} = \boldsymbol{x}^{(0)} + \boldsymbol{\delta x}$, $\lambda^{(1)} = \lambda^{(0)} + \delta\lambda$. Comment on the difference between this method and the method of inverse iteration in Section 5.8.

5.12 Suppose that $A \in \mathbb{R}^{n\times n}_{\mathrm{sym}}$ and that Jacobi's method has produced an orthogonal matrix R and a symmetric matrix B such that $B = R^{\mathrm{T}}AR$. Suppose also that $|b_{ij}| < \varepsilon$ for all $i \neq j$. Show that, for each $j = 1, 2, \ldots, n$, there is at least one eigenvalue λ of A such that

$$|\lambda - b_{jj}| < \varepsilon\sqrt{n}\,.$$

5.13 Suppose that $A \in \mathbb{R}^{n\times n}_{\mathrm{sym}}$ and that the Householder reduction and QR algorithm have produced an orthogonal matrix Q and a tridiagonal matrix T such that $T = Q^{\mathrm{T}}AQ$. Suppose also that $|t_{n,n-1}| < \varepsilon$. Show that there is at least one eigenvalue λ of A such that

$$|\lambda - t_{nn}| < \varepsilon\,.$$

6

Polynomial interpolation

6.1 Introduction

It is time to take a break from solving equations. In this chapter we consider the problem of polynomial interpolation; it involves finding a polynomial that agrees exactly with some information that we have about a real-valued function f of a single real variable x. This information may be in the form of values $f(x_0), \dots, f(x_n)$ of the function f at some finite set of points $\{x_0, \dots, x_n\}$ on the real line, and the corresponding polynomial is then called the **Lagrange interpolation polynomial**[1] or, provided that f is differentiable, it may include values of the derivative of f at these points, in which case the associated polynomial is referred to as a **Hermite interpolation polynomial**.[2]

Why should we be interested in constructing Lagrange or Hermite interpolation polynomials? If the function values $f(x)$ are known for all x in a closed interval of the real line, then the aim of polynomial

[1] Joseph-Louis Lagrange (25 January 1736, Turin, Sardinia–Piedmont (now in Italy) – 10 April 1813, Paris, France) made fundamental contributions to the calculus of variations. He succeeded Euler as Director of Mathematics at the Berlin Academy of Sciences in 1766. During his stay in Berlin Lagrange worked on astronomy, the stability of the solar system, mechanics, dynamics, fluid mechanics, probability, number theory, and the foundations of calculus. In 1787 he moved to Paris and became a member of the Académie des Sciences. Napoleon named Lagrange to the Legion of Honour and as a Count of the Empire in 1808, and on 3 April 1813, a week before his death, he received the Grand Croix of the Ordre Impérial de la Réunion.

[2] Charles Hermite (24 December 1822, Dieuze, Lorraine, France – 14 January 1901, Paris, France). Hermite did not enjoy formal examinations and had to spend five years to complete his undergraduate degree. He contributed to the theory of elliptic functions and their application to the general polynomial equation of the fifth degree. In 1873 he published the first proof that e is a transcendental number. Using methods similar to those of Hermite, Lindemann established in 1882 that π was also transcendental. A number of mathematical entities bear Hermite's name: Hermite orthogonal polynomials, Hermite's differential equation, Hermite's formula of interpolation and Hermitian matrices.

interpolation is to approximate the function f by a polynomial over this interval. Given that any polynomial can be completely specified by its (finitely many) coefficients, storing the interpolation polynomial for f in a computer will be, generally, more economical than storing f itself.

Frequently, it is the case, though, that the function values $f(x)$ are only known at a finite set of points $x_0, \dots, x_n$, perhaps as the results of some measurements. The aim of polynomial interpolation is then to attempt to reconstruct the unknown function f by seeking a polynomial p_n whose graph in the (x, y)-plane passes through the points with coordinates $(x_i, f(x_i))$, $i = 0, \dots, n$. Of course, in general, the resulting polynomial p_n will differ from f (unless f itself is a polynomial of the same degree as p_n), so an error will be incurred. In this chapter we shall also establish results which provide bounds on the size of this error.

6.2 Lagrange interpolation

Given that n is a nonnegative integer, let $\mathcal{P}_n$ denote the set of all (real-valued) polynomials of degree $\leq n$ defined over the set $\mathbb{R}$ of real numbers. The simplest interpolation problem can be stated as follows: given x_0 and y_0 in $\mathbb{R}$, find a polynomial $p_0 \in \mathcal{P}_0$ such that $p_0(x_0) = y_0$. The solution to this is, trivially, $p_0(x) \equiv y_0$. The purpose of this section is to explore the following more general problem.

> Let $n \geq 1$, and suppose that $x_i, i = 0, 1, \dots, n$, are *distinct* real numbers (*i.e.*, $x_i \neq x_j$ for $i \neq j$) and $y_i, i = 0, 1, \dots, n$, are real numbers; we wish to find $p_n \in \mathcal{P}_n$ such that $p_n(x_i) = y_i$, $i = 0, 1, \dots, n$.

To prove that this problem has a unique solution, we begin with a useful lemma.

Lemma 6.1 *Suppose that $n \geq 1$. There exist polynomials $L_k \in \mathcal{P}_n$, $k = 0, 1, \dots, n$, such that*

$$L_k(x_i) = \begin{cases} 1, & i = k, \\ 0, & i \neq k, \end{cases} \tag{6.1}$$

for all $i, k = 0, 1, \dots, n$. Moreover,

$$p_n(x) = \sum_{k=0}^{n} L_k(x) y_k \tag{6.2}$$

satisfies the above interpolation conditions; in other words, $p_n \in \mathcal{P}_n$ and $p_n(x_i) = y_i$, $i = 0, 1, \dots, n$.

Proof For each fixed k, $0 \le k \le n$, L_k is required to have n zeros: x_i, $i = 0, 1, \ldots, n$, $i \ne k$; thus, $L_k(x)$ is of the form

$$L_k(x) = C_k \prod_{\substack{i=0\\ i\ne k}}^{n} (x - x_i)\,, \tag{6.3}$$

where $C_k \in \mathbb{R}$ is a constant to be determined. It is easy to find the value of C_k by recalling that $L_k(x_k) = 1$; using this in (6.3) yields

$$C_k = \prod_{\substack{i=0\\ i\ne k}}^{n} \frac{1}{x_k - x_i}\,.$$

On inserting this expression for C_k into (6.3) we get

$$L_k(x) = \prod_{\substack{i=0\\ i\ne k}}^{n} \frac{x - x_i}{x_k - x_i}\,. \tag{6.4}$$

As the function p_n defined by (6.2) is a linear combination of the polynomials $L_k \in \mathcal{P}_n$, $k = 0, 1, \ldots, n$, also $p_n \in \mathcal{P}_n$. Finally, $p_n(x_i) = y_i$ for $i = 0, 1, \ldots, n$ is a trivial consequence of using (6.1) in (6.2). □

Remark 6.1 *Although the statement of Lemma 6.1 required that $n \ge 1$, the trivial case of $n = 0$ mentioned at the beginning of the section can also be included by defining, for $n = 0$, $L_0(x) \equiv 1$, and observing that the function p_0 defined by*

$$p_0(x) = L_0(x) y_0 \; (\equiv y_0\,)$$

is the unique polynomial in $\mathcal{P}_0$ that satisfies $p_0(x_0) = y_0$.

We note that, implicitly, the polynomials L_k, $k = 0, 1, \ldots, n$, depend on the polynomial degree n, $n \ge 0$. To highlight this fact, a more accurate but cumbersome notation would have involved writing, for example, $L_k^n(x)$ instead of $L_k(x)$; this would have made it clear that $L_k^n(x)$ differs from $L_k^m(x)$ when the polynomial degrees n and m differ. For the sake of notational simplicity, we have chosen to write $L_k(x)$; the implied value of n will always be clear from the context.

Theorem 6.1 (Lagrange's Interpolation Theorem) *Assume that $n \ge 0$. Let x_i, $i = 0, \ldots, n$, be* distinct *real numbers and suppose that $y_i, i = 0, \ldots, n$, are real numbers. Then, there exists a unique polynomial $p_n \in \mathcal{P}_n$ such that*

$$p_n(x_i) = y_i\,, \quad i = 0, \ldots, n\,. \tag{6.5}$$

Proof In view of Remark 6.1, for $n = 0$ the proof is trivial. Let us therefore suppose that $n \geq 1$. It follows immediately from Lemma 6.1 that the polynomial $p_n \in \mathcal{P}_n$ defined by

$$p_n(x) = \sum_{k=0}^{n} L_k(x) y_k$$

satisfies the conditions (6.5), thus showing the *existence* of the required polynomial. It remains to show that p_n is the *unique* polynomial in $\mathcal{P}_n$ satisfying the interpolation property

$$p_n(x_i) = y_i\,, \qquad i = 0, 1, \ldots, n\,.$$

Suppose, otherwise, that there exists $q_n \in \mathcal{P}_n$, different from p_n, such that $q_n(x_i) = y_i$, $i = 0, 1, \ldots, n$. Then, $p_n - q_n \in \mathcal{P}_n$ and $p_n - q_n$ has $n + 1$ distinct roots, x_i, $i = 0, 1, \ldots, n$; since a polynomial of degree n cannot have more than n distinct roots, unless it is identically 0, it follows that

$$p_n(x) - q_n(x) \equiv 0\,,$$

which contradicts our assumption that p_n and q_n are distinct. Hence, there exists only one polynomial $p_n \in \mathcal{P}_n$ which satisfies (6.5). □

Definition 6.1 *Suppose that $n \geq 0$. Let x_i, $i = 0, \ldots, n$, be distinct real numbers, and y_i, $i = 0, \ldots, n$, real numbers. The polynomial p_n defined by*

$$\boxed{p_n(x) = \sum_{k=0}^{n} L_k(x) y_k\,,} \tag{6.6}$$

with $L_k(x)$, $k = 0, 1, \ldots, n$, defined by (6.4) when $n \geq 1$, and $L_0(x) \equiv 1$ when $n = 0$, is called the **Lagrange interpolation polynomial** *of degree n for the set of points $\{(x_i, y_i)\colon i = 0, \ldots, n\}$. The numbers x_i, $i = 0, \ldots, n$, are called the* **interpolation points**.

Frequently, the real numbers y_i are given as the values of a real-valued function f, defined on a closed real interval $[a, b]$, at the (distinct) interpolation points $x_i \in [a, b]$, $i = 0, \ldots, n$.

Definition 6.2 *Let $n \geq 0$. Given the real-valued function f, defined and continuous on a closed real interval $[a, b]$, and the (distinct) interpolation points $x_i \in [a, b]$, $i = 0, \ldots, n$, the polynomial p_n defined by*

$$p_n(x) = \sum_{k=0}^{n} L_k(x) f(x_k) \tag{6.7}$$

is the **Lagrange interpolation polynomial of degree** n **with interpolation points** x_i, $i = 0, \dots, n$, **for the function** f.

Example 6.1 *We shall construct the Lagrange interpolation polynomial of degree* 2 *for the function* $f\colon x \mapsto \mathrm{e}^x$ *on the interval* $[-1, 1]$, *with interpolation points* $x_0 = -1$, $x_1 = 0$, $x_2 = 1$.

As $n = 2$, we have that

$$L_0(x) = \frac{(x - x_1)(x - x_2)}{(x_0 - x_1)(x_0 - x_2)} = \tfrac{1}{2}x(x-1)\,.$$

Similarly, $L_1(x) = 1 - x^2$ and $L_2(x) = \frac{1}{2}x(x+1)$. Therefore,

$$p_2(x) = \tfrac{1}{2}x(x-1)\,\mathrm{e}^{-1} + (1 - x^2)\,\mathrm{e}^0 + \tfrac{1}{2}x(x+1)\,\mathrm{e}^1\,.$$

Thus, after some simplification, $p_2(x) = 1 + x \sinh 1 + x^2(\cosh 1 - 1)$. ◇

Although the values of the function f and those of its Lagrange interpolation polynomial coincide at the interpolation points, $f(x)$ may be quite different from $p_n(x)$ when x is *not* an interpolation point. Thus, it is natural to ask just how large the difference $f(x) - p_n(x)$ is when $x \neq x_i$, $i = 0, \dots, n$. Assuming that the function f is sufficiently smooth, an estimate of the size of the **interpolation error** $f(x) - p_n(x)$ is given in the next theorem.

Theorem 6.2 *Suppose that* $n \geq 0$, *and that* f *is a real-valued function, defined and continuous on the closed real interval* $[a, b]$, *such that the derivative of* f *of order* $n + 1$ *exists and is continuous on* $[a, b]$. *Then, given that* $x \in [a, b]$, *there exists* $\xi = \xi(x)$ *in* (a, b) *such that*

$$f(x) - p_n(x) = \frac{f^{(n+1)}(\xi)}{(n+1)!}\pi_{n+1}(x)\,, \tag{6.8}$$

where

$$\pi_{n+1}(x) = (x - x_0) \dots (x - x_n)\,. \tag{6.9}$$

Moreover

$$|f(x) - p_n(x)| \leq \frac{M_{n+1}}{(n+1)!}|\pi_{n+1}(x)|\,, \tag{6.10}$$

where

$$M_{n+1} = \max_{\zeta \in [a,b]} |f^{(n+1)}(\zeta)|\,.$$

Proof When $x = x_i$ for some i, $i = 0, 1, \ldots, n$, both sides of (6.8) are zero, and the equality is trivially satisfied. Suppose then that $x \in [a, b]$ and $x \neq x_i$, $i = 0, 1, \ldots, n$. For such a value of x, let us consider the auxiliary function $t \mapsto \varphi(t)$, defined on the interval $[a, b]$ by

$$\varphi(t) = f(t) - p_n(t) - \frac{f(x) - p_n(x)}{\pi_{n+1}(x)}\pi_{n+1}(t)\,. \tag{6.11}$$

Clearly $\varphi(x_i) = 0$, $i = 0, 1, \ldots, n$, and $\varphi(x) = 0$. Thus, φ vanishes at $n+2$ points which are all distinct in $[a, b]$. Consequently, by Rolle's Theorem, Theorem A.2, $\varphi'(t)$, the first derivative of φ with respect to t, vanishes at $n+1$ points in (a, b), one between each pair of consecutive points at which φ vanishes.

In particular, if $n = 0$, we then deduce the existence of $\xi = \xi(x)$ in the interval (a, b) such that $\varphi'(\xi) = 0$. Since $p_0(x) \equiv f(x_0)$ and $\pi_1(t) = t - x_0$, it follows from (6.11) that

$$0 = \varphi'(\xi) = f'(\xi) - \frac{f(x) - p_0(x)}{\pi_1(x)}\,,$$

and hence (6.8) in the case of $n = 0$.

Now suppose that $n \geq 1$. As $\varphi'(t)$ vanishes at $n+1$ points in (a, b), one between each pair of consecutive points at which φ vanishes, applying Rolle's Theorem again, we see that φ'' vanishes at n distinct points. Our assumptions about f are sufficient to apply Rolle's Theorem $n+1$ times in succession, showing that $\varphi^{(n+1)}$ vanishes at some point $\xi \in (a, b)$, the exact value of ξ being dependent on the value of x. By differentiating $n+1$ times the function φ with respect to t, and noting that p_n is a polynomial of degree n or less, it follows that

$$0 = \varphi^{(n+1)}(\xi) = f^{(n+1)}(\xi) - \frac{f(x) - p_n(x)}{\pi_{n+1}(x)}(n+1)!\,.$$

Hence

$$f(x) - p_n(x) = \frac{f^{(n+1)}(\xi)}{(n+1)!}\pi_{n+1}(x)\,.$$

In order to prove (6.10), we note that as $f^{(n+1)}$ is a continuous function on $[a, b]$ the same is true of $|f^{(n+1)}|$. Therefore, the function $x \mapsto |f^{(n+1)}(x)|$ is bounded on $[a, b]$ and achieves its maximum there; so (6.10) follows from (6.8). □

It is perhaps worth noting that since the location of ξ in the interval $[a,b]$ is unknown (to the extent that the exact dependence of ξ on x is not revealed by the proof of Theorem 6.2), (6.8) is of little practical value; on the other hand, given the function f, an upper bound on the maximum value of $f^{(n+1)}$ over $[a,b]$ is, at least in principle, possible to obtain, and thereby we can provide an upper bound on the size of the interpolation error by means of inequality (6.10).

6.3 Convergence

An important theoretical question is whether or not a sequence (p_n) of interpolation polynomials for a continuous function f converges to f as $n \to \infty$. This question needs to be made more specific, as p_n depends on the distribution of the interpolation points x_j, $j = 0, 1, \ldots, n$, not just on the value of n. Suppose, for example, that we agree to choose equally spaced points, with

$$x_j = a + j(b-a)/n\,, \qquad j = 0, 1, \ldots, n\,, \quad n \geq 1\,.$$

The question of convergence then clearly depends on the behaviour of M_{n+1} as n increases. In particular, if

$$\lim_{n\to\infty} \frac{M_{n+1}}{(n+1)!} \max_{x\in[a,b]} |\pi_{n+1}(x)| = 0\,,$$

then, by (6.10),

$$\lim_{n\to\infty} \max_{x\in[a,b]} |f(x) - p_n(x)| = 0\,, \tag{6.12}$$

and we say that the sequence of interpolation polynomials (p_n), with equally spaced points on $[a,b]$, converges to f as $n \to \infty$, uniformly on the interval $[a,b]$.

You may now think that if all derivatives of f exist and are continuous on $[a,b]$, then (6.12) will hold. Unfortunately, this is not so, since the sequence

$$\left(M_{n+1} \max_{x\in[a,b]} |\pi_{n+1}(x)| \right)$$

may tend to ∞, as $n \to \infty$, faster than the sequence $(1/(n+1)!)$ tends to 0.

In order to convince you of the existence of such ‘pathological’ functions, we consider the sequence of Lagrange interpolation polynomials

Table 6.1. *Runge phenomenon: n denotes the degree of the interpolation polynomial p_n to f, with equally spaced points on* $[-5, 5]$. 'Max error' *signifies* $\max_{x\in[-5,5]} |f(x) - p_n(x)|$.

Degree n	Max error
2	0.65
4	0.44
6	0.61
8	1.04
10	1.92
12	3.66
14	7.15
16	14.25
18	28.74
20	58.59
22	121.02
24	252.78

p_n, $n = 0, 1, 2, \ldots$, with equally spaced interpolation points on the interval $[-5, 5]$, to

$$f(x) = \frac{1}{1 + x^2}, \qquad x \in [-5, 5].$$

This example is due to Runge,[1] and the characteristic behaviour exhibited by the sequence of interpolation polynomials p_n in Table 6.1 is referred to as the **Runge phenomenon**: Table 6.1 shows the maximum difference between $f(x)$ and $p_n(x)$ for $-5 \le x \le 5$, for values of n from 2 up to 24. The numbers indicate clearly that the maximum error increases exponentially as n increases. Figure 6.1 shows the interpolation polynomial p_{10}, using the equally spaced interpolation points $x_j = -5 + j$, $j = 0, 1, \ldots, 10$. The sizes of the local maxima near ± 5 grow exponentially as the degree n increases.

Note that, in many ways, the function f is well behaved; all its deriva-

[1] Carle David Tolmé Runge (30 August 1856, Bremen, Germany – 3 January 1927, Göttingen, Germany) studied mathematics and physics at the University of Munich. His doctoral dissertation in 1880 was in the area of differential geometry. Gradually, his research interests shifted to more applied topics: he devised a numerical procedure for the solution of algebraic equations where the roots were expressed as infinite series of rational functions of the coefficients, and in 1887 he started to work on the wavelengths of the spectral lines of elements. In 1904 Runge became Professor of Applied Mathematics in Göttingen. He was a fit and active man: on his 70th birthday he entertained his grandchildren by performing handstands. A few months later he suffered a fatal heart attack.

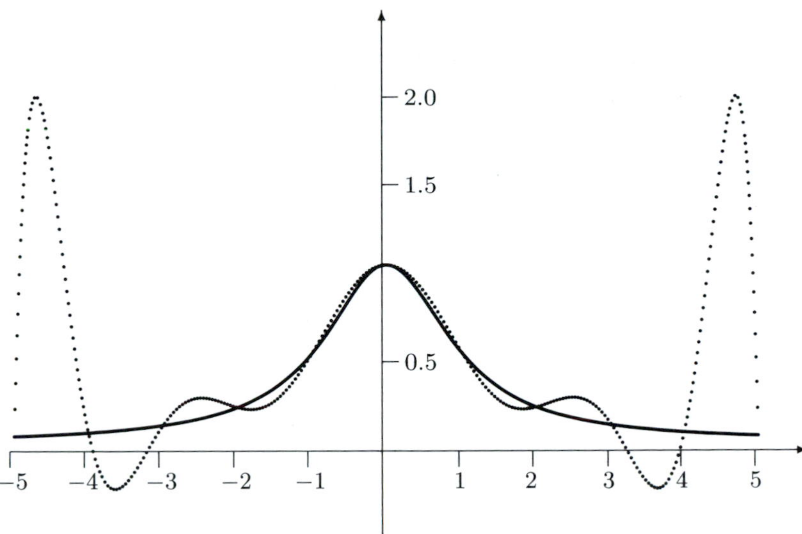

Fig. 6.1. Polynomial interpolation of $f\colon x \mapsto 1/(1+x^2)$ for $x \in [-5, 5]$. The continuous curve is f; the dotted curve is the associated Lagrange interpolation polynomial p_{10} of degree 10, using equally spaced interpolation points.

tives are continuous and bounded for all $x \in [-5, 5]$. The apparent divergence of the sequence of Lagrange interpolation polynomials (p_n) is related to the fact that, when extended to the complex plane, the Taylor series of the complex-valued function $f\colon z \mapsto 1/(1+z^2)$ converges in the open unit disc of radius 1 but not in any disc of larger radius centred at $z = 0$, given that f has poles on the imaginary axis at $z = \pm\imath$. Some further insight into this problem is given in Exercise 11, and a similar difficulty in numerical integration is discussed in Section 7.4.

6.4 Hermite interpolation

The idea of Lagrange interpolation can be generalised in various ways; we shall consider here one simple extension where a polynomial p is required to take given values and derivative values at the interpolation points. Given the distinct interpolation points $x_i, i = 0, \ldots, n$, and two sets of real numbers y_i, $i = 0, \ldots, n$, and z_i, $i = 0, \ldots, n$, with $n \geq 0$, we need to find a polynomial $p_{2n+1} \in \mathcal{P}_{2n+1}$ such that

$$p_{2n+1}(x_i) = y_i\,, \qquad p'_{2n+1}(x_i) = z_i\,, \qquad i = 0, \ldots, n\,.$$

The construction is similar to that of the Lagrange interpolation polynomial, but now requires two sets of polynomials H_k and K_k with $k = 0, \ldots, n$; these will be defined in the proof of the next theorem.

Theorem 6.3 (Hermite Interpolation Theorem) *Let $n \geq 0$, and suppose that x_i, $i = 0, \ldots, n$, are* distinct *real numbers. Then, given two sets of real numbers y_i, $i = 0, \ldots, n$, and z_i, $i = 0, \ldots, n$, there is a unique polynomial p_{2n+1} in $\mathcal{P}_{2n+1}$ such that*

$$p_{2n+1}(x_i) = y_i\,, \qquad p'_{2n+1}(x_i) = z_i\,, \qquad i = 0, \ldots, n\,. \tag{6.13}$$

Proof Let us begin by supposing that $n \geq 1$. As in the case of Lagrange interpolation, we start by constructing a set of auxiliary polynomials; we consider the polynomials H_k and K_k, $k = 0, 1, \ldots, n$, defined by

$$\begin{aligned} H_k(x) &= [L_k(x)]^2(1 - 2L'_k(x_k)(x - x_k))\,, \\ K_k(x) &= [L_k(x)]^2(x - x_k)\,, \end{aligned} \tag{6.14}$$

where

$$L_k(x) = \prod_{\substack{i=0 \\ i \neq k}}^{n} \frac{x - x_i}{x_k - x_i}\,.$$

Clearly H_k and K_k, $k = 0, 1, \ldots, n$, are polynomials of degree $2n+1$. It is easy to see that $H_k(x_i) = K_k(x_i) = 0$, $H'_k(x_i) = K'_k(x_i) = 0$ whenever $i, k \in \{0, 1, \ldots, n\}$ and $i \neq k$; moreover, a straightforward calculation verifies their values when $i = k$, showing that

$$\begin{aligned} H_k(x_i) &= \begin{cases} 1, & i = k\,, \\ 0, & i \neq k\,, \end{cases} \qquad H'_k(x_i) = 0\,, \qquad i, k = 0, 1, \ldots, n\,, \\ K_k(x_i) &= 0\,, \qquad K'_k(x_i) = \begin{cases} 1, & i = k\,, \\ 0, & i \neq k\,, \end{cases} \qquad i, k = 0, 1, \ldots, n\,. \end{aligned}$$

We deduce that

$$p_{2n+1}(x) = \sum_{k=0}^{n} [H_k(x) y_k + K_k(x) z_k]$$

satisfies the conditions (6.13), and p_{2n+1} is clearly an element of $\mathcal{P}_{2n+1}$.

To show that this is the only polynomial in $\mathcal{P}_{2n+1}$ satisfying these conditions, we suppose otherwise; then, there exists a polynomial q_{2n+1} in $\mathcal{P}_{2n+1}$, distinct from p_{2n+1}, such that

$$q_{2n+1}(x_i) = y_i \quad \text{and} \quad q'_{2n+1}(x_i) = z_i\,, \qquad i = 0, 1, \ldots, n\,.$$

Consequently, $p_{2n+1} - q_{2n+1}$ has $n+1$ distinct zeros; therefore, Rolle's Theorem implies that, in addition to the $n+1$ zeros x_i, $i = 0, 1, \ldots, n$, $p'_{2n+1} - q'_{2n+1}$ vanishes at another n points which interlace the x_i. Hence $p'_{2n+1} - q'_{2n+1} \in \mathcal{P}_{2n}$ has $2n+1$ zeros, which means that $p'_{2n+1} - q'_{2n+1}$ is identically zero, so that $p_{2n+1} - q_{2n+1}$ is a constant function. However, $(p_{2n+1} - q_{2n+1})(x_i) = 0$ for $i = 0, 1, \ldots, n$, and hence $p_{2n+1} - q_{2n+1} \equiv 0$, contradicting the hypothesis that p_{2n+1} and q_{2n+1} are distinct. Thus, p_{2n+1} is unique.

When $n = 0$, we define $H_0(x) \equiv 1$ and $K_0(x) \equiv x - x_0$, which correspond to taking $L_0(x) \equiv 1$ in (6.14). Clearly, p_1 defined by

$$p_1(x) = H_0(x)y_0 + K_0(x)z_0 = y_0 + (x - x_0)z_0$$

is the unique polynomial in $\mathcal{P}_1$ such that $p_1(x_0) = y_0$ and $p'_1(x_0) = z_0$. □

Definition 6.3 *Let $n \geq 0$, and suppose that x_i, $i = 0, \ldots, n$, are distinct real numbers and y_i, z_i, $i = 0, \ldots, n$, are real numbers. The polynomial p_{2n+1} defined by*

$$p_{2n+1}(x) = \sum_{k=0}^{n} [H_k(x)y_k + K_k(x)z_k] \qquad (6.15)$$

where $H_k(x)$ and $K_k(x)$ are defined by (6.14), is called the **Hermite interpolation polynomial** *of degree $2n+1$ for the set of values given in $\{(x_i, y_i, z_i)\colon i = 0, \ldots, n\}$.*

Example 6.2 *We shall construct a cubic polynomial p_3 such that*

$$p_3(0) = 0\,, \quad p_3(1) = 1\,, \quad p'_3(0) = 1 \quad \textit{and} \quad p'_3(1) = 0\,.$$

Here $n = 1$, and since $p_3(0) = p'_3(1) = 0$ the polynomial simplifies to

$$p_3(x) = H_1(x) + K_0(x)\,.$$

We easily find that, with $n = 1$, $x_0 = 0$ and $x_1 = 1$,

$$L_0(x) = 1 - x\,, \quad L_1(x) = x\,,$$

and then,

$$\begin{aligned} H_1(x) &= [L_1(x)]^2(1 - 2L'_1(x_1)(x - x_1)) = x^2(3 - 2x)\,, \\ K_0(x) &= [L_0(x)]^2(x - x_0) = (1 - x)^2 x\,. \end{aligned}$$

These yield the required Hermite interpolation polynomial,

$$p_3(x) = -x^3 + x^2 + x\,.$$

◇

Definition 6.4 *Suppose that f is a real-valued function, defined on the closed interval $[a,b]$ of $\mathbb{R}$, and that f is continuous and differentiable on this interval. Suppose, further, that $n \geq 0$ and that x_i, $i = 0,\ldots,n$, are distinct points in $[a,b]$. Then, the polynomial p_{2n+1} defined by*

$$p_{2n+1}(x) = \sum_{k=0}^{n} [H_k(x) f(x_k) + K_k(x) f'(x_k)] \tag{6.16}$$

is the **Hermite interpolation polynomial of degree $2n+1$ with interpolation points x_i, $i = 0,\ldots,n$, for f.** *It satisfies the conditions*

$$p_{2n+1}(x_i) = f(x_i)\,, \quad p'_{2n+1}(x_i) = f'(x_i)\,, \quad i = 0,\ldots,n\,.$$

Pictorially, the graph of p_{2n+1} touches the graph of the function f at the points x_i, $i = 0,\ldots,n$.

To conclude this section we state a result, analogous to Theorem 6.2, concerning the error in Hermite interpolation.

Theorem 6.4 *Suppose that $n \geq 0$ and let f be a real-valued function, defined, continuous and $2n+2$ times differentiable on the interval $[a,b]$, such that $f^{(2n+2)}$ is continuous on $[a,b]$. Further, let p_{2n+1} denote the Hermite interpolation polynomial of f defined by (6.16). Then, for each $x \in [a,b]$ there exists $\xi = \xi(x)$ in (a,b) such that*

$$f(x) - p_{2n+1}(x) = \frac{f^{(2n+2)}(\xi)}{(2n+2)!}[\pi_{n+1}(x)]^2\,, \tag{6.17}$$

where π_{n+1} is as defined in (6.9). Moreover,

$$|f(x) - p_{2n+1}(x)| \leq \frac{M_{2n+2}}{(2n+2)!}[\pi_{n+1}(x)]^2\,, \tag{6.18}$$

where $M_{2n+2} = \max_{\zeta\in[a,b]} |f^{(2n+2)}(\zeta)|$.

Proof The inequality (6.18) is a straightforward consequence of (6.17). In order to prove (6.17), we observe that it is trivially true if $x = x_i$

for some i, $i = 0, \dots, n$; thus, it suffices to consider $x \in [a, b]$ such that $x \neq x_i$, $i = 0, \dots, n$. For such x, let us define the function $t \mapsto \psi(t)$ by

$$\psi(t) = f(t) - p_{2n+1}(t) - \frac{f(x) - p_{2n+1}(x)}{[\pi_{n+1}(x)]^2}[\pi_{n+1}(t)]^2 .$$

Then, $\psi(x_i) = 0$ for $i = 0, \dots, n$, and also $\psi(x) = 0$. Hence, by Rolle's Theorem, $\psi'(t)$ vanishes at $n + 1$ points which lie strictly between each pair of consecutive points from the set $\{x_0, \dots, x_n, x\}$. Also $\psi'(x_i) = 0$, $i = 0, \dots, n$; hence ψ' vanishes at a total of $2n + 2$ distinct points in $[a, b]$. Applying Rolle's Theorem repeatedly, we find eventually that $\psi^{(2n+2)}$ vanishes at some point ξ in (a, b), the location of ξ being dependent on the position of x. This gives the required result on computing $\psi^{(2n+2)}(t)$ from the definition of ψ above and noting that $\psi^{(2n+2)}(\xi) = 0$ and $p_{2n+1}^{(2n+2)}(t) \equiv 0$. □

6.5 Differentiation

From the Lagrange interpolation polynomial p_n, defined by (6.7), which is an approximation to f, it is easy to obtain the polynomial p_n', which is an approximation to the derivative f'. The polynomial p_n' is given by

$$p_n'(x) = \sum_{k=0}^{n} L_k'(x)\, f(x_k) , \qquad n \geq 1 . \tag{6.19}$$

The degree of the polynomial p_n' is clearly at most $n - 1$; p_n' is a linear combination of the derivatives of the polynomials $L_k \in \mathcal{P}_n$, the coefficients being the values of f at the interpolation points x_k, $k = 0, 1, \dots, n$.

In order to find an expression for the difference between $f'(x)$ and the approximation $p_n'(x)$, we might simply differentiate (6.8) to give

$$f'(x) - p_n'(x) = \frac{\mathrm{d}}{\mathrm{d}x} \left(\frac{f^{(n+1)}(\xi(x))}{(n+1)!} \, \pi_{n+1}(x) \right) .$$

However, the result is not helpful: on application of the chain rule, the right-hand side involves the derivative $\mathrm{d}\xi/\mathrm{d}x$; the value of ξ depends on x, but not in any simple manner. In fact, it is not *a priori* clear that the function $x \mapsto \xi(x)$ is continuous, let alone differentiable. An alternative approach is given by the following theorem.

Theorem 6.5 *Let $n \geq 1$, and suppose that f is a real-valued function defined and continuous on the closed real interval $[a, b]$, such that the derivative of order $n+1$ of f is continuous on $[a, b]$. Suppose further that*

x_i, $i = 0, 1, \ldots, n$, are distinct points in $[a, b]$, and that $p_n \in \mathcal{P}_n$ is the Lagrange interpolation polynomial for f defined by these points. Then, there exist distinct points η_i, $i = 1, \ldots, n$, in (a, b), and corresponding to each x in $[a, b]$ there exists a point $\xi = \xi(x)$ in (a, b), such that

$$f'(x) - p_n'(x) = \frac{f^{(n+1)}(\xi)}{n!} \pi_n^*(x) , \tag{6.20}$$

where

$$\pi_n^*(x) = (x - \eta_1) \ldots (x - \eta_n) .$$

Proof Since $f(x_i) - p_n(x_i) = 0$, $i = 0, 1, \ldots, n$, there exists a point η_i in (x_{i-1}, x_i) at which $f'(\eta_i) - p_n'(\eta_i) = 0$, for each $i = 1, \ldots, n$. This defines the points η_i, $i = 1, \ldots, n$. Now the proof closely follows that of Theorem 6.2.

When $x = \eta_i$ for some $i \in \{1, \ldots, n\}$, both sides of (6.20) are zero. Suppose then that x is distinct from all the η_i, $i = 1, \ldots, n$, and define the function $t \mapsto \chi(t)$ by

$$\chi(t) = f'(t) - p_n'(t) - \frac{f'(x) - p_n'(x)}{\pi_n^*(x)} \pi_n^*(t) .$$

This function vanishes at every point η_i, $i = 1, \ldots, n$, and also at the point $t = x$. By successively applying Rolle's Theorem we deduce that $\chi^{(n)}$ vanishes at some point ξ. The result then follows as in the proof of Theorem 6.2. □

Corollary 6.1 *Under the conditions of Theorem 6.5,*

$$|f'(x) - p_n'(x)| \leq \frac{M_{n+1}}{n!} |\pi_n^*(x)| \leq \frac{(b-a)^n M_{n+1}}{n!}$$

for all x in $[a, b]$, where $M_{n+1} = \max_{x \in [a,b]} |f^{(n+1)}(x)|$.

In particular, we deduce that if f and all its derivatives are defined and continuous on the closed interval $[a, b]$, and

$$\lim_{n \to \infty} \frac{(b-a)^n M_{n+1}}{n!} = 0 ,$$

then $\lim_{n \to \infty} \max_{x \in [a,b]} |f'(x) - p_n'(x)| = 0$, showing the convergence of the sequence of interpolation polynomials (p_n') to f', uniformly on $[a, b]$.

The discussion in the last few paragraphs may give the impression that numerical differentiation is a straightforward procedure. In practice, however, things are much more complicated since the function values $f(x_i)$, $i = 0, 1, \ldots, n$, will be polluted by rounding errors.

Example 6.3 *Consider, for example, a real-valued function f that is defined, continuous and differentiable on the closed interval $[-h, h]$ of the real line, where $h > 0$. Suppose that f has been sampled at the points $x_0 = -h$ and $x_1 = h$, and that $f(\pm h)$ are known, but only up to rounding errors $\varepsilon_\pm$, respectively. Consider the Lagrange interpolation polynomial $p_1 \in \mathcal{P}_1$ for f that passes through the points $(-h, f(-h))$ and $(h, f(h))$; clearly,*

$$p_1(x) = \frac{f(h) - f(-h)}{2h}(x + h) + f(-h)\,.$$

Differentiating this with respect to x yields

$$p_1'(x) \equiv \frac{f(h) - f(-h)}{2h}\,.$$

Now, p_1' is a polynomial of degree 0, representing an approximation to $f'(x)$ at any $x \in [-h, h]$, and in particular to $f'(0)$. Unfortunately, in the presence of rounding errors only $f(-h)+\varepsilon_-$ and $f(h)+\varepsilon_+$ are available, with $\varepsilon_\pm$ unknown; thus, we can only calculate

$$\frac{(f(h) + \varepsilon_+) - (f(-h) + \varepsilon_-)}{2h}\,. \tag{6.21}$$

Rewriting this in the form

$$\frac{f(h) - f(-h)}{2h} + \frac{\varepsilon_+ - \varepsilon_-}{2h}\,,$$

we see that even though the first fraction converges to $f'(0)$ as the spacing $2h$ between the interpolation points $-h$ and h tends to 0, for $\varepsilon_+ - \varepsilon_-$ nonzero and fixed the second fraction will tend to infinity as $h \to 0$. Thus, if h is too small in comparison with $|\varepsilon_+ - \varepsilon_-|$, our approximation to $f'(0)$ will be polluted by a large error of size $|\varepsilon_+ - \varepsilon_-|/(2h)$, whereas if h is very large in comparison with $|\varepsilon_+ - \varepsilon_-|$, then $|\varepsilon_+ - \varepsilon_-|/(2h)$ will be small, but $(f(h) - f(-h))/(2h)$ may be a poor approximation to the value $f'(0)$. These observations indicate the existence of an 'optimal' h, depending on the size of the rounding error, for which the error between $f'(0)$ and the approximation (6.21) is smallest. (See Exercise 12 for further details.) ◇

Convergence, as $h \to 0$, of the expression $p_1'(x) \equiv (f(h) - f(-h))/(2h)$ to $f'(0)$ in the last example should not be confused with convergence, as $n \to \infty$, of the sequence of polynomials (p_n') to the function f' discussed just prior to the example. In the former case, the polynomial degree is fixed and the spacing between the two interpolation points, $x_0 = -h$

and $x_1 = h$, tends to 0; in the latter case, the degree of the polynomial p'_n tends to infinity and consequently the spacing between the increasing number of consecutive interpolation points shrinks. Nevertheless, Example 6.3 illustrates the issue that caution should be exercised in the course of numerical differentiation when rounding errors are present.

6.6 Notes

The interpolation polynomial (6.6) was discovered by Edward Waring (1736–1798) in 1776, rediscovered by Euler in 1783 and published by Joseph-Louis Lagrange (1736–1813) in his *Leçons élémentaires sur les mathématiques*, Paris, 1795.

Lagrange's interpolation theorem is a purely algebraic result, and it also holds in number fields different from the field of real numbers considered in this chapter. In particular, it holds if the numbers x_i and y_i, $i = 0, 1, \ldots, n$, are complex, and the polynomial p_n has complex coefficients. Theorem 6.2 is due to Augustin-Louis Cauchy (1789–1857). The interpolation polynomial (6.15) was discovered by Charles Hermite (1822–1901).

Before modern computers came into general use about 1960, the evaluation of a standard mathematical function for a given value of x required the use of published tables of the function, in book form. If x was not one of the tabulated values, the required result was obtained by interpolation, using tabulated values close to x. The tabulated values were given at equally spaced points, so that usually $x_j = jh$, where h is a fixed increment. In this case the Lagrange formula can be simplified; as this sort of interpolation had to be done frequently, various devices were used to make the calculations easy and quick. Older books, such as F.B. Hildebrand's *Introduction to Numerical Analysis*, published in 1956, contain extensive discussions of such special methods of interpolation, some of which date back to the time of Newton, but are now mainly of historical interest. A notable early contribution to the development of mathematical tables is the work of Henry Briggs (1560–1630), Savilian Professor of Geometry and fellow of Merton College in Oxford, entitled *Arithmetica logarithmica*, published in 1624. It contained extensive calculations of the logarithms of thirty thousand numbers to 14 decimal digits; these were the numbers from 1 to 20000 and from 90000 to 100000. It also contained tables of the sin function to 15 decimal digits, and of the tan and sec functions to 10 decimal digits.

Exercises

6.1 Construct the Lagrange interpolation polynomial p_1 of degree 1, for a continuous function f defined on the interval $[-1,1]$, using the interpolation points $x_0 = -1$, $x_1 = 1$. Show further that if the second derivative of f exists and is continuous on $[0,1]$, then

$$|f(x) - p_1(x)| \le \frac{M_2}{2}(1-x^2) \le \frac{M_2}{2}, \quad x \in [-1,1],$$

where $M_2 = \max_{x\in[-1,1]} |f''(x)|$. Give an example of a function f, and a point x, for which equality is achieved.

6.2 (i) Write down the Lagrange interpolation polynomial of degree 1 for the function $f\colon x \mapsto x^3$, using the points $x_0 = 0$, $x_1 = a$. Verify Theorem 6.2 by direct calculation, showing that in this case ξ is unique and has the value $\xi = \frac{1}{3}(x+a)$.

(ii) Repeat the calculation for the function $f\colon x \mapsto (2x-a)^4$; show that in this case there are two possible values for ξ, and give their values.

6.3 Given the distinct points x_i, $i = 0,1,\ldots,n+1$, and the points y_i, $i = 0,1,\ldots,n+1$, let q be the Lagrange polynomial of degree n for the set of points $\{(x_i,y_i)\colon i = 0,1,\ldots,n\}$ and let r be the Lagrange polynomial of degree n for the points $\{(x_i,y_i)\colon i = 1,2,\ldots,n+1\}$. Define

$$p(x) = \frac{(x-x_0)r(x) - (x-x_{n+1})q(x)}{x_{n+1}-x_0}.$$

Show that p is the Lagrange polynomial of degree $n+1$ for the points $\{(x_i,y_i)\colon i = 0,1,\ldots,n+1\}$.

6.4 Let $n \ge 1$. The points x_j are equally spaced in $[-1,1]$, so that

$$x_j = \frac{2j-n}{n}, \quad j = 0,\ldots,n.$$

With the usual notation

$$\pi_{n+1}(x) = (x-x_0)\ldots(x-x_n),$$

show that

$$\pi_{n+1}(1-1/n) = -\frac{(2n)!}{2^n n^{n+1} n!}.$$

Using Stirling's formula

$$N! \sim \sqrt{2\pi} N^{N+1/2} \mathrm{e}^{-N}, \qquad N \to \infty,$$

verify that

$$\pi_{n+1}(1 - 1/n) \sim -\frac{2^{n+1/2}\mathrm{e}^{-n}}{n}$$

for large values of n.

6.5 Let $n \geq 1$. Suppose that x_i, $i = 0, 1, \ldots, n$, are distinct real numbers, and y_i, u_i, $i = 0, 1, \ldots, n$, are real numbers. Suppose, further, that there exists $p_{2n+1} \in \mathcal{P}_{2n+1}$ such that $p_{2n+1}(x_i) = y_i$ for all $i = 0, 1, \ldots, n$, and $p''_{2n+1}(x_i) = u_i$, $i = 0, 1, \ldots, n$. Attempt to prove that p_{2n+1} is the unique polynomial with these properties, by adapting the uniqueness proofs in Sections 6.2 and 6.4, using Rolle's Theorem; explain where the proof fails. Show that there is no polynomial $p_5 \in \mathcal{P}_5$ such that $p_5(-1) = 1$, $p_5(0) = 0$, $p_5(1) = 1$, $p''_5(-1) = 0$, $p''_5(0) = 0$, $p''_5(1) = 0$, but that if the first condition is replaced by $p_5(-1) = -1$, then there is an infinite number of such polynomials. Give an explicit expression for the general form of these polynomials.

6.6 Suppose that $n \geq 1$. The function f and its derivatives of order up to and including $2n + 1$ are continuous on $[a, b]$. The points x_i, $i = 0, 1, \ldots, n$, are distinct and lie in $[a, b]$. Construct polynomials $l_0(x)$, $h_i(x)$, $k_i(x)$, $i = 1, \ldots, n$, of degree $2n$ such that the polynomial

$$p_{2n}(x) = l_0(x)f(x_0) + \sum_{i=1}^{n}[h_i(x)f(x_i) + k_i(x)f'(x_i)]$$

satisfies the conditions

$$p_{2n}(x_i) = f(x_i)\,, \quad i = 0, 1, \ldots, n\,,$$

and

$$p'_{2n}(x_i) = f'(x_i)\,, \quad i = 1, \ldots, n\,.$$

Show also that for each value of x in $[a, b]$ there is a number η, depending on x, such that

$$f(x) - p_{2n}(x) = \frac{(x - x_0)\prod_{i=1}^{n}(x - x_i)^2}{(2n+1)!} f^{(2n+1)}(\eta)\,.$$

6.7 Suppose that $n \geq 2$. The function f and its derivatives of order up to and including $2n$ are continuous on $[a, b]$. The points x_i, $i = 0, 1, \ldots, n$, are distinct and lie in $[a, b]$. Explain how to construct polynomials $l_0(x)$, $l_n(x)$, $h_i(x)$, $k_i(x)$, $i = 1, \ldots, n-1$, of degree $2n-1$ such that the polynomial

$$p_{2n-1}(x) = l_0(x)f(x_0) + l_n(x)f(x_n) + \sum_{i=1}^{n-1} [h_i(x)f(x_i) + k_i(x)f'(x_i)]$$

satisfies the conditions $p_{2n-1}(x_i) = f(x_i)$, $i = 0, 1, \ldots, n$, and $p'_{2n-1}(x_i) = f'(x_i)$, $i = 1, \ldots, n-1$. It is not necessary to give explicit expressions for these polynomials.

Show also that for each value of x in $[a, b]$ there is a number η, depending on x, such that

$$f(x) - p_{2n-1}(x) = \frac{(x - x_0)(x - x_n) \prod_{i=1}^{n-1} (x - x_i)^2}{(2n)!} f^{(2n)}(\eta).$$

6.8 By considering the symmetry of the graph of the polynomial

$$q(x) = x(x^2 - 1)(x^2 - 4)(x - 3),$$

show that the maximum of $|q(x)|$ over the interval $[0, 1]$ is attained at the point $x = \frac{1}{2}$.

The values of the function $f \colon x \mapsto \sin x$ are given at the points $x_i = i\pi/8$, for all integer values of i. For a general value of x, an approximation $u(x)$ to $f(x)$ is calculated by first defining k to be the integer part of $8x/\pi$, so that $x_k \leq x \leq x_{k+1}$, and then evaluating the Lagrange polynomial of degree 5 using the six interpolation points $(x_j, f(x_j))$, $j = k-2, \ldots, k+3$. Show that, for all values of x,

$$|\sin x - u(x)| \leq \frac{225\,\pi^6}{16^6 \times 6!} < 0.00002.$$

6.9 Let $n \geq 1$. The interpolation points x_j, $j = 0, 1, \ldots, 2n-1$, are distinct, and $x_{n+j} = x_j + \varepsilon$ for each $j = 0, \ldots, n-1$. The Lagrange polynomial of degree $2n-1$ for the function f using these points is denoted by p_{2n-1}. Show that the terms involving $f(x_j)$ and $f(x_{n+j})$ in p_{2n-1} may be written

$$\frac{\varphi_j(x)\,\varphi_j(x - \varepsilon)}{\varepsilon\,\varphi_j(x_j)} \left\{ \frac{x - x_j}{\varphi_j(x_j + \varepsilon)} f(x_j + \varepsilon) - \frac{x - x_j - \varepsilon}{\varphi_j(x_j - \varepsilon)} f(x_j) \right\},$$

where

$$\varphi_j(x) = \prod_{\substack{i=0 \\ i \neq j}}^{n-1} (x - x_i).$$

Find the limit of this expression as $\varepsilon \to 0$, and deduce that $p_{2n-1} - q_{2n-1} \to 0$ as $\varepsilon \to 0$, where q_{2n-1} is the Hermite interpolation polynomial for f, using the points x_i, $i = 0, \ldots, n-1$.

6.10 Construct the Hermite interpolation polynomial of degree 3 for the function $f\colon x \mapsto x^5$, using the points $x_0 = 0$, $x_1 = a$, and show that it has the form $p_3(x) = 3a^2x^3 - 2a^3x^2$. Verify Theorem 6.4 by direct calculation, showing that in this case ξ is unique and has the value $\xi = \frac{1}{5}(x + 2a)$.

6.11 The complex function $z \mapsto f(z)$ of the complex variable z is holomorphic in the region D of the complex plane; the boundary of D is the simple closed contour C. The interpolation points x_j, $j = 0, 1, \ldots, n$, with $n \geq 1$, and the point x all lie in D. Determine the residues of the function g defined by

$$g(z) = \frac{f(z)}{z - x} \prod_{j=0}^{n} \frac{x - x_j}{z - x_j}$$

at its poles in D, and deduce that

$$f(x) - p_n(x) = \frac{1}{2\pi\imath} \int_C \frac{f(z)}{z - x} \prod_{j=0}^{n} \frac{x - x_j}{z - x_j} \mathrm{d}z\,,$$

where p_n is the Lagrange interpolation polynomial for the function f using the interpolation points x_j, $j = 0, 1, \ldots, n$.

Now, suppose that the real number x and the interpolation points x_j, $j = 0, 1, \ldots, n$, all lie in the real interval $[a, b]$, and that D consists of all the points z such that $|z - t| < K$ for all $t \in [a, b]$, where K is a constant with $K > |b - a|$. Show that the length of the contour C is $2(b - a) + 2\pi K$, and that

$$|f(x) - p_n(x)| < \frac{(b - a + \pi K)M}{\pi} \left(\frac{b - a}{K}\right)^{n+1},$$

where M is such that $|f(z)| \leq M$ on C. Deduce that the sequence (p_n) converges to f, uniformly on $[a, b]$.

Show that these conditions are not satisfied by the function $f\colon x \mapsto 1/(1 + x^2)$ for x in the interval $[-5, 5]$. For what values of a are the conditions satisfied by f for x in the interval $[-a, a]$?

6.12 With the same notation as in Example 6.3, let

$$E(h) = \frac{(f(h) + \varepsilon_+) - (f(-h) + \varepsilon_-)}{2h} - f'(0)\,.$$

Suppose that $f'''(x)$ exists and is continuous at all $x \in [-h, h]$.

By expanding $f(h)$ and $f(-h)$ into Taylor series about the point 0, show that there exists $\xi \in (-h, h)$ such that

$$E(h) = \frac{1}{6}h^2 f'''(\xi) + \frac{\varepsilon_+ - \varepsilon_-}{2h}\,.$$

Hence deduce that

$$|E(h)| \leq \frac{1}{6}h^2 M_3 + \frac{\varepsilon}{h}$$

where $M_3 = \max_{x\in[-h,h]} |f'''(x)|$ and $\varepsilon = \max(|\varepsilon_+|, |\varepsilon_-|)$. Show further that the right-hand side of the last inequality achieves its minimum value when

$$h = \left(\frac{3\varepsilon}{M_3}\right)^{1/3}.$$

7

Numerical integration – I

7.1 Introduction

The problem of evaluating definite integrals arises both in mathematics and beyond, in many areas of science and engineering. At some point in our mathematical education we all learned to calculate simple integrals such as

$$\int_0^1 \mathrm{e}^x \mathrm{d}x \qquad \text{or} \qquad \int_0^\pi \cos x \, \mathrm{d}x$$

using a table of integrals, so you will know that the values of these are $\mathrm{e} - 1$ and 0 respectively; but how about the innocent-looking

$$\int_0^1 \mathrm{e}^{x^2} \mathrm{d}x \qquad \text{and} \qquad \int_0^\pi \cos(x^2) \mathrm{d}x \,,$$

or the more exotic

$$\int_1^{2000} \exp(\sin(\cos(\sinh(\cosh(\tan^{-1}(\log(x))))))) \, \mathrm{d}x ?$$

Please try to evaluate these using a table of integrals and see how far you can get! It is not so simple, is it? Of course, you could argue that the last example was completely artificial. Still, it illustrates the point that it is relatively easy to think of a continuous real-valued function f defined on a closed interval $[a, b]$ of the real line such that the definite integral

$$\int_a^b f(x) \, \mathrm{d}x \tag{7.1}$$

is very hard to reduce to an entry in the table of integrals by means of the usual tricks of variable substitution and integration by parts. If you have access to the computer package Maple, you may try to type

```
evalf(int(exp(sin(cos(sinh(cosh(arctan(log(x))))))), x=1..2000));
```

at the Maple command line. In about the same time as it will take you to correctly type the command at the keyboard, as if by magic, the result 1514.780678 will pop up on the screen. How was this number arrived at?

The purpose of this chapter, and its continuation, Chapter 10, is to answer this question. Specifically, we shall address the problem of evaluating (7.1) approximately, by applying the results of Chapter 6 on polynomial interpolation to derive formulae for numerical integration (also called numerical quadrature rules). We shall also explain how one can estimate the associated approximation error. What does polynomial interpolation have to do with evaluating definite integrals? The answer will be revealed in the next section which is about a class of quadrature formulae bearing the names of two English mathematicians: Newton and Cotes.[1]

7.2 Newton–Cotes formulae

Let f be a real-valued function, defined and continuous on the closed real interval $[a, b]$, and suppose that we have to evaluate the integral

$$\int_a^b f(x)\mathrm{d}x\,.$$

Since polynomials are easy to integrate, the idea, roughly speaking, is to approximate the function f by its Lagrange interpolation polynomial p_n of degree n, and integrate p_n instead. Thus,

$$\int_a^b f(x)\mathrm{d}x \approx \int_a^b p_n(x)\mathrm{d}x\,. \tag{7.2}$$

For a positive integer n, let $x_i,\ i = 0, 1, \ldots, n$, denote the interpolation

[1] Roger Cotes (10 July 1682, Burbage, Leicestershire, England – 5 June 1716, Cambridge, Cambridgeshire, England) was a fellow of Trinity College in Cambridge. At the age of 26 he became the first Plumian Professor of Astronomy and Experimental Philosophy. Even though he only published one paper in his lifetime, entitled ‘Logometria’, Cotes made important contributions to the theory of logarithms and integral calculus, particularly interpolation and table construction. In reference to Cotes’ early death, Newton said: *If he had lived we might have known something.*

points; for the sake of simplicity, we shall assume that these are equally spaced, that is,

$$x_i = a + ih\,, \qquad i = 0, 1, \ldots, n\,,$$

where

$$h = (b-a)/n\,.$$

The Lagrange interpolation polynomial of degree n for the function f, with these interpolation points, is of the form

$$p_n(x) = \sum_{k=0}^{n} L_k(x) f(x_k) \quad \text{where} \quad L_k(x) = \prod_{\substack{i=0 \\ i\neq k}}^{n} \frac{x - x_i}{x_k - x_i}\,.$$

Inserting the expression for p_n into the right-hand side of (7.2) yields

$$\int_a^b f(x)\mathrm{d}x \approx \sum_{k=0}^{n} w_k f(x_k)\,, \tag{7.3}$$

where

$$w_k = \int_a^b L_k(x)\mathrm{d}x\,, \qquad k = 0, 1, \ldots, n\,. \tag{7.4}$$

The values w_k, $k = 0, 1, \ldots, n$, are referred to as the **quadrature weights**, while the interpolation points x_k, $k = 0, 1, \ldots, n$, are called the **quadrature points**. The numerical quadrature rule (7.3), with quadrature weights (7.4) and equally spaced quadrature points, is called the **Newton–Cotes formula** of order n. In order to illustrate the general idea, we consider two simple examples.

Trapezium rule. In this case we take $n = 1$, so that $x_0 = a$, $x_1 = b$; the Lagrange interpolation polynomial of degree 1 for the function f is simply

$$\begin{aligned} p_1(x) &= L_0(x)f(a) + L_1(x)f(b) \\ &= \frac{x-b}{a-b} f(a) + \frac{x-a}{b-a} f(b) \\ &= \frac{1}{b-a}[(b-x)f(a) + (x-a)f(b)]\,. \end{aligned}$$

Integrating $p_1(x)$ from a to b yields

$$\int_a^b f(x)\mathrm{d}x \approx \frac{b-a}{2}[f(a) + f(b)]\,.$$

This numerical integration formula is called the trapezium rule. The terminology stems from the fact that the expression on the right is the area of the trapezium with vertices $(a, 0)$, $(b, 0)$, $(a, f(a))$, $(b, f(b))$.

Simpson's rule.[1] A slightly more sophisticated quadrature rule is obtained by taking $n = 2$. In this case $x_0 = a$, $x_1 = (a+b)/2$ and $x_2 = b$, and the function f is approximated by a quadratic Lagrange interpolation polynomial.

The quadrature weights are calculated from

$$\begin{aligned} w_0 &= \int_a^b L_0(x)\mathrm{d}x \\ &= \int_a^b \frac{(x-x_1)(x-x_2)}{(x_0-x_1)(x_0-x_2)}\mathrm{d}x \\ &= \int_{-1}^1 \frac{t(t-1)}{2}\,\frac{b-a}{2}\mathrm{d}t \\ &= \frac{b-a}{6}, \end{aligned}$$

where it is convenient to make the change of variable

$$x = \frac{b-a}{2}t + \frac{b+a}{2}.$$

Similarly, $w_1 = \frac{4}{6}(b-a)$, and it is easy to see that $w_2 = w_0$ by symmetry. This gives

$$\int_a^b f(x)\mathrm{d}x \approx \frac{b-a}{6}\left[f(a) + 4f\left(\frac{a+b}{2}\right) + f(b)\right],$$

a numerical integration formula known as Simpson's rule.

It is very important to notice that the weights w_k defined in (7.4) depend only on n and k, not on the function f. Their values can therefore

[1] Thomas Simpson (20 August 1710, Market Bosworth, Leicestershire, England – 14 May 1761, Market Bosworth, Leicestershire, England) was a weaver by training who taught mathematics in the London coffee-houses. His two-volume work entitled *The Doctrine and Application of Fluxions* published in 1750 contains some of the work that Cotes hoped to publish with Cambridge University Press but was prevented by his premature death. In 1796 fellow mathematician Charles Hutton gave the following description of Simpson: *It has been said that Mr Simpson frequented low company, with whom he used to guzzle porter and gin: but it must be observed that the misconduct of his family put it out of his power to keep the company of gentlemen, as well as to procure better liquor.* On a related subject: in his *New Stereometry of Wine Barrels* (*Nova stereometria doliorum vinariorum* (1615)), the astronomer Johannes Kepler (1571–1630) approximated the volumes of many three-dimensional solids, each of which was formed by revolving a two-dimensional region around an axis line. For each of these volumes of revolution, he subdivided the solid into many thin slices the sum of whose volumes then approximated the desired total volume.

be calculated in advance, as in the trapezium rule and Simpson's rule. The evaluation of the approximation to the integral (7.1) is then a trivial matter; it is only necessary to compute $f(x_k)$ at each of the quadrature points x_k, $k = 0, 1, \ldots, n$, multiply by the known weights w_k for $k = 0, 1, \ldots, n$, and form the sum on the right-hand side of (7.3).

7.3 Error estimates

Our next task is to estimate the size of the error in the numerical integration formula (7.3), that is, the error that has been committed by integrating the interpolating Lagrange polynomial of f instead of f itself. The error in (7.3) is defined by

$$E_n(f) = \int_a^b f(x)\mathrm{d}x - \sum_{k=0}^{n} w_k f(x_k)\,.$$

The next theorem provides a useful bound on $E_n(f)$ under the additional hypothesis that the function f is sufficiently smooth.

Theorem 7.1 *Let $n \geq 1$. Suppose that f is a real-valued function, defined and continuous on the interval $[a, b]$, and let $f^{(n+1)}$ be defined and continuous on $[a, b]$. Then,*

$$|E_n(f)| \leq \frac{M_{n+1}}{(n+1)!} \int_a^b |\pi_{n+1}(x)|\,\mathrm{d}x\,, \tag{7.5}$$

where $M_{n+1} = \max_{\zeta \in [a,b]} |f^{(n+1)}(\zeta)|$ and $\pi_{n+1}(x) = (x - x_0) \ldots (x - x_n)$.

Proof Recalling the definition of the weights w_k from (7.4), we can write $E_n(f)$ as follows:

$$\begin{aligned} E_n(f) &= \int_a^b f(x)\,\mathrm{d}x - \int_a^b \left(\sum_{k=0}^{n} L_k(x) f(x_k) \right) \mathrm{d}x \\ &= \int_a^b [f(x) - p_n(x)]\,\mathrm{d}x\,. \end{aligned}$$

Thus,

$$|E_n(f)| \leq \int_a^b |f(x) - p_n(x)|\,\mathrm{d}x\,.$$

The desired error estimate (7.5) follows by inserting (6.8) into the right-hand side of this inequality. □

Let us use this theorem to estimate the size of the error which arises from applying the trapezium rule to the integral $\int_a^b f(x)\,dx$. In this case, with $n = 1$ and $\pi_2(x) = (x-a)(x-b)$, the bound (7.5) reduces to

$$\begin{aligned} |E_1(f)| &\le \frac{M_2}{2}\int_a^b |(x-a)(x-b)|\,dx \\ &= \frac{M_2}{2}\int_a^b (b-x)(x-a)\,dx \\ &= \frac{(b-a)^3}{12} M_2\,. \end{aligned} \tag{7.6}$$

An analogous but slightly more tedious calculation shows that, for Simpson's rule,

$$\begin{aligned} |E_2(f)| &\le \frac{M_3}{6}\int_a^b |(x-a)(x-(a+b)/2)(x-b)|\,dx \\ &= \frac{(b-a)^4}{196} M_3\,. \end{aligned} \tag{7.7}$$

Unfortunately, (7.7) gives a considerable overestimate of the error in Simpson's rule; in particular it does not bring out the fact that $E_2(f) = 0$ whenever f is a polynomial of degree 3. The next theorem will allow us to give a sharper bound on the error in Simpson's rule which illustrates this fact. More generally, it is quite easy to prove that when n is odd the Newton–Cotes formula (7.3) (with w_k defined by (7.4)) is exact for all polynomials of degree n, while when n is even it is also exact for all polynomials of degree $n+1$ (see Exercise 2 at the end of the chapter).

Theorem 7.2 *Suppose that f is a real-valued function, defined and continuous on the interval $[a,b]$, and that $f^{\mathrm{iv}} = f^{(4)}$, the fourth derivate of f, is continuous on $[a,b]$. Then,*

$$\int_a^b f(x)\,dx - \frac{b-a}{6}\left[f(a) + 4f((a+b)/2) + f(b)\right] = -\frac{(b-a)^5}{2880} f^{\mathrm{iv}}(\xi)\,, \tag{7.8}$$

for some ξ in (a,b).

Proof Making the change of variable

$$x = \frac{a+b}{2} + \frac{b-a}{2}t\,, \qquad t \in [-1,1]\,,$$

and defining the function $t \mapsto F(t)$ by $F(t) = f(x)$, we see that

$$\int_a^b f(x)\mathrm{d}x - \frac{b-a}{6}\left[f(a) + 4f((a+b)/2) + f(b)\right]$$
$$= \frac{b-a}{2}\left(\int_{-1}^{1} F(\tau)\mathrm{d}\tau - \frac{1}{3}[F(-1) + 4F(0) + F(1)]\right). \quad (7.9)$$

We now introduce the function $t \mapsto G(t)$ by

$$G(t) = \int_{-t}^{t} F(\tau)\,\mathrm{d}\tau - \frac{t}{3}[F(-t) + 4F(0) + F(t)], \quad t \in [-1, 1];$$

the right-hand side of (7.9) is then simply $\frac{1}{2}(b-a)G(1)$.

The remainder of the proof is devoted to showing that $\frac{1}{2}(b-a)G(1)$ is, in turn, equal to the right-hand side of (7.8) for some ξ in (a, b). To do so, we define

$$H(t) = G(t) - t^5 G(1), \quad t \in [-1, 1],$$

and apply Rolle's Theorem repeatedly to the function H. Noting that $H(0) = H(1) = 0$, we deduce that there exists $\zeta_1 \in (0, 1)$ such that $H'(\zeta_1) = 0$. But it is easy to show that $H'(0) = 0$, so there exists $\zeta_2 \in (0, \zeta_1)$ such that $H''(\zeta_2) = 0$. Again we see that $H''(0) = 0$, so there exists $\zeta_3 \in (0, \zeta_2)$ such that $H'''(\zeta_3) = 0$. Now,

$$G'''(t) = -\frac{t}{3}[F'''(t) - F'''(-t)],$$

and therefore

$$H'''(\zeta_3) = -\frac{\zeta_3}{3}[F'''(\zeta_3) - F'''(-\zeta_3)] - 60\zeta_3^2 G(1).$$

Applying the Mean Value Theorem to the function F''' this shows that there exists $\zeta_4 \in (-\zeta_3, \zeta_3)$ such that

$$\begin{aligned} H'''(\zeta_3) &= -\frac{\zeta_3}{3}[2\zeta_3 F^{iv}(\zeta_4)] - 60\zeta_3^2 G(1) \\ &= -\frac{2\zeta_3^2}{3}[F^{iv}(\zeta_4) + 90G(1)]. \end{aligned}$$

Since $H'''(\zeta_3) = 0$ and $\zeta_3 \neq 0$, this means that

$$G(1) = -\frac{1}{90}F^{iv}(\zeta_4) = -\frac{(b-a)^4}{1440} f^{iv}(\xi),$$

and the required result follows. □

Theorem 7.2 yields the following bound on the error in Simpson's rule:

$$|E_2(f)| \leq \frac{(b-a)^5}{2880} M_4 \,. \tag{7.10}$$

This is a considerable improvement on the earlier bound (7.7); when f is a polynomial of degree 3, the bound correctly shows that $E_2(f) = 0$.

There is a great variety of quadrature rules constructed in the same way as the Newton–Cotes formulae. For example, it may sometimes be useful to involve quadrature points outside the interval of integration, as in

$$\int_0^1 f(x)\,\mathrm{d}x \approx c_{-1} f(-1) + c_0 f(0) + c_1 f(1)\,. \tag{7.11}$$

The coefficients are determined similarly as in (7.4), but now $x_{-1} = -1$, $x_0 = 0$, $x_1 = 1$ and

$$L_{-1}(x) = \tfrac{1}{2}x(x-1)\,, \quad L_0(x) = 1 - x^2\,, \quad L_1(x) = \tfrac{1}{2}x(x+1)\,.$$

Hence,

$$\begin{aligned} c_{-1} &= \int_0^1 L_{-1}(x)\,\mathrm{d}x \\ &= \int_0^1 \frac{x(x-1)}{2}\,\mathrm{d}x \\ &= -\tfrac{1}{12}\,. \end{aligned}$$

In a similar way we find that $c_0 = \frac{2}{3}$, $c_1 = \frac{5}{12}$.

The quadrature rule (7.11) is then exact when f is any polynomial of degree 2 or less. More generally, for any three times continuously differentiable function f, Theorem 7.1 extends in an obvious way to give

$$\begin{aligned} &\left| \int_0^1 f(x)\,\mathrm{d}x + \tfrac{1}{12} f(-1) - \tfrac{2}{3} f(0) - \tfrac{5}{12} f(1) \right| \\ &\qquad \leq \frac{M_3}{6} \int_0^1 |(x+1)x(x-1)|\,\mathrm{d}x \\ &\qquad \leq \frac{M_3}{24}\,; \end{aligned}$$

but there is an important difference. To justify this estimate we now need a condition on f outside the interval of integration: we must require that f and f''' are continuous on $[-1, 1]$, and M_3 is the maximum of $|f'''(x)|$ on $[-1, 1]$. More generally, the conditions must hold on an interval which contains the interval of integration, and also all the quadrature points.

Table 7.1. *I_n is the result of the Newton–Cotes formula of degree n for the approximation of the integral (7.12)*

n	I_n
1	0.38462
2	6.79487
3	2.08145
4	2.37401
5	2.30769
6	3.87045
7	2.89899
8	1.50049
9	2.39862
10	4.67330
11	3.24477
12	−0.31294
13	1.91980
14	7.89954
15	4.15556

7.4 The Runge phenomenon revisited

By looking at the right-hand side of the error bound (7.5) we may be led to believe that by increasing n, that is by approximating the integrand by Lagrange interpolation polynomials of increasing degree and integrating these exactly, we shall reduce the size of the quadrature error $E_n(f)$. However, this is not always the case, even for very smooth functions f. An example of this behaviour uses the same function as in Section 6.3; Table 7.1 gives the results of applying Newton–Cotes formulae of increasing degree to the evaluation of the integral

$$\int_{-5}^{5} \frac{1}{1+x^2}\,\mathrm{d}x\,. \tag{7.12}$$

These results do not evidently converge as n increases, and in fact they eventually increase without bound. This behaviour is related to the fact that the weights w_j in the Newton–Cotes formula are not all positive when $n > 8$. We shall return to this point in Theorem 10.2.

A better approach to improving accuracy is to divide the interval $[a, b]$ into an increasing number of subintervals of decreasing size, and then to use a numerical integration formula of fixed order n on each of the subintervals. Quadrature rules based on this approach are called composite formulae; in the next section we shall describe two examples.[1]

7.5 Composite formulae

We shall consider only some very simple composite quadrature rules: the composite trapezium rule and the composite Simpson rule.

Suppose that f is a function, defined and continuous on a nonempty closed interval $[a, b]$ of the real line. In order to construct an approximation to

$$\int_a^b f(x)\,\mathrm{d}x\,,$$

we now select an integer $m \geq 2$ and divide the interval $[a, b]$ into m equal subintervals, each of width $h = (b - a)/m$, so that

$$\int_a^b f(x)\,\mathrm{d}x = \sum_{i=1}^{m} \int_{x_{i-1}}^{x_i} f(x)\,\mathrm{d}x\,, \tag{7.13}$$

where

$$x_i = a + ih = a + \frac{i}{m}(b - a)\,, \quad i = 0, 1, \ldots, m\,.$$

Each of the integrals is then evaluated by the trapezium rule,

$$\int_{x_{i-1}}^{x_i} f(x)\,\mathrm{d}x \approx \frac{1}{2}h[f(x_{i-1}) + f(x_i)]\,; \tag{7.14}$$

summing these over $i = 1, 2, \ldots, m$ leads to the following definition.

Definition 7.1 (Composite trapezium rule)

$$\int_a^b f(x)\,\mathrm{d}x \approx h\left[\frac{1}{2}f(x_0) + f(x_1) + \cdots + f(x_{m-1}) + \frac{1}{2}f(x_m)\right]. \tag{7.15}$$

[1] The historical roots of composite formulae may be traced back to the work of Kepler cited in the footnote to Simpson's method earlier on in this chapter, although the idea of computing volumes of two- and three-dimensional geometrical objects by subdivision was already present in the work of Archimedes of Syracuse (287 BC, Syracuse (now in Italy) – 212 BC, Syracuse (now in Italy)). Archimedes' long-lost book known as the Palimpsest, containing his geometrical studies, resurfaced at an auction at Christie's of New York in 1998 and is now in the care of the Walters Art Gallery in Baltimore, Maryland, USA: `http://www.thewalters.org/archimedes/frame.html`

The error in the composite trapezium rule can be estimated by using the error bound (7.6) for the trapezium rule on each individual subinterval $[x_{i-1}, x_i]$, $i = 1, 2, \ldots, m$. For this purpose, let us define

$$\begin{aligned} \mathcal{E}_1(f) &= \int_a^b f(x)\,\mathrm{d}x - h\left[\tfrac{1}{2}f(x_0) + f(x_1) + \cdots + f(x_{m-1}) + \tfrac{1}{2}f(x_m)\right] \\ &= \sum_{i=1}^m \left[\int_{x_{i-1}}^{x_i} f(x)\,\mathrm{d}x - \tfrac{1}{2}h\left[f(x_{i-1}) + f(x_i)\right]\right]. \end{aligned}$$

Applying (7.6) to each of the terms under the summation sign we obtain

$$\begin{aligned} |\mathcal{E}_1(f)| &\le \frac{1}{12}h^3 \sum_{i=1}^m \left(\max_{\zeta\in[x_{i-1},x_i]} |f''(\zeta)|\right) \\ &\le \frac{(b-a)^3}{12m^2} M_2\,, \end{aligned} \tag{7.16}$$

where $M_2 = \max_{\zeta\in[a,b]} |f''(\zeta)|$.

For Simpson's rule, let us suppose that the interval $[a, b]$ has been divided into $2m$ intervals by the points $x_i = a + ih$, $i = 0, 1, \ldots, 2m$, with $m \ge 2$ and

$$h = \frac{b-a}{2m}\,,$$

and let us apply Simpson's rule on each of the intervals $[x_{2i-2}, x_{2i}]$, $i = 1, 2, \ldots, m$, giving

$$\begin{aligned} \int_a^b f(x)\,\mathrm{d}x &= \sum_{i=1}^m \int_{x_{2i-2}}^{x_{2i}} f(x)\,\mathrm{d}x \\ &\approx \sum_{i=1}^m \frac{2h}{6}\left[f(x_{2i-2}) + 4f(x_{2i-1}) + f(x_{2i})\right]. \end{aligned}$$

This leads to the following definition.

Definition 7.2 (Composite Simpson rule)

$$\begin{aligned} \int_a^b f(x)\,\mathrm{d}x \approx\ & \frac{h}{3}[f(x_0) + 4f(x_1) + 2f(x_2) + 4f(x_3) + \cdots \\ & + 2f(x_{2m-2}) + 4f(x_{2m-1}) + f(x_{2m})]\,. \end{aligned} \tag{7.17}$$

A schematic view of the pattern in which the coefficients 1, 4 and 2 appear in the composite Simpson rule is shown in Figure 7.1.

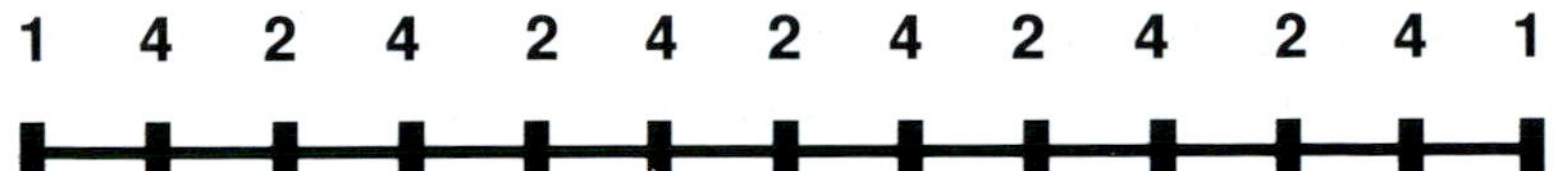

Fig. 7.1. Quadrature weights for the composite Simpson rule: the integers $1, 4, 2, 4, \ldots, 1$, when multiplied by $h/3$, where $h = (b-a)/2m$, provide the quadrature weights. This figure corresponds to taking $m = 6$.

In order to estimate the error in the composite Simpson rule, we proceed in the same way as for the composite trapezium rule. Let us define

$$\begin{aligned}\mathcal{E}_2(f) &= \int_a^b f(x)\,\mathrm{d}x - \sum_{i=1}^{m} \frac{h}{3}[f(x_{2i-2}) + 4f(x_{2i-1}) + f(x_{2i})] \\ &= \sum_{i=1}^{m} \left[\int_{x_{2i-2}}^{x_{2i}} f(x)\,\mathrm{d}x - \frac{h}{3}[f(x_{2i-2}) + 4f(x_{2i-1}) + f(x_{2i})] \right].\end{aligned}$$

Applying (7.10) to each individual term in the sum and recalling that $b - a = 2mh$ we obtain the following error bound:

$$|\mathcal{E}_2(f)| \le \frac{(b-a)^5}{2880m^4} M_4\,, \tag{7.18}$$

where $M_4 = \max_{\zeta \in [a,b]} |f^{iv}(\zeta)|$.

The composite rules (7.15) and (7.17) provide greater accuracy than the basic formulae considered in Section 7.2; this is clearly seen by comparing the error bounds (7.16) and (7.18) for the two composite rules with (7.6) and (7.10), the error estimates for the basic trapezium rule and Simpson rule respectively. The inequalities (7.16) and (7.18) indicate that, as long as the function f is sufficiently smooth, the errors in the composite rules can be made arbitrarily small by choosing a sufficiently large number of subintervals.

7.6 The Euler–Maclaurin expansion

We have seen in (7.16) that the error in the composite trapezium rule is bounded by a term involving $1/m^2$, where m is the number of subdivisions of the interval $[a,b]$; the **Euler**[1]–**Maclaurin**[2] expansion expresses this error as a series in powers of $1/m^2$, and makes it possible to improve accuracy by extrapolation methods.

We first define a sequence of polynomials.

Definition 7.3 *Consider the sequence of polynomials q_r, $r = 1, 2, \ldots$, defined by their properties, as follows:*

(i) *q_r is a polynomial of degree r;*
(ii) *for each positive integer r, $q'_{r+1} = q_r$;*
(iii) *q_r is an odd function if r is odd, and an even function if r is even;*
(iv) *if $r > 1$ is odd, then $q_r(-1) = 0$ and $q_r(1) = 0$;*
(v) *$q_1(t) = -t$.*

Using these conditions it is easy to construct the polynomials q_r in succession. From (v) and (ii) we get

$$q_2(t) = -\tfrac{1}{2}t^2 + A_2\,, \qquad q_3(t) = -\tfrac{1}{6}t^3 + A_2 t + A_3\,,$$

where A_2 and A_3 are constants. From (iii) we see that $A_3 = 0$; then, from (iv) it follows that $A_2 = \frac{1}{6}$. Hence,

$$q_2(t) = -\tfrac{1}{2}t^2 + \tfrac{1}{6}\,, \qquad q_3(t) = -\tfrac{1}{6}t^3 + \tfrac{1}{6}t\,.$$

We can then go on to construct q_4 and q_5, and so on.

[1] Leonhard Euler (15 April 1707, Basel, Switzerland – 18 September 1783, St Petersburg, Russia) was the most prolific mathematical writer of all times, who made fundamental contributions to many branches of mathematics despite being totally blind for the last third of his life. Euler and his wife Katharina had 13 children: he claimed to have made his greatest discoveries while he was holding a baby in his arms and the other children were playing around his feet. Euler studied the calculus of variations, differential geometry, number theory, differential equations, continuum mechanics, astronomy, lunar theory, the three-body problem, elasticity, acoustics, the wave theory of light, hydraulics, and music. In his *Theory of the Motions of Rigid Bodies* published in 1765 he laid the foundation of analytical mechanics. Euler integrated Leibniz's differential calculus and Newton's method of fluxions into mathematical analysis. We owe him the concepts of beta and gamma functions and the notion of integrating factor for differential equations; he is responsible for the notation e for the base of natural logarithm, $f(x)$ for a function, π for pi, $\sum$ for summation, i for the square root of -1, and Δ_y and Δ_y^2 for the first and second finite differences.

[2] Colin Maclaurin (February 1698, Kilmodan, Argyllshire, Scotland – 14 June 1746, Edinburgh, Scotland) became a student at the University of Glasgow at the age of 11 and completed his studies at the age of 14. In 1719, at the age of 21, he became Fellow of the Royal Society. His major work of 763 pages in two volumes, entitled *A Treatise of Fluxions*, was the first systematic exposition of Newton's ideas. Notable is Maclaurin's work on elliptic integrals, maxima and minima, and the attraction of ellipsoids.

Theorem 7.3 *Suppose that the function g is defined and continuous on the interval $[-1,1]$ and has a continuous derivative of order $2k$ over this interval. Then,*

$$\int_{-1}^{1} g(t)\,\mathrm{d}t - [g(-1)+g(1)] = \int_{-1}^{1} -t\,g'(t)\,\mathrm{d}t$$
$$= \sum_{r=1}^{k} q_{2r}(1)[g^{(2r-1)}(1) - g^{(2r-1)}(-1)] - \int_{-1}^{1} q_{2k}(t)g^{(2k)}(t)\,\mathrm{d}t\,. \quad (7.19)$$

Proof We observe that $\int_{-1}^{1} g(t)\,\mathrm{d}t - [g(-1)+g(1)]$ is the error in the approximation of $\int_{-1}^{1} g(t)\mathrm{d}t$ by the trapezium rule. Integration by parts gives

$$\int_{-1}^{1} -t\,g'(t)\,\mathrm{d}t \quad = \quad -[g(-1)+g(1)] + \int_{-1}^{1} g(t)\,\mathrm{d}t\,,$$

which establishes the first equality in (7.19). By repeated integration by parts in the other direction, and using the fact that $q_1(t) = -t$, we then have

$$\begin{aligned}\int_{-1}^{1} -t\,g'(t)\,\mathrm{d}t &= q_2(1)g'(1) - q_2(-1)g'(-1) - \int_{-1}^{1} q_2(t)g''(t)\mathrm{d}t \\ &= \Big[q_2(t)g'(t) - q_3(t)g''(t) + \cdots + q_{2k}(t)g^{(2k-1)}(t)\Big]_{-1}^{1} \\ &\quad - \int_{-1}^{1} q_{2k}(t)g^{(2k)}(t)\,\mathrm{d}t\,.\end{aligned}$$

The required result follows from properties (iii) and (iv) of the q_r. □

Theorem 7.4 (Euler–Maclaurin expansion) *Suppose that the real-valued function f is defined and continuous on the interval $[a,b]$ and has a continuous derivative of order $2k$ on this interval. Consider the subdivision of $[a,b]$ into $m \ge 1$ closed intervals $[x_{i-1},x_i]$, $i = 1,\dots,m$, where $x_i = a+ih$, $i = 0,1,\dots,m$, and $h = (b-a)/m$. Writing $T(m)$ for the result of approximating the integral $I = \int_a^b f(x)\mathrm{d}x$ by the composite trapezium rule with the m subintervals $[x_{i-1},x_i]$, $i = 1,\dots,m$,*

$$\begin{aligned}I - T(m) &= \sum_{r=1}^{k} c_r h^{2r}[f^{(2r-1)}(b) - f^{(2r-1)}(a)] \\ &\quad - \left(\frac{h}{2}\right)^{2k} \sum_{i=1}^{m} \int_{x_{i-1}}^{x_i} q_{2k}(t) f^{(2k)}(x)\,\mathrm{d}x\,, \qquad (7.20)\end{aligned}$$

where $t = t(x) = -1 + \frac{2}{h}(x - x_{i-1})$ *for* $x \in [x_{i-1}, x_i]$, $i = 1, \ldots, m$, *and* $c_r = q_{2r}(1)/2^{2r}$ *for* $r = 1, \ldots, k$.

Proof We express the integral as a sum of integrals over the m subintervals $[x_{i-1}, x_i]$, $i = 1, \ldots, m$, as in (7.13). In the interval $[x_{i-1}, x_i]$ we change the variable by writing $x = x_{i-1} + h(t+1)/2$, so that

$$\int_{x_{i-1}}^{x_i} f(x)\mathrm{d}x = \frac{h}{2}\int_{-1}^{1} g(t)\mathrm{d}t\,,$$

where $f(x) = g(t)$. According to Theorem 7.3, then,

$$\begin{aligned}
&\int_{x_{i-1}}^{x_i} f(x)\mathrm{d}x - \frac{h}{2}[f(x_{i-1}) + f(x_i)] \\
&\quad = \frac{h}{2}\left\{\int_{-1}^{1} g(t)\mathrm{d}t - [g(-1) + g(1)]\right\} \\
&\quad = \frac{h}{2}\left\{\sum_{r=1}^{k} q_{2r}(1)[g^{(2r-1)}(1) - g^{(2r-1)}(-1)]\right. \\
&\qquad\qquad \left. - \int_{-1}^{1} q_{2k}(t)g^{(2k)}(t)\mathrm{d}t\right\}.
\end{aligned}$$

On noting that $g^{(\ell)}(t) = (h/2)^{\ell} f^{(\ell)}(x)$, $\ell = 1, 2, \ldots, 2k$, $\mathrm{d}t = (2/h)\,\mathrm{d}x$, summation over all the subintervals $[x_{i-1}, x_i]$, for $i = 1, \ldots, m$, gives the required result. The important point is the symmetry of the polynomials q_r, which ensures that $q_{2r}(1) = q_{2r}(-1)$, so that all the derivatives of f at the internal points x_i cancel in the course of summation, leaving only the derivatives at a and b. □

Remark 7.1 *By successively computing the polynomials* $q_r(t)$, *we can determine the values of* $c_r = q_{2r}(1)/2^{2r}$, $r = 1, 2, 3, \ldots$. *For example,*

$$c_1 = -\tfrac{1}{12}\,,\ c_2 = \tfrac{1}{720}\,,\ c_3 = -\tfrac{1}{30240}\,,\ c_4 = \tfrac{1}{1209600}\,,\ c_5 = -\tfrac{1}{47900160}\,, \ldots.$$

It can be shown that $c_r = -\frac{B_{2r}}{(2r)!}$ *for all* $r = 1, 2, 3, \ldots$, *where* B_{2r} *are the Bernoulli numbers*[1] *with even index, which can be determined from*

[1] Jacob Bernoulli the elder (27 December 1654, Basel, Switzerland – 16 August 1705, Basel, Switzerland) was one of the first mathematicians to recognise the significance of the work of Newton and Leibniz on differential and integral calculus. Bernoulli contributed to the theory of infinite series, mechanics, calculus of variations, mechanics, and is also known in probability theory for his Law of Large Numbers.

the Taylor series expansion

$$\tfrac{x}{2}\coth\left(\tfrac{x}{2}\right) = \sum_{r=0}^{\infty} \frac{B_{2r}\, x^{2r}}{(2r)!}\,.$$

Easier still, typing `c[6]=-bernoulli(12)/12!;` *at the Maple command line gives* $c_6 = \frac{691}{1307674368000}$*;* $c_7, c_8, \ldots$ *can be found in the same way.*

An interesting consequence of Theorem 7.4 concerns the numerical integration of smooth periodic functions. Suppose that f is a continuous function defined on $(-\infty, \infty)$ such that all derivatives of f, up to and including order $2k$, are defined and continuous on $(-\infty, \infty)$, and f is periodic on $(-\infty, \infty)$ with period $b - a$; *i.e.*, $f(x + b - a) - f(x) = 0$ for all $x \in \mathbb{R}$. Hence, by successive differentiation of this equality and taking $x = a$ we deduce that, in particular,

$$f^{(2r-1)}(b) - f^{(2r-1)}(a) = 0 \qquad \text{for } r = 1, 2, \ldots, k\,.$$

Therefore, according to (7.20), we have that

$$I - T(m) = \mathcal{O}(h^{2k})\,.$$

The fact that for $k \gg 1$ this integration error is much smaller than the $\mathcal{O}(h^2)$ error that will be observed in the case of a nonperiodic function indicates that the composite trapezium rule is particularly well suited for the numerical integration of smooth periodic functions.

A second application of the Euler–Maclaurin expansion concerns extrapolation methods. This subject will be discussed in the next section.

7.7 Extrapolation methods

In general the calculation of the higher derivatives involved in the Euler–Maclaurin expansion (7.20) is not possible. However, the existence of the expansion allows us to eliminate successive terms by repeated calculation of the trapezium rule approximation.

For example, the case $k = 2$ of (7.20) may be written in the form

$$\int_a^b f(x)\mathrm{d}x - T(m) = C_1 h^2 + \mathcal{O}(m^{-4})\,,$$

where $C_1 = c_1[f'(b) - f'(a)]$ and $h = (b - a)/m$. This also means that

$$\int_a^b f(x)\mathrm{d}x - T(2m) = C_1 (h/2)^2 + \mathcal{O}(m^{-4})\,.$$

We can eliminate the term in h^2 from these two equalities, giving

$$\int_a^b f(x)\mathrm{d}x = \frac{4T(2m) - T(m)}{3} + \mathcal{O}(h^4)\,.$$

The same elimination process could be used for any two values of m, from the calculation of $T(m_1)$ and $T(m_2)$; the advantage of using m and $2m$ is that in the computation of $T(2m)$ half the required values of $f(x_i)$ are already known from $T(m)$, and we do not have to calculate them again. This process of eliminating the term in h^2 from the expansion of the error is known as **Richardson extrapolation**[1] or h^2 **extrapolation**. It is easy to extend the process to higher-order terms. For example,

$$\int_a^b f(x)\mathrm{d}x - T(m) = C_1h^2 + C_2h^4 + C_3h^6 + \mathcal{O}(h^8)\,.$$

Hence

$$\int_a^b f(x)\mathrm{d}x - \frac{4T(2m) - T(m)}{3} = -\tfrac{1}{4}C_2h^4 - \tfrac{5}{16}C_3h^6 + \mathcal{O}(h^8)\,,$$

which leads to

$$\int_a^b f(x)\mathrm{d}x - \frac{16T_1(2m) - T_1(m)}{15} = \mathcal{O}(h^6)\,,$$

where

$$T_1(m) = \frac{4T(2m) - T(m)}{3}\,.$$

Therefore,

$$T_2(m) = \frac{16T_1(2m) - T_1(m)}{15}$$

approximates the integral $\int_a^b f(x)\mathrm{d}x$ to accuracy $\mathcal{O}(h^6)$. Adopting the notational convention

$$T_0(m) = T(m)$$

and proceeding recursively,

[1] Lewis Fry Richardson (11 October 1881, Newcastle upon Tyne, Northumberland, England – 30 September 1953, Kilmun, Argyllshire, Scotland) studied mathematics, physics, chemistry, botany and zoology at the Durham College of Science, and subsequently Natural Science at King's College in Cambridge. He worked in the National Physical Laboratory and the Meteorological Office, and was the first to apply numerical mathematics, in particular the method of finite differences, to predicting the weather in *Weather Prediction by Numerical Process* (1922). The *Richardson number*, a quantity involving gradients of temperature and wind velocity is named after him.

Table 7.2. *Romberg table.*

m	$T(m)$	$T_1(m)$	$T_2(m)$	$T_3(m)$	$T_4(m)$
4	$T(4)$	$T_1(4)$	$T_2(4)$	$T_3(4)$	$T_4(4)$
8	$T(8)$	$T_1(8)$	$T_2(8)$	$T_3(8)$	...
16	$T(16)$	$T_1(16)$	$T_2(16)$	...	
32	$T(32)$	$T_1(32)$	...		
64	$T(64)$	...			
...	...				

$$T_k(m) = \frac{4^k T_{k-1}(2m) - T_{k-1}(m)}{4^k - 1}, \qquad k = 1, 2, 3, \dots, \tag{7.21}$$

will approximate $\int_a^b f(x)\mathrm{d}x$ to accuracy $\mathcal{O}(h^{2k+2})$, provided of course that $f^{(2k+2)}$ exists and is continuous on the closed interval $[a, b]$. This extrapolation process is known as the **Romberg**[1] integration method.

The intermediate results in Romberg's method are often arranged in the form of a table, known as the Romberg table. For example, if we start with $m = 4$ subdivisions of the closed interval $[a, b]$, each of length $h = (b - a)/4$, and proceed by doubling the number of subdivisions in each step (and thereby halving the spacing h between the quadrature points from the previous step), then the associated Romberg table is as shown in Table 7.2, where we took, successively, $m = 4, 8, 16, 32, 64$ subdivisions of the interval $[a, b]$ of length $h = (b - a)/m$ each. After $T_0(4) = T(4), \dots, T_0(64) = T(64)$ have been computed, we calculate $T_1(4), \dots, T_1(32)$ using (7.21) with $k = 1$, then we compute $T_2(4), \dots, T_2(16)$ using (7.21) with $k = 2$, then $T_3(4), T_3(8)$ using (7.21) with $k = 3$, and finally $T_4(4)$ using (7.21) with $k = 4$. Provided that the integrand is sufficiently smooth, the numbers in the $T(m)$ column approximate the integral to within an error $\mathcal{O}(h^2)$; the numbers in the $T_1(m)$ column to within $\mathcal{O}(h^4)$, those in the $T_2(m)$ column to $\mathcal{O}(h^6)$, those in the $T_3(m)$ column to $\mathcal{O}(h^8)$, and those in the $T_4(m)$ column to within $\mathcal{O}(h^{10})$.

[1] Werner Romberg, Emeritus Professor at the Institute of Applied Mathematics at the University of Heidelberg in Germany. The extrapolation process was proposed in his paper Vereinfachte numerische Integration [German], *Norske Vid. Selsk. Forh., Trondheim* **28**, 30–36, 1955.

An example is shown in Table 7.3. This gives the results of calculating the integral

$$\int_0^1 \frac{e^{-2x}}{1+4x}\mathrm{d}x$$

by Romberg's method; first the trapezium rule is used successively with $m = 4, 8, 16, 32$ and 64 equal subdivisions of the interval $[0, 1]$ of length $h = (b-a)/m$ each. There are then four stages of extrapolation: Stage 1 involves computing $T_1(m)$ for $m = 4, 8, 16, 32$; Stage 2 computes $T_2(m)$ for $m = 4, 8, 16$; Stage 3 calculates $T_3(m)$ for $m = 4, 8$; and Stage 4 then computes $T_4(m)$ for $m = 4$. Not only does the extrapolation give an accurate result, but the consistency of the numerical values in the last two columns gives a good deal of confidence in quoting the result 0.220458 correct to six decimal digits. Note that none of the individual composite trapezium rule calculations in the $T(m)$ column gives a result correct to more than three decimal digits – not even $T(64)$ which uses 64 equal subdivisions of $[0, 1]$.

Table 7.3. *Romberg table for the calculation of* $\int_0^1 (e^{-2x}/(1+4x))\mathrm{d}x$.

m	$T(m)$	$T_1(m)$	$T_2(m)$	$T_3(m)$	$T_4(m)$
4	0.248802	0.221038	0.220470	0.220458	0.220458
8	0.227979	0.220505	0.220459	0.220458	
16	0.222374	0.220461	0.220458		
32	0.220940	0.220458			
64	0.220579				

The success of Romberg integration is only justified if the integrand f satisfies the hypotheses of the Euler–Maclaurin Theorem. As an illustration of this, Table 7.4 shows the result of the same calculation, but for the integral

$$\int_0^1 x^{1/3}\mathrm{d}x\,.$$

The function $x \mapsto x^{1/3}$ is not differentiable at $x = 0$, so the required conditions are not satisfied for any extrapolation. The numerical results bear this out; they are quite close to the correct value, $3/4$, but the behaviour of the extrapolation does not give any confidence in the accuracy of the result. In fact the extrapolation has not given much improvement on $T(64)$. The calculation of integrals involving this sort of singularity requires special methods which we shall not discuss here.

We have reached the end of this chapter, but do not despair: the story about numerical integration rules will continue. In Chapter 10 we shall discuss a class of quadrature formulae, generally referred to as Gaussian quadrature rules, which are distinct from the Newton–Cotes formulae considered here. Before doing so, however, in Chapters 8 and 9 we make a brief excursion into the realm of approximation theory.

Table 7.4. *Romberg table for the calculation of* $\int_0^1 x^{1/3}\mathrm{d}x$.

m	$T(m)$	$T_1(m)$	$T_2(m)$	$T_3(m)$	$T_4(m)$
4	0.708055	0.741448	0.746950	0.748819	0.749534
8	0.733100	0.746606	0.748790	0.749531	
16	0.743230	0.748653	0.749520		
32	0.747297	0.749465			
64	0.748923				

7.8 Notes

The material presented in this chapter is classical. For further details on the theory and practice of numerical integration, we refer to the following texts:

- Philip J. Davis and Philip Rabinowitz, *Methods of Numerical Integration,* Second Edition, Computer Science and Applied Mathematics, Academic Press, Orlando, FL, 1984;
- Vladimir Ivanovich Krylov, *Approximate Calculation of Integrals,* translated from Russian by Arthur H. Stroud, ACM Monograph Series, Macmillan, New York, 1962;
- Hermann Engels, *Numerical Quadrature and Cubature,* Computational Mathematics and Applications, Academic Press, London, 1980.

The first of these is a standard text and contains a huge bibliography of more than 1500 entries. Concerning the implementation of numerical integration rules into mathematical software, the reader is referred to

- Arnold R. Krommer and Christoph W. Ueberhuber, *Computational Integration*, SIAM, Philadelphia, 1998.

It includes a comprehensive overview of computational integration techniques based on both numerical and symbolical methods, and an exposition of some more recent number-theoretical, pseudorandom and lattice algorithms; these topics are beyond the scope of the present text.

Exercises

7.1 With the usual notation for the Newton–Cotes quadrature formula and using the equally spaced quadrature points $x_k = a+kh$ for $k = 0, 1, \dots, n$ and $n \geq 1$, show that $w_k = w_{n-k}$ for $k = 0, 1, \dots, n$.

7.2 By considering the polynomial $[x-(a+b)/2]^{n+1}$, $n \geq 1$, and the result of Exercise 1, or otherwise, show that the Newton–Cotes formula using $n+1$ points x_k, $k = 0, 1, \dots, n$, is exact for all polynomials of degree $n+1$ whenever n is even.

7.3 A quadrature formula on the interval $[-1, 1]$ uses the quadrature points $x_0 = -\alpha$ and $x_1 = \alpha$, where $0 < \alpha \leq 1$:

$$\int_{-1}^{1} f(x)\mathrm{d}x \approx w_0 f(-\alpha) + w_1 f(\alpha)\,.$$

The formula is required to be exact whenever f is a polynomial of degree 1. Show that $w_0 = w_1 = 1$, independent of the value of α. Show also that there is one particular value of α for which the formula is exact also for all polynomials of degree 2. Find this α, and show that, for this value, the formula is also exact for all polynomials of degree 3.

7.4 The Newton–Cotes formula with $n = 3$ on the interval $[-1, 1]$ is

$$\int_{-1}^{1} f(x)\,\mathrm{d}x \approx w_0 f(-1) + w_1 f(-1/3) + w_2 f(1/3) + w_3 f(1)\,.$$

Using the fact that this formula is to be exact for all polynomials of degree 3, or otherwise, show that

$$\begin{aligned} 2w_0 + 2w_1 &= 2\,, \\ 2w_0 + \tfrac{2}{9}w_2 &= \tfrac{2}{3}\,, \end{aligned}$$

and hence find the values of the weights w_0, w_1, w_2 and w_3.

7.5 For each of the functions $1, x, x^2, \dots, x^6$, find the difference between $\int_{-1}^{1} f(x)\mathrm{d}x$ and (i) Simpson's rule, (ii) the formula derived in Exercise 4.

Deduce that for every polynomial of degree 5 formula (ii) is more accurate than formula (i). Find a polynomial of degree 6 for which formula (i) is more accurate than formula (ii).

7.6 Write down the errors in the approximation of

$$\int_0^1 x^4 \, \mathrm{d}x \quad \text{and} \quad \int_0^1 x^5 \, \mathrm{d}x$$

by the trapezium rule and Simpson's rule. Hence find the value of the constant C for which the trapezium rule gives the correct result for the calculation of

$$\int_0^1 (x^5 - Cx^4) \, \mathrm{d}x \,,$$

and show that the trapezium rule gives a more accurate result than Simpson's rule when $\frac{15}{14} < C < \frac{85}{74}$.

7.7 Determine the values of $c_j, j = -1, 0, 1, 2$, such that the quadrature rule

$$Q(f) = c_{-1}f(-1) + c_0 f(0) + c_1 f(1) + c_2 f(2)$$

gives the correct value for the integral

$$\int_0^1 f(x) \, \mathrm{d}x$$

when f is any polynomial of degree 3. Show that, with these values of the weights c_j, and under appropriate conditions on the function f,

$$\left| \int_0^1 f(x) \, \mathrm{d}x - Q(f) \right| \leq \tfrac{11}{720} M_4 \,.$$

Give suitable conditions for the validity of this bound, and a definition of the quantity M_4.

7.8 Writing $T(m)$ for the composite trapezium rule defined in (7.15) and $S(2m)$ for the composite Simpson's rule defined in (7.17), show that

$$S(2m) = \tfrac{4}{3} T(2m) - \tfrac{1}{3} T(m) \,.$$

7.9 Suppose that the function f has a continuous fourth derivative on the interval $[a, b]$, and that $T(m)$ denotes the composite trapezium rule approximation to $\int_a^b f(x) \mathrm{d}x$, using m subintervals. Show that

$$\frac{T(m) - T(2m)}{T(2m) - T(4m)} \to 4 \quad \text{as } m \to \infty \,.$$

Using the information in Table 7.3 evaluate this expression for $m = 4, 8, 16$.

7.10 With the same notation as in Exercise 9, suppose that the fourth derivative of f is not continuous on $[a, b]$, but that

$$\int_a^b f(x)\mathrm{d}x - T(m) = A/m^\alpha + E(m),$$

where $\alpha > 0$ and A are constants and $\lim_{m\to\infty} m^\alpha E(m) = 0$. Determine

$$\lim_{m\to\infty} \frac{T(m) - T(2m)}{T(2m) - T(4m)}.$$

Suggest a value of α which is consistent with the values of $T(m)$ given in Table 7.4.

7.11 The function f has a continuous fourth derivative on the interval $[-1, 1]$. Construct the Hermite interpolation polynomial of degree 3 for f using the interpolation points $x_0 = -1$ and $x_1 = 1$. Deduce that

$$\int_{-1}^1 f(x)\mathrm{d}x - [f(-1) + f(1)] = \tfrac{1}{3}[f'(-1) - f'(1)] + E,$$

where

$$|E| \le \tfrac{2}{45} \max_{x\in[-1,1]} |f^{iv}(x)|.$$

7.12 Construct the polynomials q_4, q_5, q_6 and q_7 given by Definition 7.3. Hence show that, in the notation of Theorem 7.4,

$$c_1 = -1/12, \quad c_2 = 1/720, \quad c_3 = -1/30240.$$

7.13 Using the relations

$$\begin{aligned} 2\sin\tfrac{1}{2}x \sum_{j=1}^m \sin jx &= \cos\tfrac{1}{2}x - \cos(m+\tfrac{1}{2})x, \\ 2\sin\tfrac{1}{2}x \sum_{j=1}^m \cos jx &= \sin(m+\tfrac{1}{2})x - \sin\tfrac{1}{2}x, \end{aligned}$$

where m is a positive integer, show that the composite trapezium rule (7.15) with m subintervals will give the exact result for each of the integrals

$$\int_{-\pi}^{\pi} \cos rx\,\mathrm{d}x, \qquad \int_{-\pi}^{\pi} \sin rx\,\mathrm{d}x,$$

for any integer value of r which is not a multiple of m.

What values are given by the composite trapezium rule for these integrals when $r = mk$ and k is a positive integer?

8

Polynomial approximation in the ∞-norm

8.1 Introduction

In Chapter 6 we considered the problem of interpolating a function by polynomials of a certain degree. Here we shall discuss other types of approximation by polynomials, the overall objective being to find the polynomial of given degree n which provides the 'best approximation' from $\mathcal{P}_n$ to a given function in a sense that will be made precise below.

8.2 Normed linear spaces

In order to be able to talk about 'best approximation' in a rigorous manner we need to recall from Chapter 2 the concept of *norm*; this will allow us to compare various approximations quantitatively and select the one which has the smallest approximation error. The definition given in Section 2.7 applies to a linear space consisting of functions in the same way as to the finite-dimensional linear spaces considered in Chapter 2.

Definition 8.1 *Suppose that $\mathcal{V}$ is a linear space over the field $\mathbb{R}$ of real numbers. A nonnegative function $\|\cdot\|$ defined on $\mathcal{V}$ whose value at $f \in \mathcal{V}$ is denoted by $\|f\|$ is called a* **norm** *on $\mathcal{V}$ if it satisfies the following axioms:*

❶ $\|f\| = 0$ *if, and only if,* $f = 0$ *in* $\mathcal{V}$;
❷ $\|\lambda f\| = |\lambda|\,\|f\|$ *for all* $\lambda \in \mathbb{R}$, *and all* f *in* $\mathcal{V}$;
❸ $\|f+g\| \le \|f\| + \|g\|$ *for all* f *and* g *in* $\mathcal{V}$ *(the triangle inequality).*

A linear space $\mathcal{V}$, equipped with a norm, is called a **normed linear space**.

Throughout this chapter $[a,b]$ will denote a nonempty, bounded and closed interval of $\mathbb{R}$ with $a < b$, and (a,b) will signify a nonempty, bounded open interval of $\mathbb{R}$.

Example 8.1 *The set* $\mathrm{C}[a,b]$ *of real-valued functions* f*, defined and continuous on the interval* $[a,b]$*, is a normed linear space with norm*

$$\|f\|_\infty = \max_{x\in[a,b]} |f(x)| \,. \tag{8.1}$$

The norm $\|\cdot\|_\infty$ is called the ∞**-norm** or **maximum norm**; it can be thought of as an analogue of the ∞-norm for vectors introduced in Chapter 2. Thus, for the sake of notational simplicity, here we shall use the same symbol $\|\cdot\|_\infty$ as in Chapter 2, tacitly assuming in what follows that $\|f\|_\infty$ signifies the ∞-norm of a continuous function f, defined on a bounded closed interval of the real line (rather than the ∞-norm of an n-component vector). The choice of the interval $[a,b]$ over which the norm is taken will always be clear from the context and will not be explicitly highlighted in our notation. ◇

Example 8.2 *Suppose that* w *is a real-valued function, defined, continuous, positive and integrable on the interval* (a,b)*. The set* $\mathrm{C}[a,b]$ *of real-valued functions* f*, defined and continuous on* $[a,b]$*, is a normed linear space equipped with the norm*

$$\|f\|_2 = \left(\int_a^b w(x)|f(x)|^2\,\mathrm{d}x\right)^{1/2} . \tag{8.2}$$

The norm $\|\cdot\|_2$ is called the **2-norm**. The function w is called a **weight function**. The assumptions on w allow for singular weight functions, such as $w\colon x \in (0,1) \mapsto x^{-1/2}$ which is continuous, positive and integrable on the open interval $(0,1)$, but is not continuous on the closed interval $[0,1]$. The norm (8.2) can be thought of as an analogue of the 2-norm for vectors introduced in Chapter 2; thus, for the sake of simplicity, we use the same notation, $\|\cdot\|_2$, as there. As for the ∞-norm, we shall not explicitly indicate in our notation the interval over which the norm is taken. The implied choice of interval $[a,b]$ and weight function w will be clear from the context. ◇

The next lemma provides a comparison of the ∞-norm with the 2-norm, defined by (8.1) and (8.2), respectively, on $\mathrm{C}[a,b]$.

Lemma 8.1 *(i) Suppose that the real-valued weight function w is defined, continuous, positive and integrable on the interval (a,b). Then, for any function $f \in \mathrm{C}[a,b]$,*

$$\|f\|_2 \leq W\,\|f\|_\infty\,, \qquad \textit{where } W = \left[\int_a^b w(x)\mathrm{d}x\right]^{1/2}.$$

(ii) Given any two positive numbers ε (however small) and M (however large), there exists a function $f \in \mathrm{C}[a,b]$ such that

$$\|f\|_2 < \varepsilon\,, \qquad \|f\|_\infty > M\,.$$

Proof The proof is left as an exercise (see Exercise 1). □

The definitions (2.34) and (2.33) of the vector norms $\|\cdot\|_\infty$ and $\|\cdot\|_2$ on $\mathbb{R}^n$ imply that

$$n^{-1/2}\|\boldsymbol{v}\|_\infty \leq \|\boldsymbol{v}\|_2 \leq n^{1/2}\|\boldsymbol{v}\|_\infty \quad \forall \boldsymbol{v} \in \mathbb{R}^n\,, \tag{8.3}$$

which means that, to all intents and purposes, these two norms are interchangeable.[1] Lemma 8.1 indicates that a similar chain of inequalities cannot possibly hold for the norms (8.1) and (8.2) on $\mathrm{C}[a,b]$, and the choice between them may therefore significantly influence the outcome of the analysis.

Stimulated by the first axiom of *norm*, we shall think of $f \in \mathrm{C}[a,b]$ as being well approximated by a polynomial p on $[a,b]$ if $\|f-p\|$ is small, where $\|\cdot\|$ is either $\|\cdot\|_\infty$ or $\|\cdot\|_2$ defined, respectively, by (8.1) or (8.2). In the light of Lemma 8.1, it should come as no surprise that the mathematical tools for the analysis of smallness of $\|f-p\|_\infty$ are quite different from those that ensure smallness of $\|f-p\|_2$. We have therefore chosen to discuss these two matters separately: the present chapter focuses on the ∞-norm (8.1), while Chapter 9 explores the use of the 2-norm (8.2).

Despite the fundamental differences between the norms (8.1) and (8.2) which we have alluded to above, there is a common underlying feature which is independent of the choice of norm: if *no limitation is imposed*

[1] The chain of inequalities (8.3) is, in fact, just a particular manifestation of the following general result from linear algebra. Suppose that $\mathcal{V}$ is a finite-dimensional linear space and let $\|\cdot\|'$ and $\|\cdot\|''$ be two norms on $\mathcal{V}$; then, there exist positive real numbers m and M such that

$$m\|v\|' \leq \|v\|'' \leq M\|v\|' \qquad \forall v \in \mathcal{V}\,.$$

on the degree of the approximating polynomial p, then the approximation error $f-p$ can be made *arbitrarily small* in both norms. This is a central result in the theory of polynomial approximation and is formulated in the next theorem.

Theorem 8.1 (Weierstrass Approximation Theorem[1]) *Suppose that f is a real-valued function, defined and continuous on a bounded closed interval $[a,b]$ of the real line; then, given any $\varepsilon > 0$, there exists a polynomial p such that*

$$\|f - p\|_\infty \le \varepsilon.$$

Further, if w is a real-valued function, defined, continuous, positive and integrable on (a,b), then an analogous result holds in the 2-norm over the interval $[a,b]$ with weight function w.

This is an important theorem in classical analysis, and several proofs are known. It is evidently sufficient to consider only the interval $[0,1]$; a simple change of variable will then extend the proof to any bounded closed interval $[a,b]$. For a real-valued function f, defined and continuous on the interval $[0,1]$, Bernstein's proof uses the polynomial

$$p_n(x) = \sum_{k=0}^{n} p_{nk}(x) f(k/n)\,, \qquad x \in [0,1]\,,$$

where the **Bernstein polynomials** $p_{nk}(x)$ are defined by

$$p_{nk}(x) = \left(\begin{array}{c} n \\ k \end{array}\right) x^k (1-x)^{n-k}\,, \qquad x \in [0,1]\,.$$

It can then be shown that, for any $\varepsilon > 0$, there exists $n = n(\varepsilon)$ such that $\|f - p_n\|_\infty < \varepsilon$. The second part of the theorem is a direct consequence of this result, using part (i) of Lemma 8.1.

The details of the proof are given in Exercise 12. For an alternative proof, the reader is referred to Theorem 6.3 in M.J.D. Powell, *Approximation Theory and Methods*, Cambridge University Press, 1996.

[1] Karl Theodor Wilhelm Weierstrass (31 October 1815, Ostenfelde, Bavaria, Germany – 19 February 1897, Berlin, Germany) is frequently referred to as the father of modern mathematical analysis. He made fundamental contributions to the theory of series, functions of real variables, elliptic functions, converging infinite products, the calculus of variations, and the theory of bilinear and quadratic forms. Weierstrass' students included Cantor, Frobenius, Gegenbauer, Hölder, Hurwitz, Killing, Klein, Kneser, Sofia Kovalevskaya, Lie, Mertens, Minkowski, Mittag-Leffler, Schwarz and Stolz.

8.3 Best approximation in the ∞-norm

According to the Weierstrass Approximation Theorem any function f in $\mathrm{C}[a,b]$ can be approximated arbitrarily well from the set of *all* polynomials. Clearly, if instead of the set of all polynomials we restrict ourselves to the set of polynomials $\mathcal{P}_n$ of degree n or less, with n *fixed*, then it is no longer true that, for any $f \in \mathrm{C}[a,b]$ and any $\varepsilon > 0$, there exists $p_n \in \mathcal{P}_n$ such that

$$\|f - p_n\|_\infty < \varepsilon\,.$$

Consider, for example, the function $x \mapsto \sin x$ defined on the interval $[0,\pi]$ and fix $n = 0$; then $\|f - q\|_\infty \geq 1/2$ for any $q \in \mathcal{P}_0$, and therefore there is no q in $\mathcal{P}_0$ such that $\|f - q\|_\infty < 1/2$. A similar situation will arise if $\mathcal{P}_0$ is replaced by $\mathcal{P}_n$, with the polynomial degree n fixed.[1]

It is therefore relevant to enquire just how well a given function f in $\mathrm{C}[a,b]$ may be approximated by polynomials of a fixed degree $n \geq 0$. This question leads us to the following approximation problem.

(A) Given that $f \in \mathrm{C}[a,b]$ and $n \geq 0$, fixed, find $p_n \in \mathcal{P}_n$ such that

$$\|f - p_n\|_\infty = \inf_{q \in \mathcal{P}_n} \|f - q\|_\infty\,;$$

such a polynomial p_n is called a **polynomial of best approximation of degree n to the function f in the ∞-norm**.

The next theorem establishes the existence of a polynomial of best approximation, showing, in particular, that the infimum of $\|f - q\|_\infty$ over $q \in \mathcal{P}_n$ is attained. We shall consider the question of uniqueness of the polynomial of best approximation later on, in Theorem 8.5.

Theorem 8.2 *Given that $f \in \mathrm{C}[a,b]$, there exists a polynomial $p_n \in \mathcal{P}_n$ such that $\|f - p_n\|_\infty = \min_{q \in \mathcal{P}_n} \|f - q\|_\infty$.*

Proof Let us define the function $(c_0, \ldots, c_n) \in \mathbb{R}^{n+1} \mapsto E(c_0, \ldots, c_n)$ of $n+1$ real variables by

$$E(c_0, \ldots, c_n) = \|f - q_n\|_\infty, \quad \text{where} \quad q_n(x) = c_0 + \cdots + c_n x^n.$$

[1] This is due to the fact that, for any fixed n, $\mathcal{P}_n$ is a closed subset of $\mathrm{C}[a,b]$; *i.e.*, if f does not belong to $\mathcal{P}_n$, there exists $\varepsilon > 0$ such that

$$\inf_{q \in \mathcal{P}_n} \|f - q\|_\infty > \varepsilon\,.$$

On the other hand, by the Weierstrass Theorem, the set of *all* polynomials is dense in $\mathrm{C}[a,b]$: any continuous function f can be represented as a limit of a uniformly convergent sequence of polynomials (of, in general, increasing degree) on $[a,b]$.

We shall first show that E is continuous; this will imply that E attains its bounds on any bounded closed set in $\mathbb{R}^{n+1}$. We shall then construct a nonempty bounded closed set $\mathcal{S} \subset \mathbb{R}^{n+1}$ such that the lower bound of E on $\mathcal{S}$ is the same as its lower bound over the whole of $\mathbb{R}^{n+1}$.

To show that E is continuous at each point $(c_0, \dots, c_n) \in \mathbb{R}^{n+1}$, consider any $(\delta_0, \dots, \delta_n) \in \mathbb{R}^{n+1}$ and define the polynomial $\eta_n \in \mathcal{P}_n$ by $\eta_n(x) = \delta_0 + \cdots + \delta_n x^n$. We see from the triangle inequality that

$$\begin{aligned} E(c_0 + \delta_0, \dots, c_n + \delta_n) &= \|f - (q_n + \eta_n)\|_\infty \\ &\le \|f - q_n\|_\infty + \|\eta_n\|_\infty \\ &= E(c_0, \dots, c_n) + \|\eta_n\|_\infty \,. \end{aligned}$$

Now, for any given positive number ε, choose $\delta = \varepsilon/(1 + \cdots + K^n)$, where $K = \max\{|a|, |b|\}$. Consider any $(\delta_0, \dots, \delta_n) \in \mathbb{R}^{n+1}$ such that $|\delta_i| \le \delta$ for all $i = 0, \dots, n$. Then,

$$\begin{aligned} E(c_0 + \delta_0, \dots, c_n + \delta_n) - E(c_0, \dots, c_n) &\le \|\eta_n\|_\infty \\ &\le \max_{x \in [a,b]} \left(|\delta_0| + |\delta_1||x| + \cdots + |\delta_n||x|^n\right) \\ &\le \delta(1 + \cdots + K^n) \\ &= \varepsilon \,. \end{aligned} \tag{8.4}$$

Similarly,

$$\begin{aligned} E(c_0, \dots, c_n) &= \|f - q_n\|_\infty = \|f - (q_n + \eta_n) + \eta_n\|_\infty \\ &\le \|f - (q_n + \eta_n)\|_\infty + \|\eta_n\|_\infty \\ &\le E(c_0 + \delta_0, \dots, c_n + \delta_n) + \varepsilon \,, \end{aligned}$$

and therefore

$$E(c_0, \dots, c_n) - E(c_0 + \delta_0, \dots, c_n + \delta_n) \le \varepsilon \,. \tag{8.5}$$

From (8.4) and (8.5) we deduce that

$$|E(c_0 + \delta_0, \dots, c_n + \delta_n) - E(c_0, \dots, c_n)| \le \varepsilon$$

for all $(\delta_0, \dots, \delta_n) \in \mathbb{R}^{n+1}$ such that $|\delta_i| \le \delta$, $i = 0, \dots, n$, where now $\delta = \varepsilon/(1 + \cdots + K^n)$ and $K = \max\{|a|, |b|\}$. Hence E is continuous at $(c_0, \dots, c_n) \in \mathbb{R}^{n+1}$. Since $(c_0, \dots, c_n)$ is an arbitrary point in $\mathbb{R}^{n+1}$, it follows that E is continuous on the whole of $\mathbb{R}^{n+1}$.

Let us denote by $\mathcal{S}$ the set of all points $(c_0, \dots, c_n)$ in $\mathbb{R}^{n+1}$ such that $E(c_0, \dots, c_n) \le \|f\|_\infty + 1$. The set $\mathcal{S}$ is evidently bounded and closed in $\mathbb{R}^{n+1}$; further, $\mathcal{S}$ is nonempty since $E(0, \dots, 0) = \|f\|_\infty \le \|f\|_\infty + 1$, so that $(0, \dots, 0) \in \mathcal{S}$. Hence the continuous function E attains its lower bound over the set $\mathcal{S}$; let us denote this lower bound by d and let $(c_0^*, \dots, c_n^*)$ denote the point in $\mathcal{S}$ where it is attained.

Since $(0, \dots, 0) \in \mathcal{S}$, it follows that

$$d = \min_{(c_0,\dots,c_n)\in\mathcal{S}} E(c_0, \dots, c_n) \le E(0, \dots, 0) = \|f\|_\infty \,.$$

According to the definition of $\mathcal{S}$,

$$E(c_0, \dots, c_n) > \|f\|_\infty + 1 \qquad \forall\, (c_0, \dots, c_n) \in \mathbb{R}^{n+1} \setminus \mathcal{S}\,.$$

Hence, if $(c_0, \dots, c_n) \notin \mathcal{S}$, then $E(c_0, \dots, c_n) > d + 1 > d$. Thus, the lower bound d of the function E over the set $\mathcal{S}$ is the same as the lower bound of E over all values of $(c_0, \dots, c_n) \in \mathbb{R}^{n+1}$. The lower bound d is attained at a point $(c_0^*, \dots, c_n^*)$ in $\mathcal{S}$; letting $p_n^*(x) = c_0^* + \dots + c_n^* x^n$, we find that $d = \|f - p_n^*\|_\infty$ and therefore p_n^* is the required polynomial of best approximation of degree n to the function f in the ∞-norm. $\square$

Due to the nonconstructive nature of its proof, the last theorem does not actually tell us how to find a polynomial of best approximation of degree n for a given function $f \in \mathrm{C}[a, b]$. Therefore, our goal is now to devise a constructive characterisation of the property '*p_n is a polynomial of best approximation of degree n to the function f in the ∞-norm*'. Before doing so, however, let us simplify our terminology.

Writing the polynomial $q \in \mathcal{P}_n$ in the form

$$q_n(x) = c_0 + \dots + c_n x^n \,,$$

we want to choose the coefficients c_j, $j = 0, \dots, n$, so that they minimise the function E: $(c_0, \dots, c_n) \mapsto E(c_0, \dots, c_n)$ defined by

$$\begin{aligned} E(c_0, \dots, c_n) &= \|f - q\|_\infty \\ &= \max_{x\in[a,b]} |f(x) - c_0 - \dots - c_n x^n| \end{aligned}$$

over $\mathbb{R}^{n+1}$. Since the polynomial of best approximation is to *minimise* (over $q \in \mathcal{P}_n$) the *maximum* absolute value of the error $f(x) - q(x)$ (over $x \in [a, b]$), it is often referred to as the **minimax polynomial**; from now on, for the sake of brevity, we shall use the latter terminology.

Before we embark on the constructive characterisation of the minimax polynomial of a continuous function, let us consider a simple example which illustrates some of its key properties.

Example 8.3 *Suppose that $f \in \mathrm{C}[0, 1]$, and that f is strictly monotonic increasing on $[0, 1]$. We wish to find the minimax polynomial p_0 of degree zero for f on $[0, 1]$.*

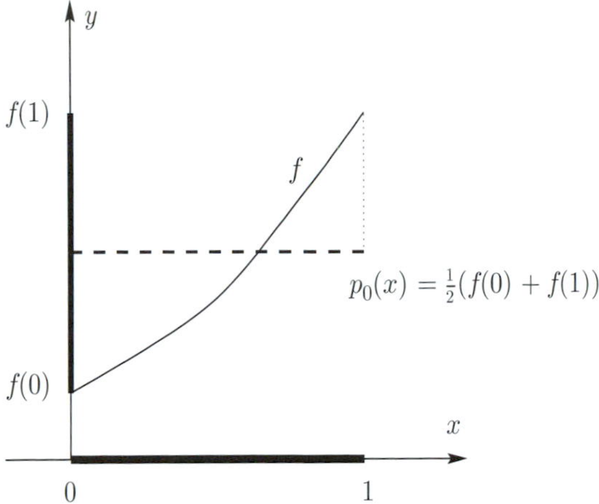

Fig. 8.1. Minimax approximation $p_0 \in \mathcal{P}_0$ of a strictly monotonic increasing continuous function f defined on the interval $[0, 1]$.

The polynomial p_0 will be of the form $p_0(x) \equiv c_0$, and we need to determine $c_0 \in \mathbb{R}$ so that

$$\|f - p_0\|_\infty = \max_{x \in [0,1]} |f(x) - c_0|$$

is minimal. Since f is monotonic increasing, $f(x) - c_0$ attains its minimum at $x = 0$ and its maximum at $x = 1$; therefore $|f(x) - c_0|$ reaches its maximum value at one of the endpoints of $[0, 1]$, *i.e.*,

$$E(c_0) = \max_{x \in [0,1]} |f(x) - c_0| = \max\{|f(0) - c_0|, |f(1) - c_0|\} .$$

Clearly,

$$E(c_0) = \begin{cases} f(1) - c_0 & \text{if } c_0 < \frac{1}{2}\,(f(0) + f(1)) , \\ c_0 - f(0) & \text{if } c_0 \geq \frac{1}{2}\,(f(0) + f(1)) . \end{cases}$$

Drawing the graph of the function $c_0 \in \mathbb{R} \mapsto E(c_0) \in \mathbb{R}$ shows that the minimum is attained when $c_0 = \frac{1}{2}\,(f(0) + f(1))$. Consequently, the desired minimax polynomial of degree 0 for the function f is

$$p_0(x) \equiv \tfrac{1}{2}\,(f(0) + f(1)) , \qquad x \in [0, 1] .$$

The function f and its minimax approximation $p_0 \in \mathcal{P}_0$ are depicted in Figure 8.1.

More generally, if $f \in \mathrm{C}[a, b]$ (not necessarily monotonic), and ξ and η denote two points in $[a, b]$ where f attains its global minimum and maximum values, respectively, then the minimax polynomial of degree zero to f on $[a, b]$ is

$$p_0(x) \equiv \tfrac{1}{2}\left(f(\xi) + f(\eta)\right), \qquad x \in [a, b].$$

◇

This example shows that the minimax polynomial p_0 of degree *zero* for $f \in \mathrm{C}[a, b]$ has the property that the approximation error $f - p_0$ attains its extrema at *two* points, $x = \xi$ and $x = \eta$, with the error

$$f(x) - p_0(x) = \frac{1}{2}\left(f(x) - f(\xi)\right) + \frac{1}{2}\left(f(x) - f(\eta)\right)$$

being *negative at one point*, $x = \xi$, and *positive at the other*, $x = \eta$. We shall prove that a property of this kind holds in general; the precise formulation of the general result is given in Theorem 8.4 which is, due to the oscillating nature of the approximation error, usually referred to as the Oscillation Theorem: it gives a complete characterisation of the minimax polynomial and provides a method for its construction. We begin with a preliminary result due to de la Vallée Poussin.[1]

Theorem 8.3 (De la Vallée Poussin's Theorem) *Let $f \in \mathrm{C}[a, b]$ and $r \in \mathcal{P}_n$. Suppose that there exist $n + 2$ points $x_0 < \cdots < x_{n+1}$ in the interval $[a, b]$, such that $f(x_i) - r(x_i)$ and $f(x_{i+1}) - r(x_{i+1})$ have opposite signs, for $i = 0, \ldots, n$. Then,*

$$\min_{q \in \mathcal{P}_n} \|f - q\|_\infty \geq \min_{i=0,1,\ldots,n+1} |f(x_i) - r(x_i)|. \tag{8.6}$$

Proof The condition on the signs of $f(x_i) - r(x_i)$ is usually expressed by saying that $f - r$ has alternating signs at the points x_i, $i = 0, 1, \ldots, n+1$. Let us denote the right-hand side of (8.6) by μ. Clearly, $\mu \geq 0$; when $\mu = 0$ the statement of the theorem is trivially true, so we shall assume that $\mu > 0$. Suppose that (8.6) is false; then, for a minimax polynomial approximation $p_n \in \mathcal{P}_n$ to the function f we have[2]

$$\|f - p_n\|_\infty = \min_{q \in \mathcal{P}_n} \|f - q\|_\infty < \mu.$$

[1] Charles Jean Gustave Nicolas, Baron de la Vallée Poussin (14 August 1866, Louvain, Belgium – 2 March 1962, Louvain, Belgium) made important contributions to approximation theory and number theory, proving in 1892 that the number of primes less than n is, asymptotically as $n \to \infty$, $n/\ln n$.

[2] Recall from Theorem 8.2 that such p_n exists.

Therefore,

$$|p_n(x_i) - f(x_i)| < |r(x_i) - f(x_i)|, \quad i = 0, 1, \ldots, n+1.$$

Now,

$$r(x_i) - p_n(x_i) = [r(x_i) - f(x_i)] - [p_n(x_i) - f(x_i)], \quad i = 0, 1, \ldots, n+1.$$

Since the first term on the right always exceeds the second term in absolute value, it follows that $r(x_i) - p_n(x_i)$ and $r(x_i) - f(x_i)$ have the same sign for $i = 0, 1, \ldots, n+1$. Hence $r - p_n$, which is a polynomial of degree n, changes sign $n+1$ times. Thus, the assumption that (8.6) is false has led to a contradiction, and the proof is complete. □

Theorem 8.3 gives a clue to formulating a constructive characterisation of the *minimax polynomial*: indeed, we shall show that if the quantities $|f(x_i) - r(x_i)|$, $i = 0, 1, \ldots, n+1$, in Theorem 8.3 are all equal to $\|f - r\|_\infty$, then $r \in \mathcal{P}_n$ is, in fact, a minimax polynomial of degree n for the function f on the interval $[a, b]$.

Theorem 8.4 (The Oscillation Theorem) *Suppose that $f \in \mathrm{C}[a, b]$. A polynomial $r \in \mathcal{P}_n$ is a minimax polynomial for f on $[a, b]$ if, and only if, there exists a sequence of $n+2$ points x_i, $i = 0, 1, \ldots, n+1$, such that $a \le x_0 < \cdots < x_{n+1} \le b$,*

$$|f(x_i) - r(x_i)| = \|f - r\|_\infty, \quad i = 0, 1, \ldots, n+1,$$

and

$$f(x_i) - r(x_i) = -[f(x_{i+1}) - r(x_{i+1})], \quad i = 0, \ldots, n.$$

The statement of the theorem is often expressed by saying that $f - r$ attains its maximum absolute value with alternating signs at the points x_i. The points x_i, $i = 0, 1, \ldots, n+1$, in the Oscillation Theorem are referred to as **critical points**.

Proof of theorem If $f \in \mathcal{P}_n$, then the result is trivially true, with $r = f$ and any sequence of $n+2$ distinct points x_i, $i = 0, 1, \ldots, n+1$, contained in $[a, b]$. Thus, we shall suppose throughout the proof that $f \notin \mathcal{P}_n$, *i.e.*, f is such that there is no polynomial $p \in \mathcal{P}_n$ whose restriction to $[a, b]$ is identically equal to f.

The sufficiency of the condition stated in the theorem is easily shown. Suppose that the sequence of points x_i, $i = 0, 1, \ldots, n+1$, exists with the given properties. Define

$$L = \|f - r\|_\infty \qquad \text{and} \qquad E_n(f) = \min_{q \in \mathcal{P}_n} \|f - q\|_\infty .$$

From De la Vallée Poussin's Theorem, Theorem 8.3, it follows that $E_n(f) \geq L$. By the definition of $E_n(f)$ we also see that $E_n(f) \leq \|f - r\|_\infty = L$. Hence $E_n(f) = L$, and the given polynomial r is a minimax polynomial.

For the necessity of the condition, suppose that the given polynomial $r \in \mathcal{P}_n$ is a minimax polynomial for f on $[a, b]$. As $x \mapsto |f(x) - r(x)|$ is a continuous function on the bounded closed interval $[a, b]$, there exists a point in $[a, b]$ at which $|f(x) - r(x)|$ attains its maximum value, $L > 0$; let

$$x_0 = \min\{x \in [a, b]\colon |f(x) - r(x)| = L\} .$$

We may assume without loss of generality that $f(x_0) - r(x_0) = L > 0$. If $x_0 = b$, then $-L < f(x) - r(x) \leq L$ for all $x \in [a, b]$. As f is continuous on $[a, b]$, there exists $\delta \in (0, L)$ such that $-L + \delta \leq f(x) - r(x) \leq L$ for all $x \in [a, b]$. Let $\varepsilon \in (0, \delta)$ and define $r^*(x) = r(x) + \varepsilon$. Then $r^* \in \mathcal{P}_n$ and $\|f - r^*\|_\infty < L = \|f - r\|_\infty$; hence, $r^* \in \mathcal{P}_n$ is a better approximation to f on $[a, b]$ than $r \in \mathcal{P}_n$ is. The contradiction implies that $x_0 < b$.

Now, we shall prove the existence of the next critical point, $x_1 \in (x_0, b]$ such that $f(x_1) - r(x_1) = -L$. Suppose otherwise, for contradiction; then, $-L < f(x) - r(x) \leq L$ for all x in $[a, b]$. Thus, by the continuity of f, there exists $\delta \in (0, L)$ such that $-L + \delta \leq f(x) - r(x) \leq L$ for all $x \in [a, b]$. Let us define $r^* \in \mathcal{P}_n$ by

$$r^*(x) = r(x) + \varepsilon ,$$

where $0 < \varepsilon < \min\{\delta, L\} = \delta$. Then, for all $x \in [a, b]$,

$$f(x) - r^*(x) = f(x) - r(x) - \varepsilon \geq -L + \delta - \varepsilon > -L$$

and

$$f(x) - r^*(x) = f(x) - r(x) - \varepsilon \leq L - \varepsilon < L ,$$

which means that

$$\|f - r^*\|_\infty < L = \|f - r\|_\infty .$$

Hence, $r^* \in \mathcal{P}_n$ is a better approximation to f on $[a, b]$ than $r \in \mathcal{P}_n$ is. This, however, contradicts our hypothesis that r is a polynomial of best approximation to f on $[a, b]$ from $\mathcal{P}_n$, and implies the existence of

$$x_1 = \inf\{x \in (x_0, b]\colon f(x) - r(x) = -L\} .$$

Consequently, $f(x_1) - r(x_1) = -L$ and $x_1 \in (x_0, b]$, as required; thus if $n = 0$, the proof is complete.

Let us, therefore, suppose that $n \geq 1$, and successively define the critical points

$$x_i = \inf\{x \in (x_{i-1}, b]\colon f(x) - r(x) = (-1)^i L\}\,, \qquad i = 1, \dots, m\,,$$

continuing either until $x_m = b$ or until we find an $x_m < b$ such that $|f(x) - r(x)| < L$ for all $x \in (x_m, b]$. Now, *either* $m \geq n+1$, and then the proof is complete as we will have found $n+2$ critical points, $x_0 < x_1 < \dots < x_{n+1}$ in $[a, b]$, with the required properties, *or* $1 \leq m \leq n$.

To complete the proof of the theorem, we shall show that the second alternative, $1 \leq m \leq n$, leads to a contradiction, and is, therefore, not possible. Let us suppose, for this purpose, that $1 \leq m \leq n$, and let $\eta_0 = a$. Further, observe that, due to the definition of the points x_i, $i = 0, 1, \dots, m$,

$$\exists \eta_i \in (x_{i-1}, x_i) \quad \forall x \in [\eta_i, x_i) \quad |f(x) - r(x)| < L\,, \quad i = 1, \dots, m\,,$$

and define $\eta_{m+1} = b$.

It follows from the choice of the η_i, $i = 0, 1, \dots, m+1$, that the following properties hold:

(a) $|f(x) - r(x)| \leq L$ for all $x \in [\eta_i, \eta_{i+1}]$ and all $i = 0, 1, \dots, m$;
(b) for each $i = 0, 1, \dots, m$ there exists $x \in [\eta_i, \eta_{i+1}]$ (say, $x = x_i$), such that $f(x) - r(x) = (-1)^i L$;
(c) there exist no $i \in \{0, 1, \dots, m\}$ and $x \in [\eta_i, \eta_{i+1}]$ such that $f(x) - r(x) = (-1)^{i+1} L$;
(d) $|f(\eta_i) - r(\eta_i)| < L$ for all $i = 1, \dots, m$.

Now, let

$$v(x) = \prod_{i=1}^{m} (\eta_i - x)\,,$$

and define

$$r^*(x) = r(x) + \varepsilon v(x)\,,$$

where $\varepsilon > 0$ is a fixed real number, to be chosen below. Since, by hypothesis, $1 \leq m \leq n$, it follows that $r^* \in \mathcal{P}_n$. Let us consider the behaviour of the difference

$$f(x) - r^*(x) = f(x) - r(x) - \varepsilon v(x)$$

on each of the intervals $[\eta_i, \eta_{i+1}]$, $i = 0, 1, \ldots, m$ (whose union is $[a, b]$). We shall prove that, for $\varepsilon > 0$ sufficiently small,

$$|f(x) - r^*(x)| < L = \|f - r\|_\infty$$

for all x in $[\eta_i, \eta_{i+1}]$ and all $i = 0, 1, \ldots, m$; *i.e.*, $\|f - r^*\|_\infty < \|f - r\|_\infty$, contradicting the fact that $r \in \mathcal{P}_n$ is a minimax polynomial for f on $[a, b]$, and refuting the hypothesis that $1 \le m \le n$.

Take, for example, the interval $[\eta_0, \eta_1]$. For each x in $[\eta_0, \eta_1)$ we have $v(x) > 0$ and therefore, by the definition of $r^*(x)$ and property (a) above,

$$f(x) - r^*(x) \le L - \varepsilon v(x) < L\,, \qquad x \in [\eta_0, \eta_1).$$

Further, as $v(\eta_1) = 0$, it follows from (d) that

$$f(\eta_1) - r^*(\eta_1) = f(\eta_1) - r(\eta_1) < L\,.$$

Therefore, $f(x) - r^*(x) < L$ for each x in $[\eta_0, \eta_1]$. For a lower bound on $f(x) - r^*(x)$, note that by (a) and (c), $f(x) - r(x) > -L$ for all x in $[\eta_0, \eta_1]$. As $f - r$ is a continuous function on $[\eta_0, \eta_1]$, there exists $\delta_1 \in (0, L)$ such that $f(x) - r(x) \ge -L + \delta_1$ for all x in $[\eta_0, \eta_1]$. Thus, for $0 < \varepsilon < \min\{L, \delta_1, \varepsilon_1\}$, where

$$\varepsilon_1 = \frac{\delta_1}{\max_{x \in [\eta_0, \eta_1]} |v(x)|}\,,$$

we have that

$$f(x) - r^*(x) \ge -L + \delta_1 - \varepsilon |v(x)| > -L\,, \qquad x \in [\eta_0, \eta_1)\,.$$

Further, by (d) above,

$$f(\eta_1) - r^*(\eta_1) = f(\eta_1) - r(\eta_1) > -L\,.$$

Hence, $f(x) - r^*(x) > -L$ for all $x \in [\eta_0, \eta_1]$, for $0 < \varepsilon < \min\{L, \delta_1, \varepsilon_1\}$. Combining the upper and lower bounds on $f(x) - r^*(x)$, we deduce that

$$|f(x) - r^*(x)| < L = \|f - r\|_\infty\,, \qquad x \in [\eta_0, \eta_1]\,.$$

Arguing in the same manner on each of the other intervals $[\eta_i, \eta_{i+1}]$, $i = 1, \ldots, m$, with $0 < \varepsilon < \min\{L, \delta_{i+1}, \varepsilon_{i+1}\}$, $i = 1, \ldots, m$, and δ_{i+1} and ε_{i+1} defined analogously to δ_1 and ε_1 above, we conclude that

$$|f(x) - r^*(x)| < L = \|f - r\|_\infty\,, \qquad x \in [\eta_i, \eta_{i+1}]\,, \quad i = 0, 1, \ldots, m\,,$$

and hence, for $0 < \varepsilon < \min\{L, \delta_1, \varepsilon_1, \ldots, \delta_{m+1}, \varepsilon_{m+1}\}$,

$$\|f - r^*\|_\infty < L = \|f - r\|_\infty\,.$$

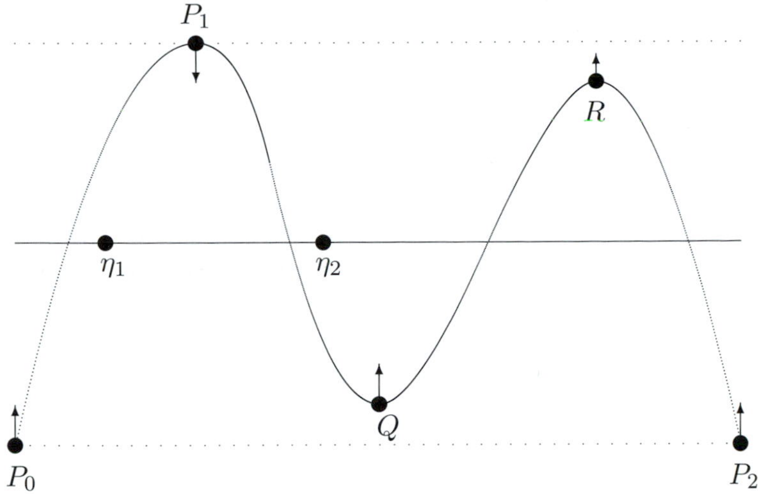

Fig. 8.2. The Oscillation Theorem: the difference $f(x) - r(x)$, where r is a cubic approximation to a continuous function f, and the effect of replacing $r(x)$ by $r^*(x) = r(x) - \varepsilon v(x)$, where $v(x) = (\eta_1 - x)(\eta_2 - x)$.

As r^* is in $\mathcal{P}_n$, the last inequality contradicts our assumption that r is a polynomial of best approximation to f on $[a, b]$ from $\mathcal{P}_n$. The contradiction rules out the possibility that $1 \le m \le n$. Since $m \ge 1$, it follows that $m \ge n + 1$, and the proof is complete. □

In the proof of the Oscillation Theorem we supposed, without loss of generality, that $f(x_0) - r(x_0) = L > 0$, where $L = \|f - r\|_\infty$. When $f(x_0) - r(x_0) = -L < 0$ the proof is analogous, except we then define $r^*(x) = r(x) - \varepsilon$ to prove the existence of the critical point $x_1 \in (x_0, b]$ and, in the discussion of the case $1 \le m \le n$, we let

$$r^*(x) = r(x) - \varepsilon v(x)\,,$$

with $v(x)$ and $\varepsilon > 0$ defined as before.

A typical situation is illustrated in Figure 8.2, which represents the difference $f - r$, where r is a polynomial approximation of degree 3 to a continuous function f. Here $|f - r|$ attains its maximum value with alternate signs at the points P_0, P_1 and P_2, so that $m = 2 < n = 3$. Let x_0, x_1 and x_2 denote the x-coordinates of P_0, P_1, P_2, respectively. Clearly, $f(x_0) - r(x_0) = -L < 0$, where $L = \|f - r\|_\infty$. Also, the two points η_1 and η_2 are as shown, $v(x) = (\eta_1 - x)(\eta_2 - x)$, and the effect of replacing r by $r^*(x) = r(x) - \varepsilon v(x)$, with $\varepsilon > 0$, is indicated by the arrows. Since $f - r^* = f - r + \varepsilon v(x)$ and v is negative for $x \in (\eta_1, \eta_2)$ and positive outside (η_1, η_2), $|f - r^*|$ will be smaller than $|f - r|$ at each of the points P_i, $i = 0, 1, 2$. There are two other local extrema for the error function $f - r$: a minimum at Q and a maximum at R. Since both these points are to the right of η_2, where $v(x) > 0$, we shall have $f - r^* > f - r$ at both of Q and R, and $|f - r^*| > |f - r|$ at R. The magnitude of the extra term $\varepsilon v(x)$ must therefore be limited by the need to avoid the new difference $f - r^*$ becoming too large at R. We can achieve this by selecting $\varepsilon > 0$ sufficiently small. In this illustration the polynomial $r \in \mathcal{P}_3$ is not a minimax approximation to f on the given interval, since we can construct a better approximation r^* which is also in $\mathcal{P}_3$.

We can now apply the Oscillation Theorem to prove that the minimax polynomial is unique.

Theorem 8.5 (Uniqueness Theorem) *Suppose that $[a, b]$ is a bounded closed interval of the real line. Each $f \in \mathrm{C}[a, b]$ has a unique minimax polynomial $p_n \in \mathcal{P}_n$ on $[a, b]$.*

Proof Suppose that $q_n \in \mathcal{P}_n$ is also a minimax polynomial for f, and that p_n and q_n are distinct. Then,

$$\|f - p_n\|_\infty = \|f - q_n\|_\infty = E_n(f),$$

where, as in the proof of the Oscillation Theorem, we have used the notation

$$E_n(f) = \min_{q \in \mathcal{P}_n} \|f - q\|_\infty .$$

This implies, by the triangle inequality, that

$$\begin{aligned} \|f - \tfrac{1}{2}(p_n + q_n)\|_\infty &= \|\tfrac{1}{2}(f - p_n) + \tfrac{1}{2}(f - q_n)\|_\infty \\ &\leq \tfrac{1}{2}\|f - p_n\|_\infty + \tfrac{1}{2}\|f - q_n\|_\infty \\ &= E_n(f) . \end{aligned}$$

Therefore $\frac{1}{2}(p_n + q_n) \in \mathcal{P}_n$ is also a minimax polynomial approximation to f on $[a, b]$. By the Oscillation Theorem there exists a sequence of $n + 2$ critical points x_i, $i = 0, 1, \ldots, n + 1$, at which

$$\left|f(x_i) - \tfrac{1}{2}\left(p_n(x_i) + q_n(x_i)\right)\right| = E_n(f), \quad i = 0, 1, \ldots, n + 1 .$$

This is equivalent to

$$|\,(f(x_i) - p_n(x_i)) + (f(x_i) - q_n(x_i))\,| = 2E_n(f)\,.$$

Now

$$|f(x_i) - p_n(x_i)| \le \max_{x\in[a,b]} |f(x) - p_n(x)| = \|f - p_n\|_\infty = E_n(f)\,,$$

and, for the same reason,

$$|f(x_i) - q_n(x_i)| \le E_n(f)\,.$$

It therefore follows[1] that

$$f(x_i) - p_n(x_i) = f(x_i) - q_n(x_i)\,, \qquad i = 0, 1, \ldots, n+1\,.$$

Thus, the difference $p_n - q_n$ vanishes at $n+2$ distinct points. As $p_n - q_n$ is a polynomial of degree n or less, it follows that $p_n - q_n$ is identically zero. This, however, contradicts our initial hypothesis that p_n and q_n are distinct, and implies the uniqueness of the minimax polynomial $p_n \in \mathcal{P}_n$ for $f \in \mathrm{C}[a,b]$. □

As an application of the Oscillation Theorem, we consider the construction of the minimax approximation $p_1 \in \mathcal{P}_1$ of degree 1 to a function $f \in \mathrm{C}[a,b]$ on the interval $[a,b]$, where we assume that f has a continuous and strictly monotonic increasing derivative f' on this interval.

We seek the minimax polynomial $p_1 \in \mathcal{P}_1$ in the form $p_1(x) = c_1x + c_0$. The difference $f(x) - (c_1x + c_0)$ attains its extrema either at the endpoints of the interval $[a,b]$ or at points where its derivative $f'(x) - c_1$ is zero. Since f' is strictly monotonic increasing it can only take the value c_1 at one point at most. Therefore the endpoints of the interval, a and b, are critical points. Let us denote by d the third critical point whose location inside (a,b) remains to be determined. Since the critical point $x = d$ is an internal extremum of $f(x) - (c_1x + c_0)$, it follows that

$$(f(x) - (c_1x + c_0))'|_{x=d} = 0\,.$$

By the Oscillation Theorem, with $x_0 = a$, $x_1 = d$, $x_2 = b$, we have the

[1] We use the following elementary result: if P and Q are two real numbers and E is a nonnegative real number such that $|P + Q| = 2E$, $|P| \le E$ and $|Q| \le E$, then $P = Q$. This follows by noting that $(P - Q)^2 = 2P^2 + 2Q^2 - (P + Q)^2 \le 2E^2 + 2E^2 - 4E^2 = 0$, and hence $P - Q = 0$. In the proof of the theorem we apply this with $P = f(x_i) - p_n(x_i)$, $Q = f(x_i) - q_n(x_i)$ and $E = E_n(f)$.

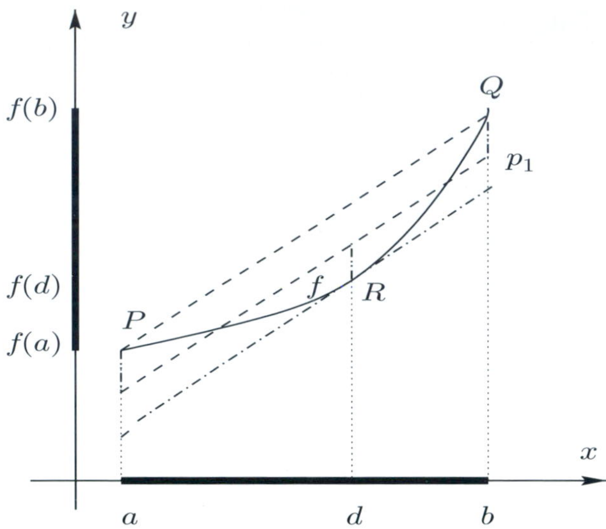

Fig. 8.3. Construction of minimax polynomial of degree 1.

equations

$$\left.\begin{aligned} f(a) - (c_1 a + c_0) &= A\,, \\ f(d) - (c_1 d + c_0) &= -A\,, \\ f(b) - (c_1 b + c_0) &= A\,, \end{aligned}\right\} \tag{8.7}$$

where either $A = L$ or $A = -L$, with $L = \max_{x\in[a,b]} |f(x) - p_1(x)|$. Along with the condition

$$f'(d) = c_1 \tag{8.8}$$

this gives four equations to determine the unknowns d, c_1, c_0 and A.

Subtracting the first equation in (8.7) from the third equation, we get $f(b) - f(a) = c_1(b-a)$, whereby $c_1 = (f(b) - f(a))/(b-a)$. Now, by the Mean Value Theorem, Theorem A.3, with this choice of c_1 equation (8.8) has at least one solution, d, in the open interval (a, b). In fact, the value of d is uniquely determined by (8.8), as f' is continuous and strictly monotonic increasing. Next, c_0 can be determined by adding the second equation in (8.7) to the first. Having calculated both c_1 and c_0 we insert them into the first equation in (8.7) to obtain A; finally $L = |A|$.

The construction of the minimax polynomial p_1 is illustrated in Figure 8.3; R is the point at which the tangent to the curve $y = f(x)$ is parallel to the chord PQ; the graph of $p_1(x)$ is parallel to these two lines, and lies half-way between them.

Table 8.1. *The first seven Chebyshev Polynomials:* $T_0, T_1, \ldots, T_6$.

$T_0(x)$	$=$	1
$T_1(x)$	$=$	x
$T_2(x)$	$=$	$2x^2 - 1$
$T_3(x)$	$=$	$4x^3 - 3x$
$T_4(x)$	$=$	$8x^4 - 8x^2 + 1$
$T_5(x)$	$=$	$16x^5 - 20x^3 + 5x$
$T_6(x)$	$=$	$32x^6 - 48x^4 + 18x^2 - 1$

8.4 Chebyshev polynomials

There are very few functions for which it is possible to write down in simple closed form the minimax polynomial. One such problem of practical importance concerns the approximation of a power of x by a polynomial of lower degree. The minimax approximation in this case is given in terms of Chebyshev polynomials.[1]

Definition 8.2 *The* **Chebyshev polynomial** T_n *of degree* n *is defined, for* $x \in [-1, 1]$, *by*

$$T_n(x) = \cos(n \cos^{-1} x), \qquad n = 0, 1, 2, \ldots .$$

Despite its unusual form, T_n is a polynomial in disguise. For example, $T_0(x) \equiv 1$, $T_1(x) = x$ for all $x \in [-1, 1]$, and so on. In order to show that this is true in general, we recall the trigonometric identity

$$\cos{(n+1)\vartheta} + \cos{(n-1)\vartheta} = 2 \cos\vartheta \cos n\vartheta ,$$

and set $\vartheta = \cos^{-1} x$, with $x \in [-1, 1]$, to obtain the recurrence relation

$$T_{n+1}(x) = 2xT_n(x) - T_{n-1}(x) , \qquad n = 1, 2, 3, \ldots , \qquad x \in [-1, 1] .$$

Since T_0 and T_1 have already been shown to be polynomials on $[-1, 1]$, we deduce from this recurrence relation, by induction, that T_n is a polynomial of degree n on $[-1, 1]$ for each $n \geq 0$. A list of the first seven Chebyshev polynomials is given in Table 8.1.

[1] Pafnuty Lvovich Chebyshev (16 May 1821, Okatovo, Russia – 8 December 1894, St Petersburg, Russia). In 1850 Chebyshev proved the Bertrand conjecture, that there is always at least one prime between n and $2n$ for $n \geq 2$. He also came close to proving the Prime Number Theorem which states that the number of primes less than n is, asymptotically as $n \to \infty$, $n/\ln n$. The proof was completed, independently, by Dirichlet and de la Vallée Poussin two years after Chebyshev's death. Chebyshev made important contributions to probability theory, orthogonal functions and the theory of integrals.

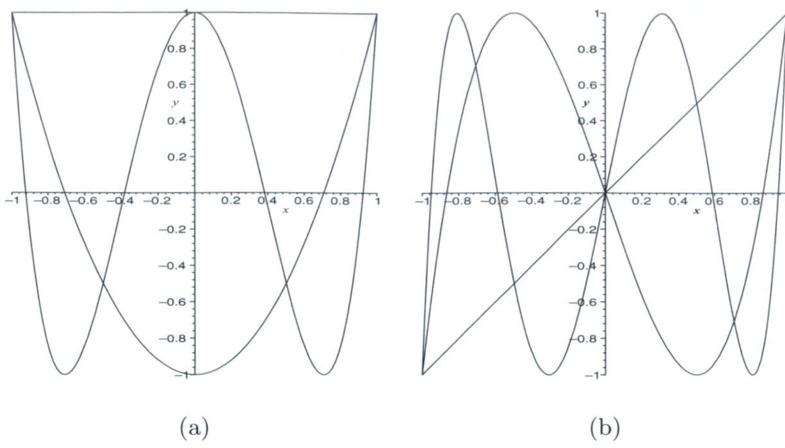

Fig. 8.4. The first three Chebyshev polynomials of (a) even degree, T_0, T_2, T_4, and (b) odd degree T_1, T_3, T_5, plotted on the interval $[-1, 1]$.

The polynomials T_0, T_2, T_4, and T_1, T_3, T_5, are depicted in Figure 8.4. We see that the even-degree Chebyshev polynomials are even functions; (*i.e.*, $T_{2k}(-x) = T_{2k}(x)$ for all $x \in [-1, 1]$) and the odd-degree ones are odd functions (*i.e.*, $T_{2k+1}(-x) = -T_{2k+1}(x)$ for all $x \in [-1, 1]$). They all map the interval $[-1, 1]$ into itself.[1]

The proof of the next lemma is straightforward and is left as an exercise (see Exercise 10).

Lemma 8.2 *The Chebyshev polynomials have the following properties:*

(i) $T_{n+1}(x) = 2xT_n(x) - T_{n-1}(x)$, $x \in [-1, 1]$, $n = 1, 2, 3, \ldots$;

(ii) *for* $n \geq 1$, T_n *is a polynomial in* x *of degree* n *on the interval* $[-1, 1]$, *with leading coefficient* $2^{n-1}x^n$;

(iii) T_n *is an even function on* $[-1, 1]$ *if* n *is even, and an odd function on* $[-1, 1]$ *if* n *is odd,* $n \geq 0$;

(iv) *for* $n \geq 1$, *the zeros of* T_n *are at*

$$x_j = \cos \frac{(2j-1)\pi}{2n}, \qquad j = 1, \ldots, n;$$

[1] In Maple, typing `plot(orthopoly[T](7,x), x=-1..1, y=-1..1);` will, for example, plot the graph of the Chebyshev polynomial T_7 of degree 7 in x; T_8, T_9, etc., can be obtained similarly. Incidentally, you may be wondering why T_n and not C_n is used to denote the Chebyshev polynomial of degree n. The reasons are largely historical: in some older books and articles Chebyshev's Russian surname has been transliterated from the Cyrillic original as Tchebyshev, following the French and German transliterations Tchebychef and Tschebyscheff, respectively.

they are all real and distinct, and lie in $(-1, 1)$*;*

(v) $|T_n(x)| \le 1$ *for all* $x \in [-1, 1]$ *and all* $n \ge 0$*;*

(vi) *for* $n \ge 1$, $T_n(x) = \pm 1$, *alternately at the* $n + 1$ *points* $x_k = \cos(k\pi/n)$, $k = 0, 1, \ldots, n$.

We can now apply the Oscillation Theorem to construct the minimax polynomial of degree n for $f \colon x \mapsto x^{n+1}$ on the interval $[-1, 1]$.

Theorem 8.6 *Suppose that* $n \ge 0$. *The polynomial* $p_n \in \mathcal{P}_n$ *defined by*

$$p_n(x) = x^{n+1} - 2^{-n} T_{n+1}(x) , \qquad x \in [-1, 1] ,$$

is the minimax approximation of degree n *to the function* $x \mapsto x^{n+1}$ *on the interval* $[-1, 1]$.

Proof By part (ii) of Lemma 8.2, $p_n \in \mathcal{P}_n$. Since

$$x^{n+1} - p_n(x) = 2^{-n} T_{n+1}(x) ,$$

by parts (v) and (vi) of Lemma 8.2, the difference $x^{n+1} - p_n(x)$ does not exceed 2^{-n} in the interval $[-1, 1]$, and attains this value with alternating signs at the $n + 2$ points $x_k = \cos(k\pi/(n + 1))$, $k = 0, 1, \ldots, n + 1$. Therefore, by the Oscillation Theorem, p_n is the (unique) minimax polynomial approximation from $\mathcal{P}_n$ to the function $x \mapsto x^{n+1}$ over $[-1, 1]$. □

A polynomial of degree n whose leading coefficient, the coefficient of x^n, is equal to 1, is called a **monic polynomial** of degree n. For example, the polynomial $r \in \mathcal{P}_{n+1}$ defined by $r(x) = x^{n+1} - q(x)$ with $q \in \mathcal{P}_n$, is a monic polynomial of degree $n + 1$.

Corollary 8.1 *Suppose that* $n \ge 0$. *Among all monic polynomials of degree* $n + 1$ *the polynomial* $2^{-n} T_{n+1}$ *has the smallest* ∞*-norm on the interval* $[-1, 1]$.

Proof Let $\mathcal{P}^1_{n+1}$ denote the set of all monic polynomials of degree $n+1$. Any $r \in \mathcal{P}^1_{n+1}$ can be regarded as the difference between the function $x \mapsto x^{n+1}$ and a polynomial of lower degree, *i.e.*, $r(x) = x^{n+1} - q(x)$ with $q \in \mathcal{P}_n$. Hence, by Theorem 8.6,

$$\begin{aligned} \min_{r \in \mathcal{P}^1_{n+1}} \|r\|_\infty &= \min_{q \in \mathcal{P}_n} \|x^{n+1} - q\|_\infty \\ &= \|x^{n+1} - (x^{n+1} - 2^{-n} T_{n+1})\|_\infty \\ &= \|2^{-n} T_{n+1}\|_\infty ; \end{aligned}$$

the minimum is, therefore, achieved when $r \in \mathcal{P}^1_{n+1}$ is the monic polynomial $2^{-n}T_{n+1}$.

□

8.5 Interpolation

We close the body of this chapter with another application of Chebyshev polynomials: it concerns the 'optimal' choice of interpolation points in Lagrange interpolation. In Chapter 6 the error between an $n+1$ times continuously differentiable function f, defined on a closed interval $[a,b]$ of the real line, and its Lagrange interpolation polynomial p_n of degree n, $n \geq 0$, with interpolation points $\xi_0, \ldots, \xi_n$, was shown to have the form

$$f(x) - p_n(x) = \frac{f^{(n+1)}(\eta)}{(n+1)!}\,\pi_{n+1}(x)\,, \tag{8.9}$$

where $\eta = \eta(x) \in (a,b)$ and

$$\pi_{n+1}(x) = (x - \xi_0)\ldots(x - \xi_n)\,. \tag{8.10}$$

Clearly, π_{n+1} is a monic polynomial of degree $n+1$.

In a practical application the values ξ_i and $f(\xi_i)$, $i = 0, 1, \ldots, n$, may be already given. However, in a situation where $[a,b] = [-1,1]$ and the ξ_i, $i = 0, 1, \ldots, n$, can be freely chosen in the interval $[-1,1]$, Corollary 8.1 suggests that they should be taken as the zeros of the Chebyshev polynomial T_{n+1}, for then π_{n+1} will have the smallest ∞-norm on the interval $[-1,1]$ among all monic polynomials. This observation motivates the following result.

Theorem 8.7 *Suppose that f is a real-valued function, defined and continuous on the closed real interval $[a,b]$, and such that the derivative of f of order $n+1$ is continuous on $[a,b]$. Let $p_n \in \mathcal{P}_n$ denote the Lagrange interpolation polynomial of f, with interpolation points*

$$\xi_j = \tfrac{1}{2}(b-a)\cos\frac{(j+\frac{1}{2})\pi}{n+1} + \tfrac{1}{2}(b+a)\,, \quad j = 0, 1, \ldots, n\,;$$

then

$$\|f - p_n\|_\infty \leq \frac{(b-a)^{n+1}}{2^{2n+1}(n+1)!} M_{n+1}$$

where $M_{n+1} = \max_{\zeta\in[a,b]} |f^{(n+1)}(\zeta)|$.

Proof Let $\tau_j = \cos\left((j+\frac{1}{2})\pi/(n+1)\right)$, $j = 0, 1, \ldots, n$, denote the zeros of the polynomial $T_{n+1}(t)$ (in the interval $(-1,1)$). Hence,

$$\prod_{j=0}^{n}(t-\tau_j) = 2^{-n}T_{n+1}(t)\,, \qquad t \in [-1,1]\,.$$

Let us define the points ξ_j, $j = 0, 1, \ldots, n$, as in the statement of the theorem. Clearly $\xi_j \in (a,b)$ is the image of $\tau_j \in (-1,1)$ under the linear transformation $t \mapsto x = \frac{1}{2}(b-a)t + \frac{1}{2}(b+a)$; we note in passing that the inverse of this mapping is $x \mapsto t(x) = (2x-a-b)/(b-a)$; thus,

$$\prod_{j=0}^{n}(x-\xi_j) = \left(\frac{b-a}{2}\right)^{n+1}\prod_{j=0}^{n}(t(x)-\tau_j) = \left(\frac{b-a}{2}\right)^{n+1} 2^{-n}T_{n+1}(t(x))\,.$$

The required bound now follows from (8.9), since $|T_{n+1}(t(x))| \le 1$ for all $x \in [a,b]$, and therefore $|\pi_{n+1}(x)| \le (b-a)^{n+1}2^{-2n-1}$. □

The De la Vallée Poussin Theorem, Theorem 8.3, suggests the notion of a **near-minimax** polynomial, which is a polynomial $p_n \in \mathcal{P}_n$ such that the difference $f(x) - p_n(x)$ changes sign at $n+1$ points ξ_j, $j = 0, 1, \ldots, n$, with $a < \xi_0 < \cdots < \xi_n < b$; for the difference $f(x) - p_n(x)$ then attains a local maximum or minimum with alternating signs in each of the intervals $[a, \xi_0), (\xi_0, \xi_1), \ldots, (\xi_n, b]$. The positions of these alternating local maxima and minima are then the points x_i, $i = 0, 1, \ldots, n+1$, required by Theorem 8.3, and we therefore know that the ∞-norm of the error of the minimax polynomial lies between the least and greatest of the absolute values of these local maxima and minima. In particular, we should expect that if the sizes of these local maxima and minima are not greatly different, then the error of the near-minimax approximation should not be very much larger than the error of the minimax approximation.

Given any set of points ξ_i, $i = 0, 1, \ldots, n$, with $a < \xi_0 < \cdots < \xi_n < b$, the polynomial $\pi_{n+1}(x) = (x-\xi_0)\ldots(x-\xi_n)$ changes sign at the $n+1$ points ξ_j, $j = 0, 1, \ldots, n$. Let us assume that $f \in \mathrm{C}[a,b]$, $f^{(n+1)}$ exists and is continuous on $[a,b]$, and $f^{(n+1)}$ has the same sign on the whole of (a,b). It then follows that the product $f^{(n+1)}(\eta)\pi_{n+1}(x)$ has exactly $n+1$ sign-changes in the open interval (a,b) for any $\eta \in (a,b)$. Thus, according to (8.9), the Lagrange interpolation polynomial p_n of degree n for the function f, with interpolation points ξ_j, $j = 0, 1, \ldots, n$, contained in the open interval (a,b), is a near-minimax polynomial from $\mathcal{P}_n$ for f on $[a,b]$.

We have therefore just shown that if $f^{(n+1)}$ exists and is continuous on the closed interval $[a,b]$, and has the same sign on the open interval

(a, b), then the polynomial constructed by interpolating at the points ξ_j, $j = 0, 1, \ldots, n$, obtained by linearly mapping the $n+1$ zeros of the Chebyshev polynomial $T_{n+1}(t)$ from $(-1, 1)$ to (a, b), is a near-minimax approximation from $\mathcal{P}_n$ for the function $f \in \mathrm{C}[a, b]$ on the interval $[a, b]$. Notice that if we use equally spaced interpolation points, so that $\xi_j = a + j(b-a)/n$, $j = 0, 1, \ldots, n$, $n \geq 1$, we shall not obtain a near-minimax approximation, since the interpolation error now changes sign at only $n-1$ points, the interpolation points which are internal to (a, b).

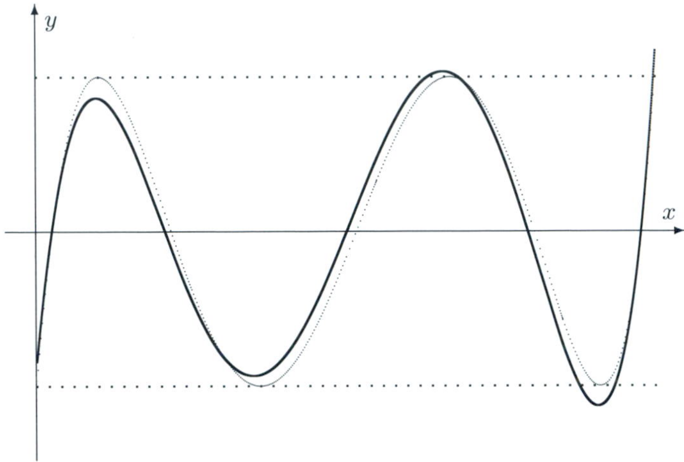

Fig. 8.5. Comparison of two polynomial approximations to e^{2x} on [0,1]: the thin curve is the error of the minimax approximation; the thick curve is the error of the polynomial obtained by interpolation at the Chebyshev points.

As an illustration, Figure 8.5 shows the errors of two approximations of degree 4 to the function $f(x) = e^{2x}$ over the interval $[0, 1]$. One of these is the minimax approximation, and the other is obtained by interpolation at the zeros of $T_5(t)$. It is clear that they are quite close; in fact the ∞-norms of the errors are 0.0015 and 0.0017 respectively.

In the next chapter we shall show that the least squares polynomial approximation to a continuous real-valued function is also near-minimax in this sense.

An alternative and very easy way of constructing polynomial approximations to many simple, smooth, functions is to truncate their Taylor series expansion. For example,

$$e^{kx} = 1 + kx + \cdots + \frac{k^n x^n}{n!} + \cdots ,$$

so we obtain a polynomial approximation $p_n(x)$ by taking the terms of this series up to the one involving x^n. Then, clearly,

$$e^{kx} - p_n(x) = \sum_{r=n+1}^{\infty} \frac{k^r x^r}{r!}.$$

Over the interval $[0,1]$, for example, this difference is nonnegative and monotonic increasing; it does not change sign at all. Hence the polynomial $p_n \in \mathcal{P}_n$ thus constructed is quite certainly not a near-minimax approximation for $x \mapsto e^{kx}$ on $[0,1]$. Nevertheless, $\max_{x\in[0,1]} |e^{kx} - p_n(x)|$ can be made arbitrarily small by choosing n sufficiently large.

8.6 Notes

For further details on the topics presented in this chapter, we refer to

- M.J.D. Powell, *Approximation Theory and Methods*, Cambridge University Press, Cambridge, 1996.

The Weierstrass Theorem is discussed in Chapter 6 of that book, and is stated in its Theorem 6.3. Although the proof presented by Powell uses the Bernstein polynomials, it is different from the more elementary but slightly lengthier argument proposed in Exercise 12 here: it relies on a proof of Bohman and Korovkin based on properties of monotone operators; see, also, p. 66 in Chapter 3 of

- E.W. Cheney, *Introduction to Approximation Theory*, McGraw–Hill, New York, 1966.

The notes contained on pp. 224–233 of Cheney's book are particularly illuminating.

The proof of the Weierstrass Theorem as proposed in Exercise 12, including the definition of what we today call Bernstein polynomials, stem from a paper of Sergei Natanovich Bernstein (1880–1968), entitled 'Démonstration du théorème de Weierstrass fondée sur le calcul des probabilités', *Comm. Soc. Math. Kharkow* **13**, 1–2, 1912/13.

Weierstrass' main contributions to approximation theory, as well as those of other mathematicians (including Picard, Volterra, Runge, Lebesgue, Mittag-Leffler, Fejér, Landau, de la Vallée Poussin, Bernstein), are reviewed in the extensive historical survey by Allan Pinkus, Weierstrass and approximation theory, *J. Approx. Theory* **107**, 1–66, 2000. Further details about the history of the subject can be found at the history of approximation theory website maintained by Allan Pinkus and Carl de Boor: `http://www.cs.wisc.edu/~deboor/HAT/`

The second part of Theorem 8.1 concerning the approximability of a continuous function by polynomials in the 2-norm is not usually presented as part of the classical Weierstrass Theorem which is posed in the ∞-norm. Here, we have chosen to state these results together in order to highlight the analogy, as well as to motivate the use of the 2-norm in polynomial approximation in the next chapter, Chapter 9.

In both Cheney's and Powell's books minimax approximation is treated in the more general framework of Haar systems. An $(n+1)$-dimensional linear subspace $\mathcal{A}$ of $\mathrm{C}[a,b]$ is said to satisfy the *Haar condition* if, for every nonzero p in $\mathcal{A}$, the number of roots of the equation $p(x) = 0$ in the interval $[a,b]$ is less than $n+1$. The concept of Haar system is due to Alfred Haar (1885–1933), Die Minkowskische Geometrie und die Annäherung an stetige Funktionen, *Math. Ann.* **78**, 294–311, 1918; this paper contains Haar's Theorem which characterises finite-dimensional Haar systems in spaces of continuous functions. The *Characterisation Theorem*, formulated as Theorem 7.2 in Powell's book, shows that the Oscillation Theorem, Theorem 8.4 of the present chapter, remains valid in a more general setting when the set of polynomials $\{1, x, \dots, x^n\}$ is replaced by an $(n+1)$-dimensional Haar system of functions contained in $\mathrm{C}[a,b]$.

Exercises

8.1 Give a proof of Lemma 8.1.

8.2 Suppose that the real-valued function f is continuous and even on the interval $[-a,a]$, that is, $f(x) = f(-x)$ for all $x \in [-a,a]$. By using the Uniqueness Theorem, or otherwise, show that the minimax polynomial approximation of degree n is an even function. Deduce that the minimax polynomial approximation of degree $2n$ is also the minimax polynomial approximation of degree $2n+1$. What does this imply about the sequence of critical points for the minimax polynomial p_{2n}?

8.3 State and prove similar results to those in Exercise 2, for the case where f is an odd function, that is, $f(x) = -f(-x)$ for all $x \in [-a,a]$.

8.4 (i) Construct the minimax polynomial $p_2 \in \mathcal{P}_2$ on the interval $[-1,1]$ for the function g defined by $g(x) = \sin x$.

(ii) Construct the minimax polynomial $p_3 \in \mathcal{P}_3$ on the interval $[-1, 1]$ for the function h defined by $h(x) = \cos x^2$.
(Use the results of Exercises 2 and 3.)

8.5 The function H is defined by $H(x) = 1$ if $x > 0$, $H(x) = -1$ if $x < 0$, and $H(0) = 0$. Show that for any $n \geq 0$ and any $p_n \in \mathcal{P}_n$, $\|H - p_n\|_\infty \geq 1$ on the interval $[-1, 1]$. Construct the polynomial, of degree 0, of best approximation to H on the interval $[-1, 1]$, and show that it is unique. (Note that since H is discontinuous most of the theorems in this chapter are not applicable.)

Show that the polynomial of best approximation, of degree 1, to H on $[-1, 1]$ is not unique, and give an expression for its most general form.

8.6 Suppose that $t_1 < t_2 < \cdots < t_k$ are k distinct points in the interval $[a, b]$; for any function f defined on $[a, b]$, write $Z_k(f) = \max_{i=1}^{k} |f(x_i)|$. Explain why $Z_k(\cdot)$ is not a norm on the space of functions which are continuous on $[a, b]$; show that it is a norm on the space of polynomials of degree n, provided that $k > n$.

In the case $k = 3$, with $t_1 = 0$, $t_2 = \frac{1}{2}$, $t_3 = 1$, where we wish to approximate the function $f\colon x \mapsto \mathrm{e}^x$ on the interval $[0, 1]$, explain graphically, or otherwise, why the polynomial p_1 of degree 1 which minimises $Z_3(f - p_1)$ satisfies the conditions

$$f(0) - p_1(0) = -[f(\tfrac{1}{2}) - p_1](\tfrac{1}{2}) = f(1) - p_1(1)\,.$$

Hence construct this polynomial p_1. Now suppose that $k = 4$, with $t_1 = 0$, $t_2 = \frac{1}{3}$, $t_3 = \frac{2}{3}$, $t_4 = \frac{1}{3}$; use a similar method to construct the polynomial of degree 1 which minimises $Z_4(f - p_1)$.

8.7 Among all polynomials $p_n \in \mathcal{P}_n$ of the form

$$p_n(x) = Ax^n + \sum_{k=0}^{n-1} a_k x^k,$$

where A is a fixed nonzero real number, find the polynomial of best approximation for the function $f(x) \equiv 0$ on the closed interval $[-1, 1]$.

8.8 Find the minimax polynomial $p_n \in \mathcal{P}_n$ on the interval $[-1, 1]$ for the function f defined by

$$f(x) = \sum_{k=0}^{n+1} a_k x^k,$$

where $a_{n+1} \neq 0$.

8.9 Construct the minimax polynomial $p_1 \in \mathcal{P}_1$ on the interval $[-1, 2]$ for the function f defined by $f(x) = |x|$.

8.10 Give a proof of Lemma 8.2.

8.11 Give an example of a continuous real-valued function f defined on the closed interval $[a, b]$ such that the set of critical points for the minimax approximation of f by polynomials from $\mathcal{P}_1$ does not contain either of the points a and b.

8.12 For each nonnegative integer n, and $x \in [0, 1]$, define the Bernstein polynomials $p_{nk} \in \mathcal{P}_n$ by

$$p_{nk}(x) = \frac{n!}{k!(n-k)!} x^k (1-x)^{n-k}, \quad k = 0, \dots, n.$$

Show that

$$(1 - x + tx)^n = \sum_{k=0}^{n} p_{nk}(x) t^k ;$$

by differentiating this relation successively with respect to t and putting $t = 1$, show that, for any $x \in [0, 1]$,

$$\begin{aligned} \sum_{k=0}^{n} p_{nk}(x) &= 1, \\ \sum_{k=0}^{n} k p_{nk}(x) &= nx, \\ \sum_{k=0}^{n} k(k-1) p_{nk}(x) &= n(n-1)x^2, \end{aligned}$$

and deduce that

$$\sum_{k=0}^{n} (x - k/n)^2 p_{nk}(x) = \frac{x(1-x)}{n}, \qquad x \in [0, 1].$$

Define M to be the upper bound of $|f(x)|$ on $[0, 1]$. Given $\varepsilon > 0$, we can choose $\delta > 0$ such that $|f(x) - f(y)| < \varepsilon/2$ for any x and y in $[0, 1]$ such that $|x - y| < \delta$. Now define the polynomial $p_n \in \mathcal{P}_n$ by

$$p_n(x) = \sum_{k=0}^{n} f(k/n) p_{nk}(x),$$

and choose a fixed value of x in $[0,1]$; show that

$$|f(x) - p_n(x)| \le \sum_{k=0}^{n} |f(x) - f(k/n)| p_{nk}(x) \,.$$

Using the notation

$$\sum_{k=0}^{n} = \sum_{1} + \sum_{2}$$

where $\sum_1$ denotes the sum over those values of k for which $|x - k/n| < \delta$, and $\sum_2$ denotes the sum over those values of k for which $|x - k/n| \ge \delta$, show that

$$\sum_{1} |f(x) - f(k/n)| p_{nk}(x) < \varepsilon/2 \,.$$

Show also that

$$\sum_{2} |f(x) - f(k/n)| p_{nk}(x) \le (2M/\delta^2) \sum_{k=0}^{n} (x - k/n)^2 p_{nk}(x) \,.$$

Now, choose $N_0 = M/(\delta^2 \varepsilon)$, and show that

$$|f(x) - p_n(x)| < \varepsilon \qquad \forall\, x \in [0,1] \,,$$

if $n \ge N_0$. Deduce that

$$\|f - p_n\|_\infty < \varepsilon, \quad \text{if } n \ge N_0 \,,$$

where $\|\cdot\|_\infty$ denotes the ∞-norm on the interval $[0,1]$.

9

Approximation in the 2-norm

9.1 Introduction

In Chapter 8 we discussed the idea of best approximation of a continuous real-valued function by polynomials of some fixed degree in the ∞-norm. Here we consider the analogous problem of best approximation in the 2-norm. Why, you might ask, is it necessary to consider best approximation in the 2-norm when we have already developed a perfectly adequate theory of best approximation in the ∞-norm? As our first example in Section 9.3 will demonstrate, the choice of norm can significantly influence the outcome of the problem of best approximation: the polynomial of best approximation of a certain fixed degree to a given continuous function in one norm need not bear any resemblance to the polynomial of best approximation of the same degree in another norm. Ultimately, in a practical situation, the choice of norm will be governed by the sense in which the given continuous function has to be well approximated.

As will become apparent, best approximation in the 2-norm is closely related to the notion of orthogonality and this in turn relies on the concept of *inner product*. Thus, we begin the chapter by recalling from linear algebra the definition of *inner product space*.

Throughout the chapter $[a, b]$ will denote a bounded closed interval of the real line, and (a, b) will signify a bounded open interval of the real line, with $a < b$.

9.2 Inner product spaces

Definition 9.1 *Let $\mathcal{V}$ be a linear space over the field of real numbers. A real-valued function $\langle \cdot, \cdot \rangle$, defined on the Cartesian product $\mathcal{V} \times \mathcal{V}$, is called an* **inner product** *on $\mathcal{V}$ if it satisfies the following axioms:*

❶ $\langle f+g, h\rangle = \langle f, h\rangle + \langle g, h\rangle$ *for all f, g and h in $\mathcal{V}$;*
❷ $\langle \lambda f, g\rangle = \lambda\langle f, g\rangle$ *for all λ in $\mathbb{R}$, and all f, g in $\mathcal{V}$;*
❸ $\langle f, g\rangle = \langle g, f\rangle$ *for all f and g in $\mathcal{V}$;*
❹ $\langle f, f\rangle > 0$ *if $f \neq 0$, $f \in \mathcal{V}$.*

A linear space with an inner product is called an **inner product space**.

Example 9.1 *The n-dimensional Euclidean space $\mathbb{R}^n$ is an inner product space with*

$$\langle \boldsymbol{x}, \boldsymbol{y}\rangle = \sum_{i=1}^{n} x_i y_i\,, \qquad \boldsymbol{x},\, \boldsymbol{y} \in \mathbb{R}^n\,,$$

where $\boldsymbol{x} = (x_1, \ldots, x_n)^{\mathrm{T}}$ and $\boldsymbol{y} = (y_1, \ldots, y_n)^{\mathrm{T}}$. We can also write this in a more compact form as $\langle \boldsymbol{x}, \boldsymbol{y}\rangle = \boldsymbol{x}^{\mathrm{T}}\boldsymbol{y}$.

Definition 9.2 *Suppose that $\mathcal{V}$ is an inner product space, and f and g are two elements of $\mathcal{V}$ such that $\langle f, g\rangle = 0$; we shall then say that f is* **orthogonal** *to g.*

Due to the third axiom of inner product, if f is orthogonal to g, then g is orthogonal to f; therefore, if $\langle f, g\rangle = 0$, we shall simply say that f and g are orthogonal. Our next example shows that Definition 9.2 is a direct generalisation of the usual geometrical notion of orthogonality.

Example 9.2 *According to Example 9.1, with $n = 2$, the formula $\langle \boldsymbol{y}, \boldsymbol{z}\rangle = \boldsymbol{y}^{\mathrm{T}}\boldsymbol{z}$, where $\boldsymbol{y} = (y_1, y_2)^{\mathrm{T}}$ and $\boldsymbol{z} = (z_1, z_2)^{\mathrm{T}}$ are two-component vectors, defines an inner product in $\mathbb{R}^2$.*

The vectors $\boldsymbol{y}$ and $\boldsymbol{z}$ have respective lengths $\sqrt{y_1^2 + y_2^2} = \|\boldsymbol{y}\|_2$ and $\sqrt{z_1^2 + z_2^2} = \|\boldsymbol{z}\|_2$, where $\|\cdot\|_2$ denotes the 2-norm for vectors in $\mathbb{R}^2$. Let $\alpha \in [0, 2\pi)$ denote the angle, measured in an anticlockwise direction, between the positive x_1-coordinate direction and $\boldsymbol{y}$; similarly, let $\beta \in [0, 2\pi)$ be the angle between the positive x_1-coordinate direction and $\boldsymbol{z}$. Then,

$$\boldsymbol{y} = \|\boldsymbol{y}\|_2(\cos\alpha, \sin\alpha) \quad \textit{and} \quad \boldsymbol{z} = \|\boldsymbol{z}\|_2(\cos\beta, \sin\beta)\,.$$

Now,

$$\begin{aligned}\langle \boldsymbol{y}, \boldsymbol{z}\rangle &= \boldsymbol{y}^{\mathrm{T}}\boldsymbol{z}\\ &= \|\boldsymbol{y}\|_2\|\boldsymbol{z}\|_2\,(\cos\alpha\cos\beta + \sin\alpha\sin\beta)\\ &= \|\boldsymbol{y}\|_2\|\boldsymbol{z}\|_2\cos(\alpha-\beta)\\ &= \|\boldsymbol{y}\|_2\|\boldsymbol{z}\|_2\cos(\vartheta_{yz})\,,\end{aligned}$$

where $\vartheta_{yz} = |\alpha-\beta|$ *is the angle between the vectors* $\boldsymbol{y}$ *and* $\boldsymbol{z}$*. The vector* $\boldsymbol{y}$ *is orthogonal to* $\boldsymbol{z}$ *if, and only if,* ϑ_{yz} *is* $\pi/2$ *or* $3\pi/2$*; either way,* $\cos(\vartheta_{yz}) = 0$*, and hence* $\langle \boldsymbol{y}, \boldsymbol{z}\rangle = 0$*. We note in passing that if* $\boldsymbol{y} = \boldsymbol{z}$*, then* $\vartheta_{yz} = 0$ *and therefore*

$$\langle \boldsymbol{y}, \boldsymbol{y}\rangle = \|\boldsymbol{y}\|_2^2\,.$$

This last observation motivates our next definition.

Definition 9.3 *Suppose that* $\mathcal{V}$ *is an inner product space over the field of real numbers, with inner product* $\langle\cdot,\cdot\rangle$*. For* f *in* $\mathcal{V}$*, we define the* ***induced norm***

$$\|f\| = \langle f, f\rangle^{1/2}\,. \tag{9.1}$$

Although our terminology and our notation appear to imply that (9.1) defines a norm on $\mathcal{V}$, this is by no means obvious. In order to show that $f \mapsto \langle f, f\rangle^{1/2}$ is indeed a norm, we begin with the following result which is a direct generalisation of the Cauchy–Schwarz inequality (2.35) from Chapter 2.

Lemma 9.1 (Cauchy–Schwarz inequality)

$$|\langle f, g\rangle| \le \|f\|\,\|g\| \qquad \forall\, f,\, g \in \mathcal{V}\,. \tag{9.2}$$

Proof The proof is analogous to that of (2.35). Recalling the definition of $\|\cdot\|$ from (9.1) and noting the first three axioms of inner product, we find that, for $f, g \in \mathcal{V}$,

$$0 \le \|\lambda f + g\|^2 = \lambda^2\|f\|^2 + 2\lambda\langle f, g\rangle + \|g\|^2 \qquad \forall\,\lambda \in \mathbb{R}\,. \tag{9.3}$$

Denoting, for $f, g \in \mathcal{V}$ fixed, the quadratic polynomial in λ on the right-hand side by $A(\lambda)$, the condition for $A(\lambda)$ to be nonnegative for all λ in $\mathbb{R}$ is that $[2\langle f, g\rangle]^2 - 4\|f\|^2\|g\|^2 \le 0$; this gives the inequality (9.2). □

Now, putting $\lambda = 1$ in (9.3) and using (9.2) on the right yields

$$\|f + g\| \le \|f\| + \|g\| \qquad \forall\, f,\, g \in \mathcal{V}\,.$$

Consequently, $\|\cdot\|$ obeys the triangle inequality, the third axiom of norm. The first two axioms of norm, namely that

- $\|f\| \geq 0$ for all $f \in \mathcal{V}$, and $\|f\| = 0$ if, and only if, $f = 0$ in $\mathcal{V}$, and
- $\|\lambda f\| = |\lambda| \, \|f\|$ for all $\lambda \in \mathbb{R}$ and all $f \in \mathcal{V}$,

follow directly form (9.1) and from the last three axioms of inner product stated in Definition 9.1.

We have thus shown the following result.

Theorem 9.1 *An inner product space $\mathcal{V}$ over the field $\mathbb{R}$ of real numbers, equipped with the induced norm $\|\cdot\|$, is a normed linear space over $\mathbb{R}$.*

We conclude this section with a relevant example of an inner product space, whose induced norm is the 2-norm considered at the beginning of Chapter 8.

Example 9.3 *The set $\mathrm{C}[a,b]$ of continuous real-valued functions defined on the closed interval $[a,b]$ is an inner product space with*

$$\langle f, g\rangle = \int_a^b w(x) f(x)\, g(x) \mathrm{d}x\,, \tag{9.4}$$

where w is a **weight function**, *defined, positive, continuous and integrable on the open interval (a,b). The norm $\|\cdot\|_2$, induced by this inner product and given by*

$$\|f\|_2 = \left(\int_a^b w(x)|f(x)|^2 \mathrm{d}x\right)^{1/2}, \tag{9.5}$$

is referred to as the 2-norm on $\mathrm{C}[a,b]$ (see Example 8.2). For the sake of simplicity, we have chosen not to distinguish in terms of our notation between the 2-norm on $\mathrm{C}[a,b]$ defined above and the 2-norm for vectors introduced in Chapter 2; it will always be clear from the context which of the two is intended.

Clearly, it is not necessary to demand the continuity of the function f on the closed interval $[a,b]$ to ensure that $\|f\|_2$ is finite. For example, $f \colon x \mapsto \operatorname{sgn}\left(x - \frac{1}{2}(a+b)\right)$, $x \in [a,b]$, has finite 2-norm, despite the fact that it has a jump discontinuity at $x = \frac{1}{2}(a+b)$.

Motivated by this observation, and the desire to develop a theory of approximation in the 2-norm whose range of applicability extends beyond the linear space of continuous functions on a bounded closed interval, we denote by $\mathrm{L}^2_w(a,b)$ the set of all real-valued functions f

defined on (a,b) such that $w(x)|f(x)|^2$ is integrable[1] on (a,b); the set $\mathrm{L}^2_w(a,b)$ is equipped with the inner product (9.4) and the induced 2-norm (9.5). Obviously, $\mathrm{C}[a,b]$ is a proper subset of $\mathrm{L}^2_w(a,b)$.

In this broader context, $\|\cdot\|_2$ is frequently referred to as the L^2-norm; for the sake of simplicity we shall continue to call it the 2-norm. As before, w is assumed to be a real-valued function, defined, positive, continuous and integrable on the open interval (a,b). When $w(x) \equiv 1$ on (a,b), we shall write $\mathrm{L}^2(a,b)$ instead of $\mathrm{L}^2_w(a,b)$.

We are now ready to consider best approximation in the 2-norm.

9.3 Best approximation in the 2-norm

The problem of best approximation in the 2-norm can be formulated as follows:

(B) Given that $f \in \mathrm{L}^2_w(a,b)$, find $p_n \in \mathcal{P}_n$ such that

$$\|f - p_n\|_2 = \inf_{q\in\mathcal{P}_n} \|f - q\|_2\,;$$

such p_n is called a **polynomial of best approximation of degree** n **to the function** f **in the 2-norm on** (a,b).

The existence and uniqueness of p_n will be shown in Theorem 9.2. However, we shall first consider some simple examples.

Example 9.4 *Suppose that $\varepsilon > 0$ and let $f(x) = 1 - \mathrm{e}^{-x/\varepsilon}$ with x in $[0,1]$. For $\varepsilon = 10^{-2}$, the function f is depicted in Figure 9.1. We shall construct the polynomial of best approximation of degree 0 in the 2-norm, with weight function $w(x) \equiv 1$, to f on $(0,1)$, and compare it with the minimax polynomial of degree 0 to f on $[0,1]$.*

The best approximation to f by a polynomial of degree 0 in the 2-norm on the interval $(0,1)$, with weight function $w(x) \equiv 1$, is determined by minimising $\|f - c\|_2$ over all $c \in \mathbb{R}$; equivalently, we need to minimise

$$\int_0^1 (f(x) - c)^2 \mathrm{d}x = \int_0^1 |f(x)|^2 \mathrm{d}x - 2c\int_0^1 f(x)\mathrm{d}x + c^2$$

[1] Strictly speaking, the integral in the definition of $\|\cdot\|_2$ should now be thought of as a Lebesgue integral, with the convention that any two functions in $\mathrm{L}^2_w(a,b)$ which differ only on a set of zero measure are identified. Readers who are unfamiliar with the concept of Lebesgue integral can safely ignore this footnote. For the definition of *set of measure zero* see Section 11.2 in Chapter 11.

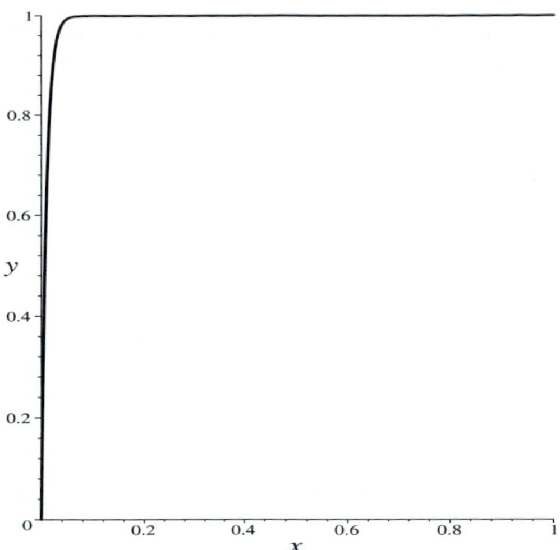

Fig. 9.1. Graph of the function $f\colon x \mapsto 1 - \mathrm{e}^{-x/\varepsilon}$ for $x \in [0,1]$ and $\varepsilon = 10^{-2}$.

over all $c \in \mathbb{R}$. The right-hand side is a quadratic polynomial in c; its minimum, as a function of c, is achieved for

$$c = \int_0^1 f(x)\,\mathrm{d}x = 1 - \varepsilon + \varepsilon \mathrm{e}^{-1/\varepsilon}\,.$$

Consequently, the polynomial of degree 0 of best approximation to f in the 2-norm on the interval $(0,1)$ with respect to the weight function $w(x) \equiv 1$ is

$$p_0^{(2\text{-norm})}(x) \equiv 1 - \varepsilon + \varepsilon \mathrm{e}^{-1/\varepsilon}\,, \qquad x \in [0,1]\,.$$

On the other hand, since $f \in \mathrm{C}[0,1]$ and f is strictly monotonic increasing on $[0,1]$, its minimax approximation of degree 0 on the interval $[0,1]$ is simply the arithmetic mean of $f(0)$ and $f(1)$:

$$p_0^{(\infty\text{-norm})}(x) \equiv \tfrac{1}{2}(1 - \mathrm{e}^{-1/\varepsilon})\,, \qquad x \in [0,1]\,.$$

Clearly, for $0 < \varepsilon \ll 1$, $p_0^{(\infty\text{-norm})}(x) \approx 1/2$, while $p_0^{(2\text{-norm})}(x) \approx 1$.

An even more dramatic discrepancy is observed between the polynomials of best approximation in the 2-norm and the ∞-norm when

$$f(x) = 1 - \varepsilon^{-1/2}\mathrm{e}^{-x/\varepsilon}\,, \qquad x \in [0,1]\,.$$

Here, for $0 < \varepsilon \ll 1$, $p_0^{(2\text{-norm})}(x) \approx 1$, as before. On the other hand,

$$p_0^{(\infty\text{-norm})}(x) \equiv 1 - \tfrac{1}{2}\varepsilon^{-1/2}(1 + \mathrm{e}^{-1/\varepsilon})\,, \qquad x \in [0,1]\,,$$

which tends to $-\infty$ as $\varepsilon \to 0+$. These examples indicate that the polynomial of best approximation from $\mathcal{P}_n$ to a function in the 2-norm can be vastly different from the minimax approximation from $\mathcal{P}_n$ to the same function. ◇

Given $f \in \mathrm{L}^2_w(a,b)$, we shall assume for the moment the existence of an associated polynomial of best approximation in the 2-norm; later on we shall prove that such a polynomial exists and is unique. In order to motivate the general discussion that will follow, it is helpful to begin with a straightforward approach to a simple example.

Let us suppose that we wish to construct the polynomial of best approximation $p_n \in \mathcal{P}_n$, $n \geq 0$, to a function $f \in \mathrm{L}^2_w(0,1)$ on the interval $(0,1)$ in the 2-norm; for simplicity, we shall assume that the weight function $w(x) \equiv 1$. Writing the polynomial p_n as

$$p_n(x) = c_0 + c_1 x + \cdots + c_n x^n\,,$$

we want to choose the coefficients c_j, $j = 0, \ldots, n$, so as to minimise the 2-norm of the error, $e_n = f - p_n$,

$$\|e_n\|_2 = \|f - p_n\|_2 = \left(\int_0^1 |f(x) - p_n(x)|^2 \mathrm{d}x\right)^{1/2}.$$

Since the 2-norm is nonnegative and the function $\xi \in \mathbb{R}_+ \mapsto \xi^{1/2}$ is monotonic increasing, this problem is equivalent to one of minimising the square of the norm; thus, instead, we shall minimise the expression

$$\begin{aligned} E(c_0, c_1, \ldots, c_n) &= \int_0^1 [f(x) - p_n(x)]^2 \mathrm{d}x \\ &= \int_0^1 [f(x)]^2 \mathrm{d}x - 2\sum_{j=0}^{n} c_j \int_0^1 f(x) x^j \mathrm{d}x \\ &\quad + \sum_{j=0}^{n}\sum_{k=0}^{n} c_j c_k \int_0^1 x^{k+j} \mathrm{d}x\,, \end{aligned}$$

by treating it as a function of $(c_0, \ldots, c_n)$. At the minimum, the partial derivatives of E with respect to the c_j, $j = 0, \ldots, n$, are equal to zero. This leads to a system of $(n+1)$ linear equations for the coefficients $c_0, \ldots, c_n$:

$$\sum_{k=0}^{n} M_{jk} c_k = b_j\,, \qquad j = 0, \ldots, n\,, \tag{9.6}$$

where

$$M_{jk} = \int_0^1 x^{k+j}\mathrm{d}x = \frac{1}{k+j+1}\,,$$
$$b_j = \int_0^1 f(x)x^j\mathrm{d}x\,.$$

Equivalently, recalling that the inner product associated with the 2-norm (in the case of $w(x) \equiv 1$) is defined by

$$\langle g, h\rangle = \int_0^1 g(x)h(x)\mathrm{d}x\,,$$

M_{jk} and b_j can be written as

$$M_{jk} = \langle x^k, x^j\rangle\,, \qquad b_j = \langle f, x^j\rangle\,. \tag{9.7}$$

By solving the system of linear equations (9.6) for $c_0, \ldots, c_n$, we obtain the coefficients of the polynomial of best approximation of degree n to the function f in the 2-norm on the interval $(0, 1)$. We can proceed in the same manner on any interval (a, b) with any positive, continuous and integrable weight function w defined on (a, b).

This approach is straightforward for small values of n, but soon becomes impractical as n increases. The source of the computational difficulties is the fact that the matrix M is the Hilbert matrix, discussed in Section 2.8. The Hilbert matrix is well known to be ill-conditioned for large n, so any solution to (9.6), computed with a fixed number of decimal digits, loses all accuracy due to accumulation of rounding errors. Fortunately, an alternative method is available, and is discussed in the next section.

9.4 Orthogonal polynomials

In the previous section we described a method for constructing the polynomial of best approximation $p_n \in \mathcal{P}_n$ to a function f in the 2-norm; it was based on seeking p_n as a linear combination of the polynomials x^j, $j = 0, \ldots, n$, which form a basis for the linear space $\mathcal{P}_n$. The approach was not entirely satisfactory because it gave rise to a system of linear equations with a full matrix that was difficult to invert. The central idea of the alternative approach that will be described in this section is to expand p_n in terms of a different basis, chosen so that the resulting system of linear equations has a diagonal matrix; solving this linear system is then a trivial exercise. Of course, the nontrivial ingredient of this alternative approach is to find a suitable basis for $\mathcal{P}_n$ which achieves the objective that the matrix of the linear system is diagonal. The expression for M_{jk} in (9.7) gives us a clue how to proceed.

Suppose that φ_j, $j = 0, \ldots, n$, form a basis for $\mathcal{P}_n$, $n \geq 0$; let us seek the polynomial of best approximation as

$$p_n(x) = \gamma_0\varphi_0(x) + \cdots + \gamma_n\varphi_n(x)\,,$$

where $\gamma_0, \ldots, \gamma_n$ are real numbers to be determined. By the same process as in the previous section, we arrive at a system of linear equations of the form (9.6):

$$\sum_{k=0}^{n} M_{jk}\gamma_k = \beta_j\,, \qquad j = 0, \ldots, n\,,$$

where now

$$M_{jk} = \langle \varphi_k, \varphi_j \rangle \qquad \text{and} \qquad \beta_j = \langle f, \varphi_j \rangle\,,$$

with the inner product $\langle \cdot\,, \cdot \rangle$ defined by

$$\langle g, h \rangle = \int_a^b w(x)g(x)h(x)\,\mathrm{d}x\,,$$

and the weight function w assumed to be positive, continuous and integrable on the interval (a, b).

Thus, $M = (M_{jk})$ will be a diagonal matrix provided that the basis functions φ_j, $j = 0, \ldots, n$, for the linear space $\mathcal{P}_n$ are chosen so that $\langle \varphi_k, \varphi_j \rangle = 0$, for $j \neq k$; in other words, φ_k is required to be orthogonal to φ_j for $j \neq k$, in the sense of Definition 9.2. This observation motivates the following definition.

Definition 9.4 *Given a weight function w, defined, positive, continuous and integrable on the interval (a, b), we say that the sequence of polynomials φ_j, $j = 0, 1, \ldots$, is a* **system of orthogonal polynomials** *on the interval (a, b) with respect to w, if each φ_j is of exact degree j, and if*

$$\int_a^b w(x)\varphi_k(x)\varphi_j(x)\mathrm{d}x \quad \begin{cases} = 0 & \textit{for all } k \neq j\,, \\ \neq 0 & \textit{when } k = j\,. \end{cases}$$

Next, we show that a system of orthogonal polynomials exists on any interval (a, b) and for any weight function w which satisfies the conditions in Definition 9.4. We proceed inductively.

Let $\varphi_0(x) \equiv 1$, and suppose that φ_j has already been constructed for $j = 0, \ldots, n$, with $n \geq 0$. Then,

$$\int_a^b w(x)\varphi_k(x)\varphi_j(x)\mathrm{d}x = 0\,, \qquad k \in \{0, \ldots, n\} \setminus \{j\}\,.$$

Let us now define the polynomial

$$q(x) = x^{n+1} - a_0\varphi_0(x) - \cdots - a_n\varphi_n(x)\,,$$

where

$$a_j = \frac{\int_a^b w(x)\,x^{n+1}\varphi_j(x)\mathrm{d}x}{\int_a^b w(x)[\varphi_j(x)]^2\mathrm{d}x}\,, \qquad j = 0, \ldots, n\,.$$

It then follows that

$$\begin{aligned} \int_a^b w(x)q(x)\varphi_j(x)\mathrm{d}x &= \int_a^b w(x)x^{n+1}\varphi_j(x)\mathrm{d}x \\ &\quad -a_j\int_a^b w(x)[\varphi_j(x)]^2\mathrm{d}x \\ &= 0 \qquad \text{for } 0 \leq j \leq n\,, \end{aligned}$$

where we have used the orthogonality of the sequence φ_j, $j = 0, \ldots, n$. Thus, with this choice of the numbers a_j we have ensured that q is orthogonal to all the previous members of the sequence, and φ_{n+1} can now be defined as any nonzero-constant multiple of q. This procedure for constructing a system of orthogonal polynomials is usually referred to as **Gram–Schmidt orthogonalisation**.[1]

Example 9.5 *We shall construct a system of orthogonal polynomials* $\{\varphi_0, \varphi_1, \varphi_2\}$ *on the interval* $(0, 1)$ *with respect to the weight function* $w(x) \equiv 1$.

We put $\varphi_0(x) \equiv 1$, and we seek φ_1 in the form

$$\varphi_1(x) = x - c_0\varphi_0(x)$$

such that $\langle \varphi_1, \varphi_0 \rangle = 0$; that is,

$$\langle x, \varphi_0 \rangle - c_0\langle \varphi_0, \varphi_0 \rangle = 0\,.$$

[1] Jørgen Pedersen Gram (27 June 1850, Nustrup, Denmark – 29 April 1916, Copenhagen, Denmark); Erhard Schmidt (13 January 1876, Dorpat, Russia (now Tartu, Estonia) – 6 December 1959, Berlin, Germany).

Hence,

$$c_0 = \frac{\langle x, \varphi_0 \rangle}{\langle \varphi_0, \varphi_0 \rangle} = \tfrac{1}{2}$$

and therefore,

$$\varphi_1(x) = x - \tfrac{1}{2}\varphi_0(x) = x - \tfrac{1}{2}\,.$$

By construction, $\langle \varphi_1, \varphi_0 \rangle = \langle \varphi_0, \varphi_1 \rangle = 0$.

We now seek φ_2 in the form

$$\varphi_2(x) = x^2 - (d_1\varphi_1(x) + d_0\varphi_0(x))$$

such that $\langle \varphi_2, \varphi_1 \rangle = 0$ and $\langle \varphi_2, \varphi_0 \rangle = 0$. Thus,

$$\begin{aligned} \langle x^2, \varphi_1 \rangle - d_1\langle \varphi_1, \varphi_1 \rangle - d_0\langle \varphi_0, \varphi_1 \rangle &= 0\,, \\ \langle x^2, \varphi_0 \rangle - d_1\langle \varphi_1, \varphi_0 \rangle - d_0\langle \varphi_0, \varphi_0 \rangle &= 0\,. \end{aligned}$$

As $\langle \varphi_0, \varphi_1 \rangle = 0$ and $\langle \varphi_1, \varphi_0 \rangle = 0$, we have that

$$\begin{aligned} d_1 &= \frac{\langle x^2, \varphi_1 \rangle}{\langle \varphi_1, \varphi_1 \rangle} = 1\,, \\ d_0 &= \frac{\langle x^2, \varphi_0 \rangle}{\langle \varphi_0, \varphi_0 \rangle} = \tfrac{1}{3}\,, \end{aligned}$$

and therefore

$$\varphi_2(x) = x^2 - x + \tfrac{1}{6}\,. \tag{9.8}$$

Clearly, $\langle \varphi_k, \varphi_j \rangle = 0$ for $j \neq k$, $j, k \in \{0, 1, 2\}$, and φ_j is of exact degree j, $j = 0, 1, 2$. Thus we have found the required system $\{\varphi_0, \varphi_1, \varphi_2\}$ of orthogonal polynomials on the interval $(0, 1)$ with respect to the given weight function w.

By continuing this procedure, we can construct a system of orthogonal polynomials $\{\varphi_0, \varphi_1, \ldots, \varphi_n\}$, with respect to the weight function $w(x) \equiv 1$ on the interval $(0, 1)$, for any $n \geq 1$. For example, when $n = 3$, we shall find $\{\varphi_0, \varphi_1, \varphi_2, \varphi_3\}$, with φ_0, φ_1, φ_2, as above, and

$$\varphi_3(x) = x^3 - \tfrac{3}{2}x^2 + \tfrac{3}{5}x - \tfrac{1}{20}\,.$$

Having generated a system of orthogonal polynomials on the interval $(0, 1)$ with respect to the weight function $w(x) \equiv 1$, by performing the linear mapping $x \mapsto (b - a)x + a$ we may obtain a system of orthogonal polynomials on any open interval (a, b) with respect to the weight function $w(x) \equiv 1$. For example, when $(a, b) = (-1, 1)$, the mapping $x \mapsto 2x - 1$ leads to the system of Legendre polynomials on $(-1, 1)$.

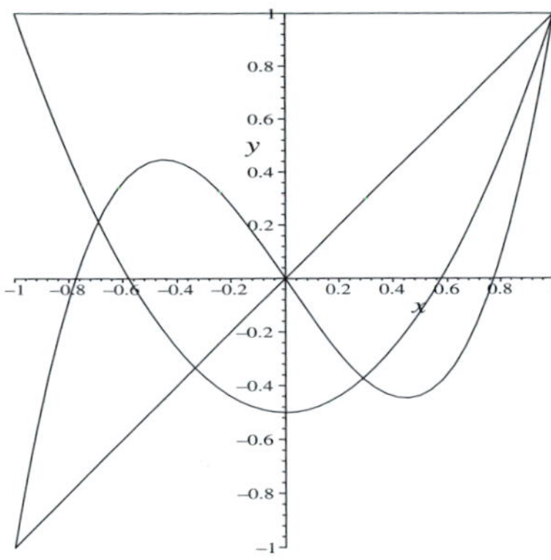

Fig. 9.2. The first four Legendre polynomials on the interval $(-1,1)$.

Example 9.6 (Legendre polynomials) *We wish to construct a system of orthogonal polynomials on $(a,b) = (-1,1)$ with respect to the weight function $w(x) \equiv 1$.*

On replacing x by

$$\frac{x-a}{b-a} = \tfrac{1}{2}(x+1)\,, \qquad x \in (a,b) = (-1,1)\,,$$

in $\varphi_0(x)$, $\varphi_1(x)$, $\varphi_2(x)$, $\varphi_3(x)$ from Example 9.5, we obtain, on normalising each of these polynomials so that its value at $x=1$ is equal to 1, the polynomials $\varphi_0, \varphi_1, \varphi_2, \varphi_3$, defined by

$$\begin{aligned}
\varphi_0(x) &= 1\,,\\
\varphi_1(x) &= x\,,\\
\varphi_2(x) &= \tfrac{3}{2}x^2 - \tfrac{1}{2}\,,\\
\varphi_3(x) &= \tfrac{5}{2}x^3 - \tfrac{3}{2}x\,.
\end{aligned}$$

These are the first four elements of the system of Legendre polynomials, orthogonal on the interval $(-1,1)$ with respect to the weight function $w(x) \equiv 1$. They are depicted in Figure 9.2. An alternative normalisation would have been to divide each φ_j by $\|\varphi_j\|_2$ so as to ensure that the 2-norm of the resulting scaled polynomial is equal to 1. ◇

Example 9.7 *The Chebyshev polynomials T_n: $x \mapsto \cos(n\cos^{-1}x)$, $n = 0,1,\ldots$, introduced in Section 8.4, form an orthogonal system on the interval $(-1,1)$ with respect to the positive, continuous and integrable weight function $w(x) = (1-x^2)^{-1/2}$.*

The proof of this is simple: let $\langle \cdot , \cdot \rangle$ denote the inner product in $\mathrm{L}^2_w(-1,1)$ with $w = (1-x^2)^{-1/2}$. By using the change of independent variable

$$t \in (0,\pi) \mapsto x = \cos t \in (-1,1)\,,$$

we have

$$\begin{aligned}\langle T_m, T_n\rangle &= \int_{-1}^{1} \frac{1}{\sqrt{1-x^2}}\,(\cos m \cos^{-1} x)\,(\cos n \cos^{-1} x)\,\mathrm{d}x\\ &= \int_0^{\pi} \cos mt\,\cos nt\,\mathrm{d}t\\ &= \tfrac{1}{2}\int_0^{\pi} \{\cos(m+n)t + \cos(m-n)t\}\,\mathrm{d}t\\ &= \begin{cases} 0 & \text{when } m \neq n\,,\\ \frac{\pi}{2} & \text{when } m = n\,,\end{cases}\end{aligned}$$

for any pair of nonnegative integers m and n. ◇

We are now ready to prove the existence and uniqueness of the polynomial of best approximation in the 2-norm. In particular, the next theorem shows that the infimum of $\|f-q\|_2$ over $q \in \mathcal{P}_n$ in problem (B) is attained and can be replaced by a minimum over $q \in \mathcal{P}_n$.

Theorem 9.2 *Given that $f \in \mathrm{L}^2_w(a,b)$, there exists a unique polynomial $p_n \in \mathcal{P}_n$ such that $\|f-p_n\|_2 = \min_{q\in\mathcal{P}_n} \|f-q\|_2$.*

Proof In order to simplify the notation, we recall the definition of the inner product $\langle \cdot , \cdot \rangle$:

$$\langle g, h\rangle = \int_a^b w(x)g(x)h(x)\mathrm{d}x\,,$$

and note that the induced 2-norm, $\|\cdot\|_2$, is defined by

$$\|g\|_2 = \langle g, g\rangle^{1/2}.$$

Suppose that φ_j, $j = 0,\dots,n$, is a system of orthogonal polynomials with respect to the weight function w on (a,b). Let us normalise the polynomials φ_j by defining a new system of orthogonal polynomials,

$$\psi_j(x) = \frac{\varphi_j(x)}{\|\varphi_j\|_2}\,, \qquad j = 0,\dots,n\,.$$

Then,

$$\langle \psi_k, \psi_j\rangle = \begin{cases} 1\,, & j = k\,,\\ 0\,, & j \neq k\,.\end{cases}$$

Such a system of polynomials is said to be **orthonormal**. The polynomials ψ_j, $j = 0, \ldots, n$, are linearly independent and form a basis for the linear space $\mathcal{P}_n$; therefore, each element $q \in \mathcal{P}_n$ can be expressed as a suitable linear combination,

$$q(x) = \beta_0\psi_0(x) + \cdots + \beta_n\psi_n(x)\,.$$

We wish to choose β_j, $j = 0, \ldots, n$, so as to ensure that the corresponding polynomial q minimises $\|f - q\|_2^2$ over all $q \in \mathcal{P}_n$. Let us, therefore, consider the function E: $(\beta_0, \ldots, \beta_n) \in \mathbb{R}^{n+1} \mapsto E(\beta_0, \ldots, \beta_n)$ defined by $E(\beta_0, \ldots, \beta_n) = \|f - q\|_2^2$, where $q(x) = \beta_0\psi_0(x) + \cdots + \beta_n\psi_n(x)$. Then,

$$\begin{aligned} E(\beta_0, \ldots, \beta_n) &= \langle f - q, f - q\rangle \\ &= \langle f, f\rangle - 2\langle f, q\rangle + \langle q, q\rangle \\ &= \|f\|_2^2 - 2\sum_{j=0}^{n} \beta_j \langle f, \psi_j\rangle + \sum_{j=0}^{n}\sum_{k=0}^{n} \beta_j\beta_k\langle \psi_k, \psi_j\rangle \\ &= \|f\|_2^2 - 2\sum_{j=0}^{n} \beta_j \langle f, \psi_j\rangle + \sum_{j=0}^{n} \beta_j^2 \\ &= \sum_{j=0}^{n} [\beta_j - \langle f, \psi_j\rangle]^2 + \|f\|_2^2 - \sum_{j=0}^{n} |\langle f, \psi_j\rangle|^2\,. \end{aligned}$$

The function $(\beta_0, \ldots, \beta_n) \mapsto E(\beta_0, \ldots, \beta_n)$ achieves its minimum value at $(\beta_0^*, \ldots, \beta_n^*)$, where

$$\beta_j^* = \langle f, \psi_j\rangle\,, \qquad j = 0, \ldots, n\,.$$

Hence $p_n \in \mathcal{P}_n$ defined by

$$p_n(x) = \beta_0^*\psi_0(x) + \cdots + \beta_n^*\psi_n(x)$$

is the unique polynomial of best approximation of degree n to the function $f \in \mathrm{L}_w^2(a, b)$ in the 2-norm on the interval (a, b). □

Remark 9.1 *As* $E(\beta_0^*, \ldots, \beta_n^*) = \|f - p_n\|_2^2 \geq 0$, *it follows from the proof of Theorem 9.2 that if* $f \in \mathrm{L}_w^2(a, b)$, *and* $\{\psi_0, \psi_1, \ldots\}$ *is an orthonormal system of polynomials in* $\mathrm{L}_w^2(a, b)$, *then*

$$\sum_{j=0}^{n} |\langle f, \psi_j\rangle|^2 \leq \|f\|_2^2$$

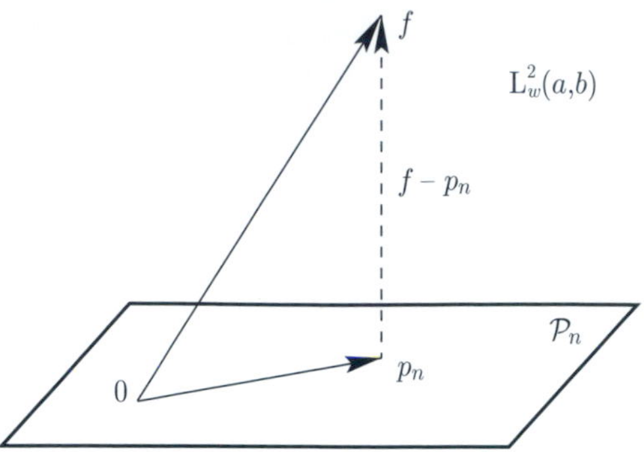

Fig. 9.3. Illustration of the orthogonality property $\langle f - p_n, q\rangle = 0$ for all q in $\mathcal{P}_n$, expressing the fact that if $p_n \in \mathcal{P}_n$ is a polynomial of best approximation to $f \in \mathrm{L}^2_w(a,b)$ in the 2-norm, then the error $f - p_n$ is orthogonal, in $\mathrm{L}^2_w(a,b)$, to all elements of the linear space $\mathcal{P}_n$. The 0 in the figure denotes the zero element of the linear space $\mathcal{P}_n$ (and, simultaneously, that of $\mathrm{L}^2_w(a,b)$), namely the function that is identically zero on the interval (a,b).

for each $n \geq 0$. This result is known as **Bessel's inequality**.[1]

The next theorem, in conjunction with the use of orthogonal polynomials, will be our key tool for constructing the polynomial of best approximation in the 2-norm.

Theorem 9.3 *A polynomial $p_n \in \mathcal{P}_n$ is the polynomial of best approximation of degree n to a function $f \in \mathrm{L}^2_w(a,b)$ in the 2-norm if, and only if, the difference $f - p_n$ is orthogonal to every element of $\mathcal{P}_n$, i.e.,*

$$\langle f - p_n, q\rangle = 0 \qquad \forall\, q \in \mathcal{P}_n\,. \tag{9.9}$$

A geometrical illustration of the property (9.9) is given in Figure 9.3.

Proof of theorem Suppose that (9.9) holds. Then,

$$\langle f - p_n, p_n - q\rangle = 0 \qquad \forall\, q \in \mathcal{P}_n,$$

given that $p_n - q \in \mathcal{P}_n$ for each q in $\mathcal{P}_n$. Therefore,

$$\begin{aligned}\|f - p_n\|_2^2 &= \langle f - p_n, f - p_n\rangle \\ &= \langle f - p_n, f - q\rangle + \langle f - p_n, q - p_n\rangle \\ &= \langle f - p_n, f - q\rangle \qquad \forall\, q \in \mathcal{P}_n\,.\end{aligned}$$

[1] Friedrich Wilhelm Bessel (22 July 1784, Minden, Westphalia, Holy Roman Empire (now Germany) – 17 March 1846, Königsberg, Prussia (now Kaliningrad, Russia)).

Hence, by the Cauchy–Schwarz inequality (9.2),

$$\|f - p_n\|_2^2 \leq \|f - p_n\|_2 \, \|f - q\|_2 \qquad \forall q \in \mathcal{P}_n .$$

This implies that

$$\|f - p_n\|_2 \leq \|f - q\|_2 \qquad \forall q \in \mathcal{P}_n .$$

On choosing $q = p_n$ on the right-hand side, equality will hold and therefore

$$\|f - p_n\|_2 = \min_{q \in \mathcal{P}_n} \|f - q\|_2 .$$

Conversely, suppose that p_n is the polynomial of best approximation to $f \in \mathrm{L}^2_w(a,b)$. We have seen in the proof of Theorem 9.2 that p_n can be written in terms of the orthonormal polynomials ψ_k, $k = 0, \dots, n$, as

$$p_n(x) = \beta_0^* \psi_0(x) + \cdots + \beta_n^* \psi_n(x) ,$$

where

$$\beta_k^* = \langle f, \psi_k \rangle , \qquad k = 0, \dots, n . \tag{9.10}$$

On recalling that $\langle \psi_k, \psi_j \rangle = \delta_{jk}$, $j, k \in \{0, \dots, n\}$, where δ_{jk} is the Kronecker delta, we deduce from (9.10) that

$$\begin{aligned} \langle f - p_n, \psi_j \rangle &= \langle f, \psi_j \rangle - \sum_{k=0}^{n} \beta_k^* \langle \psi_k, \psi_j \rangle \\ &= \langle f, \psi_j \rangle - \sum_{k=0}^{n} \beta_k^* \delta_{jk} \\ &= \langle f, \psi_j \rangle - \beta_j^* = 0 , \qquad j = 0, \dots, n . \end{aligned} \tag{9.11}$$

Since $\mathcal{P}_n = \mathrm{span}\{\psi_0, \dots, \psi_n\}$, it follows from (9.11) that $\langle f - p_n, q \rangle = 0$ for all $q \in \mathcal{P}_n$, as required. □

An equivalent, but slightly more explicit, form of writing (9.9) is

$$\int_a^b w(x)(f(x) - p_n(x))q(x) \, \mathrm{d}x = 0 \qquad \forall q \in \mathcal{P}_n .$$

Theorem 9.3 provides a simple method for determining the polynomial of best approximation $p_n \in \mathcal{P}_n$ to a function $f \in \mathrm{L}^2_w(a,b)$ in the 2-norm. First, proceeding as described in the discussion following Definition 9.4, we construct the system of orthogonal polynomials φ_j, $j = 0, \dots, n$, on the interval (a, b) with respect to the weight function w, if this system is not already known. We normalise the polynomials φ_j, $j = 0, \ldots, n$, by setting

$$\psi_j = \frac{\varphi_j}{\|\varphi_j\|_2}, \qquad j = 0, \ldots, n,$$

to obtain the system of orthonormal polynomials ψ_j, $j = 0, \ldots, n$, on (a, b). We then evaluate the coefficients $\beta_j^* = \langle f, \psi_j \rangle$, $j = 0, \ldots, n$, and form $p_n(x) = \beta_0^* \psi_0(x) + \cdots + \beta_n^* \psi_n(x)$.

We may avoid the necessity of determining the normalised polynomials ψ_j by writing

$$\begin{aligned} p_n(x) &= \beta_0^* \psi_0(x) + \cdots + \beta_n^* \psi_n(x) \\ &= \beta_0^* \langle \varphi_0, \varphi_0 \rangle^{-1/2} \varphi_0(x) + \cdots + \beta_n^* \langle \varphi_n, \varphi_n \rangle^{-1/2} \varphi_n(x) \\ &= \gamma_0 \varphi_0(x) + \cdots + \gamma_n \varphi_n(x), \end{aligned} \tag{9.12}$$

where

$$\gamma_j = \frac{\langle f, \varphi_j \rangle}{\langle \varphi_j, \varphi_j \rangle}, \qquad j = 0, \ldots, n. \tag{9.13}$$

Thus, as indicated at the beginning of the section, with this approach to the construction of the polynomial of best approximation in the 2-norm, we obtain the coefficients γ_j explicitly and there is no need to solve a system of linear equations with a full matrix.

Example 9.8 *We shall construct the polynomial of best approximation of degree 2 in the 2-norm to the function $f\colon x \mapsto \mathrm{e}^x$ over $(0, 1)$ with weight function $w(x) \equiv 1$.*

We already know a system of orthogonal polynomials φ_0, φ_1, φ_2 on this interval from Example 9.5; thus, we seek $p_2 \in \mathcal{P}_2$ in the form

$$p_2(x) = \gamma_0 \varphi_0(x) + \gamma_1 \varphi_1(x) + \gamma_2 \varphi_2(x), \tag{9.14}$$

where, according to (9.13),

$$\gamma_j = \frac{\int_0^1 \mathrm{e}^x \varphi_j(x) \mathrm{d}x}{\int_0^1 \varphi_j^2(x) \mathrm{d}x}, \qquad j = 0, 1, 2.$$

Recalling from Example 9.5 that

$$\varphi_0(x) \equiv 1, \quad \varphi_1(x) = x - \tfrac{1}{2}, \quad \varphi_2(x) = x^2 - x + \tfrac{1}{6},$$

we then have that

$$\left.\begin{aligned}\gamma_0 &= \frac{\mathrm{e}-1}{1} = \mathrm{e}-1\,,\\ \gamma_1 &= \frac{3/2-\mathrm{e}/2}{1/12} = 18-6\mathrm{e}\,,\\ \gamma_2 &= \frac{7\mathrm{e}/6-19/6}{1/180} = 210\mathrm{e}-570\,.\end{aligned}\right\} \tag{9.15}$$

Substituting the values of γ_0, γ_1 and γ_2 into (9.14), we deduce that the polynomial of best approximation of degree 2 for the function $f\colon x \mapsto \mathrm{e}^x$ in the 2-norm is

$$p_2(x) = (210\mathrm{e} - 570)x^2 + (588 - 216\mathrm{e})x + (39\mathrm{e} - 105)\,.$$

The approximation error is

$$\|f - p_2\|_2 = 0.005431\,,$$

to six decimal digits. ◇

We conclude this section by giving a property of orthogonal polynomials that will be required in the next chapter.

Theorem 9.4 *Suppose that φ_j, $j = 0, 1, \ldots$, is a system of orthogonal polynomials on the interval (a, b) with respect to the positive, continuous and integrable weight function w on (a, b). It is understood that φ_j is a polynomial of exact degree j. Then, for $j \geq 1$, the zeros of the polynomial φ_j are real and distinct, and lie in the interval (a, b).*

Proof Suppose that ξ_i, $i = 1, \ldots, k$, are the points in the open interval (a, b) at which $\varphi_j(x)$ changes sign. Let us note that $k \geq 1$, because for $j \geq 1$, by orthogonality of $\varphi_j(x)$ to $\varphi_0(x) \equiv 1$, we have that

$$\int_a^b w(x)\varphi_j(x)\mathrm{d}x = 0\,.$$

Thus, the integrand, being a continuous function that is not identically zero on (a, b), must change sign on (a, b); however, w is positive on (a, b), so φ_j must change sign at least once on (a, b). Therefore $k \geq 1$.

Let us define

$$\pi_k(x) = (x - \xi_1)\ldots(x - \xi_k)\,. \tag{9.16}$$

Now the function $\varphi_j(x)\pi_k(x)$ does not change sign in the interval (a, b), since at each point where $\varphi_j(x)$ changes sign $\pi_k(x)$ changes sign also. Hence,

$$\int_a^b w(x)\varphi_j(x)\pi_k(x)\mathrm{d}x \neq 0\,.$$

However, φ_j is orthogonal to every polynomial of lower degree with respect to the weight function w, so the degree of the polynomial π_k must be at least j; thus, $k \geq j$. On the other hand, k cannot be greater than j, since a polynomial of exact degree j cannot change sign more than j times. Therefore $k = j$; *i.e.*, the points $\xi_i \in (a, b)$, $i = 1, \ldots, j$, are the zeros (and all the zeros) of $\varphi_j(x)$. □

9.5 Comparisons

We can show that the polynomial of best approximation in the 2-norm for a function $f \in \mathrm{C}[a, b]$ is also a near-best approximation in the ∞-norm for f on $[a, b]$ in the sense defined in Section 8.5.

Theorem 9.5 *Let $n \geq 0$ and assume that f is defined and continuous on the interval $[a, b]$, and $f \notin \mathcal{P}_n$. Let p_n be the polynomial of best approximation of degree n to f in the 2-norm on $[a, b]$, where the weight function w is positive, continuous and integrable on (a, b). Then, the difference $f - p_n$ changes sign at no less than $n + 1$ distinct points in the interval (a, b).*

Proof The proof is very similar to that of Theorem 9.4; we shall give an outline and leave the details as an exercise.

As $\langle f - p_n, 1 \rangle = 0$, *i.e.*,

$$\int_a^b w(x)(f(x) - p_n(x))\mathrm{d}x = 0\,,$$

and $w(x) > 0$ for all $x \in (a, b)$, it follows that $f - p_n$ changes sign in (a, b). Let ξ_j, $j = 1, \ldots, k$, denote distinct points in (a, b) where $f - p_n$ changes sign. We shall prove that $k \geq n + 1$.

Define the polynomial $\pi_k(x)$ as in (9.16); then, $w(x)[f(x) - p_n(x)]\pi_k(x)$ does not change sign in (a, b), and so its integral over (a, b) is not zero. Therefore, $\langle f - p_n, \pi_k \rangle \neq 0$. On the other hand, according to Theorem 9.3, $f - p_n$ is orthogonal to every polynomial of degree n or less. Hence the degree of $\pi_k(x)$ must be greater than n, and so $k \geq n + 1$. □

We return to the example illustrated by Figure 8.5, and consider the difference $f - p_n$ for the function $f\colon x \mapsto \mathrm{e}^{2x}$ on the interval $(0, 1)$. Figure 9.4 shows this difference for two polynomial approximations of degree 4: the minimax approximation of Section 8.5 and the best approximation in the 2-norm with weight function $w(x) \equiv 1$. It is clear that the

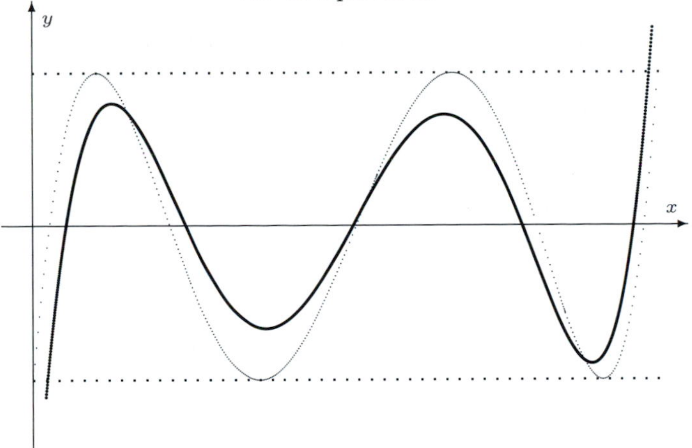

Fig. 9.4. The difference $e^{2x} - p_4(x)$ for two polynomial approximations of degree 4 on $[0, 1]$. Thin curve – minimax approximation; thick curve – best approximation in the 2-norm with weight function $w(x) \equiv 1$.

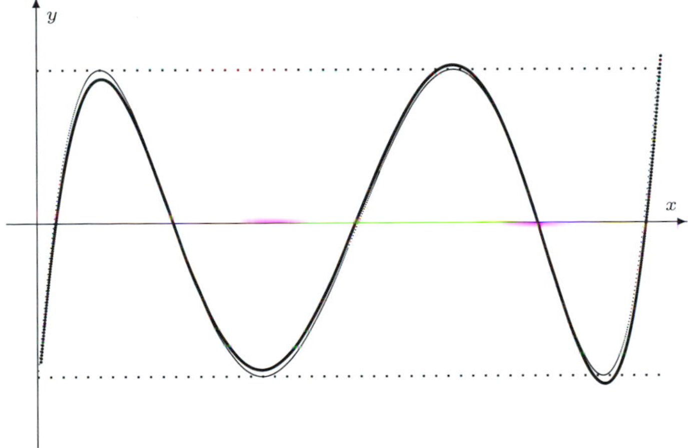

Fig. 9.5. The difference $e^{2x} - p_4(x)$ for two polynomial approximations of degree 4 on $[0, 1]$. Thin curve – minimax approximation; thick curve – best approximation in the 2-norm with weight function $w(x) = [x(1-x)]^{-1/2}$.

error of the 2-norm approximation has the right number of alternating local maxima and minima, and is a near-minimax approximation from $\mathcal{P}_4$ to f on $[0, 1]$; but the extrema at the ends of the interval are significantly larger than the internal extrema. If we use a weight function w

which gives greater weight near the ends of the interval, it seems likely that the extrema of the error might be more nearly equal. This can be achieved by using the weight function $w(x) = [x(1-x)]^{-1/2}$, so that the orthogonal polynomials are the Chebyshev polynomials adapted to the interval $(0, 1)$. Figure 9.5 shows the corresponding difference $f - p_n$, and we now see that the two best approximations, in the ∞-norm and the weighted 2-norm, are very close.

Polynomials of best approximation in the 2-norm have a special property which is often useful. Suppose that we have constructed the best polynomial approximation, p_n, of degree n, in the 2-norm, but that p_n does not achieve the required accuracy. To construct the best polynomial approximation of degree $n+1$ all we need is to calculate γ_{n+1} from

$$\gamma_{n+1} = \frac{\langle f - p_n, \varphi_{n+1} \rangle}{\|\varphi_{n+1}\|_2^2}$$

and then let $p_{n+1}(x) = p_n(x) + \gamma_{n+1}\varphi_{n+1}(x)$. By noting that

$$\langle f - p_{n+1}, \varphi_j \rangle = 0\,, \qquad j = 0, 1, \ldots, n+1\,,$$

it follows that p_{n+1} is best least squares approximation to f from $\mathcal{P}_{n+1}$. If we are constructing the minimax approximation of degree $n+1$, or using Lagrange interpolation with equally spaced points, the work involved in constructing p_n is lost, and the construction of p_{n+1} must begin completely afresh.

9.6 Notes

We give some pointers to the vast literature on orthogonal polynomials. The following are classical sources on the subject.

- Géza Freund, *Orthogonal Polynomials,* Pergamon Press, Oxford, New York, 1971.
- Paul Névai, *Orthogonal Polynomials,* Memoirs of the American Mathematical Society, no. 213, American Mathematical Society, Providence, RI, 1979.
- Gábor Szegő, *Orthogonal Polynomials,* Colloquium publications (American Mathematical Society), 23, American Mathematical Society, Providence, RI, 1959.

Tables of orthogonal polynomials are found in

- M. ABRAMOWITZ AND I.A. STEGUN (Editors), 'Orthogonal polynomials', Ch. 22 in *Handbook of Mathematical Functions with Formulas, Graphs, and Mathematical Tables*, ninth printing, Dover, New York, pp. 771–802, 1972.

Computational aspects of the theory of orthogonal polynomials are discussed in the edited volume

- W. GAUTSCHI, G.H. GOLUB, AND G. OPFER (Editors), *Applications and Computation of Orthogonal Polynomials*, Conference at the Mathematical Research Institute, Oberwolfach, Germany, March 22–28, 1998, Birkhäuser, Basel, 1999.

A recent survey of the theory and application of orthogonal polynomials in numerical computations is contained in

- W. GAUTSCHI, Orthogonal polynomials: applications and computation, *Acta Numerica* **5** (A. Iserles, ed.), Cambridge University Press, Cambridge, pp. 45–119, 1996.

Finally, we refer to the books of Powell and Cheney, cited in the Notes at the end of the previous chapter, concerning the application of orthogonal polynomials in the field of best least squares approximation.

Exercises

9.1 Construct orthogonal polynomials of degrees 0, 1 and 2 on the interval $(0,1)$ with the weight function $w(x) = -\ln x$.

9.2 Let the polynomials φ_j, $j = 0, 1, \ldots$, form an orthogonal system on the interval $(-1,1)$ with respect to the weight function $w(x) \equiv 1$. Show that the polynomials $\varphi_j((2x - a - b)/(b - a))$, $j = 0, 1, \ldots$, represent an orthogonal system for the interval (a, b) and the same weight function. Hence obtain the polynomials in Example 9.5 from the Legendre polynomials in Example 9.6.

9.3 Suppose that the polynomials φ_j, $j = 0, 1, \ldots$, form an orthogonal system on the interval $(0,1)$ with respect to the weight function $w(x) = x^\alpha$, $\alpha > 0$. Find, in terms of φ_j, a system of orthogonal polynomials for the interval $(0, b)$ and the same weight function.

9.4 Show, by induction or otherwise, that, for $0 \le k \le n$,

$$\left(\frac{\mathrm{d}}{\mathrm{d}x}\right)^k (1-x^2)^n = (1-x^2)^{n-k} q_k(x)\,,$$

where q_k is a polynomial of degree k. Deduce that all the derivatives of the function $(1-x^2)^n$ of order less than n vanish at $x = \pm 1$.

Define $\varphi_j(x) = (\mathrm{d}/\mathrm{d}x)^j (1-x^2)^j$, and show by repeated integration by parts that

$$\int_{-1}^{1} \varphi_k(x)\varphi_j(x)\mathrm{d}x = 0\,, \qquad 0 \le k < j\,.$$

Hence verify the expressions in Example 9.6 for the Legendre polynomials of degrees 0, 1, 2 and 3.

9.5 Show, by induction or otherwise, that, for $0 \le k \le j$,

$$\left(\frac{\mathrm{d}}{\mathrm{d}x}\right)^k x^j \mathrm{e}^{-x} = x^{j-k} q_k(x)\mathrm{e}^{-x}\,,$$

where $q_k(x)$ is a polynomial of degree k.

The function φ_j is defined for $j \ge 0$ by

$$\varphi_j(x) = \mathrm{e}^x \frac{\mathrm{d}^j}{\mathrm{d}x^j}(x^j \mathrm{e}^{-x})\,.$$

Show that, for each $j \ge 0$, φ_j is a polynomial of degree j, and that these polynomials form an orthogonal system on the interval $(0, \infty)$ with respect to the weight function $w(x) = \mathrm{e}^{-x}$. Write down the polynomials with $j = 0, 1, 2$ and 3.

9.6 Suppose that φ_j, $j = 0, 1, \ldots$, form a system of orthogonal polynomials with weight function $w(x)$ on the interval (a, b). Show that, for some value of the constant C_j, $\varphi_{j+1}(x) - C_j x \varphi_j(x)$ is a polynomial of degree j, and hence that

$$\varphi_{j+1}(x) - C_j x \varphi_j(x) = \sum_{k=0}^{j} \alpha_{jk}\varphi_k(x)\,, \qquad \alpha_{jk} \in \mathbb{R}\,.$$

Use the orthogonality properties to show that $\alpha_{jk} = 0$ for $k < j-1$, and deduce that the polynomials satisfy a recurrence relation of the form

$$\varphi_{j+1}(x) - (C_j x + D_j)\varphi_j(x) + E_j \varphi_{j-1}(x) = 0\,, \qquad j \ge 1\,.$$

9.7 In the notation of Exercise 6 suppose that the normalisation of the polynomials is so chosen that for each j the coefficient of x^j in $\varphi_j(x)$ is positive. Show that $C_j > 0$ for all j. By considering

$$\int_a^b w(x)\varphi_j(x)[\varphi_j(x) - C_{j-1}x\varphi_{j-1}(x)]\mathrm{d}x$$

show that

$$\int_a^b w(x)x\varphi_{j-1}(x)\varphi_j(x)\mathrm{d}x > 0\,,$$

and deduce that $E_j > 0$ for all j. Hence show that for all positive values of j the zeros of φ_j and φ_{j-1} interlace. (See the proof of Theorem 5.8.)

9.8 Using the weight function w on the interval (a, b) apply a similar argument to that for Theorem 8.6 to find the best polynomial approximation p_n of degree n in the 2-norm to the function x^{n+1}. Show that

$$\|x^{n+1} - p_n\|_2^2 = \int_a^b w(x)\varphi_{n+1}^2\mathrm{d}x/[c_{n+1}^{n+1}]^2\,,$$

where c_{n+1}^{n+1} is the coefficient of x^{n+1} in $\varphi_{n+1}(x)$.

Write down the best polynomial approximation of degree 2 to the function x^3 in the 2-norm with $w(x) \equiv 1$ on the interval $(-1, 1)$, and evaluate the 2-norm of the error.

9.9 Suppose that the weight w is an even function on the interval $(-a, a)$, and that a system of orthogonal polynomials φ_j, $j = 0, \ldots, n$, on the interval $(-a, a)$ is constructed by the Gram–Schmidt process. Show that, if j is even, then φ_j is an even function, and that, if j is odd, then φ_j is an odd function.

Now suppose that the best polynomial approximation of degree n in the 2-norm to the function f on the interval $(-a, a)$ is expressed in the form

$$p_n(x) = \gamma_0\varphi_0(x) + \cdots + \gamma_n\varphi_n(x)\,.$$

Show that if f is an even function, then all the odd coefficients γ_{2j-1} are zero, and that if f is an odd function, then all the even coefficients γ_{2j} are zero.

9.10 The function $H(x)$ is defined by $H(x) = 1$ if $x > 0$, and $H(-x) = -H(x)$. Construct the best polynomial approximations of degrees 0, 1 and 2 in the 2-norm to this function over the interval $(-1, 1)$ with weight function $w(x) \equiv 1$. (It may not appear very useful to consider a polynomial approximation to a discontinuous function, but representations of such functions by Fourier series will be familiar to most readers. Note that the function H belongs to $\mathrm{L}^2_w(-1,1)$.)

10

Numerical integration – II

10.1 Introduction

In Section 7.2 we described the Newton–Cotes family of formulae for numerical integration. These were constructed by replacing the integrand by its Lagrange interpolation polynomial with equally spaced interpolation points and integrating this exactly. Here, we consider another family of numerical integration rules, called Gauss quadrature formulae, which are based on replacing the integrand f by its Hermite interpolation polynomial and choosing the interpolation points x_j in such a way that, after integrating the Hermite polynomial, the derivative values $f'(x_j)$ do not enter the quadrature formula. It turns out that this can be achieved by requiring that the x_j are roots of a polynomial of a certain degree from a system of orthogonal polynomials.

10.2 Construction of Gauss quadrature rules

Suppose that the function f is defined on the closed interval $[a, b]$ and that it is continuous and differentiable on this interval. Suppose, further, that w is a weight function, defined, positive, continuous and integrable on (a, b). We wish to construct quadrature formulae for the approximate evaluation of the integral

$$\int_a^b w(x) f(x) \mathrm{d}x \, .$$

For a nonnegative integer n, let x_i, $i = 0, \ldots, n$, be $n + 1$ points in the interval $[a, b]$; the precise location of these points will be determined later on. The Hermite interpolation polynomial of degree $2n + 1$ for the function f is given by the expression (see Section 6.4)

$$p_{2n+1}(x) = \sum_{k=0}^{n} H_k(x) f(x_k) + \sum_{k=0}^{n} K_k(x) f'(x_k)\,, \qquad (10.1)$$

where

$$\begin{aligned} H_k(x) &= [L_k(x)]^2 (1 - 2L'_k(x_k)(x - x_k))\,, \\ K_k(x) &= [L_k(x)]^2 (x - x_k)\,. \end{aligned} \qquad (10.2)$$

Further, for $n \geq 1$, $L_k \in \mathcal{P}_n$ is defined by

$$L_k(x) = \prod_{\substack{i=0 \\ i \neq k}}^{n} \frac{x - x_i}{x_k - x_i}\,, \qquad k = 0, 1, \ldots, n\,;$$

if $n = 0$, we let $L_0(x) \equiv 1$ and thereby $H_0(x) \equiv 1$ and $K_0(x) = x - x_0$ for this value of n. Thus, we deduce from (10.1) that

$$\begin{aligned} \int_a^b w(x) f(x) \mathrm{d}x &\approx \int_a^b w(x) p_{2n+1}(x) \mathrm{d}x \\ &= \sum_{k=0}^{n} W_k f(x_k) + \sum_{k=0}^{n} V_k f'(x_k)\,, \end{aligned} \qquad (10.3)$$

where

$$W_k = \int_a^b w(x) H_k(x) \mathrm{d}x\,, \qquad V_k = \int_a^b w(x) K_k(x) \mathrm{d}x\,.$$

There is an obvious advantage in choosing the points x_k in such a way that all the coefficients V_k are zero, for then the derivative values $f'(x_k)$ are not required. Recalling the form of the polynomial K_k and inserting it into the defining expression for V_k, we have

$$\begin{aligned} V_k &= \int_a^b w(x) [L_k(x)]^2 (x - x_k) \mathrm{d}x \\ &= C_n \int_a^b w(x) \pi_{n+1}(x) L_k(x) \mathrm{d}x\,, \end{aligned} \qquad (10.4)$$

where $\pi_{n+1}(x) = (x - x_0) \ldots (x - x_n)$ and

$$C_n = \begin{cases} \left(\prod_{i=0, i \neq k}^{n} (x_k - x_i)^{-1} \right) & \text{if } n \geq 1\,, \\ 1 & \text{if } n = 0\,. \end{cases}$$

Since π_{n+1} is of degree $n + 1$ while $L_k(x)$ is of degree n for each k, $0 \leq k \leq n$, each V_k will be zero if the polynomial π_{n+1} is orthogonal to every polynomial of lower degree with respect to the weight function

w. We can therefore construct the required quadrature formula (10.3) with $V_k = 0$, $k = 0, \ldots, n$, by choosing the points x_k, $k = 0, \ldots, n$, to be the zeros of the polynomial of degree $n+1$ in a system of orthogonal polynomials over the interval (a, b) with respect to the weight function w; we know from Theorem 9.4 that these zeros are real and distinct, and all lie in the open interval (a, b).

Having chosen the location of the points x_k, we now consider W_k:

$$\begin{aligned} W_k &= \int_a^b w(x) H_k(x) \mathrm{d}x \\ &= \int_a^b w(x) [L_k(x)]^2 (1 - 2L_k'(x_k)(x - x_k)) \mathrm{d}x \\ &= \int_a^b w(x) [L_k(x)]^2 \mathrm{d}x - 2L_k'(x_k) V_k \,. \end{aligned} \tag{10.5}$$

Since $V_k = 0$, the second term in the last line vanishes and thus we obtain the following numerical integration formula, known as the **Gauss quadrature**[1] rule:

$$\int_a^b w(x) f(x) \mathrm{d}x \approx \mathcal{G}_n(f) = \sum_{k=0}^{n} W_k f(x_k) \,, \tag{10.6}$$

where the **quadrature weights** are

$$W_k = \int_a^b w(x) [L_k(x)]^2 \mathrm{d}x \,, \tag{10.7}$$

and the **quadrature points** x_k, $k = 0, \ldots, n$, are chosen as the zeros of the polynomial of degree $n+1$ from a system of orthogonal polynomials over the interval (a, b) with respect to the weight function w. Since this quadrature rule was obtained by exact integration of the Hermite interpolation polynomial of degree $2n+1$ for f, it gives the exact result whenever f is a polynomial of degree $2n+1$ or less.

Example 10.1 *Consider the case* $n = 1$, *with the weight function* $w(x) \equiv 1$ *over the interval* $(0, 1)$.

The quadrature points x_0, x_1 are then the zeros of the polynomial φ_2 constructed in Example 9.5 and given by (9.8),

$$\varphi_2(x) = x^2 - x + \tfrac{1}{6} \,, \tag{10.8}$$

[1] Carl Friedrich Gauss, *Methodus nova integralium valores per approximationem inveniendi*, 1814.

and therefore

$$x_0 = \tfrac{1}{2} - \sqrt{\tfrac{1}{12}}\,, \qquad x_1 = \tfrac{1}{2} + \sqrt{\tfrac{1}{12}}\,.$$

Clearly, x_0 and x_1 belong to the open interval $(0,1)$, in accordance with Theorem 9.4. The weights are obtained from (10.7):

$$\begin{aligned} W_0 &= \int_0^1 \left(\frac{x - x_1}{x_0 - x_1}\right)^2 \mathrm{d}x \\ &= 3\int_0^1 (x^2 - 2x_1 x + x_1^2)\,\mathrm{d}x \\ &= 3(\tfrac{1}{3} - x_1 + x_1^2) \\ &= \tfrac{1}{2}\,, \end{aligned} \tag{10.9}$$

and $W_1 = \frac{1}{2}$ in the same way. We thus have the Gauss quadrature rule

$$\int_0^1 f(x)\mathrm{d}x \approx \tfrac{1}{2} f(\tfrac{1}{2} - \sqrt{\tfrac{1}{12}}) + \tfrac{1}{2} f(\tfrac{1}{2} + \sqrt{\tfrac{1}{12}})\,, \tag{10.10}$$

which is exact whenever f is a polynomial of degree $2 \times 1 + 1 = 3$ or less. ◇

10.3 Direct construction

The calculation of the weights and the quadrature points in a Gauss quadrature rule requires little work when the system of orthogonal polynomials is already known. If this is not known, at the very least it is necessary to construct the polynomial from the system whose roots are the quadrature points; in that case a straightforward approach, which avoids this construction, may be easier.

Suppose, for example, that we wish to find the values of A_0, A_1, x_0 and x_1 such that the quadrature rule

$$\int_0^1 f(x)\mathrm{d}x \approx A_0 f(x_0) + A_1 f(x_1) \tag{10.11}$$

is exact for all $f \in \mathcal{P}_3$.

We have to determine four unknowns, A_0, A_1, x_0 and x_1, so we need four equations; thus we take, in turn, $f(x) \equiv 1$, $f(x) = x$, $f(x) = x^2$ and $f(x) = x^3$ and demand that the quadrature rule (10.11) is exact (that is, the integral of f is equal to the corresponding approximation obtained by inserting f into the right-hand side of (10.11)). Hence,

$$1 = A_0 + A_1\,, \tag{10.12}$$

$$\begin{aligned} \tfrac{1}{2} &= A_0x_0 + A_1x_1\,, && (10.13)\\ \tfrac{1}{3} &= A_0x_0^2 + A_1x_1^2\,, && (10.14)\\ \tfrac{1}{4} &= A_0x_0^3 + A_1x_1^3\,. && (10.15) \end{aligned}$$

It remains to solve this system. To do so, we consider the quadratic polynomial π_2 defined by

$$\pi_2(x) = (x - x_0)(x - x_1)$$

whose roots are the unknown quadrature points x_0 and x_1. In expanded form, $\pi_2(x)$ can be written as

$$\pi_2(x) = x^2 + px + q\,.$$

First we shall determine p and q; then we shall find the roots x_0 and x_1 of π_2. We shall then insert the values of x_0 and x_1 into (10.13) and solve the linear system (10.12), (10.13) for A_0 and A_1.

To find p and q, we multiply (10.12) by q, (10.13) by p and (10.14) by 1, and we add up the resulting equations to deduce that

$$\begin{aligned} \tfrac{1}{3} + \tfrac{1}{2}p + q &= A_0(x_0^2 + px_0 + q) + A_1(x_1^2 + px_1 + q)\\ &= A_0\pi_2(x_0) + A_1\pi_2(x_1) = A_0 \cdot 0 + A_1 \cdot 0 = 0\,. \end{aligned}$$

Therefore,

$$\tfrac{1}{3} + \tfrac{1}{2}p + q = 0\,. \qquad (10.16)$$

Similarly, we multiply (10.13) by q, (10.14) by p and (10.15) by 1, and we add up the resulting equations to obtain

$$\begin{aligned} \tfrac{1}{4} + \tfrac{1}{3}p + \tfrac{1}{2}q &= A_0x_0(x_0^2 + px_0 + q) + A_1x_1(x_1^2 + px_1 + q)\\ &= A_0x_0\pi_2(x_0) + A_1x_1\pi_2(x_1) = A_0 \cdot 0 + A_1 \cdot 0 = 0\,. \end{aligned}$$

Thus,

$$\tfrac{1}{4} + \tfrac{1}{3}p + \tfrac{1}{2}q = 0\,. \qquad (10.17)$$

From (10.16) and (10.17) we immediately find that $p = -1$ and $q = \frac{1}{6}$. Having determined p and q, we see that

$$\pi_2(x) = x^2 - x + \tfrac{1}{6}\,,$$

in agreement with (10.8). We then find the roots of this quadratic polynomial to give x_0 and x_1 as before. With these values of x_0 and x_1 we deduce from (10.12) and (10.13) that

$$\begin{aligned} A_0 + A_1 &= 1\,,\\ A_0(\tfrac{1}{2} - \sqrt{\tfrac{1}{12}}) + A_1(\tfrac{1}{2} + \sqrt{\tfrac{1}{12}}) &= \tfrac{1}{2}\,, \end{aligned}$$

and therefore $A_0 = A_1 = \frac{1}{2}$. Thus, we conclude that the required quadrature rule is (10.10), as before.

It is easy to see that equations (10.16) and (10.17) express the condition that the polynomial $x^2 + px + q$ is orthogonal to the polynomials 1 and x respectively. This alternative approach has simply constructed a quadratic polynomial from a system of orthogonal polynomials by requiring that it is orthogonal to every polynomial of lower degree, instead of building up the whole system of orthogonal polynomials.

A straightforward calculation shows that, in general, the quadrature rule (10.10) is not exact for polynomials of degree higher than 3 (take $f(x) = x^4$, for example, to verify this).

Example 10.2 *We shall apply the quadrature rule (10.10) to compute an approximation to the integral $I = \int_0^1 \mathrm{e}^x \mathrm{d}x$.*

Using (10.10) with $f(x) = \exp(x) = \mathrm{e}^x$ yields

$$I \approx \tfrac{1}{2}\exp\left(\tfrac{1}{2} - \sqrt{\tfrac{1}{12}}\right) + \tfrac{1}{2}\exp\left(\tfrac{1}{2} + \sqrt{\tfrac{1}{12}}\right) = \sqrt{\mathrm{e}}\cosh\sqrt{\tfrac{1}{12}}\,.$$

On rounding to six decimal digits, $I \approx 1.717896$. The exact value of the integral is $I = \mathrm{e} - 1 = 1.718282$, rounding to six decimal digits. ◇

10.4 Error estimation for Gauss quadrature

The next theorem provides a bound on the error that has been committed by approximating the integral on the left-hand side of (10.6) by the quadrature rule on the right.

Theorem 10.1 *Suppose that w is a weight function, defined, integrable, continuous and positive on (a, b), and that f is defined and continuous on $[a, b]$; suppose further that f has a continuous derivative of order $2n+2$ on $[a, b]$, $n \geq 0$. Then, there exists a number η in (a, b) such that*

$$\int_a^b w(x)f(x)\mathrm{d}x - \sum_{k=0}^{n} W_k f(x_k) = K_n f^{(2n+2)}(\eta)\,, \tag{10.18}$$

and

$$K_n = \frac{1}{(2n+2)!}\int_a^b w(x)[\pi_{n+1}(x)]^2\mathrm{d}x\,.$$

Consequently, the integration formula (10.6), (10.7) will give the exact result for every polynomial of degree $2n+1$.

Proof Recalling the definition of the Hermite interpolation polynomial p_{2n+1} for the function f and using Theorem 6.4, we have

$$\begin{aligned}\int_a^b w(x)f(x)\mathrm{d}x - \sum_{k=0}^{n} W_k f(x_k) &= \int_a^b w(x)(f(x) - p_{2n+1}(x))\mathrm{d}x \\ &= \int_a^b w(x)\frac{f^{(2n+2)}(\xi(x))}{(2n+2)!}[\pi_{n+1}(x)]^2\mathrm{d}x\,. \end{aligned} \tag{10.19}$$

However, by the Integral Mean Value Theorem, Theorem A.6, the last term is equal to

$$\frac{f^{(2n+2)}(\eta)}{(2n+2)!}\int_a^b w(x)[\pi_{n+1}(x)]^2\mathrm{d}x\,,$$

for some $\eta \in (a, b)$, and hence the desired error bound. □

Note that, by virtue of Theorem 10.1, the Gauss quadrature rule gives the exact value of the integral when f is a polynomial of degree $2n+1$ or less, which is the highest possible degree that one can hope for with the $2n+2$ free parameters consisting of the quadrature weights W_k, $k = 0, \dots, n$, and the quadrature points x_k, $k = 0, \dots, n$.

A different approach leads to a proof of convergence of the Gauss formulae $\mathcal{G}_n(f)$, defined in (10.6), (10.7), as $n \to \infty$.

Theorem 10.2 *Suppose that the weight function w is defined, positive, continuous and integrable on the open interval (a, b). Suppose also that the function f is continuous on the closed interval $[a, b]$. Then,*

$$\lim_{n\to\infty} \mathcal{G}_n(f) = \int_a^b w(x)f(x)\mathrm{d}x\,.$$

Proof If we choose any positive real number ε_0 then, since f is continuous on $[a, b]$, the Weierstrass Theorem (Theorem 8.1) shows that there is a polynomial p such that

$$|f(x) - p(x)| \le \varepsilon_0 \qquad \text{for all } x \in [a, b]\,. \tag{10.20}$$

Let N be the degree of this polynomial, and write p as p_N.

Thus we deduce that

$$\begin{aligned}\int_a^b w(x)f(x)\mathrm{d}x - \mathcal{G}_n(f) &= \int_a^b w(x)[f(x) - p_N(x)]\mathrm{d}x \\ &\quad + \int_a^b w(x)p_N(x)\mathrm{d}x - \mathcal{G}_n(p_N) \\ &\quad + \mathcal{G}_n(p_N) - \mathcal{G}_n(f)\,. \end{aligned} \tag{10.21}$$

Consider the first term on the right of this equality; it follows from (10.20) that

$$\left|\int_a^b w(x)[f(x) - p_n(x)]\mathrm{d}x\right| \leq \varepsilon_0 W\,,$$

where

$$W = \int_a^b w(x)\mathrm{d}x\,.$$

For the last term on the right of (10.21),

$$\begin{aligned}
|\mathcal{G}_n(f) - \mathcal{G}_n(p_N)| &\leq \sum_{k=0}^{n} |W_k[f(x_k) - p_N(x_k)]| \\
&\leq \varepsilon_0 \sum_{k=0}^{n} W_k \\
&= \varepsilon_0 \int_a^b w(x)\mathrm{d}x \\
&= \varepsilon_0 W\,, \qquad\qquad (10.22)
\end{aligned}$$

where we have used the fact that all the quadrature weights W_k are positive (see (10.7)), and that a Gauss quadrature rule integrates a constant function exactly. Now for the middle term in (10.21), if we define N_0 to be the integer part of $\frac{1}{2}N$, we see that when $n \geq N_0$ the quadrature formula is exact for all polynomials of degree $2N_0 + 1$ or less, and hence for the polynomial p_N (given that $N \leq 2N_0 + 1 \leq 2n + 1$). Therefore,

$$\int_a^b w(x)p_N(x)\mathrm{d}x - \mathcal{G}_n(p_N) = 0 \quad \text{if } n \geq N_0\,.$$

Putting these three terms together, we see that

$$\left|\int_a^b w(x)f(x)\mathrm{d}x - \mathcal{G}_n(f)\right| \leq \varepsilon_0 W + 0 + \varepsilon_0 W \quad \text{if } n \geq N_0\,.$$

Finally, given any positive number ε, we define $\varepsilon_0 = \varepsilon/(2W)$ and find the corresponding value of $N_0 = N_0(\varepsilon)$ to deduce that

$$\left|\int_a^b w(x)f(x)\mathrm{d}x - \mathcal{G}_n(f)\right| \leq \varepsilon \quad \text{if } n \geq N_0\,,$$

which is what we were required to prove. □

The interest of this theorem is mainly theoretical, as it gives no indication of how rapidly the error tends to zero. However, it does show the importance of the fact that the weights W_k are positive. Much of the above proof would apply with little change to the Newton–Cotes formulae of Section 7.2. We saw there that for the formulae of order 1 and 2, the trapezium rule and Simpson's rule, the weights are positive. However, when $n > 8$ some of the weights in the Newton–Cotes formula of order n become negative. In this case we have $\sum_{k=0}^{n} W_k = (b-a)$, but we find that $\sum_{k=0}^{n} |W_k| \to \infty$ as $n \to \infty$, so the proof breaks down. Stronger conditions must be imposed on the function f to ensure that the Newton–Cotes formula converges to the required integral. (See the example in Section 7.4.)

10.5 Composite Gauss formulae

It is often useful to define composite Gauss formulae, just as we did for the trapezium rule and Simpson's rule in Section 7.5. Let us suppose, for the sake of simplicity, that $w(x) \equiv 1$. We divide the range $[a, b]$ into m subintervals $[x_{j-1}, x_j]$, $j = 1, 2, \ldots, m$, $m \geq 2$, each of width $h = (b-a)/m$, and write

$$\int_a^b f(x)\mathrm{d}x = \sum_{j=1}^{m} \int_{x_{j-1}}^{x_j} f(x)\mathrm{d}x \,,$$

where

$$x_j = a + jh \,, \qquad j = 0, 1, \ldots, m \,.$$

We then map each of the subintervals $[x_{j-1}, x_j]$, $j = 1, 2, \ldots, m$, onto the reference interval $[-1, 1]$ by the change of variable

$$x = \tfrac{1}{2}(x_{j-1} + x_j) + \tfrac{1}{2}ht \,, \qquad t \in [-1, 1] \,,$$

giving

$$\int_a^b f(x)\mathrm{d}x = \tfrac{1}{2}h \sum_{j=1}^{m} \int_{-1}^{1} g_j(t)\mathrm{d}t = \tfrac{1}{2}h \sum_{j=1}^{m} I_j \,,$$

where

$$g_j(t) = f\left(\tfrac{1}{2}(x_{j-1} + x_j) + \tfrac{1}{2}ht\right) \quad \text{and} \quad I_j = \int_{-1}^{1} g_j(t)\mathrm{d}t \,.$$

The composite Gauss quadrature rule is then obtained by applying the same Gauss formula to each of the integrals I_j. This gives

$$\begin{aligned}\int_a^b f(x)\mathrm{d}x &\approx \tfrac{1}{2}h\sum_{j=1}^{m}\sum_{k=0}^{n} W_k g_j(\xi_k) \\ &= \tfrac{1}{2}h\sum_{j=1}^{m}\sum_{k=0}^{n} W_k f\left(\tfrac{1}{2}(x_{j-1}+x_j)+\tfrac{1}{2}h\xi_k\right),\end{aligned} \tag{10.23}$$

where ξ_k are the quadrature points in $(-1,1)$ and W_k are the associated weights for $k=0,\ldots,n$ with $n\geq 0$.

An expression for the error of this composite formula is obtained, as in Section 7.5, by adding the expressions (10.18) for the errors in the integrals I_j. The result is

$$\mathcal{E}_{n,m} = C_n \frac{(b-a)^{2n+3}}{2^{2n+3}m^{2n+2}(2n+2)!} f^{(2n+2)}(\eta) \tag{10.24}$$

where $\eta \in (a,b)$ and

$$C_n = \int_{-1}^{1} [\pi_{n+1}(t)]^2 \mathrm{d}t\,.$$

Definition 10.1 *The* **composite midpoint rule** *is the composite Gauss formula with $w(x)\equiv 1$ and $n=0$ defined by*

$$\int_a^b f(x)\mathrm{d}x \approx h\sum_{j=1}^{m} f(a+(j-\tfrac{1}{2})h)\,. \tag{10.25}$$

This follows from the fact that when $n=0$ there is one quadrature point $\xi_0 = 0$ in $(-1,1)$, which is at the midpoint of the interval, and the corresponding quadrature weight W_0 is equal to the length of the interval $(-1,1)$, *i.e.*, $W_0 = 2$. It follows from (10.24) with $n=0$ and

$$C_0 = \int_{-1}^{1} t^2 \mathrm{d}t = \tfrac{2}{3}$$

that the error in the composite midpoint rule is

$$\mathcal{E}_{0,m} = \frac{(b-a)^3}{24m^2} f''(\eta)\,,$$

where $\eta \in (a,b)$, provided that the function f has a continuous second derivative on $[a,b]$.

10.6 Radau and Lobatto quadrature

We have now discussed two types of quadrature formulae, which have the same form, $\sum_{k=0}^{n} W_k f(x_k)$. In the Newton–Cotes formulae the (equally spaced) quadrature points x_k are given, and we were able to find the weights W_k so that the result was exact for polynomials of degree n. By allowing the quadrature points as well as the weights to be freely chosen, we constructed Gauss quadrature formulae which were exact for polynomials of degree $2n+1$. There are also many possible formulae of mixed type, where some, but not all, of the quadrature points are given, and the rest can be freely chosen. We might expect that each quadrature point which is fixed will reduce the degree of polynomial for which such a formula is exact by 1, from the maximum degree of $2n+1$.

It is often useful to be able to fix one of the endpoints of the interval as one of the quadrature points. As an example, suppose we prescribe that $x_0 = a$. Let p_{2n} be an arbitrary polynomial of degree $2n$, and write

$$p_{2n}(x) = (x-a)q_{2n-1}(x) + r\,,$$

where the quotient q_{2n-1} is a polynomial of degree $2n-1$ and the remainder r is a constant. The integral of $w\,p_{2n}$ is then

$$\int_a^b w(x)p_{2n}(x)\mathrm{d}x = \int_a^b (x-a)w(x)q_{2n-1}(x)\mathrm{d}x + r\int_a^b w(x)\mathrm{d}x\,.$$

We can now construct the usual Gauss quadrature formula for the interval $[a,b]$ with the modified weight function $(x-a)w(x)$, giving n quadrature points and n weights $x_k^*,\ W_k^*,\ k=1,\ldots,n$. This formula will be exact for all polynomials q of degree $2n-1$. Provided that the weight function w satisfies the standard conditions on (a,b), the modified weight function does also; in particular it is clearly positive on (a,b). This gives

$$\begin{aligned}\int_a^b w(x)p_{2n}(x)\mathrm{d}x &= \sum_{k=1}^{n} W_k^* q_{2n-1}(x_k^*) + r\int_a^b w(x)\mathrm{d}x \\ &= \sum_{k=1}^{n} \frac{W_k^*}{x_k^*-a} p_{2n}(x_k^*) \\ &\quad + r\left[\int_a^b w(x)\mathrm{d}x - \sum_{k=1}^{n}\frac{W_k^*}{x_k^*-a}\right].\end{aligned} \qquad (10.26)$$

The fact that $r = p_{2n}(a)$ then leads us to consider the quadrature rule

$$\int_a^b w(x)f(x)\mathrm{d}x \approx W_0 f(a) + \sum_{k=1}^{n} W_k f(x_k)\,, \tag{10.27}$$

where

$$\begin{aligned} W_k &= W_k^*/(x_k^* - a)\,, \quad k = 1,\ldots,n\,, \\ W_0 &= \int_a^b w(x)\mathrm{d}x - \sum_{k=1}^{n} W_k\,. \end{aligned} \tag{10.28}$$

By construction, this formula is exact for all polynomials of degree $2n$. It is obvious that $W_k > 0$ for $k = 1,\ldots,n$. We leave it as an exercise to show that $W_0 > 0$ also (see Exercise 5).

With only trivial changes it is easy to see how to construct a similar formula where instead of fixing $x_0 = a$ we fix $x_n = b$. These are known as **Radau quadrature formulae**. We leave it as an exercise to construct the formula corresponding to fixing both $x_0 = a$ and $x_n = b$, which is known as a **Lobatto quadrature formula**; as might be expected, this is exact for all polynomials of degree $2n - 1$ (see Exercise 7).

The formal process could evidently be generalised to allow for fixing one of the quadrature points at an internal point c, where $a < c < b$. However, this leads to the difficulty that the modified weight function

$$w^*\colon x \mapsto (x - c)w(x)$$

is not positive over the whole interval (a, b); hence we can no longer be sure that it is possible to construct a system of orthogonal polynomials, or, even if we can, that these polynomials will have all their zeros real and distinct and lying in $[a, b]$. In general, therefore, such quadrature formulae may not exist.

10.7 Note

For a detailed guide to the literature on Gauss quadrature rules and its connection to the theory of orthogonal polynomials, we refer to the books cited in the Notes at the end of Chapter 7.

Exercises

10.1 Determine the quadrature points and weights for the weight function $w\colon x \mapsto -\ln x$ on the interval $(0, 1)$, for $n = 0$ and $n = 1$.

10.2 The weights in the Gauss quadrature formula are given by (10.7), which is

$$W_k = \int_a^b w(x)[L_k(x)]^2 \mathrm{d}x \, .$$

Show that W_k can also be calculated from

$$W_k = \int_a^b w(x) L_k(x) \mathrm{d}x \, .$$

(This is a simpler way of calculating W_k than (10.7); the importance of (10.7) is that it shows that the weights are all positive.)

10.3 Suppose that f has a continuous second derivative on $[0, 1]$. Show that there is a point ξ in $(0, 1)$ such that

$$\int_0^1 x f(x) \mathrm{d}x = \tfrac{1}{2} f(\tfrac{2}{3}) + \tfrac{1}{72} f''(\xi) \, .$$

10.4 Let $n \geq 0$. Write down the quadrature points x_j, $j = 0, \ldots, n$, for the weight function $w\colon x \mapsto (1 - x^2)^{-1/2}$ on the interval $(-1, 1)$.

By induction, or otherwise, show that for positive integer values of n,

$$\sum_{j=0}^{n} \cos(2j + 1)\vartheta = \frac{\sin(2n + 2)\vartheta}{2 \sin \vartheta} \, ,$$

unless ϑ is a multiple of π. What is the value of the sum when ϑ is a multiple of π?

Deduce that

$$\sum_{j=0}^{n} T_k(x_j) = \int_{-1}^{1} (1 - x^2)^{-1/2} T_k(x) \, \mathrm{d}x \, , \qquad k = 1, \ldots, n \, ,$$

and show that

$$\sum_{j=0}^{n} T_0(x_j) = \frac{n + 1}{\pi} \int_{-1}^{1} (1 - x^2)^{-1/2} T_0(x) \, \mathrm{d}x \, ,$$

where T_n is the Chebyshev polynomial of degree n.

Deduce that the weights of the quadrature formula with weight function $w\colon x \mapsto (1 - x^2)^{-1/2}$ on the interval $(-1, 1)$ are

$$W_k = \frac{\pi}{n + 1} \, , \qquad k = 0, \ldots, n \, .$$

10.5 In the notation for the construction of the Radau quadrature formula in Section 10.6, show that $W_0 > 0$.

10.6 The **Laguerre polynomials**[1] L_j, $j = 0, 1, 2, \ldots$, are the orthogonal polynomials associated with the weight function $w\colon x \mapsto e^{-x}$ on the semi-infinite interval $(0, \infty)$, with L_j of exact degree j. Show that

$$\int_0^\infty e^{-x} x[L_j(x) - L_j'(x)]p_r(x)\mathrm{d}x = 0$$

when p_r is any polynomial of degree less than j.

In the Radau formula

$$\int_0^\infty e^{-x} p_{2n}(x)\mathrm{d}x = W_0 p_{2n}(0) + \sum_{k=1}^{n} W_k p_{2n}(x_k),$$

where one of the quadrature points is fixed at $x = 0$, show that the other quadrature points x_k, $k = 1, \ldots, n$, are the zeros of the polynomial $L_n - L_n'$. Deduce that

$$\int_0^\infty e^{-x} p_2(x)\mathrm{d}x = \tfrac{1}{2}p_2(0) + \tfrac{1}{2}p_2(2).$$

10.7 Let $n \geq 2$. Show that a polynomial p_{2n-1} of degree $2n-1$ can be written

$$p_{2n-1}(x) = (x-a)(b-x)q_{2n-3}(x) + r(x-a) + s(b-x),$$

where q_{2n-3} is a polynomial of degree $2n-3$, and r and s are constants. Hence construct the Lobatto quadrature formula

$$\int_a^b w(x)f(x)\mathrm{d}x \approx W_0 f(a) + \sum_{k=1}^{n-1} W_k f(x_k) + W_n f(b),$$

which is exact when f is any polynomial of degree $2n-1$. Show that all the weights W_k, $k = 0, 1, \ldots, n$, are positive.

10.8 Construct the Lobatto quadrature formula

$$\int_{-1}^{1} f(x) \approx A_0 f(-1) + A_1 f(x_1) + A_2 f(1)$$

for the interval $(-1, 1)$ with weight function $w(x) \equiv 1$, and with $n = 2$; write down and solve four equations to determine x_1, A_0, A_1 and A_2.

[1] Edmond Nicolas Laguerre (9 April 1834, Bar-le-Duc, France – 14 Aug 1886, Bar-le-Duc, France.)

10.9 Write I_m for the composite trapezium rule (7.15), S_m for the composite Simpson rule (7.17) and M_m for the composite mid-point rule (10.25), each with m subintervals. Show that

$$M_m = 2I_{2m} - I_m\,, \quad S_m = \frac{4I_{2m} - I_m}{3}\,, \quad S_m = \frac{2M_m + I_m}{3}\,.$$

11

Piecewise polynomial approximation

11.1 Introduction

Up to now, the focus of our discussion has been the question of approximation of a given function f, defined on an interval $[a, b]$, by a polynomial on that interval either through Lagrange interpolation or Hermite interpolation, or by seeking the polynomial of best approximation (in the ∞-norm or 2-norm). Each of these constructions was *global* in nature, in the sense that the approximation was defined by the same analytical expression on the whole interval $[a, b]$. An alternative and more flexible way of approximating a function f is to divide the interval $[a, b]$ into a number of subintervals and to look for a piecewise approximation by polynomials of low degree. Such piecewise-polynomial approximations are called **splines**, and the endpoints of the subintervals are known as the **knots**.

More specifically, a spline of degree n, $n \geq 1$, is a function which is a polynomial of degree n or less in each subinterval and has a prescribed degree of smoothness. We shall expect the spline to be at least continuous, and usually also to have continuous derivatives of order up to k for some k, $0 \leq k < n$. Clearly, if we require the derivative of order n to be continuous everywhere the spline is just a single polynomial, since if two polynomials of degree n have the same value and the same derivatives of every order up to n at a knot, then they must be the same polynomial. An important class of splines have degree n, with continuous derivatives of order up to and including $n-1$, but as we shall see later, lower degrees of smoothness are sometimes considered.

To give a flavour of the theory of splines, we concentrate here on two simple cases: linear splines and cubic splines.

11.2 Linear interpolating splines

Definition 11.1 *Suppose that f is a real-valued function, defined and continuous on the closed interval $[a,b]$. Further, let $K = \{x_0, \dots, x_m\}$ be a subset of $[a,b]$, with $a = x_0 < x_1 < \cdots < x_m = b$, $m \geq 2$. The* **linear spline** *s_L, interpolating f at the points x_i, is defined by*

$$s_L(x) = \frac{x_i - x}{x_i - x_{i-1}} f(x_{i-1}) + \frac{x - x_{i-1}}{x_i - x_{i-1}} f(x_i), \qquad x \in [x_{i-1}, x_i], \quad i = 1, 2, \dots, m. \tag{11.1}$$

The points x_i, $i = 0, 1, \dots, m$, are the **knots** *of the spline, and K is referred to as the* **set of knots**.

As the function s_L interpolates the function f at the knots, i.e., $s_L(x_i) = f(x_i)$, $i = 0, 1, \dots, m$, and over each interval $[x_{i-1}, x_i]$, for $i = 0, 1, \dots, m$, the function s_L is a linear polynomial (and therefore continuous), we conclude that s_L is a continuous piecewise linear function on the interval $[a,b]$.

Given a set of knots $K = \{x_0, \dots, x_m\}$, we shall use the notation $h_i = x_i - x_{i-1}$, and let $h = \max_i h_i$. Also, for a positive integer k, we denote by $\mathrm{C}^k[a,b]$ the set of all real-valued functions, defined and continuous on the closed interval $[a,b]$, such that all derivatives, up to and including order k, are defined and continuous on $[a,b]$.

In order to highlight the accuracy of interpolation by linear splines we state the following error bound in the ∞-norm over the interval $[a,b]$.

Theorem 11.1 *Suppose that $f \in \mathrm{C}^2[a,b]$ and let s_L be the linear spline that interpolates f at the knots $a = x_0 < x_1 < \cdots < x_m = b$; then, the following error bound holds:*

$$\|f - s_L\|_\infty \leq \frac{1}{8} h^2 \|f''\|_\infty,$$

where $h = \max_i h_i = \max_i(x_i - x_{i-1})$, and $\|\cdot\|_\infty$ denotes the ∞-norm over $[a,b]$, defined in (8.1).

Proof Consider a subinterval $[x_{i-1}, x_i]$, $1 \leq i \leq m$. According to Theorem 6.2, applied on the interval $[x_{i-1}, x_i]$,

$$f(x) - s_L(x) = \frac{1}{2} f''(\xi)(x - x_{i-1})(x - x_i), \qquad x \in [x_{i-1}, x_i],$$

where $\xi = \xi(x) \in (x_{i-1}, x_i)$. Thus,

$$|f(x) - s_{\mathrm{L}}(x)| \le \frac{1}{8} h_i^2 \max_{\zeta \in [x_{i-1}, x_i]} |f''(\zeta)| \, .$$

Hence,

$$|f(x) - s_{\mathrm{L}}(x)| \le \frac{1}{8} h^2 \|f''\|_\infty \, ,$$

for each $x \in [x_{i-1}, x_i]$ and each $i = 1, 2, \ldots, m$. This gives the required error bound. □

Figure 11.1 shows a typical example: a linear spline approximation to the function $f\colon x \mapsto \mathrm{e}^{-3x}$ over the interval $[0, 1]$, using two internal knots, $x_1 = \frac{1}{3}$, $x_2 = \frac{2}{3}$, together with the endpoints of the interval, $x_0 = 0$ and $x_3 = 1$.

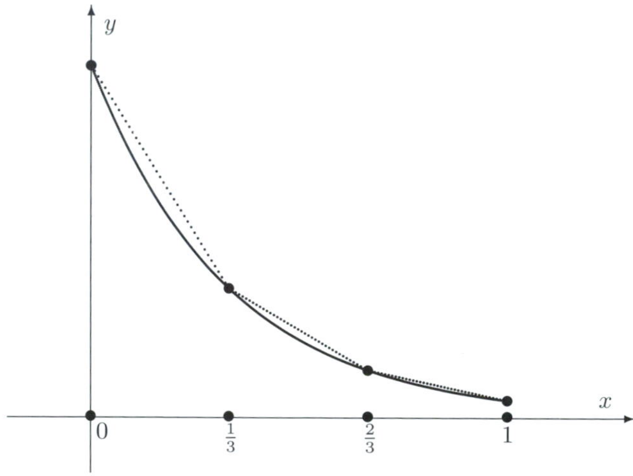

Fig. 11.1. The function $f\colon x \mapsto \mathrm{e}^{-3x}$ (full curve) and its linear spline approximation (dotted curve). The interval is $[0, 1]$, and the knots are at 0, $\frac{1}{3}$, $\frac{2}{3}$ and 1.

We conclude this section with a result that provides a characterisation of linear splines from the viewpoint of the calculus of variations.

A subset A of the real line is said to have **measure zero** if it can be contained in a countable union of open intervals of arbitrarily small total length; in other words, for every $\varepsilon > 0$ there exists a sequence of open intervals (a_i, b_i), $i = 1, 2, 3, \ldots$, such that

$$A \subset \bigcup_{i=1}^{\infty} (a_i, b_i) \quad \text{and} \quad \sum_{i=1}^{\infty} (b_i - a_i) < \varepsilon \, .$$

In particular, any finite or countable set $A \subset \mathbb{R}$ has measure zero. For example, the set of all rational numbers is countable, and therefore it has measure zero. Trivially, the empty set has measure zero.

Suppose that B is a subset of $\mathbb{R}$. We shall say that a certain property $\mathsf{P} = \mathsf{P}(x)$ holds for **almost every** x in B, if there exists a set $A \subset B$ of measure zero such that $\mathsf{P}(x)$ holds for *all* $x \in B \setminus A$.

A real-valued function v defined on the interval $[a, b]$ is said to be **absolutely continuous** on $[a, b]$ if it has finite derivative $v'(\xi)$ at almost every point ξ in $[a, b]$, v' is (Lebesgue-) integrable on $[a, b]$, and

$$\int_a^x v'(\xi)\mathrm{d}\xi = v(x) - v(a)\,, \qquad a \le x \le b\,.$$

Example 11.1 *Any $v \in \mathrm{C}^1[a, b]$ is absolutely continuous on the interval $[a, b]$. The function $x \mapsto \left|x - \frac{1}{2}(a + b)\right|$ is absolutely continuous on $[a, b]$, but it does not belong to $\mathrm{C}^1[a, b]$ as it is not differentiable at $x = \frac{1}{2}(a+b)$.*

Let us denote by $\mathrm{H}^1(a, b)$ the set of all absolutely continuous functions v defined on $[a, b]$ such that $v' \in \mathrm{L}^2(a, b)$, *i.e.*,

$$\|v'\|_2 = \left(\int_a^b |v'(\xi)|^2 \mathrm{d}\xi\right)^{1/2} < \infty\,.$$

We observe in passing that any function $v \in \mathrm{H}^1(a, b)$ is uniformly continuous on the closed interval $[a, b]$. This follows by noting that, for any pair of points $x, y \in [a, b]$,

$$\begin{aligned} |v(x) - v(y)| &= \left|\int_x^y v'(\xi)\,\mathrm{d}\xi\right| \\ &\le |x - y|^{\frac{1}{2}} \left|\int_x^y |v'(\xi)|^2 \mathrm{d}\xi\right|^{1/2} \\ &\le |x - y|^{\frac{1}{2}} \|v'\|_2\,. \end{aligned}$$

In the transition from the first line to the second we used the Cauchy–Schwarz inequality.

If $k \ge 1$, we shall denote by $\mathrm{H}^{k+1}(a, b)$ the set of all $v \in \mathrm{H}^k(a, b)$ such that $v^{(k)}$ is absolutely continuous on $[a, b]$ and $v^{(k+1)} \in \mathrm{L}^2(a, b)$. The set $\mathrm{H}^k(a, b)$ is called a **Sobolev space** of index k. We observe that

$$\mathrm{C}^k[a, b] \subset \mathrm{H}^k(a, b)$$

for any $k \ge 1$, with strict inclusion. For example, any linear spline on

$[a,b]$ belongs to $\mathrm{H}^1(a,b)$, but not to $\mathrm{C}^1[a,b]$ unless it is a linear function over the *whole of* the interval $[a,b]$.

Example 11.2 *Let $\alpha > 1/2$; the function $x \mapsto x^\alpha$ then belongs to $\mathrm{H}^1(0,1)$, although it only belongs to $\mathrm{C}^1[0,1]$ if $\alpha \geq 1$.*

As a second example, consider the function $x \mapsto x \ln x$ which belongs to $\mathrm{H}^1(0,1)$, but not to $\mathrm{C}^1[0,1]$.

The variational characterisation of linear splines stated in the next theorem expresses the fact that, among all functions $v \in \mathrm{H}^1(a,b)$ which interpolate a given continuous function f at a fixed set of knots in $[a,b]$, the linear spline s_{L} that interpolates f at these knots is the 'flattest', in the sense that its 'average slope' $\|s'_{\mathrm{L}}\|_2$ is smallest.

Theorem 11.2 *Suppose that s_{L} is the linear spline that interpolates $f \in \mathrm{C}[a,b]$ at the knots $a = x_0 < x_1 < \cdots < x_m = b$. Then, for any function v in $\mathrm{H}^1(a,b)$ that also interpolates f at these knots,*

$$\|s'_{\mathrm{L}}\|_2 \leq \|v'\|_2 .$$

Proof Let us observe that

$$\begin{aligned} \|v'\|_2^2 &= \int_a^b (v'(x) - s'_{\mathrm{L}}(x))^2 \mathrm{d}x + \int_a^b |s'_{\mathrm{L}}(x)|^2 \mathrm{d}x \\ &\quad +2 \int_a^b (v'(x) - s'_{\mathrm{L}}(x)) s'_{\mathrm{L}}(x) \mathrm{d}x . \end{aligned} \tag{11.2}$$

We shall now use integration by parts to show that the last integral is equal to 0; the desired inequality will then follow by noting that the first term on the right-hand side is nonnegative and it is equal to 0 if, and only if, $v = s_{\mathrm{L}}$. Clearly,

$$\int_a^b (v'(x) - s'_{\mathrm{L}}(x)) s'_{\mathrm{L}}(x) \mathrm{d}x = \sum_{k=1}^m \int_{x_{k-1}}^{x_k} (v'(x) - s'_{\mathrm{L}}(x)) s'_{\mathrm{L}}(x) \mathrm{d}x$$

$$\begin{aligned} &= \sum_{k=1}^m \Big[(v(x_k) - s_{\mathrm{L}}(x_k)) s'_{\mathrm{L}}(x_k-) - (v(x_{k-1}) - s_{\mathrm{L}}(x_{k-1})) s'_{\mathrm{L}}(x_{k-1}+) \\ &\qquad - \int_{x_{k-1}}^{x_k} (v(x) - s_{\mathrm{L}}(x)) s''_{\mathrm{L}}(x) \mathrm{d}x \Big] . \end{aligned} \tag{11.3}$$

Now $v(x_i) - s_{\mathrm{L}}(x_i) = f(x_i) - f(x_i) = 0$ for $i = 0, 1, \ldots, m$ and, since s_{L} is a linear polynomial over each of the open intervals (x_{k-1}, x_k), $k =$

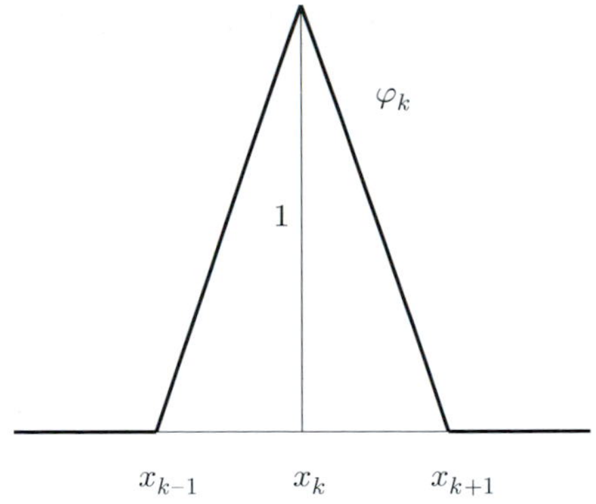

Fig. 11.2. The linear basis spline (or hat function) φ_k, $1 \le k \le m-1$.

$1, 2, \ldots, m$, it follows that s_{L}'' is identically 0 on each of these intervals. Thus, the expression in the square bracket in (11.3) is equal to 0 for each $k = 1, 2, \ldots, m$. □

Sobolev spaces play an important role in approximation theory. We shall encounter them again in Chapter 14 which is devoted to the approximation of solutions to differential equations by piecewise polynomial functions.

11.3 Basis functions for the linear spline

Suppose that s_{L} is a linear spline with knots x_i, $i = 0, 1, \ldots, m$, interpolating the function $f \in \mathrm{C}[a, b]$. Instead of specifying the value of s_{L} on each subinterval $[x_{i-1}, x_i]$, $i = 1, 2, \ldots, m$, we can express s_{L} as a linear combination of suitable 'basis functions' φ_k as follows:

$$s_{\mathrm{L}}(x) = \sum_{k=0}^{m} \varphi_k(x) f(x_k)\,, \qquad x \in [a, b]\,.$$

Here, we require that each φ_k is itself a linear spline which vanishes at every knot except x_k, and $\varphi_k(x_k) = 1$. The function φ_k is often known as the **linear basis spline** or **hat function**, and is depicted in Figure 11.2.

The formal definition of φ_k is as follows:

$$\left.\begin{aligned}
&\varphi_k(x) = \begin{cases} 0 & \text{if } x \le x_{k-1}\,, \\ (x - x_{k-1})/h_k & \text{if } x_{k-1} \le x \le x_k\,, \\ (x_{k+1} - x)/h_{k+1} & \text{if } x_k \le x \le x_{k+1}\,, \\ 0 & \text{if } x_{k+1} \le x\,, \end{cases} \\
&\text{for } k = 1, \dots, m-1, \text{ and with} \\
&\varphi_0(x) = \begin{cases} (x_1 - x)/h_0 & \text{if } a = x_0 \le x \le x_1\,, \\ 0 & \text{if } x_1 \le x \end{cases} \\
&\text{and} \\
&\varphi_m(x) = \begin{cases} 0 & \text{if } x \le x_{m-1}\,, \\ (x - x_{m-1})/h_m & \text{if } x_{m-1} \le x \le x_m = b\,. \end{cases}
\end{aligned}\right\} \qquad (11.4)$$

11.4 Cubic splines

Suppose that $f \in \mathrm{C}[a,b]$ and let $K = \{x_0, \dots, x_m\}$ be a set of $m+1$ knots in the interval $[a,b]$, $a = x_0 < x_1 < \cdots < x_m = b$. Consider the set $\mathcal{S}$ of all functions $s \in \mathrm{C}^2[a,b]$ such that

❶ $s(x_i) = f(x_i)$, $i = 0, 1, \dots, m$,
❷ s is a cubic polynomial on $[x_{i-1}, x_i]$, $i = 1, 2, \dots, m$.

Any element of $\mathcal{S}$ is referred to as an **interpolating cubic spline**. We note that, unlike linear splines which are uniquely determined by the interpolating conditions, there is more than one interpolating cubic spline $s \in \mathrm{C}^2[a,b]$ that satisfies the two conditions stated above; indeed, there are $4m$ coefficients of cubic polynomials (four on each subinterval $[x_{i-1}, x_i]$, $i = 1, 2, \dots, m$), and only $m+1$ interpolating conditions and $3(m-1)$ continuity conditions; since s belongs to $\mathrm{C}^2[a,b]$, this means that s, s' and s'' are continuous at the internal knots $x_1, \dots, x_{m-1}$. Hence, we have a total of $4m-2$ conditions for the $4m$ unknown coefficients. Depending on the choice of the remaining two conditions we can construct various interpolating cubic splines.

An important class of cubic splines is singled out by the following definition.

Definition 11.2 *The* **natural cubic spline**, *denoted by s_2, is the element of the set $\mathcal{S}$ satisfying the end conditions*

$$s_2''(x_0) = s_2''(x_m) = 0\,.$$

We shall prove that this definition is correct in the sense that the two additional conditions in Definition 11.2 uniquely determine s_2: this will be done by describing an algorithm for constructing s_2.

Construction of the natural cubic spline. Let us begin by defining $\sigma_i = s_2''(x_i)$, $i = 0, 1, \ldots, m$, and noting that s_2'' is a linear function on each subinterval $[x_{i-1}, x_i]$. Therefore, s_2'' can be expressed as

$$s_2''(x) = \frac{x_i - x}{h_i}\sigma_{i-1} + \frac{x - x_{i-1}}{h_i}\sigma_i\,, \qquad x \in [x_{i-1}, x_i]\,.$$

Integrating this twice we obtain

$$\begin{aligned} s_2(x) \;&=\; \frac{(x_i - x)^3}{6h_i}\sigma_{i-1} + \frac{(x - x_{i-1})^3}{6h_i}\sigma_i \\ &\quad + \alpha_i(x - x_{i-1}) + \beta_i(x_i - x)\,, \quad x \in [x_{i-1}, x_i]\,, \end{aligned} \tag{11.5}$$

where α_i and β_i are constants of integration. Equating s_2 with f at the knots x_{i-1}, x_i yields

$$f(x_{i-1}) = \frac{1}{6}\sigma_{i-1}h_i^2 + h_i\beta_i\,, \qquad f(x_i) = \frac{1}{6}\sigma_i h_i^2 + h_i\alpha_i\,. \tag{11.6}$$

Expressing α_i and β_i from these, inserting them into (11.5) and exploiting the continuity of s_2' at the internal knots, (*i.e.*, using that $s_2'(x_i-) = s_2'(x_i+)$, $i = 1, \ldots, m-1$), gives

$$\begin{aligned} &h_i\sigma_{i-1} + 2(h_{i+1} + h_i)\sigma_i + h_{i+1}\sigma_{i+1} \\ &\qquad = 6\left(\frac{f(x_{i+1}) - f(x_i)}{h_{i+1}} - \frac{f(x_i) - f(x_{i-1})}{h_i}\right) \end{aligned} \tag{11.7}$$

for $i = 1, \ldots, m-1$, together with

$$\sigma_0 = \sigma_m = 0\,,$$

which is a system of linear equations for the σ_i. The matrix of the system is tridiagonal and nonsingular, since the conditions of Theorem 3.4 are clearly satisfied. By solving this linear system we obtain the σ_i, $i = 0, 1, \ldots, m$, and thereby all the α_i, β_i, $i = 1, 2, \ldots, m$, from (11.6).

We have seen in a previous section, in Theorem 11.2, that a linear spline can be characterised as a minimiser of the functional $v \mapsto \|v'\|_2$ over all $v \in \mathrm{H}^1(a,b)$ which interpolate a given continuous function at the knots of the spline. Natural cubic splines have an analogous property: among all functions $v \in \mathrm{H}^2(a,b)$ which interpolate a given continuous function f at a fixed set of knots in $[a,b]$, the natural cubic spline s_2 is smoothest, in the sense that it minimises $v \mapsto \|v''\|_2$, the 'average curvature' of v.

Theorem 11.3 *Let s_2 be the natural cubic spline that interpolates a function $f \in \mathrm{C}[a,b]$ at the knots $a = x_0 < x_1 < \cdots < x_m = b$. Then, for any function v in $\mathrm{H}^2(a,b)$ that also interpolates f at the knots,*

$$\|s_2''\|_2 \le \|v''\|_2 \,.$$

The proof is analogous to that of Theorem 11.2 and is left as an exercise.

The *smoothest interpolation property* expressed by Theorem 11.3 is the source of the name *spline.*[1] A spline is a flexible thin curve-drawing aid, made of wood, metal or acrylic. Assuming that its shape is given by the equation $y = v(x)$, $x \in [a,b]$, and is constrained by requiring that it passes through a finite set of prescribed points in the plane, v will take on a shape which minimises the strain energy

$$\mathrm{E}(v) = \int_a^b \frac{|v''(x)|^2}{(1+|v'(x)|^2)^3}\,\mathrm{d}x$$

over all functions v which are constrained in the same way. If the function v is slowly varying, *i.e.*, $\max_{x\in[a,b]} |v'(x)| \ll 1$, this energy-minimisation property is very similar to the result in Theorem 11.3.

11.5 Hermite cubic splines

In the previous section we took $f \in \mathrm{C}[a,b]$ and demanded that s belonged to $\mathrm{C}^2[a,b]$; here we shall strengthen our requirements on the smoothness of the function that we wish to interpolate and assume that $f \in \mathrm{C}^1[a,b]$; simultaneously, we shall relax the smoothness requirements on the associated spline approximation s by demanding that $s \in \mathrm{C}^1[a,b]$ only.

Let $K = \{x_0, \ldots, x_m\}$ be a set of knots in the interval $[a,b]$ with $a = x_0 < x_1 < \cdots < x_m = b$ and $m \ge 2$. We define the **Hermite cubic spline** as a function $s \in \mathrm{C}^1[a,b]$ such that

❶ $s(x_i) = f(x_i)$, $s'(x_i) = f'(x_i)$ for $i = 0, 1, \ldots, m$,
❷ s is a cubic polynomial on $[x_{i-1}, x_i]$ for $i = 1, 2, \ldots, m$.

Writing the spline s on the interval $[x_{i-1}, x_i]$ as

$$s(x) = c_0 + c_1(x - x_{i-1}) + c_2(x - x_{i-1})^2 + c_3(x - x_{i-1})^3\,, \qquad x \in [x_{i-1}, x_i]\,, \quad (11.8)$$

[1] See Carl de Boor: *A Practical Guide to Splines*, Revised Edition, Springer Applied Mathematical Sciences, 27, Springer, New York, 2001.

we find that $c_0 = f(x_{i-1})$, $c_1 = f'(x_{i-1})$, and

$$\begin{aligned} c_2 &= 3\frac{f(x_i) - f(x_{i-1})}{h_i^2} - \frac{f'(x_i) + 2f'(x_{i-1})}{h_i}, \\ c_3 &= \frac{f'(x_i) + f'(x_{i-1})}{h_i^2} - 2\frac{f(x_i) - f(x_{i-1})}{h_i^3}. \end{aligned} \tag{11.9}$$

Note that the Hermite cubic spline only has a continuous first derivative at the knots, and therefore it is *not* an interpolating cubic spline in the sense of Section 11.4.

Unlike natural cubic splines, the coefficients of a Hermite cubic spline on each subinterval can be written down explicitly without the need to solve a tridiagonal system.

Concerning the size of the interpolation error, we have the following result.

Theorem 11.4 *Let $f \in \mathrm{C}^4[a, b]$, and let s be the Hermite cubic spline that interpolates f at the knots $a = x_0 < x_1 < \cdots < x_m = b$; then, the following error bound holds:*

$$\|f - s\|_\infty \le \frac{1}{384} h^4 \|f^{\mathsf{iv}}\|_\infty ,$$

where $f^{\mathsf{iv}} = f^{(4)}$ is the fourth derivative of f with respect to its argument, x, $h = \max_i h_i = \max_i(x_i - x_{i-1})$, and $\|\cdot\|_\infty$ denotes the ∞-norm on the interval $[a, b]$.

The proof is analogous to that of Theorem 11.1, except that Theorem 6.4 is used instead of Theorem 6.2.

Both the linear spline and the Hermite cubic spline are local approximations; the value of the spline at a point x between two knots x_{i-1} and x_i depends only on the values of the function and its derivative at these two knots. On the other hand, the natural cubic interpolating spline is a global approximation and, in this respect, it is more typical of a generic spline: a change in just one of the values at a knot, $f(x_k)$, will alter the right-hand side of the system of equations (11.7), so the values of all the quantities σ_i will change. Thus, the spline will change throughout the whole interval $[x_0, x_m]$. We conclude this section with an example.

Example 11.3 *Figure 11.3 shows the Hermite cubic spline approximation to the function $f\colon x \mapsto 1/(1 + x^2)$, using four equally spaced knots in the interval $[0, 5]$.*

The accuracy of this approximation is in striking contrast to the Lagrange polynomial approximation of degree 10 in Figure 6.1. The approximation over $[-5, 5]$, using seven equally spaced knots, is obviously obtained by symmetry; here we show only half the range for clarity.

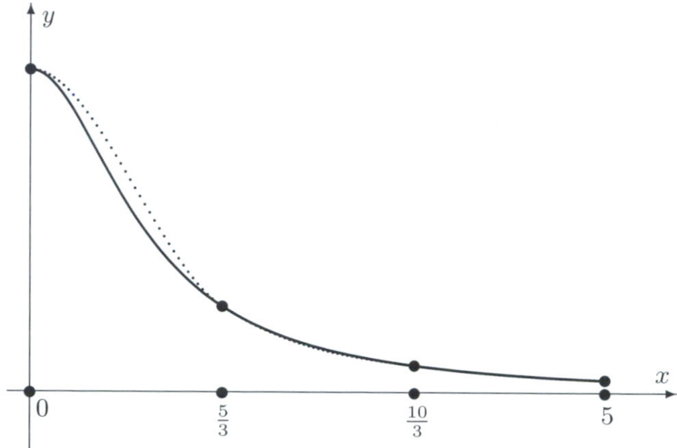

Fig. 11.3. The function $f\colon x \mapsto 1/(1+x^2)$ (full curve) and its Hermite cubic spline approximation (dotted curve). The interval is $[0, 5]$, and the knots are at 0, $\frac{5}{3}$, $\frac{10}{3}$ and 5.

As the error of this approximation is quite small, we show in Figure 11.4 graphs of the errors of three spline approximations, each using the same four knots. Note that in the first interval, $[0, \frac{5}{3}]$, the maximum error of the Hermite cubic spline is larger than that of the linear spline, but on the other two intervals it is much less. Both of these two splines are local approximations, as their values on any interval between two knots depend only on information about the function at those two knots. The natural cubic spline is a global approximation, as its value at any point depends on the values of the function at all the knots; on the first interval its error is much the same size as that of the Hermite cubic spline, but on the other two intervals its error is affected by this global coupling, and is a good deal bigger than that of the Hermite cubic spline. $\diamond$

11.6 Basis functions for cubic splines

We have seen that the family of hat functions forms a basis for the linear space of linear splines corresponding to a certain fixed set of knots; we

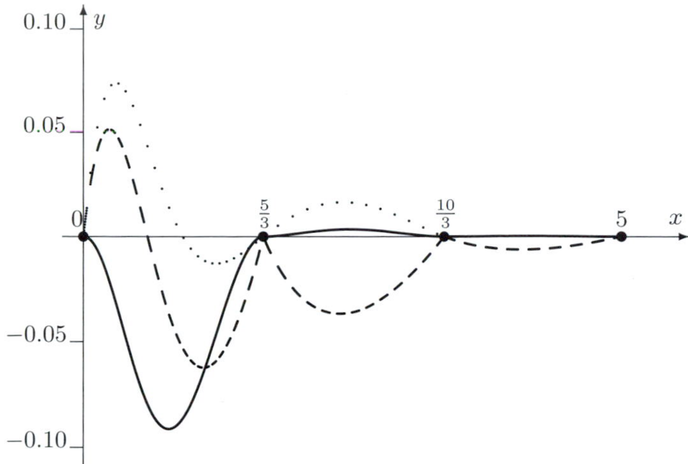

Fig. 11.4. Errors of three spline approximations to $f(x) = 1/(1 + x^2)$: Hermite cubic (full curve), natural cubic (dotted curve) and linear spline (broken curve). The interval is $[0, 5]$, and the knots are at 0, $\frac{5}{3}$, $\frac{10}{3}$ and 5.

shall now show how to construct a set of basis functions for cubic splines. The basis functions for splines are usually known as **B-splines**. Thus, the basis-splines constructed in Section 11.3 are referred to as linear B-splines. Here we shall be concerned with the construction of cubic B-splines. To simplify the notation we shall assume in this section that the knots are equally spaced, so that

$$x_k = kh\,, \qquad k = 0, 1, \ldots, n+1\,,$$

with $h > 0$.

We begin by introducing the idea of the positive part of a function.

Definition 11.3 *Suppose that* $n \geq 1$. *The* **positive part** *of the function* $x \mapsto (x-a)^n$ *is the function* $x \mapsto (x-a)^n_+$ *defined by*

$$(x-a)^n_+ = \begin{cases} (x-a)^n\,, & x \geq a\,, \\ 0\,, & x < a\,. \end{cases}$$

Clearly the function $x \mapsto (x - x_k)^n_+$ is a spline of degree n; at the knot x_k the derivatives of order up to $n-1$ are zero, but the derivative of order n is not continuous at $x = x_k$.

Figure 11.5 shows the graphs of the functions $x \mapsto x_+$ and $x \mapsto x^3_+$ on the interval $[-1, 1]$.

We shall also need the following result.

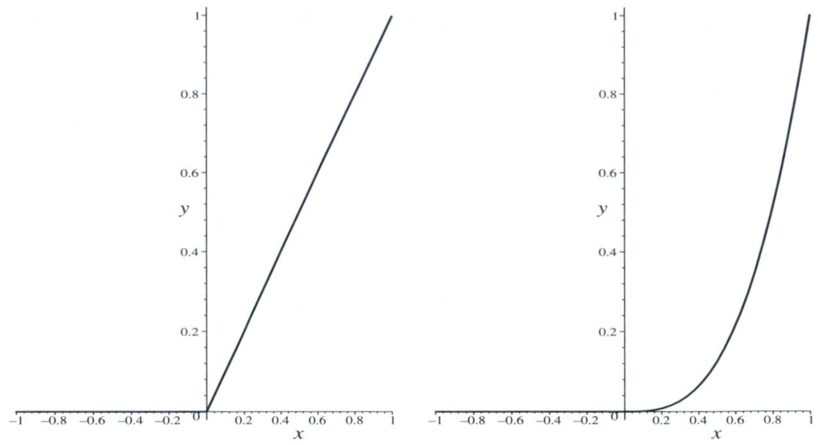

Fig. 11.5. The graph of the function $x \mapsto (x)_+^n$, for x in the interval $[-1, 1]$, with $n = 1$ (left) and $n = 3$ (right).

Lemma 11.1 *Suppose that P is a polynomial in x of degree $n \geq 1$. Then, for each $r = 1, \ldots, n$, the function $Q_{(r)}$ defined by*

$$Q_{(r)}(x) = \sum_{k=0}^{r} (-1)^k \binom{r}{k} P(x - kh)$$

is a polynomial of degree $n - r$ and $Q_{(n+1)}(x) \equiv 0$, $x \in \mathbb{R}$.

Proof It is easy to see that $Q_{(1)}(x) = P(x) - P(x - h)$, and therefore $Q_{(1)}$ is a polynomial of degree $n - 1$. Suppose now that, for some $r > 0$, $Q_{(r)}$ is a polynomial in x of degree $n - r$; then, $x \mapsto Q_{(r)}(x) - Q_{(r)}(x - h)$ is a polynomial of degree $n - r - 1$. But

$$\begin{aligned}
Q_{(r)}(x) &- Q_{(r)}(x - h) \\
&= \sum_{k=0}^{r} (-1)^k \binom{r}{k} [P(x - kh) - P(x - (k + 1)h)] \\
&= P(x) + (-1)^{r+1} P(x - (r + 1)h) \\
&\quad + \sum_{k=1}^{r} (-1)^k \left[\binom{r}{k} + \binom{r}{k - 1} \right] P(x - kh) \\
&= \sum_{k=0}^{r+1} (-1)^k \binom{r + 1}{k} P(x - kh) \\
&= Q_{(r+1)}(x) \,, \qquad (11.10)
\end{aligned}$$

from the standard properties of binomial coefficients. Hence $Q_{(r+1)}$ is a polynomial in x of degree $n-r-1$, and the result follows by induction. Finally, this shows that $Q_{(n)}$ is a polynomial of degree 0, and is therefore constant on $\mathbb{R}$. Thus, by the same argument, $Q_{(n+1)}$ is identically 0 on $\mathbb{R}$. □

Theorem 11.5 *For each $n \geq 1$, the function $S_{(n)}$ defined by*

$$S_{(n)}(x) = \sum_{k=0}^{n+1} (-1)^k \binom{n+1}{k} (x - kh)_+^n$$

is a spline of degree n with equally spaced knots kh, $k = 0, 1, \ldots, n+1$. It has a continuous derivative of order $n-1$ and is identically 0 outside the interval $(0, (n+1)h)$.

Proof The function $S_{(n)}$ is clearly a spline as stated, and $S_{(n)}(x)$ is identically 0 for $x \leq 0$. When $x \geq (n+1)h$ the arguments $x - kh$, $k = 0, 1, \ldots, n+1$, of the positive parts are all nonnegative, so that

$$S_{(n)}(x) = \sum_{k=0}^{n+1} (-1)^k \binom{n+1}{k} (x - kh)^n,$$

and this is identically zero by Lemma 11.1. □

Taking $n = 1$ we find that

$$S_{(1)}(x) = x_+ - 2(x-h)_+ + (x-2h)_+ .$$

After normalisation by $1/h$ so as to have a maximum value of 1, and shifting $x = 0$ to $x = x_{k-1}$, this yields a representation of the linear hat function φ_k from (11.4) in the form

$$\varphi_k(x) = \frac{1}{h} S_{(1)}(x - x_{k-1}),$$

which, for $1 \leq k \leq n$, is nonzero over two consecutive intervals: $(x_{k-1}, x_k]$ and $[x_k, x_{k+1})$.

In the same way we obtain a basis function for the cubic spline by taking $n = 3$:

$$S_{(3)}(x) = x_+^3 - 4(x-h)_+^3 + 6(x-2h)_+^3 - 4(x-3h)_+^3 + (x-4h)_+^3 .$$

Normalising so as to have a maximum value of 1 and shifting $x = 0$ to $x = x_{k-2}$, we get

$$\psi_k(x) = \frac{1}{4h^3} S_{(3)}(x - x_{k-2}) .$$

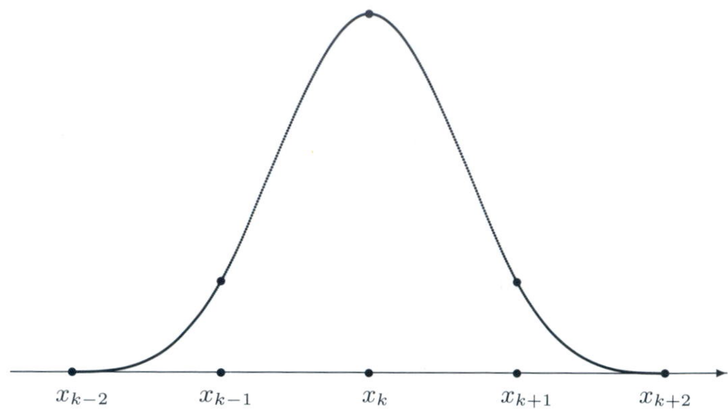

Fig. 11.6. Normalised cubic B-spline, $\psi_k(x)$, $2 \le k \le n-1$.

For $2 \le k \le n-1$, this function is nonzero over four consecutive intervals $(x_{k-2}, x_{k-1}]$, $[x_{k-1}, x_k]$, $[x_k, x_{k+1}]$ and $[x_{k+1}, x_{k+2})$, and is illustrated in Figure 11.6.

We see that both φ_k and ψ_k are nonnegative for all x; this is true for a spline basis function of any degree n, $n \ge 1$, constructed in this way, but we shall not prove it here (see Exercise 6).

For a finite set of knots $a = x_0 < x_1 < \cdots < x_{n+1} = b$ on the bounded and closed interval $[a, b]$ the normalised linear basis splines $x \mapsto \varphi_0(x)$ and $x \mapsto \varphi_{n+1}(x)$ are considered only for x in $[a, b]$, so as to avoid reference to nonexisting knots (such as x_{-1} or x_{n+2}) that lie outside $[a, b]$. A similar comment applies to the normalised cubic basis splines ψ_0, ψ_1, ψ_n and ψ_{n+1}.

11.7 Notes

There are many excellent texts covering the theory of piecewise polynomial approximation by splines. For a detailed survey of key results we refer to Chapters 18–24 of

- M.J.D. Powell, *Approximation Theory and Methods*, Cambridge University Press, Cambridge, 1996.

You may have noticed that we have given bounds on the error in linear spline approximation in Theorem 11.1, and in Hermite cubic spline approximation in Theorem 11.4, but not for the natural cubic spline. The analysis of the error in the natural cubic spline approximation is quite complicated; Powell gives full details in his book.

The following are classical texts on the theory of splines.

- J.H. Ahlberg, E.N. Nilson, and J.L. Walsh, *The Theory of Splines and Their Applications*, Mathematics in Science and Engineering, 38, Academic Press, New York, 1967.
- C. de Boor, *A Practical Guide to Splines*, Revised Edition, Springer Applied Mathematical Sciences, 27, Springer, New York, 2001.
- Larry L. Schumaker, *Spline Functions: Basic Theory,* John Wiley & Sons, New York, 1981.

The variational characterisations of splines stated in Sections 11.2 and 11.4 stem from the work of J.C. Holladay, Smoothest curve approximation, *Math. Comput.* **11**, 233–243, 1957.

Our definition of the Sobolev space $\mathrm{H}^k(a, b)$ in Section 11.2, based on the concept of absolute continuity, is specific to functions of a single variable. More generally, for functions of several real variables one needs to invoke the theory of weak differentiability or the theory of distributions to give a rigorous definition of the Sobolev space $H^k(\Omega)$ with $\Omega \subset \mathbb{R}^n$; alternatively, one can define $H^k(\Omega)$ by completion of the set of smooth functions in a suitable norm. For the sake of simplicity of exposition we have chosen to avoid such general approaches.

Exercises

11.1 An interpolating spline of degree n is required to have continuous derivatives of order up to and including $n-1$ at the knots. How many additional conditions are required to specify the spline uniquely?

11.2 (i) Suppose that f is a polynomial of degree 1. Show that the linear spline s_{L} which interpolates f at the knots x_i for $i = 0, 1, \ldots, m$ is identical to f, so that $s_{\mathrm{L}} \equiv f$.

(ii) Suppose that f is a polynomial of degree 3. Show that the Hermite cubic spline s_{H} which interpolates f at the knots x_i, $i = 0, 1, \ldots, m$, is identical to f, so that $s_{\mathrm{H}} \equiv f$.

(iii) Suppose that f is a polynomial of degree 3. Show that the natural cubic spline s_2 which interpolates f at the knots x_i, $i = 0, 1, \ldots, m$, is not in general identical to f.

11.3 Suppose that the natural cubic spline s_2 interpolates the function $f\colon x \mapsto x^3$ on the interval $[0,1]$, the knots being equally spaced, so that $x_i = ih$, $i = 0, 1, \ldots, m$, with $h = 1/m$, $m \geq 2$. Write down the equations which determine the quantities σ_i. If the two additional conditions are $\sigma_0 = \sigma_m = 0$, show that these equations are not satisfied by $\sigma_i = f''(x_i)$, $i = 1, \ldots, m-1$, so that s_2 and f are not identical. If, however, these two additional conditions are replaced by $\sigma_0 = f''(0)$, $\sigma_m = f''(1)$, show that $\sigma_i = f''(x_i)$, $i = 0, 1, \ldots, m$, and deduce that s_2 and f are identical.

11.4 A linear spline on the interval $[0,1]$ is expressed in terms of the basis functions as

$$s(x) = \sum_{k=0}^{m} \alpha_k \varphi_k(x)\,.$$

Instead of being required to interpolate the function f at the knots, the spline s is required to minimise $\|f - s\|_2$. Show that the coefficients α_k satisfy the system of equations

$$A\boldsymbol{\alpha} = \boldsymbol{b}\,,$$

where the elements of the matrix A are

$$A_{ij} = \int_0^1 \varphi_j(x)\varphi_i(x)\mathrm{d}x$$

and the elements of $\boldsymbol{b}$ are

$$b_i = \int_0^1 f(x)\varphi_i(x)\mathrm{d}x\,.$$

Now suppose that the knots are equally spaced, so that $x_k = kh$, $k = 0, 1, \ldots, m$, where $h = 1/m$, $m \geq 2$. Show that the matrix A is tridiagonal, with $A_{ii} = \frac{2}{3}h$ for $i = 1, \ldots, m-1$, and determine the other nonzero elements of A. Show also that A has the properties required for the use of the Thomas algorithm described in Section 3.3.

11.5 In the notation of Exercise 4, suppose that $f(x) = x$. Verify that the system of equations is satisfied by $\alpha_k = kh$, so that $s = f$.

Now suppose that $f(x) = x^2$. Verify that the equations are satisfied by $\alpha_k = (kh)^2 + Ch^2$, where C is a constant to be determined. Deduce that $s(x_k) = f(x_k) + Ch^2$.

11.6 In the notation of Theorem 11.5, the spline basis function $S_{(n)}$ of degree n is defined by

$$S_{(n)}(x) = \sum_{k=0}^{n+1} (-1)^k \binom{n+1}{k} (x - kh)_+^n \,.$$

Explain why, for any value of a,

$$(x-a)_+^n (x-a) = (x-a)_+^{n+1} \,.$$

Show that

$$x S_{(n)}(x) + [(n+2)h - x] S_{(n)}(x-h) = S_{(n+1)}(x) \,.$$

Hence show by induction that $S_{(n)}(x) \geq 0$ for all x.

11.7 Use the result of Exercise 6 to show by induction that each basis function $S_{(n)}$ is symmetric; that is,

$$S_{(n)}(p+x) = S_{(n)}(p-x)$$

for all x, where $p = \frac{1}{2}(n+1)h$.

12

Initial value problems for ODEs

12.1 Introduction

Ordinary differential equations frequently occur in mathematical models that arise in many branches of science, engineering and economics. Unfortunately it is seldom that these equations have solutions which can be expressed in closed form, so it is common to seek approximate solutions by means of numerical methods. Nowadays this can usually be achieved very inexpensively to high accuracy and with a reliable bound on the error between the analytical solution and its numerical approximation. In this section we shall be concerned with the construction and the analysis of numerical methods for first-order differential equations of the form

$$y' = f(x, y) \tag{12.1}$$

for the real-valued function y of the real variable x, where $y' \equiv \frac{dy}{dx}$ and f is a given real-valued function of two real variables. In order to select a particular integral from the infinite family of solution curves that constitute the general solution to (12.1), the differential equation will be considered in tandem with an **initial condition**: given two real numbers x_0 and y_0, we seek a solution to (12.1) for $x > x_0$ such that

$$y(x_0) = y_0\,. \tag{12.2}$$

The differential equation (12.1) together with the initial condition (12.2) is called an **initial value problem**.

If you believe that any initial value problem of the form (12.1), (12.2) possesses a unique solution, take a look at the following example.

Example 12.1 *Consider the differential equation $y' = |y|^\alpha$, subject to the initial condition $y(0) = 0$, where α is a fixed real number, $\alpha \in (0,1)$.*

It is a simple matter to verify that, for any nonnegative real number c,

$$y_c(x) = \begin{cases} (1-\alpha)^{\frac{1}{1-\alpha}}(x-c)^{\frac{1}{1-\alpha}}, & c \le x < \infty, \\ 0, & 0 \le x \le c, \end{cases}$$

is a solution to the initial value problem on the interval $[0,\infty)$. Consequently the existence of the solution is ensured, but not its uniqueness; in fact, the initial value problem has an infinite family of solutions $\{y_c\}$, parametrised by $c \ge 0$.

We note in passing that in contrast with the case of $\alpha \in (0,1)$, when $\alpha \ge 1$, the initial value problem $y' = |y|^\alpha$, $y(0) = 0$ has the unique solution $y(x) \equiv 0$. ◇

Example 12.1 indicates that the function f has to obey a certain growth condition with respect to its second argument so as to ensure that (12.1), (12.2) has a unique solution. The precise hypotheses on f guaranteeing the existence of a unique solution to the initial value problem (12.1), (12.2) are stated in the next theorem.

Theorem 12.1 (Picard's Theorem[1]) *Suppose that the real-valued function $(x,y) \mapsto f(x,y)$ is continuous in the rectangular region D defined by $x_0 \le x \le X_M$, $y_0 - C \le y \le y_0 + C$; that $|f(x,y_0)| \le K$ when $x_0 \le x \le X_M$; and that f satisfies the Lipschitz condition: there exists $L > 0$ such that*

$$|f(x,u) - f(x,v)| \le L|u-v| \quad \textit{for all } (x,u) \in D, \ (x,v) \in D.$$

Assume further that

$$C \ge \frac{K}{L}\left(\mathrm{e}^{L(X_M - x_0)} - 1\right). \tag{12.3}$$

Then, there exists a unique function $y \in \mathrm{C}^1[x_0, X_M]$ such that $y(x_0) = y_0$ and $y' = f(x,y)$ for $x \in [x_0, X_M]$; moreover,

$$|y(x) - y_0| \le C, \quad x_0 \le x \le X_M.$$

[1] Charles Emile Picard (24 July 1856, Paris, France – 11 December 1941, Paris, France). Although as a child he was a brilliant pupil, Picard disliked mathematics and only became interested in the subject during the vacation following his secondary studies. He was appointed to the chair of differential calculus at the Sorbonne in Paris at the age of 29 but could only take up his position a year later, as university regulations prevented anyone below the age of 30 holding a chair. Picard made important contributions to mathematical analysis and the theory of differential equations.

Proof We define a sequence of functions $(y_n)_{n=0}^{\infty}$ by

$$\begin{aligned} y_0(x) &\equiv y_0\,, \\ y_n(x) &= y_0 + \int_{x_0}^{x} f(s, y_{n-1}(s))\mathrm{d}s\,, \quad n = 1, 2, \ldots\,. \end{aligned} \tag{12.4}$$

Since f is continuous on D, it is clear that each function y_n is continuous on $[x_0, X_M]$. Further, since

$$y_{n+1}(x) = y_0 + \int_{x_0}^{x} f(s, y_n(s))\mathrm{d}s\,,$$

it follows by subtraction that

$$y_{n+1}(x) - y_n(x) = \int_{x_0}^{x} \left[f(s, y_n(s)) - f(s, y_{n-1}(s))\right] \mathrm{d}s\,. \tag{12.5}$$

We now proceed by induction, and assume that, for some positive value of n,

$$|y_n(x) - y_{n-1}(x)| \leq \frac{K}{L}\frac{[L(x - x_0)]^n}{n!}\,, \quad x_0 \leq x \leq X_M\,, \tag{12.6}$$

and that

$$\begin{aligned} |y_k(x) - y_0| &\leq \frac{K}{L}\sum_{j=1}^{k} \frac{[L(x - x_0)]^j}{j!}\,, \\ & \qquad x_0 \leq x \leq X_M\,, \quad k = 1, \ldots, n\,. \end{aligned} \tag{12.7}$$

Trivially, the hypotheses of the theorem and (12.4) imply that (12.6) and (12.7) hold for $n = 1$.

Now, (12.7) and (12.3) yield that

$$\begin{aligned} |y_k(x) - y_0| &\leq \frac{K}{L}\left(\mathrm{e}^{L(X_M - x_0)} - 1\right) \leq C\,, \\ & \qquad x_0 \leq x \leq X_M\,, \quad k = 1, \ldots, n\,. \end{aligned}$$

Therefore $(x, y_{n-1}(x)) \in D$ and $(x, y_n(x)) \in D$ for all $x \in [x_0, X_M]$. Hence, using (12.5), the Lipschitz condition and (12.6),

$$\begin{aligned} |y_{n+1}(x) - y_n(x)| &\leq L\int_{x_0}^{x} \frac{K}{L}\frac{[L(s - x_0)]^n}{n!}\mathrm{d}s \\ &= \frac{K}{L}\frac{[L(x - x_0)]^{n+1}}{(n+1)!}\,, \end{aligned} \tag{12.8}$$

for all $x \in [x_0, X_M]$. Moreover, using (12.8) and (12.7),

$$\begin{aligned}|y_{n+1}(x) - y_0| &\le |y_{n+1}(x) - y_n(x)| + |y_n(x) - y_0| \\ &\le \frac{K}{L}\frac{[L(x-x_0)]^{n+1}}{(n+1)!} + \frac{K}{L}\sum_{j=1}^{n}\frac{[L(x-x_0)]^j}{j!} \\ &= \frac{K}{L}\sum_{j=1}^{n+1}\frac{[L(x-x_0)]^{j+1}}{(j+1)!}\,, \end{aligned} \tag{12.9}$$

for all $x \in [x_0, X_M]$. Thus, (12.6) and (12.7) hold with n replaced by $n+1$, and hence, by induction, they hold for all positive integers n.

Since the infinite series $\sum_{j=1}^{\infty}(c^j/j!)$ converges (to $\mathrm{e}^c - 1$) for any value of $c \in \mathbb{R}$, and for $c = L(X_M - x_0)$ in particular, it follows from (12.6) that the infinite series

$$\sum_{j=1}^{\infty}[y_j(x) - y_{j-1}(x)]$$

converges absolutely and uniformly for $x \in [x_0, X_M]$. However,

$$y_0 + \sum_{j=1}^{n}[y_j(x) - y_{j-1}(x)] = y_n(x)\,,$$

showing that the sequence of continuous functions (y_n) converges to a limit, uniformly on $[x_0, X_M]$, and hence that the limit itself is a continuous function. Calling this limit y, we see from (12.4) that

$$\begin{aligned} y(x) &= \lim_{n\to\infty} y_{n+1}(x) \\ &= y_0 + \lim_{n\to\infty}\int_{x_0}^{x} f(s, y_n(s))\mathrm{d}s\,, \\ &= y_0 + \int_{x_0}^{x}\lim_{n\to\infty} f(s, y_n(s))\mathrm{d}s\,, \\ &= y_0 + \int_{x_0}^{x} f(s, y(s))\mathrm{d}s\,, \end{aligned} \tag{12.10}$$

where we used the uniform convergence of the sequence of functions (y_n) in the transition from line two to line three to interchange the order of the limit process and integration, and the continuity of the function f in the transition from line three to line four. As $s \mapsto f(s, y(s))$ is a continuous function of s on the interval $[x_0, X_M]$, its integral over the interval $[x_0, x]$ is a continuously differentiable function of x. Hence, by

(12.10), y is a continuously differentiable function of x on $[x_0, X_M]$; *i.e.*, $y \in \mathrm{C}^1[x_0, X_M]$. On differentiating (12.10) we deduce that

$$y' = f(x, y),$$

as required; also $y(x_0) = y_0$. We have already seen that $(x, y_n(x)) \in D$ when $x_0 \le x \le X_M$; as D is a closed set in $\mathbb{R}^2$, on letting $n \to \infty$ it then follows that also $(x, y(x)) \in D$ when $x_0 \le x \le X_M$.

To show that the solution of the initial value problem is unique, suppose, if possible, that there are two different solutions y and z. Then, by subtraction,

$$y(x) - z(x) = \int_{x_0}^{x} (f(s, y(s)) - f(s, z(s)))\,\mathrm{d}s, \qquad x \in [x_0, X_M],$$

from which it follows that

$$|y(x) - z(x)| \le L \int_{x_0}^{x} |y(s) - z(s)|\mathrm{d}s \tag{12.11}$$

for all $x \in [x_0, X_M]$. Suppose that m is the maximum value of the expression $|y(x) - z(x)|$ for $x_0 \le x \le X_M$, and that $m > 0$. Then,

$$|y(x) - z(x)| \le mL(x - x_0), \qquad x_0 \le x \le X_M.$$

Substituting this inequality into the right-hand side of (12.11) we find

$$|y(x) - z(x)| \le L^2 m \int_{x_0}^{x} (s - x_0)\,\mathrm{d}s = m\frac{[L(x - x_0)]^2}{2!}.$$

Proceeding in a similar manner, it is easy to show by induction that

$$|y(x) - z(x)| \le m\frac{[L(x - x_0)]^k}{k!}, \qquad k = 1, 2, \ldots,$$

for all $x \in [x_0, X_M]$. However, the right-hand side in the last inequality is bounded above by $m[L(X_M - x_0)]^k/k!$ for all $x \in [x_0, X_M]$, which can be made arbitrarily small by choosing k sufficiently large. Therefore, $|y(x) - z(x)|$ must be zero for all $x \in [x_0, X_M]$. Hence the solutions y and z are identical. □

In an application of this theorem it is necessary to choose a value of the constant C in Picard's Theorem so that the various hypotheses are satisfied, in particular (12.3); it is not difficult to see that if $\partial f/\partial y$ is continuous in a neighbourhood of (x_0, y_0) the conditions will be satisfied if $X_M - x_0$ is sufficiently small.

As a very simple example, consider the linear equation

$$y' = py + q\,, \tag{12.12}$$

where p and q are constants. Then, $L = |p|$, independently of C, and $K = |py_0| + |q|$. Hence, for any interval $[x_0, X_M]$, the conditions are satisfied by choosing C sufficiently large; therefore, the initial value problem has a unique continuously differentiable solution, defined for all $x \in [x_0, \infty)$.

Now, consider another example

$$y' = y^2\,, \qquad y(0) = 1\,.$$

Here for any interval $[0, X_M]$ we have $K = 1$. Choosing any positive value of C we find that

$$|u^2 - v^2| = |u + v|\,|u - v| \le L|u - v| \qquad \forall u, v \in \mathbb{R}\,,$$

where $L = 2(1 + C)$. We therefore now require the condition

$$C \ge \frac{1}{2(1+C)} \left(\mathrm{e}^{2(1+C)X_M} - 1\right)\,.$$

This is satisfied if

$$X_M \le F(C) \equiv \frac{1}{2(1+C)} \ln(1 + 2C + 2C^2)\,,$$

where ln means $\log_{\mathrm{e}}$. A sketch of the graph of the function F against C shows that F takes its maximum value near $C = 1.714$, and this gives the condition $X_M \le 0.43$ (see Figure 12.1).

Thus, we are *unable* to prove the existence of the solution over the infinite interval $[0, \infty)$. This is correct, of course, as the unique solution of the initial value problem is

$$y(x) = \frac{1}{1-x}\,, \quad 0 \le x < 1\,,$$

and this is not continuous, let alone continuously differentiable, on any interval $[0, X_M]$ with $X_M \ge 1$. The conditions of Picard's Theorem, which are sufficient but not necessary for the existence and the uniqueness of the solution, have given a rather more restrictive bound on the size of the interval over which the solution exists.

The method of proof of Picard's Theorem also suggests a possible technique for constructing approximations to the solution, by determining the functions y_n from (12.4). In practice it may be impossible, or very difficult, to evaluate the necessary integrals in closed form. We

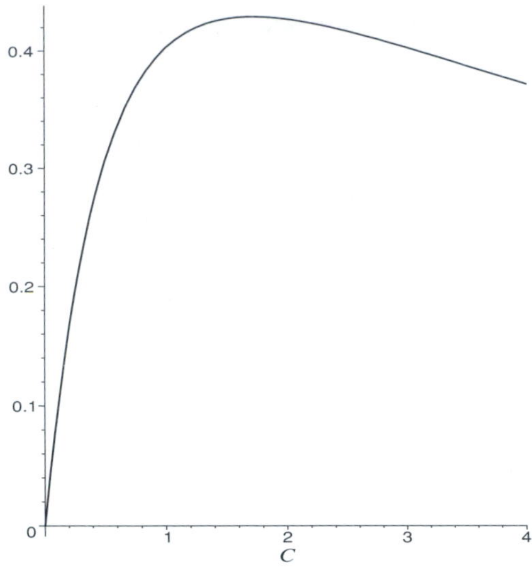

Fig. 12.1. Graph of the function $C \mapsto F(C)$ on the interval $[0, 4]$; F achieves its maximum value near $C = 1.714$ and $F(C) \leq 0.43$ for all $C \geq 0$.

leave it as an exercise (see Exercise 3) to show that for the simple linear equation (12.12), with initial condition $y(0) = 1$, the function y_n is the same as the approximation obtained from the exact solution by expanding the exponential function as a power series and retaining the terms up to the one involving x^n.

In the rest of this chapter we shall consider step-by-step numerical methods for the approximate solution of the initial value problem (12.1), (12.2). We shall suppose throughout that the function f satisfies the conditions of Picard's Theorem. Suppose that the initial value problem (12.1), (12.2) is to be solved on the interval $[x_0, X_M]$. We divide this interval by the **mesh points** $x_n = x_0 + nh$, $n = 0, 1, \ldots, N$, where $h = (X_M - x_0)/N$ and N is a positive integer. The positive real number h is called the **step size** or **mesh size**. For each n we seek a numerical approximation y_n to $y(x_n)$, the value of the analytical solution at the mesh point x_n; these values y_n are calculated in succession, for $n = 1, 2, \ldots, N$.

12.2 One-step methods

A one-step method expresses y_{n+1} in terms of the previous value y_n; later on we shall consider k-step methods, where y_{n+1} is expressed in terms of the k previous values $y_{n-k+1}, \ldots, y_n$, where $k \geq 2$. The simplest example of a one-step method for the numerical solution of the initial value problem (12.1), (12.2) is Euler's method.

Euler's method. Given that $y(x_0) = y_0$, let us suppose that we have already calculated y_n, up to some n, $0 \leq n \leq N-1$, $N \geq 1$; we define

$$y_{n+1} = y_n + hf(x_n, y_n).$$

Thus, taking in succession $n = 0, 1, \ldots, N-1$, one step at a time, the approximate values y_n at the mesh points x_n can be easily obtained. This numerical method is known as **Euler's method**.

In order to motivate the definition of Euler's method, let us observe that on expanding $y(x_{n+1}) = y(x_n + h)$ into a Taylor series about x_n, retaining only the first two terms, and writing $y'(x_n) = f(x_n, y(x_n))$, we have that

$$y(x_n + h) = y(x_n) + hf(x_n, y(x_n)) + \mathcal{O}(h^2).$$

After replacing $y(x_n)$ and $y(x_n + h)$ by their numerical approximations, denoted by y_n and y_{n+1}, respectively, and discarding the $\mathcal{O}(h^2)$ term, we arrive at Euler's method.

More generally, a one-step method may be written in the form

$$y_{n+1} = y_n + h\Phi(x_n, y_n; h), \qquad n = 0, 1, \ldots, N-1, \quad y(x_0) = y_0, \tag{12.13}$$

where $\Phi(\,\cdot\,, \,\cdot\,; \,\cdot\,)$ is a continuous function of its variables. For example, in the case of Euler's method, $\Phi(x_n, y_n; h) = f(x_n, y_n)$. More intricate examples of one-step methods will be discussed below.

In order to assess the accuracy of the numerical method (12.13), we define the **global error**, e_n, by

$$e_n = y(x_n) - y_n.$$

We also need the concept of **truncation error**, T_n, defined by

$$T_n = \frac{y(x_{n+1}) - y(x_n)}{h} - \Phi(x_n, y(x_n); h). \tag{12.14}$$

The next theorem provides a bound on the magnitude of the global error in terms of the truncation error.

Theorem 12.2 *Consider the general one-step method (12.13) where, in addition to being a continuous function of its arguments, Φ is assumed to satisfy a Lipschitz condition with respect to its second argument, that is, there exists a positive constant L_Φ such that, for $0 \le h \le h_0$ and for all (x, u) and (x, v) in the rectangle*

$$D = \{(x, y) \colon x_0 \le x \le X_M,\ |y - y_0| \le C\},$$

we have that

$$|\Phi(x, u; h) - \Phi(x, v; h)| \le L_\Phi |u - v|. \tag{12.15}$$

Then, assuming that $|y_n - y_0| \le C$, $n = 1, 2, \ldots, N$, it follows that

$$|e_n| \le \frac{T}{L_\Phi}\left(\mathrm{e}^{L_\Phi(x_n - x_0)} - 1\right), \qquad n = 0, 1, \ldots, N, \tag{12.16}$$

where $T = \max_{0 \le n \le N-1} |T_n|$.

Proof Rewriting (12.14) as

$$y(x_{n+1}) = y(x_n) + h\Phi(x_n, y(x_n); h) + hT_n$$

and subtracting (12.13) from this, we obtain

$$e_{n+1} = e_n + h[\Phi(x_n, y(x_n); h) - \Phi(x_n, y_n; h)] + hT_n.$$

Then, since $(x_n, y(x_n))$ and (x_n, y_n) belong to D, the Lipschitz condition (12.15) implies that

$$|e_{n+1}| \le |e_n| + hL_\Phi |e_n| + h|T_n|, \qquad n = 0, 1, \ldots, N-1. \tag{12.17}$$

That is,

$$|e_{n+1}| \le (1 + hL_\Phi)|e_n| + h|T_n|, \qquad n = 0, 1, \ldots, N-1.$$

It easily follows by induction that

$$|e_n| \le \frac{T}{L_\Phi}[(1 + hL_\Phi)^n - 1], \qquad n = 0, 1, \ldots, N,$$

since $e_0 = 0$. Observing that $1 + hL_\Phi \le \exp(hL_\Phi)$ gives (12.16). □

Let us apply this general result in order to obtain a bound on the global error in Euler's method. The truncation error for Euler's method is given by

$$\begin{aligned} T_n &= \frac{y(x_{n+1}) - y(x_n)}{h} - f(x_n, y(x_n)) \\ &= \frac{y(x_{n+1}) - y(x_n)}{h} - y'(x_n). \end{aligned} \tag{12.18}$$

Assuming that $y \in C^2[x_0, X_M]$, *i.e.*, that y is a twice continuously differentiable function of x on $[x_0, X_M]$, and expanding $y(x_{n+1})$ about the point x_n into a Taylor series with remainder (see Theorem A.4), we have that

$$y(x_{n+1}) = y(x_n) + hy'(x_n) + \frac{h^2}{2!}y''(\xi_n)\,, \qquad x_n < \xi_n < x_{n+1}\,.$$

Substituting this expansion into (12.18) gives

$$T_n = \frac{1}{2}hy''(\xi_n)\,.$$

Let $M_2 = \max_{\zeta \in [x_0, X_M]} |y''(\zeta)|$. Then, $|T_n| \le T$, $n = 0, 1, \ldots, N-1$, where $T = \frac{1}{2}hM_2$. Inserting this into (12.16) and noting that for Euler's method $\Phi(x_n, y_n; h) \equiv f(x_n, y_n)$ and therefore $L_\Phi = L$ where L is the Lipschitz constant for f, we have that

$$|e_n| \le \frac{1}{2}M_2 \left[\frac{e^{L(x_n - x_0)} - 1}{L}\right] h\,, \quad n = 0, 1, \ldots, N\,. \tag{12.19}$$

Let us highlight the practical relevance of our error analysis by focusing on a particular example.

Example 12.2 *Let us consider the initial value problem* $y' = \tan^{-1} y$, $y(0) = y_0$, *where* y_0 *is a given real number. In order to find an upper bound on the global error* $e_n = y(x_n) - y_n$, *where* y_n *is the Euler approximation to* $y(x_n)$, *we need to determine the constants* L *and* M_2 *in the inequality (12.19).*

Here $f(x, y) = \tan^{-1} y$; so, by the Mean Value Theorem (Theorem A.3),

$$|f(x, u) - f(x, v)| = \left|\frac{\partial f}{\partial y}(x, \eta)\,(u - v)\right| = \left|\frac{\partial f}{\partial y}(x, \eta)\right|\,|u - v|\,,$$

where η lies between u and v. In our case

$$\left|\frac{\partial f}{\partial y}(x, y)\right| = |(1 + y^2)^{-1}| \le 1\,,$$

and therefore $L = 1$. To find M_2 we need to obtain a bound on $|y''|$ (without actually solving the initial value problem!). This is easily achieved by differentiating both sides of the differential equation with respect to the variable x:

$$y'' = \frac{d}{dx}(\tan^{-1} y) = (1 + y^2)^{-1}\frac{dy}{dx} = (1 + y^2)^{-1}\tan^{-1} y\,.$$

Therefore $|y''(x)| \le M_2 = \frac{1}{2}\pi$. Inserting the values of L and M_2 into (12.19) and noting that $x_0 = 0$, we have

$$|e_n| \le \tfrac{1}{4}\pi\,(\mathrm{e}^{x_n} - 1)\,h\,, \quad n = 0, 1, \ldots, N\,.$$

Thus, given a tolerance TOL, specified beforehand, we can ensure that the error between the (unknown) analytical solution and its numerical approximation does not exceed this tolerance by choosing a positive step size h such that

$$h \le \frac{4}{\pi(\mathrm{e}^{X_M} - 1)}\,\mathtt{TOL}\,.$$

For such h we shall have $|y(x_n) - y_n| = |e_n| \le \mathtt{TOL}$, for $n = 0, 1, \ldots, N$, as required. Thus, at least in principle, we can calculate the numerical solution to arbitrarily high accuracy by choosing a sufficiently small step size h.

A numerical experiment shows that this error estimate is rather pessimistic. Taking, for example, $y_0 = 1$ and $X_M = 1$, our bound implies that the tolerance $\mathtt{TOL} = 0.01$ will be achieved with $h \le 0.0074$; hence, it would appear that we need $N \ge 135$. In fact, using $N = 27$ gives a result from Euler's method which is just within this tolerance, so the error estimate has predicted the use of a step size which is five times smaller than is actually required. ◇

Example 12.3 *As a more typical practical example, consider the problem*

$$y' = y^2 + g(x)\,, \quad y(0) = 2\,, \tag{12.20}$$

where

$$g(x) = \frac{x^4 - 6x^3 + 12x^2 - 14x + 9}{(1+x)^2}\,,$$

is so chosen that the solution is known, and is

$$y(x) = \frac{(1-x)(2-x)}{1+x}\,.$$

The results of some numerical calculations on the interval $x \in [0, 1.6]$ are shown in Figure 12.2. They use step sizes 0.2, 0.1 and 0.05, and show how halving the step size gives a reduction of the error also by a factor of roughly 2, in agreement with the error bound (12.19). ◇

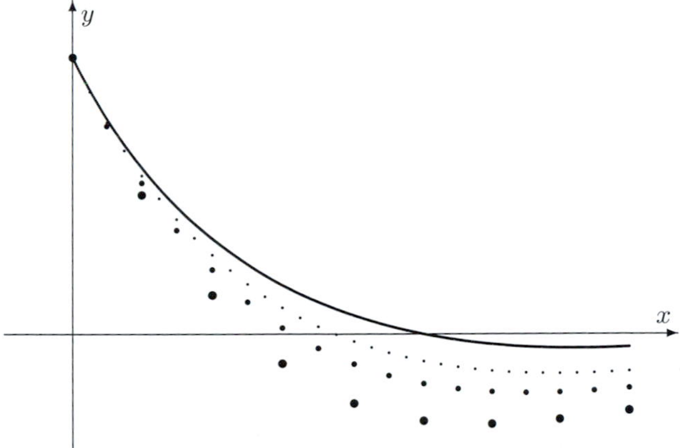

Fig. 12.2. Euler's method for the solution of (12.20). The exact solution (solid curve) and three sets of results are shown (large, medium and small dots), using respectively 8 steps of size 0.2, 16 steps of size 0.1 and 32 steps of size 0.05 on the interval $[0, 1.6]$.

12.3 Consistency and convergence

Returning to the general one-step method (12.13), we consider the choice of the function Φ. Theorem 12.2 suggests that if the truncation error 'approaches zero' as $h \to 0$, then the global error 'converges to zero' also. This observation motivates the following definition.

Definition 12.1 *The numerical method (12.13) is* **consistent** *with the differential equation (12.1) if the truncation error, defined by (12.14), is such that for any $\varepsilon > 0$ there exists a positive $h(\varepsilon)$ for which $|T_n| < \varepsilon$ for $0 < h < h(\varepsilon)$ and any pair of points $(x_n, y(x_n))$, $(x_{n+1}, y(x_{n+1}))$ on any solution curve in D.*

For the general one-step method (12.13) we have assumed that the function $\Phi(\cdot, \cdot; \cdot)$ is continuous; since y' is also a continuous function on $[x_0, X_M]$ it follows from (12.14) that, in the limit of

$$h \to 0 \text{ and } n \to \infty, \text{ with } \lim_{n\to\infty} x_n = x \in [x_0, X_M],$$

we have

$$\lim_{n\to\infty} T_n = y'(x) - \Phi(x, y(x); 0).$$

In this limit h tends to zero and n tends to infinity in such a way that x_n tends to a limit point x which lies in the interval $[x_0, X_M]$. This implies that the one-step method (12.13) is consistent if, and only if,

$$\Phi(x, y; 0) \equiv f(x, y)\,. \tag{12.21}$$

This condition is sometimes taken as the definition of consistency. We shall henceforth always assume that (12.21) holds.

Now, we are ready to state a convergence theorem for the general one-step method (12.13).

Theorem 12.3 *Suppose that the initial value problem (12.1), (12.2) satisfies the conditions of Picard's Theorem, and also that its approximation generated from (12.13) when $h \le h_0$ lies in the region D. Assume further that the function $\Phi(\,\cdot\,,\cdot\,;\cdot\,)$ is continuous on $D \times [0, h_0]$, and satisfies the consistency condition (12.21) and the Lipschitz condition*

$$|\Phi(x, u; h) - \Phi(x, v; h)| \le L_\Phi |u - v| \qquad \text{on } D \times [0, h_0]\,. \tag{12.22}$$

Then, if successive approximation sequences (y_n), generated by using the mesh points $x_n = x_0 + nh$, $n = 1, 2, \ldots, N$, are obtained from (12.13) with successively smaller values of h, each h less than h_0, we have convergence of the numerical solution to the solution of the initial value problem in the sense that

$$\lim_{n\to\infty} y_n = y(x) \qquad \text{as} \qquad x_n \to x \in [x_0, X_M] \text{ when } h \to 0 \text{ and } n \to \infty\,.$$

Proof Suppose that $h = (X_M - x_0)/N$, where N is a positive integer. We shall assume that N is sufficiently large so that $h \le h_0$. Since $y(x_0) = y_0$ and therefore $e_0 = 0$, Theorem 12.2 implies that

$$|y(x_n) - y_n| \le \left(\frac{e^{L_\Phi(X_M - x_0)} - 1}{L_\Phi}\right) \max_{0\le m\le n-1} |T_m|\,, \qquad n = 1, 2, \ldots, N\,. \tag{12.23}$$

From the consistency condition (12.21) we have

$$\begin{aligned} T_n &= \left(\frac{y(x_{n+1}) - y(x_n)}{h} - f(x_n, y(x_n))\right) \\ &\quad + (\Phi(x_n, y(x_n); 0) - \Phi(x_n, y(x_n); h))\,. \end{aligned} \tag{12.24}$$

According to the Mean Value Theorem, Theorem A.3, the expression in the first bracket is equal to $y'(\xi_n) - y'(x_n)$, where $\xi_n \in [x_n, x_{n+1}]$. By Picard's Theorem, y' is continuous on the closed interval $[x_0, X_M]$; therefore, it is uniformly continuous on this interval. Hence, for each $\varepsilon > 0$ there exists $h_1(\varepsilon)$ such that

$$|y'(\xi_n) - y'(x_n)| \le \tfrac{1}{2}\varepsilon \qquad \text{for } h < h_1(\varepsilon)\,, \quad n = 0, 1, \ldots, N-1\,.$$

Also, since $\Phi(\cdot, \cdot; \cdot)$ is a continuous function on the closed set $D \times [0, h_0]$ and is, therefore, uniformly continuous on $D \times [0, h_0]$, there exists $h_2(\varepsilon)$ such that

$$|\Phi(x_n, y(x_n); 0) - \Phi(x_n, y(x_n); h)| \leq \tfrac{1}{2}\varepsilon$$

for $h < h_2(\varepsilon)$, $n = 0, 1, \ldots, N-1$. On defining $h(\varepsilon) = \min\{h_1(\varepsilon), h_2(\varepsilon)\}$, we then have that

$$|T_n| \leq \varepsilon \qquad \text{for } h < h(\varepsilon)\,, \qquad n = 0, 1, \ldots, N-1\,.$$

Inserting this into (12.23) we deduce that

$$\begin{aligned} |y(x) - y_n| &\leq |y(x) - y(x_n)| + |y(x_n) - y_n| \\ &\leq |y(x) - y(x_n)| + \varepsilon\, \frac{\mathrm{e}^{L_\Phi (X_M - x_0)} - 1}{L_\Phi}\,. \end{aligned} \qquad (12.25)$$

Now, in the limit of $h \to 0$, $n \to \infty$ with $x_n \to x \in [x_0, X_M]$, we have $\lim_{n\to\infty} y(x_n) = y(x)$, since y is a continuous function on $[x_0, X_M]$. Further, the second term on the right-hand side of (12.25) can be made arbitrarily small, independently of h and n, by letting $\varepsilon \to 0$. Therefore, in the limit of $h \to 0$, $n \to \infty$ with $x_n \to x \in [x_0, X_M]$, we have that $\lim_{n\to\infty} y_n = y(x)$, as stated. □

We saw earlier that for Euler's method the magnitude of the truncation error T_n is bounded above by a constant multiple of the step size h, that is,

$$|T_n| \leq Kh \qquad \text{for } 0 < h \leq h_0\,,$$

where K is a positive constant, independent of h. However, there are other one-step methods (a class of which, called Runge–Kutta[1] methods, will be considered below) for which we can do better. Thus, in order to quantify the asymptotic rate of decay of the truncation error as the step size h converges to 0, we introduce the following definition.

Definition 12.2 *The numerical method (12.13) is said to have* **order of accuracy** p, *if p is the largest positive integer such that, for any sufficiently smooth solution curve $(x, y(x))$ in D of the initial value problem (12.1), (12.2), there exist constants K and h_0 such that*

$$|T_n| \leq Kh^p \qquad \textit{for } 0 < h \leq h_0$$

[1] After Carle David Tolmé Runge (30 August 1856, Bremen, Germany – 3 January 1927, Göttingen, Germany) and Martin Wilhelm Kutta (3 November 1867, Pitschen, Upper Silesia, Prussia, North Germany (now Byczyna, Poland) – 25 December 1944, Fürstenfeldbruck, Germany).

for any pair of points $(x_n, y(x_n))$, $(x_{n+1}, y(x_{n+1}))$ on the solution curve.

12.4 An implicit one-step method

A one-step method with second-order accuracy is the **trapezium rule method**

$$y_{n+1} = y_n + \tfrac{h}{2}[f(x_n, y_n) + f(x_{n+1}, y_{n+1})]\,. \tag{12.26}$$

This method is easily motivated by writing

$$y(x_{n+1}) - y(x_n) = \int_{x_n}^{x_{n+1}} y'(x)\,\mathrm{d}x\,,$$

and approximating the integral by the trapezium rule. Since the right-hand side involves the integral of the function $x \mapsto y'(x) = f(x, y(x))$ we see at once from (7.6) that the truncation error

$$T_n = \frac{y(x_{n+1}) - y(x_n)}{h} - \tfrac{1}{2}\left[f(x_n, y(x_n)) + f(x_{n+1}, y(x_{n+1}))\right]$$

of the trapezium rule method satisfies the bound

$$|T_n| \le \tfrac{1}{12}h^2 M_3\,, \quad \text{where } M_3 = \max_{x\in[x_0, X_M]} |y'''(x)|\,. \tag{12.27}$$

The important difference between this method and Euler's method is that the value y_{n+1} appears on both sides of (12.26). To calculate y_{n+1} from the known y_n therefore requires the solution of an equation, which will usually be nonlinear. This additional complication means an increase in the amount of computation required, but not usually a very large increase. The equation (12.26) is easily solved for y_{n+1} by Newton's method, assuming that the derivative $\partial f/\partial y$ can be calculated quickly; as a starting point for the Newton iteration the obvious estimate

$$y_n + hf(x_n, y_n)$$

will usually be close, and a couple of iterations will then suffice.

Methods of this type, which require the solution of an equation to determine the new value y_{n+1}, are known as **implicit methods**.

Writing the trapezium rule method in the standard form (12.13) we see that

$$\begin{aligned} h\Phi(x_n, y_n; h) &= \tfrac{h}{2}[f(x_n, y_n) + f(x_{n+1}, y_{n+1})] \\ &= \tfrac{h}{2}[f(x_n, y_n) + f(x_{n+1}, y_n + h\Phi(x_n, y_n; h)]\,. \end{aligned} \tag{12.28}$$

Hence, the function Φ is also defined in an implicit form.

In order to employ Theorem 12.2 to estimate the error in the trapezium rule method we need a value for the Lipschitz constant L_Φ. From (12.28) we find that

$$|\Phi(x_n, u; h) - \Phi(x_n, v; h)| = \tfrac{1}{2}|f(x_n, u) - f(x_n + h, u + h\Phi(x_n, u; h)) \\ - f(x_n, v) - f(x_n + h, v + h\Phi(x_n, v; h))| .$$

Hence,

$$\begin{aligned} &|\Phi(x_n, u; h) - \Phi(x_n, v; h)| \\ &\quad \le \tfrac{1}{2}|f(x_n, u) - f(x_n, v)| \\ &\qquad + \tfrac{1}{2}|f(x_n + h, u + h\Phi(x_n, u; h)) - f(x_n + h, v + h\Phi(x_n, v; h))| \\ &\quad \le \tfrac{1}{2}L_f|u - v| \\ &\qquad + \tfrac{1}{2}L_f|u + h\Phi(x_n, u; h) - v - h\Phi(x_n, v; h)| \\ &\quad \le \tfrac{1}{2}L_f|u - v| + \tfrac{1}{2}L_f|u - v| + \tfrac{1}{2}L_f h|\Phi(x_n, u; h) - \Phi(x_n, v; h)| . \end{aligned}$$

This shows that

$$\left(1 - \tfrac{1}{2}hL_f\right)|\Phi(x_n, u; h) - \Phi(x_n, v; h)| \le L_f|u - v| ,$$

and, therefore,

$$L_\Phi \le \frac{L_f}{1 - \frac{1}{2}hL_f}, \qquad \text{provided that } \tfrac{1}{2}hL_f < 1 .$$

Consequently, (12.16) and (12.27) imply that the global error in the trapezium rule method is $\mathcal{O}(h^2)$, as h tends to 0.

Figure 12.3 depicts the results of some numerical calculations on the interval $x \in [0, 1.6]$ for the same problem as in Figure 12.2. The step sizes are 0.4 and 0.2, larger than for Euler's method; nevertheless we see a much reduced error in comparison with Euler's method, and also how the reduction in the step size h by a factor of 2 gives a reduction in the error by a factor of about 4, as predicted by our error analysis.

12.5 Runge–Kutta methods

Euler's method is only first-order accurate; nevertheless, it is simple and cheap to implement because, to obtain y_{n+1} from y_n, we only require a single evaluation of the function f, at (x_n, y_n). Runge–Kutta methods aim to achieve higher accuracy by sacrificing the efficiency of Euler's method through re-evaluating $f(\,\cdot\,,\,\cdot\,)$ at points intermediate between

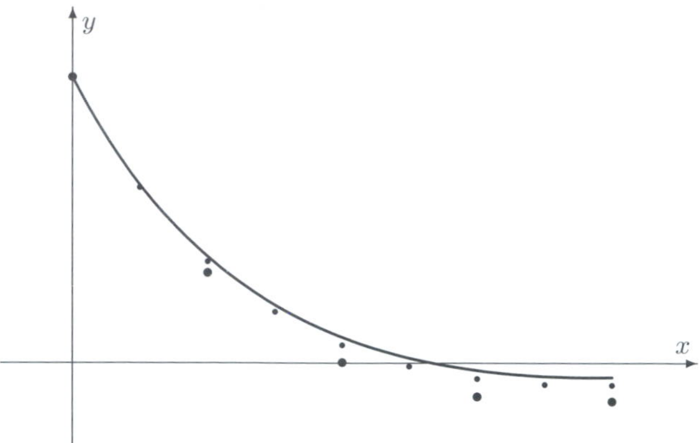

Fig. 12.3. Trapezium rule method for the solution of (12.20). The exact solution (solid curve) and two sets of results are shown (large and small dots), using respectively 4 steps of size 0.4, and 8 steps of size 0.2 on $[0, 1.6]$.

$(x_n, y(x_n))$ and $(x_{n+1}, y(x_{n+1}))$. Consider, for example, the following family of methods:

$$y_{n+1} = y_n + h(ak_1 + bk_2)\,, \tag{12.29}$$

where

$$k_1 = f(x_n, y_n)\,, \tag{12.30}$$

$$k_2 = f(x_n + \alpha h, y_n + \beta h k_1)\,, \tag{12.31}$$

and where the parameters a, b, α and β are to be determined.

Note that Euler's method is a member of this family of methods, corresponding to $a = 1$ and $b = 0$. However, we are now seeking methods that are at least second-order accurate. Clearly (12.29)–(12.31) can be written in the form (12.13) with

$$\Phi(x_n, y_n; h) = af(x_n, y_n) + bf(x_n + \alpha h, y_n + \beta h f(x_n, y_n))\,.$$

By the condition (12.21), a method from this family will be consistent if, and only if, $a + b = 1$. Further conditions on the parameters are found by attempting to maximise the order of accuracy of the method.

To determine the truncation error of the method from (12.14) we need the higher derivatives of $y(x)$, which are obtained by differentiating the function f:

$$y'(x_n) = f\,,$$

$$\begin{aligned} y''(x_n) &= f_x + f_y y' = f_x + f_y f\,, \\ y'''(x_n) &= f_{xx} + f_{xy} f + (f_{xy} + f_{yy} f) f + f_y (f_x + f_y f)\,, \end{aligned}$$

and so on; in these expressions the subscripts x and y denote partial derivatives, and all functions appearing on the right-hand sides are to be evaluated at $(x_n, y(x_n))$. We also need to expand $\Phi(x_n, y(x_n); h)$ in powers of h, giving (with the same notational conventions as before)

$$\begin{aligned} \Phi(x_n, y(x_n); h) &= af + b\left(f + \alpha h f_x + \beta h f f_y + \tfrac{1}{2}(\alpha h)^2 f_{xx}\right. \\ &\qquad \left. + \alpha\beta h^2 f f_{xy} + \tfrac{1}{2}(\beta h)^2 f^2 f_{yy} + \mathcal{O}(h^3)\right) . \end{aligned}$$

Thus, we obtain the truncation error in the form

$$\begin{aligned} T_n &= \frac{y(x_n + h) - y(x_n)}{h} - \Phi(x_n, y(x_n); h) \\ &= f + \tfrac{1}{2} h (f_x + f f_y) \\ &\quad + \tfrac{1}{6} h^2 [f_{xx} + 2 f_{xy} f + f_{yy} f^2 + f_y (f_x + f_y f)] \\ &\quad - \{af + b[f + \alpha h f_x + \beta h f f_y + \tfrac{1}{2}(\alpha h)^2 f_{xx} \\ &\qquad + \alpha\beta h^2 f f_{xy} + \tfrac{1}{2}(\beta h)^2 f^2 f_{yy}]\} + \mathcal{O}(h^3)\,. \end{aligned}$$

As $1 - a - b = 0$, the term $(1 - a - b)f$ is equal to 0. The coefficient of the term in h is

$$\tfrac{1}{2}(f_x + f f_y) - b\alpha f_x - b\beta f f_y$$

which vanishes for all functions f provided that

$$b\alpha = b\beta = \tfrac{1}{2}\,.$$

The method is therefore second-order accurate if

$$\beta = \alpha\,, \quad a = 1 - \tfrac{1}{2\alpha}\,, \quad b = \tfrac{1}{2\alpha}\,, \qquad \alpha \neq 0\,,$$

showing that there is a one-parameter family of second-order methods of this form, parametrised by $\alpha \neq 0$. The truncation error of the method then becomes

$$\begin{aligned} T_n &= h^2\{(\tfrac{1}{6} - \tfrac{\alpha}{4})(f_{xx} + f_{yy} f^2) + (\tfrac{1}{3} - \tfrac{\alpha}{2}) f f_{xy} \\ &\qquad + \tfrac{1}{6}(f_x f_y + f f_y^2)\} + \mathcal{O}(h^3)\,. \end{aligned} \tag{12.32}$$

Evidently there is no choice of the free parameter α which will make this method third-order accurate for all functions f; this can be seen, for example, by considering the initial value problem $y' = y$, $y(0) = 1$, and noting that in this case (12.32), with $f(x, y) = y$, yields

$$T_n = \tfrac{1}{6} h^2 y(x_n) + \mathcal{O}(h^3) = \tfrac{1}{6} h^2 \mathrm{e}^{x_n} + \mathcal{O}(h^3)\,.$$

Two examples of second-order Runge–Kutta methods of the form (12.29)–(12.31) are the modified Euler method and the improved Euler method.

(a) **The modified Euler method.** In this case we take $\alpha = \frac{1}{2}$ to obtain

$$y_{n+1} = y_n + h\,f\left(x_n + \frac{1}{2}h, y_n + \frac{1}{2}hf(x_n, y_n)\right).$$

(b) **The improved Euler method.** This is arrived at by choosing $\alpha = 1$ which gives

$$y_{n+1} = y_n + \frac{1}{2}h\left[f(x_n, y_n) + f(x_n + h, y_n + hf(x_n, y_n))\right].$$

For these two methods it is easily verified using (12.32) that the truncation error is of the form, respectively,

$$\begin{aligned} T_n &= \frac{1}{6}h^2\left[f_y(f_x + f_y f) + \frac{1}{4}(f_{xx} + 2f_{xy}f + f_{yy}f^2)\right] + \mathcal{O}(h^3), \\ T_n &= \frac{1}{6}h^2\left[f_y(f_x + f_y f) - \frac{1}{2}(f_{xx} + 2f_{xy}f + f_{yy}f^2)\right] + \mathcal{O}(h^3). \end{aligned}$$

A similar but more complicated analysis is used to construct Runge–Kutta methods of higher order. One of the most frequently used methods of the Runge–Kutta family is often known as the **classical fourth-order method**:

$$y_{n+1} = y_n + \frac{1}{6}h\,(k_1 + 2k_2 + 2k_3 + k_4),$$

where

$$\left.\begin{aligned} k_1 &= f(x_n, y_n), \\ k_2 &= f\left(x_n + \tfrac{1}{2}h, y_n + \tfrac{1}{2}hk_1\right), \\ k_3 &= f\left(x_n + \tfrac{1}{2}h, y_n + \tfrac{1}{2}hk_2\right), \\ k_4 &= f(x_n + h, y_n + hk_3). \end{aligned}\right\} \qquad (12.33)$$

Here k_2 and k_3 represent approximations to the derivative y' at points on the solution curve, intermediate between $(x_n, y(x_n))$ and $(x_{n+1}, y(x_{n+1}))$, and $\Phi(x_n, y_n; h)$ is a weighted average of the k_i, $i = 1, 2, 3, 4$, the weights corresponding to those of Simpson's rule (to which the classical fourth-order Runge–Kutta method reduces when $\frac{\partial f}{\partial y} \equiv 0$).

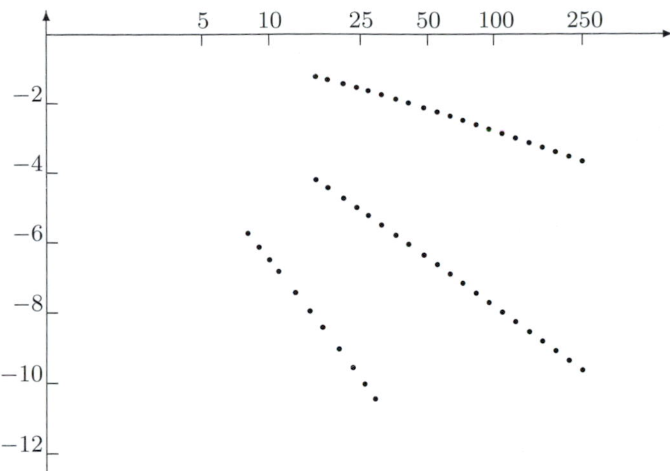

Fig. 12.4. The errors in three methods for the solution of (12.20) on the interval $[0, 1.6]$. Reading from the top, the lines (whose slopes indicate first-, second- and fourth-order convergence) represent the errors of Euler's method, the trapezium rule method, and the classical Runge–Kutta method respectively. The horizontal axis indicates the number $N = 1.6/h$, on a logarithmic scale, and the vertical axis shows $\ln |e_N| = \ln |y(1.6) - y_N|$.

To illustrate the behaviour of the one-step methods which we have discussed, Figure 12.4 shows the errors in the calculation of $y(1.6)$, where $y(x)$ is the solution to the problem (12.20) on the interval $[0, 1.6]$. The horizontal axis indicates N, the number of equally spaced mesh points used in the interval $(0, 1.6]$, on a logarithmic scale, and the vertical axis shows $\ln |e_N| = \ln |y(1.6) - y_N|$. The three methods employed are Euler's method, the trapezium rule method, and the classical Runge–Kutta method (12.33). The three lines show clearly the improved accuracy of the higher-order methods, and the rate at which the accuracy improves as N increases.

12.6 Linear multistep methods

While Runge–Kutta methods give an improvement over Euler's method in terms of accuracy, this is achieved by investing additional computational effort; in fact, Runge–Kutta methods require more evaluations of $f(\cdot, \cdot)$ than would seem necessary. For example, the fourth-order method involves four function evaluations per step. For comparison, by considering three consecutive points x_{n-1}, $x_n = x_{n-1} + h$, $x_{n+1} = x_{n-1} + 2h$, integrating the differential equation between x_{n-1} and x_{n+1}, yields

$$y(x_{n+1}) = y(x_{n-1}) + \int_{x_{n-1}}^{x_{n+1}} f(x, y(x))\mathrm{d}x\,,$$

and applying Simpson's rule to approximate the integral on the right-hand side then leads to the method

$$y_{n+1} = y_{n-1} + \frac{1}{3}h\left[f(x_{n-1}, y_{n-1}) + 4f(x_n, y_n) + f(x_{n+1}, y_{n+1})\right], \tag{12.34}$$

requiring only three function evaluations per step. In contrast with the one-step methods considered in the previous section where only a single value y_n was required to compute the next approximation y_{n+1}, here we need *two* preceding values, y_n and y_{n-1}, to be able to calculate y_{n+1}, and therefore (12.34) is *not* a one-step method.

In this section we consider a class of methods of the type (12.34) for the numerical solution of the initial value problem (12.1), (12.2), called **linear multistep methods**.

Given a sequence of equally spaced mesh points (x_n) with step size h, we consider the general **linear k-step method**

$$\sum_{j=0}^{k} \alpha_j y_{n+j} = h \sum_{j=0}^{k} \beta_j f(x_{n+j}, y_{n+j})\,, \tag{12.35}$$

where the coefficients $\alpha_0, \ldots, \alpha_k$ and $\beta_0, \ldots, \beta_k$ are real constants. In order to avoid degenerate cases, we shall assume that $\alpha_k \neq 0$ and that α_0 and β_0 are not both equal to 0. If $\beta_k = 0$, then y_{n+k} is obtained explicitly from previous values of y_j and $f(x_j, y_j)$, and the k-step method is then said to be **explicit**. On the other hand, if $\beta_k \neq 0$, then y_{n+k} appears not only on the left-hand side but also on the right, within $f(x_{n+k}, y_{n+k})$; due to this implicit dependence on y_{n+k} the method is then called **implicit**. The method (12.35) is called *linear* because it involves only linear combinations of the y_{n+j} and the $f(x_{n+j}, y_{n+j})$, $j = 0, 1, \ldots, k$; for the sake of notational simplicity, henceforth we shall often write f_n instead of $f(x_n, y_n)$.

Example 12.4 *We have already seen an example of a linear two-step method in (12.34); here we present further examples of linear multistep methods.*

(a) Euler's method is a trivial case: it is an explicit linear one-step method. The **implicit Euler method**

$$y_{n+1} = y_n + hf(x_{n+1}, y_{n+1}) \tag{12.36}$$

is an implicit linear one-step method. Another trivial example is the **trapezium rule method**, given by

$$y_{n+1} = y_n + \frac{1}{2}h\,(f_{n+1} + f_n)\,;$$

it, too, is an implicit linear one-step method.

(b) The **Adams[1]–Bashforth[2] method**

$$y_{n+4} = y_{n+3} + \frac{1}{24}h\,(55f_{n+3} - 59f_{n+2} + 37f_{n+1} - 9f_n)$$

is an example of an explicit linear four-step method, while the **Adams–Moulton[3] method**

$$y_{n+3} = y_{n+2} + \frac{1}{24}h\,(9f_{n+3} + 19f_{n+2} - 5f_{n+1} - 9f_n)$$

is an implicit linear three-step method. ◇

There are systematic ways of generating linear multistep methods, but these constructions will not be discussed here. Instead, we turn our attention to the analysis of linear multistep methods and introduce the concepts of *(zero-) stability*, *consistency* and *convergence*. The significance of these properties cannot be overemphasised: the failure of any of the three will render the linear multistep method practically useless.

12.7 Zero-stability

As is clear from (12.35) we need k starting values, $y_0, \ldots, y_{k-1}$, before we can apply a linear k-step method to the initial value problem (12.1), (12.2): of these, y_0 is given by the initial condition (12.2), but the others,

[1] John Couch Adams (5 June 1819, Laneast, Cornwall, England – 21 January 1892, Cambridge, Cambridgeshire, England) was educated at St John's College in Cambridge. In 1841 while he was still an undergraduate, he began to study the irregularities of the motion of Uranus to discover whether these can be attributed to the action of an undiscovered planet. Four years later he gave accurate information about the position of the new planet (Neptune) to the director of the Cambridge Observatory. Adams made several other contributions to astronomy.

[2] F. Bashforth: *An Attempt to Test the Theories of Capillary Action by Comparing the Theoretical and Measured Forms of Drops of Fluid. With an Explanation of the Method of Integration in Constructing Tables Which Give the Theoretical Form of Such Drops, by J.C. Adams*, Cambridge University Press, 1883.

[3] F.R. Moulton: *New Methods in Exterior Ballistics*, University of Chicago Press, 1926.

$y_1, \dots, y_{k-1}$, have to be computed by other means: say, by using a suitable one-step method (*e.g.* a Runge–Kutta method). At any rate, the starting values will contain numerical errors and it is important to know how these will affect further approximations y_n, $n \geq k$, which are calculated by means of (12.35). Thus, we wish to consider the 'stability' of the numerical method with respect to 'small perturbations' in the starting conditions.

Definition 12.3 *A linear k-step method (for the ordinary differential equation $y' = f(x, y)$) is said to be* **zero-stable** *if there exists a constant K such that, for any two sequences (y_n) and (z_n) that have been generated by the same formulae but different starting values $y_0, y_1, \dots, y_{k-1}$ and $z_0, z_1, \dots, z_{k-1}$, respectively, we have*

$$|y_n - z_n| \leq K \max\{|y_0 - z_0|, |y_1 - z_1|, \dots, |y_{k-1} - z_{k-1}|\} \qquad (12.37)$$

for $x_n \leq X_M$, and as h tends to 0.

We shall prove later on that whether or not a method is zero-stable can be determined by merely considering its behaviour when applied to the trivial differential equation $y' = 0$, corresponding to (12.1) with $f(x, y) \equiv 0$; it is for this reason that the concept of stability formulated in Definition 12.3 is referred to as *zero*-stability. While Definition 12.3 is expressive in the sense that it conforms with the intuitive notion of stability whereby 'small perturbations at input give rise to small perturbations at output', it would be a very tedious exercise to verify the zero-stability of a linear multistep method using Definition 12.3 alone. Thus, we shall next formulate an algebraic equivalent of zero-stability, known as the **Root Condition**, which will simplify this task. Before doing so, however, we introduce some notation.

Given the linear k-step method (12.35) we consider its **first** and **second characteristic polynomials**, respectively

$$\rho(z) = \sum_{j=0}^{k} \alpha_j z^j,$$

$$\sigma(z) = \sum_{j=0}^{k} \beta_j z^j,$$

where, as before, we assume that

$$\alpha_k \neq 0, \quad \alpha_0^2 + \beta_0^2 \neq 0.$$

Before stating the main theorem of this section, we recall a classical result from the theory of kth-order linear recurrence relations.

Lemma 12.1 *Consider the kth-order homogeneous linear recurrence relation*

$$\alpha_k y_{n+k} + \cdots + \alpha_1 y_{n+1} + \alpha_0 y_n = 0\,, \qquad n = 0, 1, 2, \ldots\,, \tag{12.38}$$

with $\alpha_k \neq 0$, $\alpha_0 \neq 0$, $\alpha_j \in \mathbb{R}$, $j = 0, 1, \ldots, k$, and the corresponding **characteristic polynomial**

$$\rho(z) = \alpha_k z^k + \cdots + \alpha_1 z + \alpha_0\,.$$

Let z_r, $1 \leq r \leq \ell$, $\ell \leq k$, be the distinct roots of the polynomial ρ, and let $m_r \geq 1$ denote the multiplicity of z_r, with $m_1 + \cdots + m_\ell = k$. If a sequence (y_n) of complex numbers satisfies (12.38), then

$$y_n = \sum_{r=1}^{\ell} p_r(n) z_r^n\,, \qquad \text{for all } n \geq 0\,, \tag{12.39}$$

where $p_r(\,\cdot\,)$ is a polynomial in n of degree $m_r - 1$, $1 \leq r \leq \ell$. In particular, if all roots are simple, that is $m_r = 1$, $1 \leq r \leq k$, then the p_r, $r = 1, \ldots, k$, are constants.

Proof We give a sketch of the proof.[1] Let us first consider the case when all of the (distinct) roots $z_1, z_2, \ldots, z_k$ are simple. As, by assumption, $\alpha_0 \neq 0$, none of the roots is equal to 0. It is then easy to verify by direct substitution that, since $\rho(z_r) = 0$, $r = 1, 2, \ldots, k$, each of the sequences $(y_n) = (z_r^n)$, $r = 1, 2, \ldots, k$, satisfies (12.38).

In order to prove that any solution (y_n) of (12.38) can be expressed as a linear combination of the sequences $(z_1^n), (z_2^n), \ldots, (z_k^n)$, we first show that these k sequences are linearly independent. To do so, let us suppose that

$$C_1 z_1^n + C_2 z_2^n + \cdots + C_k z_k^n = 0\,, \qquad \text{for all } n = 0, 1, 2, \ldots.$$

Then, in particular,

$$\begin{aligned}
C_1 \quad + C_2 \quad + \cdots + C_k \quad &= 0\,,\\
C_1 z_1 + C_2 z_2 + \cdots + C_k z_k &= 0\,,\\
\cdots\cdots\cdots\cdots\cdots\cdots\cdots &\\
C_1 z_1^{k-1} + C_2 z_2^{k-1} + \cdots + C_k z_k^{k-1} &= 0\,.
\end{aligned}$$

[1] For details, see, for example, pp. 213–214 of P. Henrici, *Discrete Variable Methods in Ordinary Differential Equations*, Wiley, New York, 1962.

The matrix of this system of k simultaneous linear equations for the k unknowns $C_1, C_2, \ldots, C_k$ has the determinant

$$\mathcal{D} = \begin{vmatrix} 1 & 1 & \ldots & 1 \\ z_1 & z_2 & \ldots & z_k \\ \ldots & \ldots & \ldots & \ldots \\ z_1^{k-1} & z_2^{k-1} & \ldots & z_k^{k-1} \end{vmatrix},$$

known as the Vandermonde determinant, and $\mathcal{D} = \prod_{r<s}(z_s - z_r)$. Since the roots are distinct, $\mathcal{D} \neq 0$, so the matrix of the system is nonsingular. Therefore $C_1 = C_2 = \cdots = C_k = 0$ is the unique solution, which then means that the sequences $(z_1^n), (z_2^n), \ldots, (z_k^n)$ are linearly independent.

Now, suppose that (y_n) is any solution of (12.38); as $\mathcal{D} \neq 0$, there exists a unique set of k constants, $C_1, C_2, \ldots, C_k$, such that

$$y_m = C_1 z_1^m + C_2 z_2^m + \cdots + C_k z_k^m, \qquad m = 0, 1, \ldots, k-1. \quad (12.40)$$

Substituting these equalities into (12.38) for $n = 0$, we conclude that

$$\begin{aligned} 0 &= \alpha_k y_k + \alpha_{k-1}(C_1 z_1^{k-1} + \cdots + C_k z_k^{k-1}) + \cdots \\ &\quad + \alpha_0(C_1 z_1^0 + \cdots + C_k z_k^0) \\ &= \alpha_k y_k + C_1(\rho(z_1) - \alpha_k z_1^k) + \cdots + C_k(\rho(z_k) - \alpha_k z_k^k) \\ &= \alpha_k(y_k - (C_1 z_1^k + \cdots + C_k z_k^k)). \end{aligned}$$

As $\alpha_k \neq 0$, it follows that

$$y_k = C_1 z_1^k + \cdots + C_k z_k^k,$$

which, together with (12.40), proves (12.39) for $0 \le n \le k$ in the case of simple roots. Next, we select $n = 1$ in (12.38) and proceed in the same manner as in the case of $n = 0$ discussed above to show that (12.39) holds for $0 \le n \le k+1$. Continuing in the same way, we deduce by induction that in the case of simple roots (12.39) holds for all $n \ge 0$.

In the case when $\rho(z)$ has repeated roots, the proof is similar, except that instead of (z_r^n), $r = 1, 2, \ldots, n$, the following k sequences are used:

$$\left.\begin{array}{l} (z_r^n), \\ (n z_r^n), \\ \cdots\cdots\cdots \\ (n(n-1)\ldots(n-m_r+2) z_r^n), \quad r = 1, 2, \ldots, \ell. \end{array}\right\} \quad (12.41)$$

These can be shown to satisfy (12.38) by direct substitution on noting that $\rho(z_r) = \rho'(z_r) = \cdots = \rho^{(m_r-1)}(z_r) = 0$, given that z_r is a root of

$\rho(z)$ of multiplicity m_r, $r = 1, 2, \ldots, \ell$. The linear independence of the sequences (12.41) follows as before, except instead of $\prod_{r<s}(z_s - z_r)$, the value of the corresponding determinant is now

$$\mathcal{D}_1 = \prod_{1 \le r < s \le \ell} (z_r - z_s)^{m_r + m_s} \prod_{r=1}^{\ell} (m_r - 1)!!$$

where $0!! = 1$, $m!! = m!\,(m-1)! \ldots 1!$ for $m = 1, 2, \ldots$. As the roots $z_1, z_2, \ldots, z_\ell$ are distinct, we have that $\mathcal{D}_1 \neq 0$, and therefore the sequences (12.41) are linearly independent. The rest of the argument is identical as in the case of simple roots.[1] □

Now, we are ready to state the main result of this section.

Theorem 12.4 (Root Condition) *A linear multistep method is zero-stable for any initial value problem of the form (12.1), (12.2), where f satisfies the hypotheses of Picard's Theorem, if, and only if, all roots of the first characteristic polynomial of the method are inside the closed unit disc in the complex plane, with any which lie on the unit circle being simple.*

The algebraic condition contained in this theorem, namely that *the roots of the first characteristic polynomial lie in the closed unit disc and those on the unit circle are simple*, is often called the **Root Condition.**

Proof of theorem Necessity. Consider the method (12.35), applied to $y' = 0$:

$$\alpha_k y_{n+k} + \cdots + \alpha_1 y_{n+1} + \alpha_0 y_n = 0\,. \tag{12.42}$$

According to Lemma 12.1, every solution of this kth-order linear recurrence relation has the form

$$y_n = \sum_{r=1}^{\ell} p_r(n) z_r^n\,, \tag{12.43}$$

where z_r is a root, of multiplicity $m_r \geq 1$, of the first characteristic polynomial ρ of the method, and the polynomial p_r has degree $m_r - 1$, $1 \le r \le \ell$, $\ell \le k$. Clearly, if $|z_r| > 1$ for some r, then there are starting values $y_0, y_1, \ldots, y_{k-1}$ for which the corresponding solution grows like

[1] We warn the reader that in certain mathematical texts the notation $m!!$ is, instead, used to mean $m \cdot (m-2) \ldots 5 \cdot 3 \cdot 1$ for m odd and $m \cdot (m-2) \ldots 6 \cdot 4 \cdot 2$ for m even.

$|z_r|^n$, and if $|z_r| = 1$ and the multiplicity is $m_r > 1$, then there is a solution growing like $n^{m_r - 1}$. In either case there are solutions that grow unboundedly as $n \to \infty$, *i.e.*, as $h \to 0$ with nh fixed. Considering starting values $y_0, y_1, \ldots, y_{k-1}$ which give rise to such an unbounded solution (y_n), and starting values $z_0 = z_1 = \cdots = z_{k-1} = 0$ for which the corresponding solution of (12.42) is (z_n) with $z_n = 0$ for all n, we see that (12.37) cannot hold. To summarise, if the Root Condition is violated, then the method is not zero-stable.

Sufficiency. The proof that the Root Condition is sufficient for zero-stability is long and technical, and will be omitted here. For details, the interested reader is referred to Theorem 3.1 on page 353 of W. Gautschi, *Numerical Analysis: an Introduction*, Birkhäuser, Boston, MA, 1997. □

Example 12.5 *We shall explore the zero-stability of the methods from Example 12.4 using the Root Condition.*

(a) The Euler method and the implicit Euler method have first characteristic polynomial $\rho(z) = z - 1$ with simple root $z = 1$, so both methods are zero-stable. The same is true of the trapezium rule method.

(b) The Adams–Bashforth and Adams–Moulton methods considered in Example 12.4 have first characteristic polynomials, respectively, $\rho(z) = z^3(z-1)$ and $\rho(z) = z^2(z-1)$. These have multiple root $z = 0$ and simple root $z = 1$, and therefore both methods are zero-stable.

(c) The three-step method

$$\begin{aligned} 11y_{n+3} + 27y_{n+2} - 27y_{n+1} - 11y_n \\ = 3h\left(f_{n+3} + 9f_{n+2} + 9f_{n+1} + f_n\right) \end{aligned} \tag{12.44}$$

is *not* zero-stable. Indeed, the corresponding first characteristic polynomial $\rho(z) = 11z^3 + 27z^2 - 27z - 11$ has roots at $z_1 = 1$, $z_2 \approx -0.32$, $z_3 = -3.14$, so $|z_3| > 1$.

(d) The first characteristic polynomial of the three-step method

$$y_{n+3} + y_{n+2} - y_{n+1} - y_n = 2h(f_{n+2} + f_{n+1})$$

is $\rho(z) = z^3 + z^2 - z - 1 = (z+1)(z^2 - 1)$, which has roots $z_{1/2} = -1$, $z_3 = 1$. The first of these is a double root lying on the unit circle; therefore, the method is *not* zero-stable. ◇

12.8 Consistency

In this section we consider the accuracy of the linear k-step method (12.35). For this purpose, as in the case of one-step methods, we introduce the notion of truncation error. Thus, suppose that y is a solution to the ordinary differential equation (12.1). The truncation error of (12.35) is then defined as follows:

$$T_n = \frac{\sum_{j=0}^{k} [\alpha_j y(x_{n+j}) - h\beta_j f(x_{n+j}, y(x_{n+j}))]}{h\sum_{j=0}^{k} \beta_j}. \tag{12.45}$$

Of course, the definition requires implicitly that $\sigma(1) = \sum_{j=0}^{k} \beta_j \neq 0$. Again, as in the case of one-step methods, the truncation error can be thought of as the residual that is obtained by inserting the solution of the differential equation into the formula (12.35) and scaling this residual appropriately (in this case dividing through by $h\sum_{j=0}^{k} \beta_j$), so that T_n resembles $y' - f(x, y(x))$.

Definition 12.4 *The numerical method (12.35) is said to be* **consistent** *with the differential equation (12.1) if the truncation error defined by (12.45) is such that for any $\varepsilon > 0$ there exists an $h(\varepsilon)$ for which*

$$|T_n| < \varepsilon \quad \text{for } 0 < h < h(\varepsilon),$$

and any $k+1$ points $(x_n, y(x_n)), \dots, (x_{n+k}, y(x_{n+k}))$ on any solution curve in D of the initial value problem (12.1), (12.2).

Now, let us suppose that the solution to the differential equation is sufficiently smooth, and let us expand the expressions $y(x_{n+j})$ and $f(x_{n+j}, y(x_{n+j})) = y'(x_{n+j})$ into Taylor series about the point x_n. On substituting these expansions into the numerator in (12.45) we obtain

$$T_n = \frac{1}{h\sigma(1)} \left[C_0 y(x_n) + C_1 h y'(x_n) + C_2 h^2 y''(x_n) + \cdots\right] \tag{12.46}$$

where

$$\left.\begin{aligned}
C_0 &= \sum_{j=0}^{k} \alpha_j, \\
C_1 &= \sum_{j=1}^{k} j\alpha_j - \sum_{j=0}^{k} \beta_j, \\
C_2 &= \sum_{j=1}^{k} \frac{j^2}{2!}\alpha_j - \sum_{j=1}^{k} j\beta_j, \\
&\dots \\
C_q &= \sum_{j=1}^{k} \frac{j^q}{q!}\alpha_j - \sum_{j=1}^{k} \frac{j^{q-1}}{(q-1)!}\beta_j.
\end{aligned}\right\} \tag{12.47}$$

For consistency we need that, as $h \to 0$ and $n \to \infty$ with $x_n \to x \in [x_0, X_M]$, the truncation error T_n tends to 0. This requires that $C_0 = 0$ and $C_1 = 0$ in (12.46). In terms of the characteristic polynomials this consistency requirement can be restated in compact form as

$$\rho(1) = 0 \quad \text{and} \quad \rho'(1) = \sigma(1)\,(\neq 0)\,.$$

Let us observe that, according to this condition, if a linear multistep method is consistent, then it has a *simple* root on the unit circle at $z = 1$; thus, the Root Condition is not violated by this root.

Definition 12.5 *The numerical method (12.35) is said to have* **order of accuracy** p, *if p is the largest positive integer such that, for any sufficiently smooth solution curve in D of the initial value problem (12.1), (12.2), there exist constants K and h_0 such that*

$$|T_n| \leq Kh^p \quad \textit{for } 0 < h \leq h_0\,,$$

for any $k + 1$ points $(x_n, y(x_n)), \ldots, (x_{n+k}, y(x_{n+k}))$ on the solution curve.

Thus, we deduce from (12.46) that the method is of order of accuracy p if, and only if,

$$C_0 = C_1 = \cdots = C_p = 0 \quad \text{and} \quad C_{p+1} \neq 0\,.$$

In this case,

$$T_n = \frac{C_{p+1}}{\sigma(1)} h^p y^{(p+1)}(x_n) + \mathcal{O}(h^{p+1})\,.$$

The number $C_{p+1}/\sigma(1)$ is called the **error constant** of the method.

Example 12.6 *Let us determine all values of the real parameter b, $b \neq 0$, for which the linear multistep method*

$$y_{n+3} + (2b - 3)(y_{n+2} - y_{n+1}) - y_n = hb(f_{n+2} + f_{n+1})$$

is zero-stable. We shall show that there exists a value of b for which the order of the method is 4, and that if the method is zero-stable for some value of b, then its order cannot exceed 2.

According to the Root Condition, this linear multistep method is zero-stable if, and only if, all roots of its first characteristic polynomial

$$\rho(z) = z^3 + (2b-3)(z^2 - z) - 1$$

belong to the closed unit disc, and those on the unit circle are simple.

Clearly, $\rho(1) = 0$; upon dividing $\rho(z)$ by $z-1$ we see that $\rho(z)$ can be written in the following factorised form:

$$\rho(z) = (z-1)\rho_1(z), \qquad \text{where} \quad \rho_1(z) = z^2 - 2(1-b)z + 1.$$

Thus, the method is zero-stable if, and only if, all roots of the polynomial $\rho_1(z)$ belong to the closed unit disc, and those on the unit circle are simple and differ from 1. Suppose that the method is zero-stable. It then follows that $b \neq 0$ and $b \neq 2$, since these values of b correspond to double roots of $\rho_1(z)$ on the unit circle, respectively, $z = 1$ and $z = -1$. Further, since the product of the two roots of $\rho_1(z)$ is equal to 1, both have modulus less than or equal to 1, and neither of them is equal to ± 1, it follows that they must both be strictly complex; hence the discriminant of the quadratic polynomial $\rho_1(z)$ must be negative. That is, $4(1-b)^2 - 4 < 0$. In other words, $b \in (0, 2)$.

Conversely, suppose that $b \in (0, 2)$. Then, the roots of $\rho(z)$ are

$$z_1 = 1, \qquad z_{2/3} = 1 - b + \imath\sqrt{1-(b-1)^2}.$$

Since $|z_{2/3}| = 1$, $z_{2/3} \neq 1$ and $z_2 \neq z_3$, all roots of $\rho(z)$ lie on the unit circle and they are simple. Hence the method is zero-stable. To summarise, the method is zero-stable if, and only if, $b \in (0, 2)$.

In order to analyse the order of accuracy of the method, we note that, upon Taylor series expansion, its truncation error can be written in the form

$$\begin{aligned} T_n &= \frac{1}{\sigma(1)}\left[\left(1 - \frac{b}{6}\right)h^2 y'''(x_n) + \frac{1}{4}(6-b)h^3 y^{\mathrm{iv}}(x_n)\right. \\ &\qquad \left. + \frac{1}{120}(150 - 23b)h^4 y^{\mathrm{v}}(x_n) + \mathcal{O}(h^5)\right], \end{aligned}$$

where $\sigma(1) = 2b \neq 0$. If $b = 6$, then $T_n = \mathcal{O}(h^4)$ and so the method is of order 4. As $b = 6$ does not belong to the interval $(0, 2)$, we deduce that the method is *not* zero-stable for $b = 6$.

Since zero-stability requires $b \in (0, 2)$, in which case $1 - \frac{b}{6} \neq 0$, it follows that if the method is zero-stable, then $T_n = \mathcal{O}(h^2)$. ◇

12.9 Dahlquist's theorems

An important result connecting the concepts of zero-stability, consistency and convergence of a linear multistep method was proved by the Swedish mathematician Germund Dahlquist.

Theorem 12.5 (Dahlquist's Equivalence Theorem) *For a linear k-step method that is consistent with the ordinary differential equation (12.1) where f is assumed to satisfy a Lipschitz condition, and with consistent starting values,*[1] *zero-stability is necessary and sufficient for convergence. Moreover if the solution y has continuous derivative of order $p+1$ and truncation error $\mathcal{O}(h^p)$, then the global error of the method, $e_n = y(x_n) - y_n$, is also $\mathcal{O}(h^p)$.*

The proof of this result is long and technical; for details of the argument, see Theorem 6.3.4 on page 357 of W. Gautschi, *Numerical Analysis: an Introduction*, Birkhäuser, Boston, MA, 1997, or Theorem 5.10 on page 244 of P. Henrici, *Discrete Variable Methods in Ordinary Differential Equations*, Wiley, New York, 1962.

By virtue of Dahlquist's theorem, if a linear multistep method is not zero-stable its global error cannot be made arbitrarily small by taking the mesh size h sufficiently small for any sufficiently accurate initial data. In fact, if the Root Condition is violated, then there exists a solution to the linear multistep method which will grow by an arbitrarily large factor in a fixed interval of x, however accurate the starting conditions are. This result highlights the importance of the concept of zero-stability and indicates its relevance in practical computations.

A second theorem by Dahlquist imposes a restriction on the order of accuracy of a zero-stable linear multistep method.

Theorem 12.6 (Dahlquist's Barrier Theorem) *The order of accuracy of a zero-stable k-step method cannot exceed $k+1$ if k is odd, or $k+2$ if k is even.*

A proof of this result will be found in Section 4.2 of Gautschi's book or in Section 5.2-8 of Henrici's book, cited above.

Theorem 12.6 makes it very difficult to choose a 'best' multistep method of a given order. Suppose, for example, that we consider five-step methods. The general five-step method involves 12 parameters, of

[1] That is, with starting values $y_j = \eta_j \equiv \eta_j(h)$, $j = 0, \ldots, k-1$, which all converge to the exact initial value y_0, as $h \to 0$.

which 11 are independent: the method is obviously unaffected by multiplying all the parameters by a nonzero constant. Now it would be possible to construct a five-step method of order 10, by solving the 11 equations of the form $C_q = 0$, $q = 0, 1, \ldots, 10$, where C_q is given in (12.47). But the Barrier Theorem states that this method would not be zero-stable, and the order of a zero-stable five-step method cannot exceed 6. There is a family of stable five-step methods of order 6, involving 4 free parameters, and there is no obvious way of deciding whether any one of these methods is better than the others.

Example 12.7 (i) *The Barrier Theorem says that when $k = 1$ the order of accuracy of a zero-stable method cannot exceed 2. The trapezium rule method has order 2, and is zero-stable.*

(ii) *The two-step method*

$$y_{n+2} - y_n = h(\tfrac{1}{3} f_{n+2} + \tfrac{4}{3} f_{n+1} + \tfrac{1}{3} f_n)$$

is zero-stable, as the roots of the first characteristic polynomial, $\rho(z) = z^2 - 1$, are 1 and -1. A simple calculation shows that its order of accuracy is 4; by the Barrier Theorem, this is the highest order which could be achieved by a two-step method.

(iii) *The three-step method*

$$\begin{aligned} 11y_{n+3} + 27y_{n+2} - 27y_{n+1} - 11y_n & \\ = 3h\,(f_{n+3} + 9f_{n+2} + 9f_{n+1} + f_n) & \end{aligned}$$

has order 6. The Barrier Theorem therefore implies that this method is not zero-stable. We have already shown this in Example 12.5(c) using the Root Condition.

It is found that all the zero-stable k-step methods of highest possible order are *implicit*, with β_k nonzero.

12.10 Systems of equations

In this section we discuss the application of numerical methods to simultaneous systems of differential equations, which we shall write in the form

$$\frac{\mathrm{d}\boldsymbol{y}}{\mathrm{d}x} = \boldsymbol{f}(x, \boldsymbol{y})\,.$$

Here $\boldsymbol{y}$ is an m-component vector function of x, and $\boldsymbol{f}$ is an m-component vector function of the independent variable x and the vector variable $\boldsymbol{y}$. In component form the system becomes

$$\frac{\mathrm{d}y_j}{\mathrm{d}x} = f_j(x, y_1, \ldots, y_m), \qquad j = 1, 2, \ldots, m.$$

The system comprises m simultaneous differential equations. To single out a unique solution we need m side conditions, and we shall suppose that all these conditions are given at the same value of x, and have the form

$$\boldsymbol{y}(x_0) = \boldsymbol{y}_0,$$

or, in component form,

$$y_j(x_0) = y_{j,0}, \qquad j = 1, 2, \ldots, m,$$

where the values of $y_{j,0}$ are given. This is called an initial value problem for a system of ordinary differential equations; we may also require a solution of the system on an interval $[a, b]$, with r conditions given at one end of the interval and $m - r$ conditions at the other end. This constitutes a boundary value problem, and requires different numerical methods which are considered in the next chapter.

All the numerical methods which we have discussed apply without change to systems of differential equations; it is only necessary to realise that we are dealing with vectors. For example, the first stage of the classical Runge–Kutta method (12.33) becomes

$$\boldsymbol{k}_1 = \boldsymbol{f}(x_n, \boldsymbol{y}_n);$$

we must evaluate all the elements of the vector $\boldsymbol{k}_1$ before proceeding to the next stage to calculate $\boldsymbol{k}_2$, and so on.

The most important difference which arises in dealing with a system of differential equations is in the practical use of an *implicit* multistep method. As we have seen, this almost always requires an iterative method for the solution of an equation to determine y_{n+1}. Applying such a method to a system of differential equations now involves the solution of a system of equations, which will usually be nonlinear, to determine the elements of the vector $\boldsymbol{y}_{n+1}$. In real-life problems it is quite common to deal with systems of several hundred differential equations, and it then becomes very important to be sure that the improved efficiency of the implicit method justifies the very considerable extra work in each step of the process.

We shall not discuss the extension of our earlier analysis to deal with systems of differential equations; in almost all cases we simply need to introduce vector notation, and replace the absolute value of a number by the norm of a vector. For example, in the proof of Theorem 12.2, (12.17) becomes

$$\|\boldsymbol{e}_{n+1}\| \leq \|\boldsymbol{e}_n\| + hL_\Phi\|\boldsymbol{e}_n\| + h\|\mathbf{T}_n\|\,, \qquad n = 0, 1, \ldots, N-1\,,$$

where $\|\cdot\|$ is any norm on $\mathbb{R}^m$, with obvious definitions of the global error $\boldsymbol{e}_n$ and the truncation error $\mathbf{T}_n$. Similarly, Picard's Theorem and its proof, discussed at the beginning of the chapter in the case of a single ordinary differential equation, can be easily extended to an m-component system of differential equations by replacing the absolute value sign with a vector norm on $\mathbb{R}^m$ throughout.

12.11 Stiff systems

The phenomenon of stiffness usually appears only in a system of differential equations, but we begin by discussing an almost trivial example of a single equation,

$$y' = \lambda y\,, \qquad y(0) = y_0\,,$$

where λ is a constant. The solution of this equation is evidently $y(x) = y_0 \exp(\lambda x)$. When $\lambda < 0$ the absolute value of the solution is exponentially decreasing, so it is sensible to require that the absolute value of our numerical solution also decreases. It is very easy to give expressions for the result of a numerical solution using Euler's method and the implicit Euler method (12.36). They are, respectively,

$$y_n^{\mathrm{E}} = (1 + h\lambda)^n y_0\,, \qquad y_n^{\mathrm{I}} = (1 - h\lambda)^{-n} y_0\,.$$

When $\lambda < 0$ and $h > 0$, we have $(1 - h\lambda) > 1$; therefore, the sequence $(|y_n^{\mathrm{I}}|)$ decreases monotonically with increasing n. On the other hand, for $\lambda < 0$ and $h > 0$,

$$|1 + h\lambda| < 1 \qquad \text{if, and only if,} \qquad 0 < h|\lambda| < 2\,.$$

This gives the restriction $h|\lambda| < 2$ on the size of h for which the sequence $(|y_n^{\mathrm{E}}|)$ decreases monotonically; if h exceeds $2/|\lambda|$, the numerical solution obtained by Euler's method will oscillate with increasing magnitude with increasing n and fixed $h > 0$, instead of converging to zero as $n \to \infty$.

We now consider the same two methods applied to the initial value problem for a system of differential equations of the form

$$\boldsymbol{y}' = A\boldsymbol{y}\,, \qquad \boldsymbol{y}(0) = \boldsymbol{y}_0\,,$$

where A is a square matrix of order m, each of whose elements is a constant. For simplicity we assume that the eigenvalues of A are distinct, so there exists a matrix M such that $MAM^{-1} = \Lambda$ is a diagonal matrix. The system of differential equations is therefore equivalent to

$$\boldsymbol{z}' = \Lambda \boldsymbol{z}\,, \qquad \boldsymbol{z}(0) = \boldsymbol{z}_0 = M\boldsymbol{y}_0\,,$$

with $\boldsymbol{z} = M\boldsymbol{y}$. In this form the system reduces to a set of m independent equations, whose solutions are

$$z_j(x) = z_j(0)\exp(\lambda_j x)\,, \qquad j = 1, 2, \ldots, m\,,$$

where the numbers λ_j, $j = 1, 2, \ldots, m$, are the diagonal elements of the matrix Λ, and are therefore the eigenvalues of A. In particular, if all the λ_j, $j = 1, 2, \ldots, m$, are real and negative, then $\lim_{x\to+\infty} \|\boldsymbol{z}(x)\| = 0$ and since

$$\|\boldsymbol{y}(x)\| = \|M^{-1}\,\boldsymbol{z}(x)\| \le \|M^{-1}\|\,\|\boldsymbol{z}(x)\|\,,$$

also

$$\lim_{x\to+\infty} \|\boldsymbol{y}(x)\| = 0\,.$$

Here $\|\cdot\|$ is any norm on $\mathbb{R}^m$, and the norm on M^{-1} is the associated subordinate matrix norm defined in Chapter 2.

In just the same way, Euler's method applied to the system gives

$$\boldsymbol{y}_{n+1} = (I + hA)\boldsymbol{y}_n\,,$$

which leads to

$$\begin{aligned} \boldsymbol{z}_{n+1} &= M\boldsymbol{y}_{n+1} = M(I + hA)\boldsymbol{y}_n \\ &= M(I + hA)M^{-1}\boldsymbol{z}_n = (I + h\Lambda)\boldsymbol{z}_n\,. \end{aligned}$$

Thus, the result $\boldsymbol{y}_{n+1}$ of Euler's method applied to the initial value problem $\boldsymbol{y}' = A\boldsymbol{y}$, $\boldsymbol{y}(0) = \boldsymbol{y}_0$, is exactly the same as $M^{-1}\boldsymbol{z}_{n+1}$, where $\boldsymbol{z}_{n+1}$ is the result of applying Euler's method to the transformed problem $\boldsymbol{z}' = \Lambda\boldsymbol{z}$, $\boldsymbol{z}(0) = M\boldsymbol{y}_0$; an analogous remark applies to the use of the implicit Euler method.

Suppose that all the eigenvalues λ_j, $j = 1, 2, \ldots, m$, are real and negative. Then, in order to ensure that, for a fixed positive value of h,

$$\lim_{n\to\infty} \|\boldsymbol{y}_n\| = 0\,,$$

we must require that, for Euler's method, $h|\lambda_j| < 2$, $j = 1, 2, \ldots, m$; for the implicit Euler method no such condition is required. The importance of this fact is highlighted by a numerical example.

We consider the system where A is the 2×2 matrix

$$A = \begin{pmatrix} -8003 & 1999 \\ 23988 & -6004 \end{pmatrix},$$

and the initial condition is

$$\boldsymbol{y}(0) = \begin{pmatrix} 1 \\ 4 \end{pmatrix}.$$

The eigenvalues of A are $\lambda_1 = -7$ and $\lambda_2 = -14000$; the solution of the problem is

$$\boldsymbol{y}(x) = \begin{pmatrix} \mathrm{e}^{-7x} \\ 4\mathrm{e}^{-7x} \end{pmatrix}.$$

Clearly, $\lim_{x\to+\infty} \|\boldsymbol{y}(x)\| = 0$.

The numerical solution uses 12 steps of size $h = 0.004$; the results are shown in Table 12.1. The second column gives the first component of the solution, $y_1(x) = \mathrm{e}^{-7x}$, the third column shows the result from the implicit Euler method, and the last gives the result of the standard Euler method. The last column is a dramatic example of what happens when the step size h is too large; in this case $h|\lambda_2| = 56$. The numerical values given by the implicit Euler method have an error of a few units in the third decimal digit; to get the same accuracy from the Euler method would require a step size about 30 times smaller, and about 30 times as much work.

It is clear that the difficulty in the numerical example is caused by the size of the eigenvalue -14000, but what is important is its size relative to the other eigenvalue. The special constant-coefficient system $y' = Ay$ is said to be **stiff** if all the eigenvalues of A have negative real parts, and if the ratio of the largest of the real parts to the smallest of the real parts is large. Most practical problems are nonlinear, and for such problems it is quite difficult to define precisely what is meant by stiffness.[1] To begin with we may replace the system by a linearised approximation, the first terms of an expansion

$$\boldsymbol{y}'(x) = \boldsymbol{y}'(x_n) + \frac{\partial \boldsymbol{f}}{\partial x}(x_n, y(x_n))(x - x_n) + J(x_n)(\boldsymbol{y}(x) - \boldsymbol{y}(x_n)) + \cdots$$

[1] Indeed, even in the case of variable-coefficient linear systems of differential equations, stiffness can be defined in several (nonequivalent) ways; for a discussion of the pros and cons of the various definitions, we refer to Section 6.2 of J.D. Lambert, *Numerical Methods for Ordinary Differential Systems*, Wiley, Chichester, 1991.

Table 12.1. *The use of Euler's method and the implicit Euler method to solve a stiff system.*

x	$y_1(x)$	Implicit Euler	Euler
0.000	1.000	1.000	1.000
0.004	0.972	0.973	0.972
0.008	0.946	0.946	0.945
0.012	0.919	0.920	0.918
0.016	0.894	0.895	0.893
0.020	0.869	0.871	0.868
0.024	0.845	0.847	0.843
0.028	0.822	0.824	0.820
0.032	0.799	0.802	0.794
0.036	0.777	0.780	0.941
0.040	0.756	0.759	−8.430
0.044	0.735	0.738	505.769
0.048	0.715	0.718	−27776.357

where J is the Jacobian matrix of the function $\boldsymbol{f}$, whose (i,j)-entry is

$$(J(x_n))_{ij} = \frac{\partial f_i}{\partial y_j}(x_n, \boldsymbol{y}(x_n)) .$$

We can then think of the system as being stiff if the eigenvalues of the matrix $J(x_n)$ have negative real parts and if the ratio of the largest of the real parts to the smallest is large. Although this gives some indication of the sort of problems which may cause difficulty, the behaviour of nonlinear systems is much more complicated than this. It is not difficult to construct examples in which all the eigenvalues of the Jacobian matrix have negative real parts, yet the norm of the solution of the differential equation is exponentially increasing as $x \to +\infty$.

Even though any classification of nonlinear systems of differential equations into stiff and nonstiff, based only on monitoring the eigenvalues of $J(x_n)$, is somewhat simplistic, it does highlight some of the key difficulties. Stiff systems of differential equations arise in many application areas, a typical one being chemical engineering. For example, in parts of an oil refinery there may be a large number of substances undergoing chemical reactions with widely different reaction rates. These reaction rates correspond to the eigenvalues of the Jacobian matrix, and it is not unusual to find the ratio of the largest of the real parts to the smallest to be in excess of 10^{10}. For such problems it is essential to find a numerical method which imposes no restriction on the step size;

Euler's method, which might require the restriction $10^{10}h < 2$, would evidently be quite useless.

Application of the linear multistep method

$$\sum_{j=0}^{k} \alpha_j y_{n+j} = h \sum_{j=0}^{k} \beta_j f(x_{n+j}, y_{n+j})$$

to the equation $y' = \lambda y$ leads to the kth-order linear recurrence relation

$$\sum_{j=0}^{k} (\alpha_j - \lambda h \beta_j) y_{n+j} = 0 \,. \tag{12.48}$$

The characteristic polynomial of the linear recurrence relation (12.48) is

$$\pi(z; \lambda h) = \sum_{j=0}^{k} (\alpha_j - \lambda h \beta_j) z^j \,.$$

Alternatively, we can write this in terms of the first and second characteristic polynomials of the linear multistep method as

$$\pi(z; \lambda h) = \rho(z) - \lambda h \sigma(z) \,.$$

In the present context, the polynomial $\pi(\,\cdot\,; \lambda h)$ is usually referred to as the **stability polynomial** of the linear multistep method. According to Lemma 12.1, the general solution of the recurrence relation (12.48) can be expressed in terms of the distinct roots z_r, $1 \le r \le \ell$, $\ell \le k$, of $\pi(\,\cdot\,; \lambda h)$. Letting m_r denote the multiplicity of the root z_r, $1 \le r \le \ell$, $m_1 + \cdots + m_\ell = k$, we have that

$$y_n = \sum_{r=1}^{\ell} p_r(n) z_r^n \,, \tag{12.49}$$

where the polynomial $p_r(\,\cdot\,)$ has degree $m_r - 1$, $1 \le r \le \ell$.

Clearly, the roots z_r are functions of λh. For $\lambda \in \mathbb{C}$, with $\mathrm{Re}(\lambda) < 0$, the solution of the model problem

$$y' = \lambda y \,, \qquad y(0) = y_0 \,,$$

converges in $\mathbb{C}$ to 0 as $x \to \infty$. Thus, we would like to ensure that, when a linear multistep method is applied to this problem, the step size h can be chosen so that the resulting sequence of numerical approximations (y_n) exhibits an analogous behaviour as $n \to \infty$, that is, $\lim_{n\to\infty} y_n = 0$. By virtue of (12.49), this can be guaranteed by demanding that each root $z_r = z_r(\lambda h)$ has modulus less then 1.

Definition 12.6 *A linear multistep method is said to be* **absolutely stable** *for a given value of* λh *if each root* $z_r = z_r(\lambda h)$ *of the associated stability polynomial* $\pi(\,\cdot\,;\lambda h)$ *satisfies* $|z_r(\lambda h)| < 1$.

Our aim is, therefore, to single out those values of λh for which the linear multistep method is absolutely stable.

Definition 12.7 *The* **region of absolute stability** *of a linear multistep method is the set of all points* λh *in the complex plane for which the method is absolutely stable.*

Ideally, the region of absolute stability of the method should admit all values of λ, $\mathrm{Re}(\lambda) < 0$, so as to ensure that there is no limitation on the size of h, however large $|\lambda|$ may be. This leads us to the next definition.

Definition 12.8 *A linear multistep method is said to be* **A-stable** *if its region of absolute stability contains the negative (left) complex half-plane.*

Unfortunately, the condition of A-stability is extremely demanding. Dahlquist[1] has shown the following results which are collectively known as his **Second Barrier Theorem**:

(i) No *explicit* linear multistep method is A-stable;
(ii) No A-stable linear multistep method can have order greater than 2.
(iii) The second-order A-stable linear multistep method with the smallest error constant is the trapezium rule method.

The trapezium rule method is a one-step method, so the associated stability polynomial has only one root, given by

$$z = \frac{1 + \frac{1}{2}\lambda h}{1 - \frac{1}{2}\lambda h}.$$

Evidently $|z| < 1$ if $\mathrm{Re}(h\lambda) = h\,\mathrm{Re}(\lambda) < 0$, so the trapezium rule method is indeed A-stable.

To construct useful methods of higher order we need to relax the condition of A-stability by requiring that the region of absolute stability should include a large part of the negative half-plane, and certainly that it contains the whole of the negative real axis.

[1] G. Dahlquist, A special stability problem for linear multistep methods, *BIT* **3**, 27–43, 1963.

The most efficient methods of this kind in current use are the **Backward Differentiation Formulae**, or BDF methods. These are the linear multistep methods (12.35) in which $\beta_j = 0$, $0 \le j \le k-1$, $k \ge 1$, and $\beta_k \ne 0$. Thus,

$$\alpha_k y_{n+k} + \cdots + \alpha_0 y_n = h\beta_k f_{n+k} \,.$$

The coefficients are obtained by requiring that the order of accuracy of the method is as high as possible, *i.e.*, by making the coefficients C_j zero in (12.47) for $j = 0, 1, \ldots, k$. For $k = 1$ this yields the implicit Euler method (BDF1), whose order of accuracy is, of course, 1; the method is A-stable. The choice of $k = 6$ results in the sixth-order, six-step BDF method (BDF6):

$$\begin{aligned} 147y_{n+6} - 360y_{n+5} + 450y_{n+4} - 400y_{n+3} + 225y_{n+2} - 72y_{n+1} + 10y_n \\ = 60hf_{n+6} \,. \end{aligned} \tag{12.50}$$

Although the method (12.50) is not A-stable, its region of absolute stability includes the whole of the negative real axis (see Figure 12.5). For the intermediate values, $k = 2, 3, 4, 5$, we have the following kth-order, k-step BDF methods, respectively:

$$\begin{aligned} 3y_{n+2} - 4y_{n+1} + y_n &= 2hf_{n+2} \,, \\ 11y_{n+3} - 18y_{n+2} + 9y_{n+1} - 2y_n &= 6hf_{n+3} \,, \\ 25y_{n+4} - 48y_{n+3} + 36y_{n+2} - 16y_{n+1} + 3y_n &= 12hf_{n+4} \,, \\ 137y_{n+5} - 300y_{n+4} + 300y_{n+3} - 200y_{n+2} + 75y_{n+1} - 12y_n &= 60hf_{n+5} \,, \end{aligned}$$

referred to as BDF2, BDF3, BDF4 and BDF5. Their regions of absolute stability are also shown in Figure 12.5. In each case the region of absolute stability includes the negative real axis. Higher-order methods of this type cannot be used, as all BDF methods, with $k > 6$, are zero-unstable.

12.12 Implicit Runge–Kutta methods

For Runge–Kutta methods absolute stability is defined in much the same way as for linear multistep methods; *i.e.*, by applying the method in question to the model problem $y' = \lambda y$, $y(0) = y_0$, $\lambda \in \mathbb{C}$, $\mathrm{Re}(\lambda) < 0$, and demanding that the resulting sequence (y_n) converges to 0 as $n \to \infty$, with $h\lambda$ held fixed. The set of all values of $h\lambda$ in the complex plane for which the method is absolutely stable is called the region of absolute stability of the Runge–Kutta method.

Classical Runge–Kutta methods are explicit, and are unsuitable for

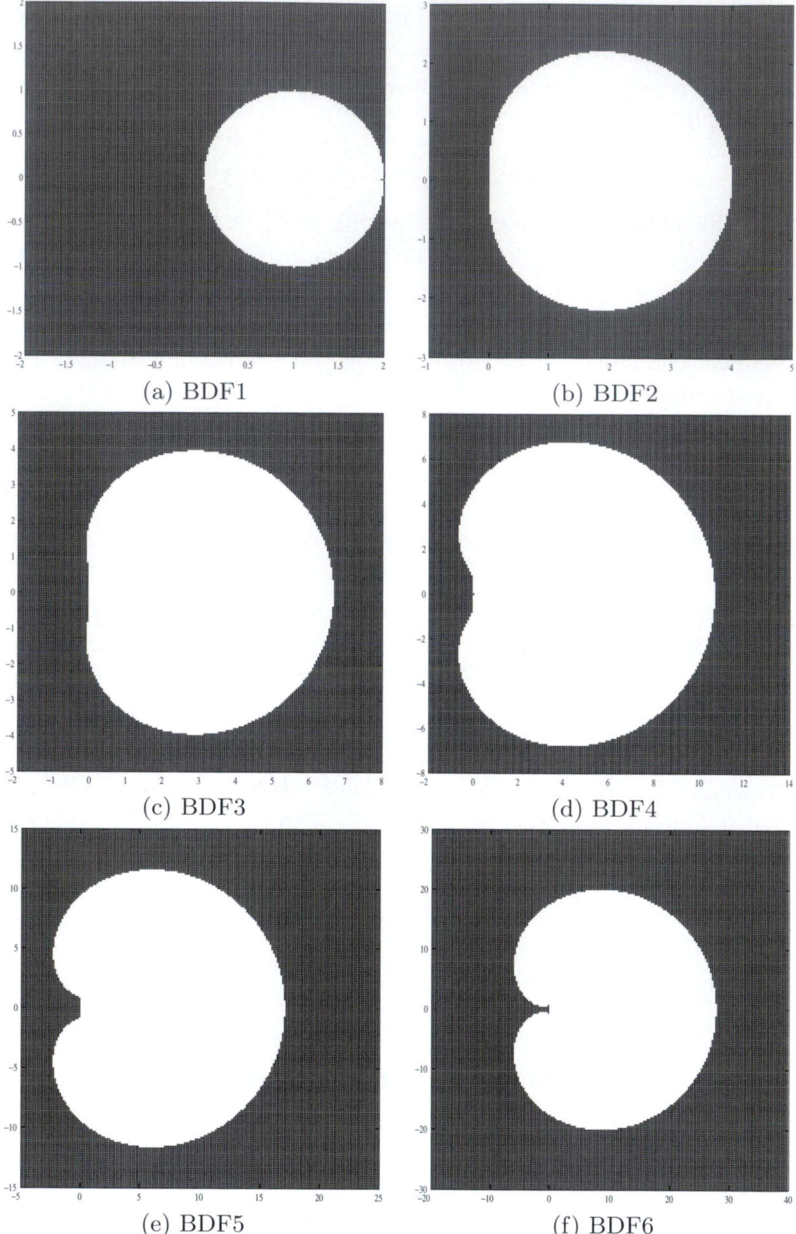

(a) BDF1 (b) BDF2

(c) BDF3 (d) BDF4

(e) BDF5 (f) BDF6

Fig. 12.5. Absolute stability regions in the complex plane for k-step Backward Differentiation Formulae, $k = 1, 2, \ldots, 6$. In each case the region of absolute stability is the set of points in the complex plane outside the white region. In each case, the region of absolute stability contains the whole of the negative real axis.

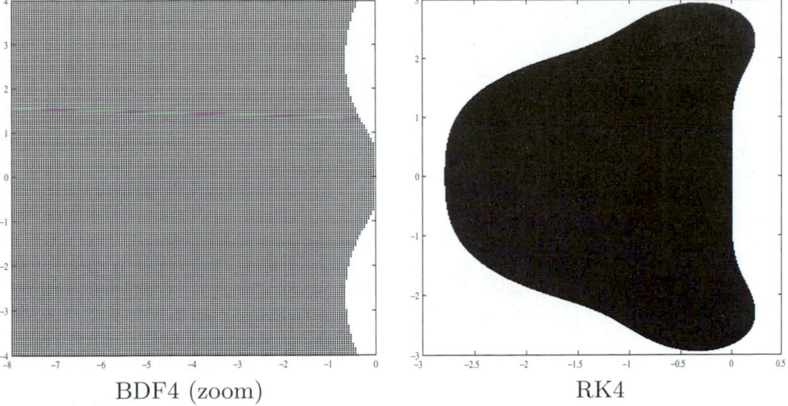

BDF4 (zoom) RK4

Fig. 12.6. The dark chequered region in the figure on the left indicates part of the absolute stability region in the complex plane for the four-step, fourth-order Backward Differentiation Formula, BDF4 (zoom into Figure 12.5(d)); here we only show the section of the region of absolute stability for BDF4 which lies in the rectangle $-8 < \mathrm{Re}(\lambda h) < 0$ and $-4 < \mathrm{Im}(\lambda h) < 4$, with $\mathrm{Re}(\lambda) < 0$, $h > 0$. The dark region in the figure on the right shows the region of absolute stability for the classical explicit fourth-order Runge–Kutta method, RK4. For BDF4, the region of absolute stability includes the whole of the negative real axis; clearly, this is not the case for RK4.

stiff systems because of their small region of absolute stability. Figure 12.6 depicts the region of absolute stability of the classical fourth-order Runge–Kutta method, together with that of the fourth-order Backward Differentiation Formula, BDF4. The contrast is striking: while the region of absolute stability of BDF4 includes most of the negative half-plane and, in particular, all of the negative real axis, for RK4 the region of absolute stability is bounded[1] (for example, along the negative real axis it does not extend to the left of, approximately, -2.8).

Motivated by the fact that BDF methods are implicit, we now go on to introduce *implicit* Runge–Kutta methods, which can also have a large region of absolute stability.

The general s-stage Runge–Kutta method is written

$$y_{n+1} = y_n + h \sum_{i=1}^{s} b_i k_i \,,$$

[1] This is not a peculiarity of RK4. It can be shown that every explicit Runge–Kutta method has bounded region of absolute stability; see, for example, Section 5.12, in J.D. Lambert's book, cited in the previous section.

where

$$k_i = f(x_n + hc_i, y_n + h\sum_{j=1}^{s} a_{ij}k_j)\,, \qquad 1 \le i \le s\,. \tag{12.51}$$

It is convenient to display the coefficients in a **Butcher tableau**

$$\begin{array}{c||ccc} c_1 & a_{11} & \dots & a_{1s} \\ \dots & \dots & \dots & \dots \\ c_s & a_{s1} & \dots & a_{ss} \\ \hline\hline & b_1 & \dots & b_s \end{array}$$

The method is then defined by the matrix $A = (a_{ij}) \in \mathbb{R}^{s\times s}$, of order s, and the two vectors $\boldsymbol{b} = (b_1, \dots, b_s)^{\mathrm{T}} \in \mathbb{R}^s$ and $\boldsymbol{c} = (c_1, \dots, c_s)^{\mathrm{T}} \in \mathbb{R}^s$. For example, the classical four-stage Runge–Kutta method is defined by the tableau

$$\begin{array}{c||cccc} 0 & & & & \\ \frac{1}{2} & \frac{1}{2} & & & \\ \frac{1}{2} & 0 & \frac{1}{2} & & \\ 1 & 0 & 0 & 1 & \\ \hline\hline & \frac{1}{6} & \frac{2}{3} & \frac{2}{3} & \frac{1}{6} \end{array}$$

The 4×4 array representing the matrix A for this method, displayed in the upper right quadrant of the tableau, follows the usual notational convention that zero elements after the last nonzero element in each row of the matrix A are omitted.

This is an explicit method, shown by the fact that the matrix A is *strictly lower triangular*, with $a_{ij} = 0$ when $1 \le i \le j \le 4$. Each value k_i can therefore be calculated in sequence, all the quantities on the right-hand side of (12.51) being known.

It is not difficult to construct s-stage implicit methods which are A-stable. For example, this can be done by choosing the coefficients c_i and b_i to be the quadrature points and weights respectively in the Gauss quadrature formula for the evaluation of

$$\int_0^1 g(x)\mathrm{d}x \approx \sum_{i=1}^{s} b_i g(c_i)\,.$$

The numbers a_{ij} can then be chosen so that the method has order $2s$, and is A-stable.

For example, the array

$$\begin{array}{c||cc}
\frac{1}{6}(3-\sqrt{3}) & \frac{1}{4} & \frac{1}{12}(3-2\sqrt{3}) \\
\frac{1}{6}(3+\sqrt{3}) & \frac{1}{12}(3+2\sqrt{3}) & \frac{1}{4} \\
\hline\hline
 & \frac{1}{2} & \frac{1}{2}
\end{array}$$

defines a 2-stage A-stable method of order 4.

However, there is a heavy price to pay for using implicit methods of this kind, as we now have to calculate all the numbers k_i, $i = 1, 2, \ldots, s$, simultaneously, not in succession. For a system of m differential equations an implicit linear multistep method requires the solution of m simultaneous equations at each step; an s-stage implicit Runge–Kutta method requires the solution of sm simultaneous equations. This is a considerable increase in cost, and the general implicit Runge–Kutta methods cannot compete in efficiency with the Backward Differentiation Formulae such as (12.50); their use is almost exclusively limited to stiff systems of ODEs.

The overall computational effort can be somewhat reduced by using **diagonally implicit Runge–Kutta** (or DIRK) methods, in which the matrix A is lower triangular, so that $a_{ij} = 0$ if $j > i$. A further improvement in efficiency is possible by requiring in addition that all the diagonal elements a_{ii} are the same; unfortunately it has proved difficult to construct such methods with order greater than 4.

12.13 Notes

In this chapter we have only been able to introduce some of the basic ideas in what has become a vast area of numerical analysis. In particular we have not discussed the practical implementation of the various methods. The questions of how to choose the step size h to obtain efficiently a prescribed accuracy, and when and how to adjust h during the course of the calculation, are dealt with in the following books.

- E. Hairer, S.P. Nørsett, and G. Wanner, *Solving Ordinary Differential Equations I: Nonstiff Problems,* Second Edition, Springer Series in Computational Mathematics, 8, Springer, Berlin, 1993.
- A. Iserles, *A First Course in the Numerical Analysis of Differential Equations*, Cambridge University Press, Cambridge, 1996.
- J.D. Lambert, *Numerical Methods for Ordinary Differential Systems,* John Wiley & Sons, Chichester, 1991.

For a study of dynamical systems and their numerical analysis, with focus on long-time behaviour, we refer to

- A.M. STUART AND A.R. HUMPHRIES, *Dynamical Systems and Numerical Analysis*, Cambridge University Press, Cambridge, 1999.

The numerical solution of stiff initial value problems for systems of ordinary differential equations is discussed in

- E. HAIRER AND G. WANNER, *Solving Ordinary Differential Equations II: Stiff and Differential-Algebraic Problems*, Springer Series in Computational Mathematics, 14, Springer, Berlin, 1991.

An extensive survey of the theory of Runge–Kutta and linear multistep methods is found in

- J.C. BUTCHER, *The Numerical Analysis of Ordinary Differential Equations. Runge–Kutta and General Linear Methods*, Wiley-Interscience, John Wiley & Sons, Chichester, 1987.

Satisfactory theoretical treatment of nonlinear systems of differential equations from the point of view of stiffness requires the development of a genuinely nonlinear stability theory which does not involve the rather dubious idea of defining stiffness through linearisation based on the 'frozen Jacobian matrix'. We close by mentioning just one concept in this direction – that of *algebraic stability*. Given a Runge–Kutta method with Butcher tableau

$$\begin{array}{c||c} \boldsymbol{c} & A \\ \hline\hline & \boldsymbol{b}^{\mathrm{T}} \end{array}$$

we define the matrices

$$B = \mathrm{diag}(b_1, b_2, \ldots, b_s) \quad \text{and} \quad M = BA + A^{\mathrm{T}}B - bb^{\mathrm{T}}\,.$$

The method is said to be **algebraically stable** if the matrices B and M are both positive semidefinite, *i.e.*, $\boldsymbol{x}^{\mathrm{T}}B\boldsymbol{x} \geq 0$ and $\boldsymbol{x}^{\mathrm{T}}M\boldsymbol{x} \geq 0$ for all $\boldsymbol{x} \in \mathbb{R}^s$. Algebraic stability can be seen to ensure that approximations to solutions of nonlinear systems of differential equations exhibit acceptable numerical behaviour. For example, the Gauss–Runge–Kutta methods discussed in the last section are algebraically stable. For further details, see, for example,

- K. DEKKER AND J.G. VERVER, *Stability of Runge–Kutta Methods for Stiff Nonlinear Differential Equations*, North-Holland, Amsterdam, 1984.

Exercises

12.1 Verify that the following functions satisfy a Lipschitz condition on the respective intervals and find the associated Lipschitz constants:

(a) $f(x,y) = 2yx^{-4}\,, \qquad x \in [1,\infty)\,;$
(b) $f(x,y) = \mathrm{e}^{-x^2}\tan^{-1} y\,, \qquad x \in [1,\infty)\,;$
(c) $f(x,y) = 2y(1+y^2)^{-1}(1+\mathrm{e}^{-|x|})\,, \qquad x \in (-\infty,\infty)\,.$

12.2 Suppose that m is a fixed positive integer. Show that the initial value problem

$$y' = y^{2m/(2m+1)}\,, \qquad y(0) = 0\,,$$

has infinitely many continuously differentiable solutions. Why does this not contradict Picard's Theorem?

12.3 Write down the solution y of the initial value problem

$$y' = py + q\,, \quad y(0) = 1\,,$$

where p and q are constants. Suppose that the method in the proof of Picard's Theorem is used to generate the sequence of approximations $y_n(x)$, $n = 0, 1, 2, \ldots$; show that $y_n(x)$ is a polynomial of degree n, and consists of the first $n+1$ terms in the series expansion of $y(x)$ in powers of x.

12.4 Show that Euler's method fails to approximate the solution $y(x) = (4x/5)^{5/4}$ of the initial value problem $y' = y^{1/5}$, $y(0) = 0$. Justify your answer.

Consider approximating the same problem with the implicit Euler method. Show that there is a solution of the form $y_n = (C_n h)^{5/4}$, $n \geq 0$, with $C_0 = 0$ and $C_1 = 1$ and $C_n > 1$ for all $n \geq 2$.

12.5 Write down Euler's method for the solution of the problem

$$y' = x\mathrm{e}^{-5x} - 5y\,, \qquad y(0) = 0$$

on the interval $[0,1]$ with step size $h = 1/N$. Denoting by y_N the resulting approximation to $y(1)$, show that $y_N \to y(1)$ as $N \to \infty$.

12.6 Consider the initial value problem

$$y' = \ln\ln(4 + y^2)\,, \qquad x \in [0,1]\,, \qquad y(0) = 1\,,$$

and the sequence $(y_n)_{n=0}^N$, $N \geq 1$, generated by the Euler method

$$y_{n+1} = y_n + h \ln\ln(4+y_n^2) , \qquad n = 0, 1, \ldots, N-1 , \qquad y_0 = 1 ,$$

using the mesh points $x_n = nh$, $n = 0, 1, \ldots, N$, with spacing $h = 1/N$.

(i) Let T_n denote the truncation error of Euler's method for this initial value problem at the point $x = x_n$. Show that $|T_n| \leq h/4$.

(ii) Verify that

$$|y(x_{n+1}) - y_{n+1}| \leq (1 + hL)|y(x_n) - y_n| + h|T_n|$$

for $n = 0, 1, \ldots, N-1$, where $L = 1/(2\ln 4)$.

(iii) Find a positive integer N_0, as small as possible, such that

$$\max_{0 \leq n \leq N} |y(x_n) - y_n| \leq 10^{-4}$$

whenever $N \geq N_0$.

12.7 Define the truncation error T_n of the trapezium rule method

$$y_{n+1} = y_n + \frac{1}{2}h\,(f_{n+1} + f_n)$$

for the numerical solution of $y' = f(x, y)$ with $y(0) = y_0$ given, where $f_n = f(x_n, y_n)$ and $h = x_{n+1} - x_n$.

By integrating by parts the integral

$$\int_{x_n}^{x_{n+1}} (x - x_{n+1})(x - x_n)y'''(x)\mathrm{d}x ,$$

or otherwise, show that

$$T_n = -\frac{1}{12}h^2 y'''(\xi_n)$$

for some ξ_n in the interval (x_n, x_{n+1}), where y is the solution of the initial value problem.

Suppose that f satisfies the Lipschitz condition

$$|f(x, u) - f(x, v)| \leq L|u - v|$$

for all real x, u, v, where L is a positive constant independent of x, and that $|y'''(x)| \leq M$ for some positive constant M independent of x. Show that the global error $e_n = y(x_n) - y_n$ satisfies the inequality

$$|e_{n+1}| \leq |e_n| + \frac{1}{2}hL\left(|e_{n+1}| + |e_n|\right) + \frac{1}{12}h^3M\,.$$

For a constant step size $h > 0$ satisfying $hL < 2$, deduce that, if $y_0 = y(x_0)$, then

$$|e_n| \leq \frac{h^2M}{12L}\left[\left(\frac{1+\frac{1}{2}hL}{1-\frac{1}{2}hL}\right)^n - 1\right].$$

12.8 Show that the one-step method defined by

$$y_{n+1} = y_n + \tfrac{1}{2}h(k_1 + k_2)\,,$$

where

$$k_1 = f(x_n, y_n)\,, \qquad k_2 = f(x_n + h, y_n + hk_1)$$

is consistent and has truncation error

$$T_n = \tfrac{1}{6}h^2\left[f_y(f_x + f_yf) - \tfrac{1}{2}(f_{xx} + 2f_{xy}f + f_{yy}f^2)\right] + \mathcal{O}(h^3)\,.$$

12.9 When the classical fourth-order Runge–Kutta method is applied to the differential equation $y' = \lambda y$, where λ is a constant, show that

$$y_{n+1} = (1 + h\lambda + \tfrac{1}{2}h^2\lambda^2 + \tfrac{1}{6}h^3\lambda^3 + \tfrac{1}{24}h^4\lambda^4)y_n\,.$$

Compare this with the Taylor series expansion of $y(x_{n+1}) = y(x_n + h)$ about the point $x = x_n$.

12.10 Consider the one-step method

$$y_{n+1} = y_n + \alpha hf(x_n, y_n) + \beta hf(x_n + \gamma h, y_n + \gamma hf(x_n, y_n))\,,$$

where α, β and γ are real parameters and $h > 0$. Show that the method is consistent if, and only if, $\alpha + \beta = 1$. Show also that the order of the method cannot exceed 2.

Suppose that a second-order method of the above form is applied to the initial value problem $y' = -\lambda y$, $y(0) = 1$, where λ is a positive real number. Show that the sequence $(y_n)_{n\geq 0}$ is bounded if, and only if, $h \leq \frac{2}{\lambda}$. Show further that, for such λ,

$$|y(x_n) - y_n| \leq \frac{1}{6}\lambda^3h^2x_n\,, \quad n \geq 0\,.$$

12.11 Find the values of α and β so that the three-step method

$$y_{n+3} + \alpha(y_{n+2} - y_{n+1}) - y_n = h\beta(f_{n+2} + f_{n+1})$$

has order of accuracy 4, and show that the resulting method is *not* zero-stable.

12.12 Consider approximating the initial value problem $y' = f(x, y)$, $y(0) = y_0$ by the linear multistep method

$$y_{n+1} + by_{n-1} + ay_{n-2} = hf(x_n, y_n)$$

on the regular mesh $x_n = nh$ where a and b are constants.

(i) For a certain (unique) choice of a and b, this method is consistent. Find these values of a and b and verify that the order of accuracy is 1.

(ii) Although the method is consistent for the choice of a and b from part (i), the numerical solution it generates will not, in general, converge to the solution of the initial value problem as $h \to 0$, because the method is not zero-stable. Show that the method is not zero-stable for these a and b, and describe quantitatively what the unstable solutions will look like for small h.

12.13 Given that α is a positive real number, consider the linear two-step method

$$y_{n+2} - \alpha y_n = \frac{h}{3}\left[f(x_{n+2}, y_{n+2}) + 4f(x_{n+1}, y_{n+1}) + f(x_n, y_n)\right],$$

on the mesh $\{x_n\colon x_n = x_0 + nh, n = 1, 2, \ldots, N\}$ of spacing h, $h > 0$. Determine the set of all α such that the method is zero-stable. Find α such that the order of accuracy is as high as possible; is the method convergent for this value of α?

12.14 Which of the following linear multistep methods for the solution of the initial value problem $y' = f(x, y)$, $y(0)$ given, are zero-stable?

(a) $y_{n+1} - y_n = hf_n$,
(b) $y_{n+1} + y_n - 2y_{n-1} = h(f_{n+1} + f_n + f_{n-1})$,
(c) $y_{n+1} - y_{n-1} = \frac{1}{3}h(f_{n+1} + 4f_n + f_{n-1})$,
(d) $y_{n+1} - y_n = \frac{1}{2}h(3f_n - f_{n-1})$,
(e) $y_{n+1} - y_n = \frac{1}{12}h(5f_{n+1} + 8f_n - f_{n-1})$.

For the methods under (a) and (c) explore absolute stability when applied to the differential equation $y' = \lambda y$ with $\lambda < 0$.

12.15 Determine the order of the linear multistep method

$$y_{n+2} - (1+a)y_{n+1} + y_n = \frac{1}{4}h\left[(3-a)f_{n+2} + (1-3a)f_n\right]$$

and investigate its zero-stability and absolute stability.

12.16 Assuming that $\sigma(z) = z^2$ is the second characteristic polynomial of a linear two-step method, find a quadratic polynomial $\rho(z)$ such that the order of the method is 2. Is this method convergent? By applying the method to $y' = \lambda y$, $y(0) = 1$, where λ is a negative real number, show that the method is absolutely stable for all $h > 0$.

12.17 Consider the θ-method

$$y_{n+1} = y_n + h\left[(1-\theta)f_n + \theta f_{n+1}\right]$$

for $\theta \in [0,1]$. Show that the method is A-stable if, and only if, $\theta \geq 1/2$.

12.18 Write down an expression for the Lagrange interpolation polynomial of degree 2 for a function $x \mapsto y(x)$, using the interpolation points x_n, $x_{n+1} = x_n + h$ and $x_{n+2} = x_n + 2h$, $h > 0$. Differentiate this polynomial to show that

$$y'(x_{n+2}) = \frac{1}{2h}\left(3y(x_{n+2}) - 4y(x_{n+1}) + y(x_n)\right) + \mathcal{O}(h^2),$$

provided that $y \in \mathrm{C}^3[x_n, x_{n+2}]$. Confirm this result by determining the truncation error of the BDF2 method

$$3y_{n+2} - 4y_{n+1} + y_n = 2hf_{n+2}\,.$$

12.19 When the general two-stage implicit Runge–Kutta method is applied to the single constant-coefficient differential equation $y' = \lambda y$, show that

$$\begin{aligned} k_1 &= [1 + \lambda h(a_{12} - a_{22})]\lambda y_n/\Delta\,, \\ k_2 &= [1 + \lambda h(a_{21} - a_{11})]\lambda y_n/\Delta\,, \end{aligned}$$

where Δ is the determinant of the matrix $I - \lambda h A$ with

$$A = \begin{pmatrix} a_{11} & a_{12} \\ a_{21} & a_{22} \end{pmatrix}.$$

For the method defined by the Butcher tableau

$$\begin{array}{c|cc}
\frac{1}{6}(3-\sqrt{3}) & \frac{1}{4} & \frac{1}{12}(3-2\sqrt{3}) \\
\frac{1}{6}(3+\sqrt{3}) & \frac{1}{12}(3+2\sqrt{3}) & \frac{1}{4} \\
\hline
 & \frac{1}{2} & \frac{1}{2}
\end{array}$$

deduce that $y_{n+1} = R(\lambda h) y_n$, where

$$R(\lambda h) = \frac{1 + \frac{1}{2}\lambda h + \frac{1}{12}\lambda^2 h^2}{1 - \frac{1}{2}\lambda h + \frac{1}{12}\lambda^2 h^2}\,.$$

By writing $R(z)$ in the factorised form $(z+p)(z+q)/(z-p)(z-q)$, deduce that this Runge–Kutta method is A-stable.

13

Boundary value problems for ODEs

13.1 Introduction

In the previous chapter we discussed numerical methods for initial value problems in which all the associated side conditions for a system of differential equations are prescribed at the same point. Now we go on to consider problems where these conditions specify values at more than one point. Typically we require the solution on an interval $[a, b]$, and some conditions are given at a, and the rest at b, although more complicated situations are possible, involving three or more points.

We shall begin with the simplest case, of a second-order equation with one condition given at a and one at b. This problem is sufficient to introduce the basic ideas, and is of a type which arises quite often in practice.

We then go on to discuss the shooting method for the solution of more general problems.

13.2 A model problem

The simplest two-point boundary problem involves the second-order differential equation

$$-y'' + r(x)y = f(x)\,, \qquad a < x < b\,, \tag{13.1}$$

with the boundary conditions

$$y(a) = A\,, \quad y(b) = B\,, \tag{13.2}$$

where A and B are given real numbers. We shall assume that r and f are given real-valued functions, defined and continuous on the bounded closed interval $[a, b]$ of the real line, and that

$$r(x) \geq 0\,, \quad a \leq x \leq b\,.$$

The reason for this condition will appear later, in Theorem 13.4.

We shall construct a numerical approximation to the solution on a uniform mesh of points

$$x_j = a + jh\,, \quad j = 0, 1, \ldots, n\,, \quad h = (b-a)/n\,, \quad n \geq 2\,,$$

so that $x_0 = a$, $x_n = b$. The second derivative is approximated using the second central difference defined below.

Definition 13.1 *The central difference δy of y is defined by*

$$\delta y(x_j) = y(x_j + \tfrac{1}{2}h) - y(x_j - \tfrac{1}{2}h)\,.$$

Higher-order differences are defined recursively by

$$\delta^{m+1} y(x_j) = \delta[\delta^m y(x_j)] = \delta^m y(x_j + \tfrac{1}{2}h) - \delta^m y(x_j - \tfrac{1}{2}h)\,.$$

In particular, the second central difference may be written

$$\begin{aligned} \delta^2 y(x_j) &= \delta y(x_j + \tfrac{1}{2}h) - \delta y(x_j - \tfrac{1}{2}h) \\ &= y(x_j + h) - 2y(x_j) + y(x_j - h)\,. \end{aligned}$$

Theorem 13.1 *(i) Suppose that $y \in \mathrm{C}^4[x-h, x+h]$, i.e., that y has continuous fourth derivative on the interval $[x-h, x+h]$. Then, there exists a number ξ in $(x-h, x+h)$ such that*

$$\frac{\delta^2 y(x)}{h^2} = y''(x) + \tfrac{1}{12}h^2 y^{\mathrm{iv}}(\xi)\,.$$

(ii) Suppose that $y \in \mathrm{C}^6[x-h, x+h]$; then, there exists a number η in $(x-h, x+h)$ such that

$$\frac{\delta^2 y(x)}{h^2} = y''(x) + \tfrac{1}{12}h^2 y^{\mathrm{iv}}(x) + \tfrac{1}{360}h^4 y^{\mathrm{vi}}(\eta)\,. \tag{13.3}$$

Proof (i) Taylor's Theorem shows that there exist numbers ξ_1 and ξ_2 in the intervals $(x-h, x)$ and $(x, x+h)$, respectively, such that

$$\left.\begin{aligned} y(x-h) &= y(x) - hy'(x) + \tfrac{1}{2}h^2 y''(x) - \tfrac{1}{6}h^3 y'''(x) + \tfrac{1}{24}h^4 y^{\mathrm{iv}}(\xi_1)\,, \\ y(x+h) &= y(x) + hy'(x) + \tfrac{1}{2}h^2 y''(x) + \tfrac{1}{6}h^3 y'''(x) + \tfrac{1}{24}h^4 y^{\mathrm{iv}}(\xi_2)\,. \end{aligned}\right\} \tag{13.4}$$

Since y^{iv} is continuous on $[x-h, x+h]$, there is a number ξ in (ξ_1, ξ_2), and thus also in $(x-h, x+h)$, such that

$$\tfrac{1}{2}(y^{\mathrm{iv}}(\xi_1) + y^{\mathrm{iv}}(\xi_2)) = y^{\mathrm{iv}}(\xi)\,.$$

The required result is now obtained by adding the two equalities (13.4) and dividing by h^2.

(ii) The proof is completely analogous, and is left to the reader as an exercise. (See Exercise 1.) □

We can now use the central difference approximation to construct the numerical solution. Writing Y_j for the numerical approximation to $y(x_j)$, we approximate the differential equation by

$$-\frac{\delta^2 Y_j}{h^2} + r_j Y_j = f_j\,, \quad j = 1, 2, \ldots, n-1\,, \tag{13.5}$$

where we have used the notation $r_j = r(x_j)$, $f_j = f(x_j)$. Now, (13.5) is a system of $n-1$ linear algebraic equations for the $n-1$ unknowns Y_j, $j = 1, 2, \ldots, n-1$, with the boundary conditions specifying the values of Y_0 and Y_n,

$$Y_0 = A\,, \qquad Y_n = B\,. \tag{13.6}$$

The system may be written in matrix form as

$$M\mathbf{Y} = \boldsymbol{g}\,,$$

where $\mathbf{Y}, \boldsymbol{g} \in \mathbb{R}^{n-1}$ and, for $n \geq 4$, the matrix $M \in \mathbb{R}^{(n-1)\times(n-1)}$ is tridiagonal. Here $\mathbf{Y} = (Y_1, \ldots, Y_{n-1})^{\mathrm{T}}$, the nonzero elements of M are

$$M_{jj} = 2/h^2 + r_j\,, \qquad M_{j\,j-1} = M_{j\,j+1} = -1/h^2\,, \tag{13.7}$$

and the elements of the column vector $\boldsymbol{g}$ on the right-hand side are

$$g_1 = f_1 + A/h^2,\;\; g_{n-1} = f_{n-1} + B/h^2,\;\; g_j = f_j\,,\;\; j = 2, 3, \ldots, n-2\,.$$

Note how the known boundary values Y_0 and Y_n have been transferred to the right-hand side, and appear in the first and last elements of $\boldsymbol{g}$. The solution of this system is very easy, using the algorithm for tridiagonal matrices described in Section 3.3. Using the fact that $r(x) \geq 0$, we see that the off-diagonal elements of M are negative, the diagonal elements are positive, and in each row the diagonal element is at least as large as the sum of absolute values of the off-diagonal elements. Theorem 3.4 and Exercise 3.5 imply that no row interchanges are needed in the calculation, and that the matrix M is nonsingular. The calculation is therefore very straightforward and efficient, and requires very little computational time, even for a mesh which may contain several hundred points.

13.3 Error analysis

Having obtained the numerical solution we must now analyse its accuracy. In the same way as for initial value problems, we begin by finding the truncation error.

Definition 13.2 *The* **truncation error** *of the central difference approximation to the problem (13.1), (13.2) is*

$$T_j = -\frac{\delta^2 y(x_j)}{h^2} + r_j y(x_j) - f_j\,, \qquad j = 1, 2, \ldots, n-1\,,$$

where y is the exact solution of (13.1), (13.2).

Theorem 13.2 *Suppose that the solution y to the boundary value problem (13.1), (13.2) has a continuous fourth derivative on $[a, b]$. Then, the truncation error may be written*

$$T_j = -\tfrac{1}{12} h^2 y^{\mathrm{iv}}(\xi_j)\,, \tag{13.8}$$

for some value of ξ_j in the interval (x_{j-1}, x_{j+1}), $j = 1, 2, \ldots, n-1$. The truncation error is bounded by T, where

$$|T_j| \le T = \tfrac{1}{12} h^2 M_4\,, \qquad j = 1, 2, \ldots, n-1\,,$$

and

$$M_4 = \max_{x\in[a,b]} |y^{\mathrm{iv}}(x)|\,. \tag{13.9}$$

Proof The expression for T_j follows from the substitution of the expression for $\delta^2 y(x_j)$ given by Theorem 13.1 into the definition of T_j, and the use of the fact that y is the solution of the differential equation. The proof of the bound for T_j is then immediate; since y^{iv} is known to be continuous on $[a, b]$ it is bounded on $[a, b]$, so M_4 exists. □

In order to simplify writing, we define

$$L(u_j) = -\frac{\delta^2 u_j}{h^2} + r_j u_j\,, \qquad j = 1, 2, \ldots, n-1,$$

for any set of real numbers $\{u_0, u_1, \ldots, u_n\}$. The **global error** in the numerical solution is defined by

$$e_j = y(x_j) - Y_j\,, \qquad j = 0, 1, \ldots, n\,.$$

Now, $y(x_j)$ and Y_j satisfy

$$\begin{aligned} L(y(x_j)) &= f_j + T_j\,, \qquad j = 1, 2, \dots, n-1\,,\\ L(Y_j) &= f_j\,, \qquad j = 1, 2, \dots, n-1\,, \end{aligned}$$

from the definition of truncation error and (13.5); hence, by subtraction,

$$L(e_j) = T_j\,, \qquad j = 1, 2, \dots, n-1\,,$$

with the boundary conditions $e_0 = e_n = 0$. We must now use the bound on T_j to derive a bound on the error e_j. This will be achieved by means of the following theorem.

Theorem 13.3 (**Maximum Principle**) *Let a_j, b_j, c_j, $j = 0, 1, \dots, n$, be positive real numbers such that $b_j \geq a_j + c_j$, and suppose that u_j, $j = 0, 1, \dots, n$, are real numbers such that*

$$-a_j u_{j-1} + b_j u_j - c_j u_{j+1} \leq 0\,, \qquad j = 1, 2, \dots, n-1\,.$$

Then, $u_j \leq K$, $j = 0, 1, \dots, n$, where $K = \max\{u_0, u_n, 0\}$.

Proof Let $u_m = \max\{u_0, u_1, \dots, u_n\}$; then if $m = 0$, $m = n$, or $u_m \leq 0$ the result is trivial. Suppose then that $1 \leq m \leq n-1$, and that $u_m > 0$. Since u_m is the maximum of the u_j, we know that

$$u_m \geq u_{m-1}\,, \qquad u_m \geq u_{m+1}\,.$$

Hence

$$\begin{aligned} b_m u_m &\leq a_m u_{m-1} + c_m u_{m+1}\\ &\leq a_m u_m + c_m u_m\\ &\leq b_m u_m\,, \end{aligned}$$

since $u_m > 0$. This means that equality holds throughout, so that $u_{m-1} = u_m = u_{m+1}$. We can then apply the same argument to both u_{m-1} and u_{m+1}, continuing until we find that either $u_m = u_n$ or $u_m = u_0$. Thus, in this case $u_0 = u_n = \max\{u_0, u_1, \dots, u_n\}$, as required. □

Theorem 13.4 *Suppose that the solution y of the boundary value problem (13.1), (13.2) has a continuous fourth derivative on $[a, b]$, and that Y_j, $j = 0, 1, \dots, n$, is the solution of the central difference approximation (13.5), (13.6). Then,*

$$\max_{0 \leq j \leq n} |y(x_j) - Y_j| \leq \tfrac{1}{96} h^2 (b-a)^2 M_4\,. \tag{13.10}$$

Proof Let $e_j = y(x_j) - Y_j$. We have already seen that $L(e_j) = T_j$, $j = 1, 2, \ldots, n-1$. Defining

$$\varphi_j = C\left\{(2j-n)^2h^2 - n^2h^2\right\}, \qquad j = 0, 1, \ldots, n, \tag{13.11}$$

where C is a constant, we see that

$$\begin{aligned} L(\varphi_j) &= -C\left\{(2j-2-n)^2 - 2(2j-n)^2 + (2j+2-n)^2\right\} + r_j\varphi_j \\ &= -8C + r_j\varphi_j, \qquad j = 1, 2, \ldots, n-1. \end{aligned}$$

Hence

$$L(e_j + \varphi_j) = T_j - 8C + r_j\varphi_j, \qquad j = 1, 2, \ldots, n-1.$$

If we choose $C = T/8$ with $T = \frac{1}{12}h^2M_4$, we see that $L(e_j + \varphi_j) \le 0$, since $|T_j| \le T$, $r_j \ge 0$ and $\varphi_j \le 0$, and L satisfies the conditions of the Maximum Principle. Now,

$$e_0 + \varphi_0 = e_n + \varphi_n = 0,$$

so that, according to Theorem 13.3, $e_j + \varphi_j \le 0$ for $j = 0, 1, \ldots, n$. However, $-Cn^2h^2 \le \varphi_j \le 0$, so we have the result

$$e_j \le Cn^2h^2 = \tfrac{1}{8}(b-a)^2T = \tfrac{1}{96}h^2(b-a)^2M_4, \qquad j = 0, 1, \ldots, n.$$

By applying the same argument to $L(-e_j + \varphi_j)$ we find that

$$-e_j \le \tfrac{1}{96}h^2(b-a)^2M_4, \qquad j = 0, 1, \ldots, n.$$

Combining these upper bounds for e_j and $-e_j$ gives the required result. □

The function φ defined by (13.11) is called a **comparison function**. An alternative proof of Theorem 13.4, based on the properties of monotone matrices, can be given by using the result in Exercise 2. Notice that the condition $r(x) \ge 0$ is used in the application of the Maximum Principle in the above proof.

This theorem shows that, provided the solution y has a continuous fourth derivative, the numerical method is **convergent**, that is

$$\max_{0 \le j \le n} |y(x_j) - Y_j| \to 0 \qquad \text{as } n \to \infty$$

(or, equivalently, as $h = (b-a)/n \to 0$). This means that we can obtain any required accuracy by choosing n sufficiently large.

13.4 Boundary conditions involving a derivative

The same differential equation (13.1) may be associated with boundary conditions involving the first derivative of the solution. Suppose, for example, that we are given real numbers $\alpha > 0$, A and B. Consider the differential equation (13.1) together with the boundary conditions

$$y'(a) - \alpha y(a) = A\,, \qquad y(b) = B\,. \tag{13.12}$$

The condition at $x = a$ may be approximated in various ways; we shall introduce an extra mesh point x_{-1} outside the interval and use the approximate version

$$\frac{Y_1 - Y_{-1}}{2h} - \alpha Y_0 = A\,.$$

This gives

$$Y_{-1} = Y_1 - 2h\alpha Y_0 - 2hA\,.$$

Writing the same central difference approximation (13.5) as before, but now for $j = 0, 1, \ldots, n-1$, we can eliminate the extra unknown Y_{-1} from the equation at $j = 0$ to give

$$\left[\frac{2(1+\alpha h)}{h^2} + r_0\right] Y_0 - \frac{2}{h^2} Y_1 = f_0 - \frac{2}{h} A\,.$$

Together with (13.5), for $j = 1, 2, \ldots, n-1$, and $Y_n = B$, we now have a system of n equations for the unknowns Y_j, $j = 0, 1, \ldots, n-1$. There are one more equation and one more unknown than before, but the new matrix is still tridiagonal, and also diagonally dominant because of the condition $\alpha > 0$. The computation is again very straightforward.

Theorem 13.5 *Suppose that $y \in \mathrm{C}^3[x-h, x+h]$; then, there exists a real number χ in $(x-h, x+h)$ such that*

$$\frac{y(x+h) - y(x-h)}{2h} = y'(x) + \tfrac{1}{6}h^2 y'''(\chi)\,. \tag{13.13}$$

Proof Taylor's Theorem shows that there exist $\chi_1 \in (x-h, x)$ and $\chi_2 \in (x, x+h)$ such that

$$\begin{aligned} y(x-h) &= y(x) - hy'(x) + \tfrac{1}{2}h^2 y''(x) - \tfrac{1}{6}h^3 y'''(\chi_1)\,, \\ y(x+h) &= y(x) + hy'(x) + \tfrac{1}{2}h^2 y''(x) + \tfrac{1}{6}h^3 y'''(\chi_2)\,. \end{aligned}$$

We subtract the first equality from the second, and the result follows as in the proof of Theorem 13.1. □

Note that the approximation to $y'(x)$ at $x = x_0$ may be written

$$\frac{\frac{1}{2}[\delta y(x_0 + \frac{1}{2}h) + \delta y(x_0 - \frac{1}{2}h)]}{h}.$$

For $j = 1, 2, \ldots, n-1$, we define the truncation error T_j as in Definition 13.2. In addition, since we shall now also incur an error in the approximation of the boundary condition at $x = a$, we define

$$T_0 = \left[\frac{2(1+\alpha h)}{h^2} + r_0\right] y(0) - \frac{2}{h^2}y(h) - f_0 + \frac{2}{h}A\,.$$

The aim of our next result is to quantify the size of the truncation error in terms of the mesh size h.

Theorem 13.6 *Suppose that the solution y to the boundary value problem (13.1), (13.2) has a continuous fourth derivative on the closed interval $[a-h, b]$. Then, the truncation error of the central difference approximation to (13.1) with boundary conditions (13.12) may be written*

$$\begin{aligned} T_j &= -\tfrac{1}{12}h^2 y^{\mathrm{iv}}(\xi_j)\,, \qquad j = 1, 2, \ldots, n-1\,, \\ T_0 &= -\tfrac{1}{12}h^2 y^{\mathrm{iv}}(\xi_0) - \tfrac{1}{3}h y'''(\chi)\,, \end{aligned}$$

for some value of ξ_j in the interval (x_{j-1}, x_{j+1}), $1 \le j \le n-1$, and some values of ξ_0 and χ in the interval (x_{-1}, x_1) where $x_{-1} = a - h$.

Proof For $j = 1, 2, \ldots, n-1$, this is the same result as in Theorem 13.2. When $j = 0$, we find that

$$\begin{aligned} T_0 &= \left[\frac{2(1+\alpha h)}{h^2} + r_0\right] y(0) - \frac{2}{h^2}y(h) - f_0 + \frac{2}{h}A \\ &= -\frac{y(h) - 2y(0) + y(-h)}{h^2} + r(0)y(0) - f(0) \\ &\quad -\frac{2}{h}\left[\frac{y(h) - y(-h)}{2h} - \alpha y(0) - A\right] \\ &= -\tfrac{1}{12}h^2 y^{\mathrm{iv}}(\xi_0) - \tfrac{2}{h}\,\tfrac{1}{6}h^2 y'''(\chi)\,, \end{aligned}$$

where we have used Theorem 13.5. □

Theorem 13.7 *Suppose that the solution y of (13.1) with the boundary conditions (13.12) has a continuous fourth derivative on the interval $[a-h, b]$; then, the numerical solution obtained from the central difference* *approximation satisfies*

$$\max_{0\le j\le n} |y(x_j) - Y_j| \le h^2 \left\{\tfrac{1}{24}(b-a)^2 M_4 + \tfrac{1}{6}(b-a)M_3\right\} .$$

Proof The proof is very similar to that of Theorem 13.4, but requires the use of a more complicated comparison function φ_j. Let us define

$$\begin{aligned} L^*(u_j) &= -\frac{\delta^2 u_j}{h^2} + r_j u_j\,, \qquad j = 1, 2, \ldots, n-1\,, \\ L^*(u_0) &= \left[\frac{2(1+\alpha h)}{h^2} + r_0\right] u_0 - \frac{2}{h^2} u_1\,, \end{aligned}$$

for any set of real numbers $\{u_0, u_1, \ldots, u_n\}$, and let

$$\varphi_j = Cj^2h^2 + Djh + E\,, \qquad j = 0, 1, \ldots, n\,,$$

where C, D and E are constants to be determined. Then, with $e_j = y(x_j) - Y_j$, as in the proof of Theorem 13.4, we see that

$$L^*(e_j) = T_j\,, \qquad j = 0, 1, \ldots, n-1\,.$$

A simple calculation shows that

$$\begin{aligned} L^*(\varphi_j) &= -2C + r_j\varphi_j\,, \qquad j = 1, 2, \ldots, n-1\,, \\ L^*(\varphi_0) &= -2C - 2D/h + [2\alpha/h + r_0]E\,. \end{aligned}$$

Hence

$$\begin{aligned} L^*(e_j + \varphi_j) &= -\tfrac{1}{12}h^2 y^{\mathrm{iv}}(\xi_j) - 2C + r_j\varphi_j\,, \quad j = 1, 2, \ldots, n-1\,, \\ L^*(e_0 + \varphi_0) &= -\tfrac{1}{12}h^2 y^{\mathrm{iv}}(\xi_0) - \tfrac{1}{3}hy'''(\chi) \\ &\quad -2C - 2D/h + [2\alpha/h + r_0]E\,. \end{aligned}$$

If we now choose

$$C = \tfrac{1}{24}h^2 M_4\,, \qquad D = \tfrac{1}{6}h^2 M_3\,, \qquad E = -C(b-a)^2 - D(b-a)\,,$$

it is easy to check that

$$\begin{aligned} \varphi_j &\le 0\,, \qquad j = 0, 1, \ldots, n\,, \\ L^*(e_j + \varphi_j) &\le 0\,, \qquad j = 0, 1, \ldots, n-1\,. \end{aligned}$$

The Maximum Principle then applies, and we deduce that

$$e_j + \varphi_j \le \max\{e_0 + \varphi_0, e_n + \varphi_n, 0\}\,, \qquad j = 0, 1, \ldots, n\,.$$

We see at once that $e_n = \varphi_n = 0$ and $\varphi_0 \le 0$, but in this case e_0 is not zero. Therefore, all we can deduce for the moment is that

$$e_j + \varphi_j \le \max\{e_0 + \varphi_0, 0\}\,, \qquad j = 0, 1, \ldots, n\,. \tag{13.14}$$

In particular,

$$e_1 + \varphi_1 \le \max\{e_0 + \varphi_0, 0\}\,. \tag{13.15}$$

However, $L^*(e_0 + \varphi_0) \le 0$; thus, by the definition of $L^*(e_0 + \varphi_0)$,

$$e_0 + \varphi_0 \le \frac{2}{2(1 + \alpha h) + h^2 r_0}\,(e_1 + \varphi_1)\,.$$

On writing $\delta = 2/(2(1 + \alpha h) + h^2 r_0)$ and noting that, since $\alpha > 0$ and $r_0 \ge 0$, we have $0 < \delta < 1$, it follows that

$$e_0 + \varphi_0 \le \delta(e_1 + \varphi_1)\,. \tag{13.16}$$

Inserting this inequality into the left-hand side of (13.15), we find that

$$e_0 + \varphi_0 \le \max\{\delta(e_0 + \varphi_0), 0\}\,.$$

If $e_0 + \varphi_0$ were positive, this inequality and the fact that $0 < \delta < 1$ would imply $e_0 + \varphi_0 \le 0$, leading to a contradiction. Therefore, $e_0 + \varphi_0 \le 0$. Returning with this information to (13.14), we deduce that $e_j + \varphi_j \le 0$ for $j = 0, 1, \dots, n$, and the rest of the proof then follows as in the proof of Theorem 13.1. □

13.5 The general self-adjoint problem

The general self-adjoint boundary value problem is

$$-\frac{\mathrm{d}}{\mathrm{d}x}\left(p(x)\frac{\mathrm{d}y}{\mathrm{d}x}\right) + r(x)y = f(x)\,, \qquad a < x < b\,, \tag{13.17}$$

where r and f are real-valued functions, defined and continuous on $[a, b]$, p is a real-valued continuously differentiable function on $[a, b]$, $r(x) \ge 0$ and $p(x) \ge c_0 > 0$. We shall consider only the case where the boundary conditions prescribe the values of y at each end,

$$y(a) = A\,, \qquad y(b) = B\,. \tag{13.18}$$

The central difference approximation to the equation (13.17) may be written

$$-\frac{\delta(p_j\,\delta Y_j)}{h^2} + r_j Y_j = f_j\,, \qquad j = 1, 2, \dots, n-1\,,$$

or, in detail,

$$-\frac{p_{j+1/2}(Y_{j+1} - Y_j) - p_{j-1/2}(Y_j - Y_{j-1})}{h^2} + r_j Y_j = f_j\,, \tag{13.19}$$

for $j = 1, 2, \ldots, n$, and is supplemented by the boundary conditions

$$Y_0 = A\,, \qquad Y_n = B\,. \tag{13.20}$$

It is easy to see that this represents a system of linear equations for the unknowns $Y_1, Y_2, \ldots, Y_{n-1}$, and that the matrix of the system is tridiagonal and diagonally dominant, just as it was in the special case (13.1), which corresponds to $p(x) \equiv 1$. The solution of the system is therefore a very simple matter.

Next, we consider the error analysis of the difference scheme (13.19), (13.20). We begin by quantifying the size of the truncation error

$$T_j = -\frac{\delta(p_j\,\delta y(x_j))}{h^2} + r_j y(x_j) - f_j\,, \qquad j = 1, 2, \ldots, n-1\,,$$

in terms of the mesh size h.

Lemma 13.1 *Suppose that $p \in \mathrm{C}^3[a,b]$ and $y \in \mathrm{C}^4[a,b]$. The truncation error T_j of the central difference approximation (13.19) then satisfies*

$$|T_j| \le T = \tfrac{1}{24}h^2 \max_{x\in[a,b]} \left\{|(py')'''(x)| + |p'y'''(x)| + 2|py^{\mathrm{iv}}(x)|\right\},$$

for $j = 1, 2, \ldots, n-1$.

Proof By expanding in Taylor series as we have done before, we find that

$$\begin{aligned} p_{j+1/2}[y(x_{j+1}) - y(x_j)] &= p_{j+1/2}[hy'_{j+1/2} + \tfrac{1}{24}h^3 y'''(\xi_1)]\,,\\ p_{j-1/2}[y(x_j) - y(x_{j-1})] &= p_{j-1/2}[hy'_{j-1/2} + \tfrac{1}{24}h^3 y'''(\xi_2)]\,, \end{aligned}$$

where $\xi_1 \in (x_j, x_{j+1})$ and $\xi_2 \in (x_{j-1}, x_j)$. The first term in the difference of these expressions gives, in the same way,

$$h[p_{j+1/2}y'(x_{j+1/2}) - p_{j-1/2}y'(x_{j-1/2})] = h[h(py')'(x_j) + \tfrac{1}{24}h^3(py')'''(\xi_3)]$$

where $\xi_3 \in (x_{j-1/2}, x_{j+1/2})$. For the other term we can write

$$\begin{aligned} &\tfrac{1}{24}h^3|p_{j+1/2}y'''(\xi_1) - p_{j-1/2}y'''(\xi_2)|\\ &\qquad = \tfrac{1}{24}h^3|(p_{j+1/2} - p_{j-1/2})y'''(\xi_1) + p_{j-1/2}[y'''(\xi_1) - y'''(\xi_2)]|\\ &\qquad \le \tfrac{1}{24}h^3\left\{|hp'(\xi_4)y'''(\xi_1)| + |p_{j-1/2}2hy^{\mathrm{iv}}(\xi_5)|\right\}, \end{aligned}$$

since $|\xi_1 - \xi_2| < 2h$. Here, $\xi_4 \in (x_{j-1/2}, x_{j+1/2})$ and ξ_5 lies between ξ_1 and ξ_2. The required bound follows immediately. □

As in the proof of Theorem 13.4, we can now derive a bound on the global error in the numerical solution in terms of the truncation error by using the Maximum Principle. The only difficulty in extending that theorem to the more general self-adjoint problem lies in the construction of a comparison function corresponding to (13.11). The general case requires some detailed analysis, which can be simplified under certain conditions on the function p, for example if p is monotonic.

Lemma 13.2 *Suppose that p and r are continuous functions defined on $[a,b]$, p is monotonic increasing on $[a,b]$, $p(x) \geq c_0 > 0$, $r(x) \geq 0$, and define*

$$L(u_j) = -\frac{\delta(p\,\delta u_j)}{h^2} + r_j u_j\,, \qquad j = 1, 2, \ldots, n-1\,,$$

for any set of real numbers $\{u_0, u_1, \ldots, u_n\}$. Further, let

$$\varphi_j = C(j^2 - n^2)h^2\,, \qquad j = 0, 1, \ldots, n\,,$$

where C is a positive constant. Then,

$$L(\varphi_j) \leq -2c_0 C\,, \quad j = 1, 2, \ldots, n-1\,.$$

Proof It follows from the definition that

$$\begin{aligned} L(\varphi_j) &= -p_{j+1/2}C(2j+1) + p_{j-1/2}C(2j-1) + C(j^2-n^2)h^2 r_j \\ &= -C[(p_{j+1/2} + p_{j-1/2}) + 2j(p_{j+1/2} - p_{j-1/2}) + h^2(n^2 - j^2)r_j] \\ &\leq -2c_0 C\,, \end{aligned}$$

for $j = 1, 2, \ldots, n-1$, as required. □

Note that we have imposed various conditions on the problem, which are usually necessary, though some can be slightly relaxed. The condition in this lemma, that p should be monotonic increasing on $[a,b]$, is only needed to simplify the subsequent proof. The main result is true much more generally. We leave it as an exercise to derive the same result under the assumption that p is monotonic decreasing on $[a,b]$.

Theorem 13.8 *Suppose that p and r are continuous functions defined on $[a,b]$, p is monotonic increasing on $[a,b]$, $p(x) \geq c_0 > 0$, $r(x) \geq 0$. Assume further that the solution y of (13.17), (13.18) has a continuous fourth derivative on $[a,b]$, that p has a continuous third derivative, and* *that Y_j, $j = 0, 1, \ldots, n$, is the solution of the central difference approximation (13.19), (13.20). Then, with T as in Lemma 13.1,*

$$\max_{0 \le j \le n} |y(x_j) - Y_j| \le \tfrac{1}{2c_0} T \,. \qquad (13.21)$$

Proof The proof of this theorem follows that of Theorem 13.4, using the bound from Lemma 13.1 on the truncation error and the comparison function φ_j from Lemma 13.2. The details are left as an exercise. □

13.6 The Sturm–Liouville eigenvalue problem

Suppose that r is a real-valued function, defined and continuous on the closed interval $[a, b]$, p is a real-valued function, defined and continuously differentiable on $[a, b]$, and $r(x) \ge 0$, $p(x) \ge c_0 > 0$ for all $x \in [a, b]$. The differential equation

$$-\frac{\mathrm{d}}{\mathrm{d}x}\left(p(x)\frac{\mathrm{d}y}{\mathrm{d}x}\right) + r(x)y = \lambda y \,, \qquad a < x < b \,, \qquad (13.22)$$

with homogeneous boundary conditions $y(a) = y(b) = 0$, has only the trivial solution $y \equiv 0$, except for an infinite sequence of positive *eigenvalues* $\lambda = \lambda_m$, $m = 1, 2, \ldots$. We shall now consider a numerical method for finding these eigenvalues and the corresponding *eigenfunctions*, $y_{(m)}(x)$, $m = 1, 2, \ldots$.

In the simple case where $p(x) \equiv 1$ and $r(x) \equiv 0$ the solution to this problem is, of course, $\lambda_m = [m\pi/(b - a)]^2$, $y_{(m)}(x) = A \sin m\pi t$, $m = 1, 2, \ldots$, where A is a nonzero constant and $t = (x - a)/(b - a)$.

Using the same finite difference approximation as in the previous section, we obtain the equations

$$-\frac{p_{j+1/2}(Y_{j+1} - Y_j) - p_{j-1/2}(Y_j - Y_{j-1})}{h^2} + r_j Y_j = \Lambda Y_j \,, \qquad j = 1, 2, \ldots, n-1 \,.$$

Together with the boundary conditions $Y_0 = Y_n = 0$, this shows that Λ is an eigenvalue of a symmetric tridiagonal matrix M whose entries are

$$M_{jj} = \frac{p_{j+1/2} + p_{j-1/2}}{h^2} + r_j \,, \qquad 1 \le j \le n-1 \,,$$

$$M_{j\,j-1} = -\frac{p_{j-1/2}}{h^2} \,, \ 2 \le j \le n \,, \qquad M_{j\,j+1} = -\frac{p_{j+1/2}}{h^2} \,, \ 1 \le j \le n-1 \,,$$

and the approximate function values Y_j are the elements of the corresponding eigenvector. This algebraic eigenvalue problem is easily solved by the method described in Chapter 5.

The boundary value problems which we have discussed so far have all had a unique solution. The eigenvalue problem (13.22) has an infinite number of solutions, and the mesh used in the numerical computation has to be chosen to adequately represent the eigenfunctions required – the computation can obviously only find a finite number of them. The matrix M has $n-1$ eigenvalues and eigenvectors and, as we shall see, it will normally give a good approximation to the first few eigenvalues, $\lambda_1, \lambda_2, \ldots$, and a much less accurate approximation to λ_{n-1}.

To analyse the error in the eigenvalue we proceed as before, by defining the truncation error

$$T_j = -\frac{p_{j+1/2}(y_{j+1}-y_j) - p_{j-1/2}(y_j - y_{j-1})}{h^2} + r_j y_j - \lambda y_j\,, \qquad j = 1, 2, \ldots, n-1\,,$$

where $y_j = y(x_j)$. These equations can now be written

$$\begin{aligned}(M - \Lambda I)\mathbf{Y} &= \mathbf{0}\,,\\ (M - \lambda I)\boldsymbol{y} &= \mathbf{T}\,,\end{aligned}$$

where

$$\begin{aligned}\mathbf{Y} &= (Y_1, \ldots, Y_{n-1})^{\mathrm{T}}\,,\\ \boldsymbol{y} &= (y_1, \ldots, y_{n-1})^{\mathrm{T}}\,,\\ \mathbf{T} &= (T_1, \ldots, T_{n-1})^{\mathrm{T}}\,.\end{aligned}$$

Theorem 5.14 of Chapter 5 applies to this problem, and shows that one of the eigenvalues, Λ_m, of the matrix M satisfies

$$|\lambda_m - \Lambda_m| \le \|\mathbf{T}\|_2 / \|\boldsymbol{y}\|_2\,. \tag{13.23}$$

In the simpler case where $p(x) \equiv 1$ and $r(x) \equiv 0$ the truncation error is

$$T_j = -\tfrac{1}{12}h^2 y^{\mathrm{iv}}(\xi_j)\,, \qquad \xi_j \in (x_{j-1}, x_{j+1})\,,$$

so the numerical method has evaluated the eigenvalue with error less than

$$\tfrac{1}{12}h^2 \left\{\sum_{j=1}^{n-1}[y^{\mathrm{iv}}(\xi_j)]^2\right\}^{1/2} \left\{\sum_{j=1}^{n-1}[y(x_j)]^2\right\}^{-1/2}.$$

Since the mth eigenfunction $y_{(m)}$ is given by

$$y(x) = y_{(m)}(x) = \sin(m\pi(x-a)/(b-a)), \qquad x \in (a,b),$$

we see that

$$y^{\text{iv}}(x) = \left[\frac{m\pi}{b-a}\right]^4 y(x), \qquad x \in (a,b).$$

This shows that, for example, the error in the tenth eigenvalue, corresponding to $m = 10$, is likely to be about 10^4 times larger than the error in the first eigenvalue; more generally, to evaluate higher eigenvalues of the equation will require the use of a smaller mesh size h.

13.7 The shooting method

The methods we have described for the linear boundary value problem may be extended to nonlinear differential equations. We shall not discuss how this is done; instead, we shall describe an alternative approach, called the **shooting method**. We shall consider the nonlinear model problem

$$y'' = f(x,y), \quad a < x < b, \qquad y(a) = A, \quad y(b) = B,$$

where we assume that the function $f(x,y)$ is continuous and differentiable, and that

$$\frac{\partial f}{\partial y}(x,y) \geq 0, \qquad a < x < b, \quad y \in \mathbb{R}.$$

The central idea of the method is to replace the boundary value problem under consideration by an initial value problem of the form

$$y'' = f(x,y), \quad a < x \leq b, \qquad y(a) = A, \quad y'(a) = t,$$

where t is to be chosen in such a way that $y(b) = B$. This can be thought of as a problem of trying to determine the angle of inclination $\tan^{-1} t$ of a loaded gun, so that, when shot from height A at the point $x = a$, the bullet hits the target placed at height B at the point $x = b$. Hence the name, shooting method.

Once the boundary value problem has been transformed into such an 'equivalent' initial value problem, any of the methods for the numerical solution of initial value problems discussed in Chapter 12 can be applied to find a numerical solution. Thus, in particular, the costly exercise of solving a large simultaneous system of nonlinear equations, arising from a direct finite difference approximation of the nonlinear boundary value problem, can be completely avoided.

If we write

$$y'(a) = t\,,$$

a numerical solution of the differential equation with the initial conditions $y(a) = A, y'(a) = t$ can be obtained by any of the methods of Chapter 12. This solution will depend on t, and we may write it as $y(x;t)$. In particular the value at $x = b$ will be a function of t,

$$y(b;t) = \psi(t)\,. \tag{13.24}$$

The solution of the nonlinear boundary value problem therefore reduces to the determination of the value of t for which the boundary condition at $x = b$ is also satisfied, *i.e.*,

$$\psi(t) - B = 0\,.$$

There are a number of well-known methods for the solution of equations of this form; Newton's method is an obvious example. Generally, we shall not, of course, have a closed form expression for the function $\psi(t)$, but this is not necessary; all that is needed is a numerical algorithm to calculate the value of $\psi(t)$ for a given value of t, and this we have. To use Newton's method we shall also need to be able to calculate the value of $\psi'(t)$, and this is easily done.

The function $y(x;t)$ is defined, for all t, as the solution of the initial value problem

$$y''(x;t) = f(x, y(x;t))\,, \qquad y(a;t) = A\,, \quad y'(a;t) = t\,, \tag{13.25}$$

where $'$ and $''$ indicate differentiation with respect to the variable x. We can differentiate these throughout with respect to t, giving

$$\frac{\partial}{\partial t} y''(x;t) = \frac{\partial f}{\partial y}(x, y(x;t))\frac{\partial y}{\partial t}(x;t)\,, \quad \frac{\partial y}{\partial t}(a;t) = 0\,, \quad \frac{\partial y'}{\partial t}(a;t) = 1\,.$$

Writing

$$w(x,t) = \frac{\partial y}{\partial t}(x;t)\,,$$

and interchanging the order of differentiation, we find that $w(x;t)$ may be obtained as the solution of the initial value problem

$$w''(x;t) = w(x;t)\,\frac{\partial f}{\partial y}(x, y(x;t))\,, \qquad w(a;t) = 0\,, \quad w'(a;t) = 1\,. \tag{13.26}$$

By virtue of (13.24), the required derivative is then given by

$$\psi'(t) = w(b,t)\,.$$

To implement this method, it is convenient to solve the two initial value problems, (13.25) and (13.26), in tandem, by writing them as a system of four simultaneous first-order differential equations:

$$\left.\begin{array}{rcl} u_1'(x;t) &=& u_2(x;t)\,,\\ u_2'(x;t) &=& f(x,u_1(x;t))\,,\\ u_3'(x;t) &=& u_4(x;t)\,,\\ u_4'(x;t) &=& u_3(x;t)\,\frac{\partial f}{\partial u_1}(x,u_1(x;t))\,, \end{array}\right\} \qquad (13.27)$$

with the initial conditions

$$u_1(a;t) = A\,,\quad u_2(a;t) = t\,,\quad u_3(a;t) = 0\,,\quad u_4(a;t) = 1\,,$$

where $u_1(x;t)$ denotes $y(x;t)$, $u_3(x;y)$ signifies $w(x;t)$, and u_2 and u_4 are defined by $u_2 = u_1' = y'$ and $u_4 = u_3' = w'$.

Having obtained a numerical solution of this system of differential equations for some chosen value of t, $t^{(k)}$ say, Newton's method gives, as the next, improved, value for t,

$$t^{(k+1)} = t^{(k)} - \frac{\psi(t^{(k)}) - B}{\psi'(t^{(k)})} = t^{(k)} - \frac{u_1(b,t^{(k)}) - B}{u_3(b,t^{(k)})}\,,\qquad k = 0,1,\ldots\,,$$

iterating until a certain number of decimal digits have converged.

Theorem 13.9 *Suppose that a numerical algorithm for the solution of the system of differential equations (13.27) gives the result $v_{i,j}(t)$, the numerical approximation to $u_i(x_j;t)$, $i = 1,2,3,4$, $j = 1,2,\ldots,n$, where the error satisfies*

$$\max_{1\le j\le n} |u_i(x_j;t) - v_{i,j}(t)| \le C(t)h^s\,,\qquad i = 1,2,3,4\,,$$

for some $s > 0$; here $C(t)$ depends on bounds on the derivatives of y and $f(x,y)$, and on t. Suppose also that the Newton iteration is performed until

$$|v_{1,n}(t^{(k)}) - B| \le \varepsilon\,.$$

Then, $v_{1,j}(t^{(k)})$ is an approximation to the solution of the boundary value problem which satisfies

$$\max_{1\le j\le n} |y(x_j) - v_{1,j}(t^k)| \le 2C(t^{(k)})h^s + \varepsilon\,.$$

Proof Suppose that the solution of the system of differential equations with $t = t^{(k)}$ is $u_i(x; t^{(k)})$, $i = 1, 2, 3, 4$, and the corresponding numerical solution is $v_{i,j}(t^{(k)})$, $i = 1, 2, 3, 4$, $j = 1, 2, \ldots, n$; then

$$|u_i(x_j; t^{(k)}) - v_{i,j}(t^{(k)})| \le C(t^{(k)})h^s .$$

Moreover $|v_{1,n}(t^{(k)}) - B| \le \varepsilon$, so that

$$\begin{aligned} |u_1(b; t^{(k)}) - B| &\le |u_1(b; t^{(k)}) - v_{1,n}(t^{(k)})| + |v_{1,n}(t^{(k)}) - B| \\ &\le C(t^{(k)})h^s + \varepsilon . \end{aligned} \tag{13.28}$$

Let us write $\eta(x; t) = y(x) - u_1(x; t)$; by subtraction we see that

$$\begin{aligned} \eta''(x; t) &= y''(x) - u_1''(x; t) \\ &= f(x, y(x)) - f(x, u_1(x, t)) \\ &= \eta(x; t)\frac{\partial f}{\partial y}(x, \xi(x; t)) , \end{aligned}$$

where $\xi(x; t)$ lies between $u_1(x; t)$ and $y(x)$.

Suppose that $\eta'(a; t) > 0$; since $\eta(a; t) = 0$, there is some interval to the right of a in which $\eta(x; t) > 0$. Then, either $\eta(x; t) > 0$ for the whole of $(a, b]$, or there is a value c such that $a < c < b$ and $\eta(c; t) = 0$. In the latter case, $\eta'(x; t)$ must vanish at some point $x = d$ between a and c. However, in the interval $[a, d]$, $\eta(x; t) > 0$ and $\partial f / \partial y \ge 0$, so that $\eta''(x; t) > 0$. Consequently, in the interval $[a, d]$, $\eta'(x; t) > \eta'(a; t) > 0$, and we have a contradiction. Thus, $\eta(x; t) > 0$ for $a < x \le b$. It then follows that $\eta''(x; t)$, and hence also $\eta'(x; t)$ are positive on the whole interval $[a, b]$, which means that $x \mapsto \eta(x; t)$ is monotonic increasing on $[a, b]$ and positive on $(a, b]$. If we had begun with the assumption that $\eta'(a; t) < 0$ an analogous argument shows that $x \mapsto \eta(x; t)$ would have been monotonic decreasing on $[a, b]$ and negative on $(a, b]$. It is left to the reader to discuss the trivial case when $\eta'(a, t) = 0$.

In any case,

$$|\eta(x; t)| \le |\eta(b; t)| , \qquad a \le x \le b ,$$

and therefore, since $y(b) = B$ and recalling (13.28),

$$|y(x) - u_1(x; t^{(k)})| \le |B - u_1(b; t^{(k)})| \le C(t^{(k)})h^s + \varepsilon .$$

Thus, finally,

$$\begin{aligned} |y(x_j) - v_{1,j}(t^{(k)})| &\le |y(x_j) - u_1(x_j; t^{(k)})| + |u_1(x_j; t^{(k)}) - v_{1,j}(t^{(k)})| \\ &\le C(t^{(k)})h^s + \varepsilon + C(t^{(k)})h^s , \qquad j = 1, 2, \ldots, n , \end{aligned}$$

and hence the desired bound. □

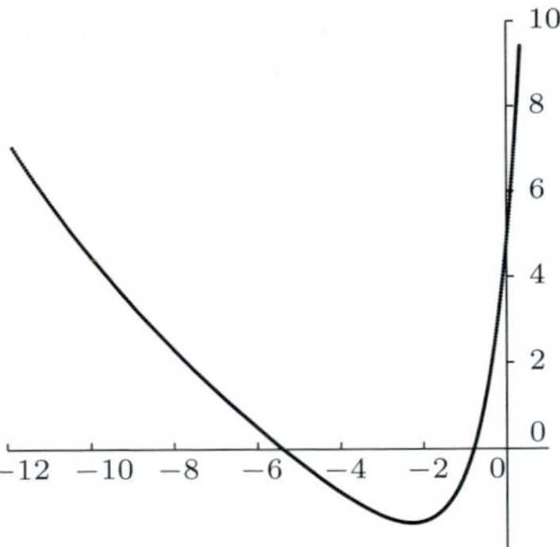

Fig. 13.1. The function $t \mapsto \psi(t)$.

The shooting method is an example of a technique which can be applied to much more general problems, including systems of differential equations of any order, with some boundary conditions specified at each end of the interval. The condition $\partial f/\partial y \geq 0$ is restrictive, and may often not be satisfied in practical problems. Note that if $f(x, y)$ is linear in y, of the form $f(x, y) = r(x)y + g(x)$, this condition is the same as the condition $r(x) \geq 0$ imposed on the model problem in (13.2). Perhaps the simplest example of a nonlinear two-point boundary value problem is

$$y'' = y^2, \qquad y(-1) = y(1) = 1, \tag{13.29}$$

where $\partial f/\partial y = 2y$, which does *not* satisfy the condition $\partial f/\partial y \geq 0$, $y \in \mathbb{R}$. In fact, problem (13.29) has two solutions, one of which is positive, and the other takes negative values around $x = 0$.

Figure 13.1 shows a graph of the corresponding function $t \mapsto \psi(t)$ defined in (13.24), over the range $-12 \leq t \leq 0$; outside this range the function ψ tends quite rapidly to $+\infty$. This shows clearly the two solutions to the boundary value problem, given by the two values of t at which $\psi(t) = 1$. The two solutions are displayed in Figure 13.2.

For the positive solution it is reasonable to suppose that the above proof could be modified so that it requires only that $\partial f/\partial y$ is positive for values of y in the neighbourhood of the solution, and the error bound would then hold, at least if h and ε were sufficiently small. The analysis of the error of the other solution, which takes negative values, will be much more difficult, as our proof relies heavily on the monotonicity of solutions of the linearised equation.

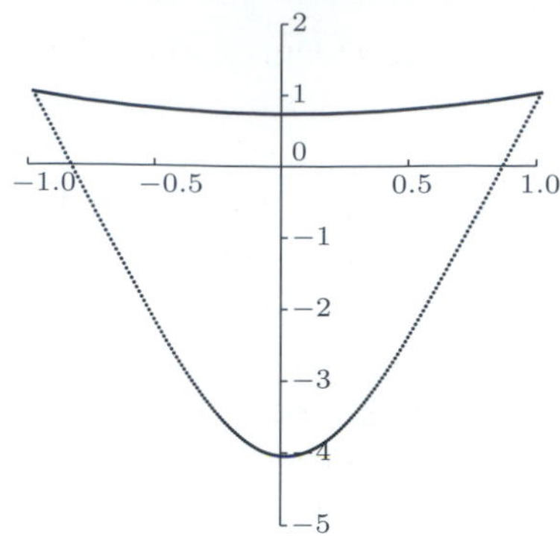

Fig. 13.2. The two solutions of the nonlinear boundary value problem (13.29).

13.8 Notes

The following books are standard texts on the subject of numerical approximation of boundary value problems:

- H.B. KELLER, *Numerical Methods for Two-Point Boundary Value Problems,* Reprint of the 1968 original published by Blaisdell, Dover, New York, 1992.
- H.B. KELLER, *Numerical Solution of Two-Point Boundary Value Problems,* SIAM, Philadelphia, fourth printing, 1990.

A more recent survey of the subject is found in

- U.M. ASCHER, R.M.M. MATTHEIJ AND R.D. RUSSELL, *Numerical Solution of Boundary Value Problems for Ordinary Differential Equations,* Corrected reprint of the 1988 original, Classics in Applied Mathematics, 13, SIAM, Philadelphia, 1995.

In practical implementations of the shooting method into mathematical software (see, for example, Appendix A in the Ascher *et al.* book), the interval $[a, b]$ is subdivided into smaller intervals on each of which the shooting method is applied with appropriately chosen initial values. The 'initial' conditions on the subintervals are then simultaneously adjusted in order to satisfy the boundary conditions and appropriate continuity conditions at the points of the subdivision. From the practical viewpoint, this extension of the basic shooting method considered in this chapter is extremely important: the various difficulties which may arise in the implementation of the basic method (such as, for example, growth of the solution to the initial value problem over the interval $[a,b]$, leading to loss of accuracy in the solution of the equation $\psi(t) = B$) are discussed, for example, in Section 2.4 of the 1992 book by Keller.

Sturm–Liouville problems originated in a paper of Jacques Charles François Sturm: Sur les équations différentielles linéaires du second ordre, *J. Math. Pures Appl.* **1**, 106–186, 1836, in Joseph Liouville's newly founded journal. Sturm's paper was followed by a series of articles by Sturm and Liouville in subsequent volumes of the journal. They examined general linear second-order differential equations, the properties of their eigenvalues, the behaviour of the eigenfunctions and the series expansion of arbitrary functions in terms of these eigenfunctions. An extensive survey of the theory and numerical analysis of Sturm–Liouville problems can be found in

- John D. Pryce, *Numerical Solution for Sturm–Liouville Problems,* Oxford University Press Monographs in Numerical Analysis, Clarendon Press, Oxford, 1993.

See also Section 11.3, page 478, of the Ascher *et al.* book cited above.

Exercises

13.1 Suppose that $y \in \mathrm{C}^6[x-h, x+h]$; show that there exists a real number η in $(x-h, x+h)$ such that

$$\frac{\delta^2 y(x)}{h^2} = y''(x) + \tfrac{1}{12}h^2 y^{\mathsf{iv}}(x) + \tfrac{1}{360}h^4 y^{\mathsf{vi}}(\eta)\,.$$

13.2 Use Theorem 3.6 to show that the matrix M in (13.7) is monotone. Use the result of Exercise 3.4 to show that $\|M^{-1}\|_\infty \le \frac{1}{8}$.

13.3 On the interval $[a,b]$ the differential equation

$$-y'' + f(x)y = g(x)$$

is approximated by

$$-\frac{\delta^2 y_j}{h^2} + \beta_{-1}y_{j-1} + \beta_0 y_j + \beta_1 y_{j+1} = \beta_{-1}g_{j-1} + \beta_0 g_j + \beta_1 g_{j+1}\,,$$

where β_{-1}, β_0 and β_1 are constants. Assuming that the solution y has the appropriate number of continuous derivatives, show that the truncation error of this approximation may be written as follows:

(i) if $\beta_{-1} + \beta_0 + \beta_1 \neq 1$, then

$$T_j = (\beta_{-1} + \beta_0 + \beta_1) y''(x_j) + Z_j^{(0)} h\,,$$

where $|Z_j^{(0)}| \leq (|\beta_{-1}| + |\beta_1|) M_3$;

(ii) if $\beta_{-1} + \beta_0 + \beta_1 = 1$ and $\beta_{-1} \neq \beta_1$, then

$$T_j = (\beta_1 - \beta_{-1}) h y'''(x_j) + Z_j^{(1)} h^2\,,$$

where $|Z_j^{(1)}| \leq [\frac{1}{2}(|\beta_{-1}| + |\beta_1|) + \frac{1}{12}] M_4$;

(iii) if $\beta_{-1} + \beta_0 + \beta_1 = 1$, $\beta_1 = \beta_{-1}$ and $\beta_1 \neq \frac{1}{12}$, then

$$T_j = (\beta_1 - \tfrac{1}{12}) h^2 y^{\mathsf{iv}}(x_j) + Z_j^{(2)} h^3\,,$$

where $|Z_j^{(2)}| \leq [\frac{1}{12}|\beta_1| + \frac{1}{360}] M_6$;

(iv) if $\beta_{-1} = \beta_1 = \frac{1}{12}$ and $\beta_0 = \frac{5}{6}$, then

$$T_j = \tfrac{1}{240} h^4 y^{\mathsf{vi}}(x_j) + Z_j^{(4)} h^6\,,$$

where $|Z_j^{(4)}| \leq \frac{1}{60480} M_8$.

13.4 The approximation of Exercise 3 is used, with the values $\beta_1 = \beta_{-1} = 1/12$, $\beta_0 = 5/6$. Use Taylor's Theorem with integral remainder (Appendix, Theorem A.5) to show that the truncation error of this approximation may be written

$$T_j = \int_{-h}^{h} G(s) y^{\mathsf{vi}}(x_j + s)\, \mathrm{d}s\,,$$

where

$$G(s) = (h - s)^5/5! - \tfrac{1}{12} h^2 (h - s)^3/3!\,, \qquad 0 \leq s \leq h\,,$$

with a similar expression for $-h \leq s \leq 0$. Show that $G(s) \leq 0$ for all $s \in [-h, h]$, and hence use the Integral Mean Value Theorem to show that the truncation error can be expressed as

$$T_j = \frac{h^6}{240} y^{\mathsf{vi}}(\xi)$$

for some value of ξ in $(x_j - h, x_j + h)$.

13.5 Suppose that the solution of (13.1), (13.2) has a continuous sixth derivative on $[a, b]$, and that Y_j is the solution of the approximation used in Exercise 4. Show that

$$|y(x_j) - Y_j| \leq \tfrac{1}{2880} h^4 (b-a)^2 M_6 , \qquad j = 0, \ldots, n ,$$

provided that

$$h^2 r(x_j) \leq 12 , \qquad j = 1, \ldots, n-1 .$$

13.6 Complete the proof of Theorem 13.7.

13.7 Show that the solution of the boundary value problem

$$-y'' + a^2 y = 0 , \quad y(-1) = 1 , \qquad y(1) = 1 ,$$

is

$$y(x) = \frac{\cosh ax}{\cosh a} .$$

Use the identity

$$\cosh(x+h) + \cosh(x-h) = 2 \cosh x \, \cosh h$$

to verify that the solution of the difference approximation (13.5) to this problem is

$$Y_j = \frac{\cosh \vartheta x_j}{\cosh \vartheta} ,$$

where

$$\vartheta = (1/h) \cosh^{-1}(1 + \tfrac{1}{2} a^2 h^2) .$$

By expanding in Taylor series, show that

$$Y_j = y(x_j) + \tfrac{1}{24} h^2 a^3 (\cosh ax \sinh a - x \sinh ax \cosh a)/(\cosh a)^2 + \mathcal{O}(h^4) .$$

Verify that this result is consistent with Theorem 13.4 when h is small.

13.8 Carry out a similar analysis as in Exercise 7 for the boundary value problem

$$-y'' - a^2 y = 0 , \qquad y(0) = 0 , \quad y(1) = 1 ,$$

and explain why in this case Theorem 13.4 cannot be used. What restriction is required on the value of a?

13.9 The eigenvalue problem

$$-y'' = \lambda y\,, \qquad y(0) = y(1) = 0\,,$$

is approximated by

$$-\frac{Y_{j+1} - 2Y_j + Y_{j-1}}{h^2} = \mu Y_j\,,\ 1 \le j \le n-1\,, \quad Y_0 = Y_n = 0\,.$$

Show that the differential equation has solution $y = \sin m\pi x$, $\lambda = m^2\pi^2$ for any positive integer m. Show also that the difference approximation has solution $Y_j = \sin m\pi x_j$, $j = 0, 1, \ldots, n$, and give an expression for the corresponding value of μ. Use the fact that

$$1 - \cos\vartheta = \tfrac{1}{2}\vartheta^2 - \tfrac{1}{24}\xi\vartheta^4, \quad |\xi| \le 1\,,$$

to show that $|\lambda - \mu| \le m^4\pi^4h^2/12$, and compare with the bound given by (13.23).

14

The finite element method

14.1 Introduction: the model problem

In Chapter 13 we explored finite difference methods for the numerical solution of two-point boundary value problems. The present chapter is devoted to the foundations of the theory of finite element methods. For the sake of simplicity the exposition will be, at least initially, confined to the second-order ordinary differential equation

$$-\frac{\mathrm{d}}{\mathrm{d}x}\left(p(x)\frac{\mathrm{d}u}{\mathrm{d}x}\right)+r(x)u=f(x)\,,\quad a<x<b\,, \tag{14.1}$$

where $p \in \mathrm{C}^1[a,b]$, $r \in \mathrm{C}[a,b]$, $f \in \mathrm{L}^2(a,b)$ and $p(x) \geq c_0 > 0$, $r(x) \geq 0$ for all $x \in [a,b]$, subject to the boundary conditions

$$u(a)=A\,,\quad u(b)=B\,. \tag{14.2}$$

Later on in the chapter, in Section 14.5, we shall also consider the ordinary differential equation

$$-\frac{\mathrm{d}}{\mathrm{d}x}\left(p(x)\frac{\mathrm{d}u}{\mathrm{d}x}\right)+q(x)\frac{\mathrm{d}u}{\mathrm{d}x}+r(x)u=f(x)\,,\quad a<x<b\,, \tag{14.3}$$

subject to the boundary conditions (14.2). Indeed, much of the material discussed here can be extended to partial differential equations; for pointers to the relevant literature we refer to the Notes at the end of the chapter.

The finite element method was proposed in a paper by Richard Courant in the early 1940s,[1] although the historical roots of the method can be traced back to earlier work by Galerkin[2] in 1915; unfortunately, the relevance of Courant's article was not recognised at the time and the idea was forgotten. In the early 1950s the method was rediscovered by engineers, but its systematic mathematical analysis began only a decade later. Since then, the finite element method has been developed into one of the most general and powerful techniques for the numerical solution of differential equations and it is widely used in engineering design and analysis.

Unlike finite difference schemes which seek to approximate the unknown analytical solution to a differential equation at a finite number of selected points, the grid points or mesh points in the computational domain, the finite element method supplies an approximation to the analytical solution in the form of a piecewise polynomial function, defined over the entire computational domain. For example, in the case of the boundary value problem (14.1), (14.2), the simplest finite element method uses a linear spline, defined over the interval $[a, b]$, to approximate the analytical solution u.

We shall consider two techniques for the construction of finite element approximations: the **Rayleigh–Ritz principle** and the **Galerkin principle**. In the case of the boundary value problem (14.1), (14.2) the approximations which stem from these two principles will be seen to coincide. We note, however, that since the Rayleigh–Ritz principle relies on the fact that the boundary value problem under consideration can be restated as a variational problem involving the minimisation of a certain quadratic functional over a function space, its use is restricted to *symmetric* boundary value problems, such as (14.1), (14.2) where (14.1) does not contain a first-derivative term; for example, the Rayleigh–Ritz principle is not applicable to (14.3), (14.2) unless $q(x) \equiv 0$. The precise sense in which the word *symmetric* is to be interpreted here will be clar-

[1] R. Courant, Variational methods for the solution of problems in equilibrium and vibrations, *Bull. Amer. Math. Soc.* **49**, 1–23, 1943; Richard Courant (8 January 1888, Lublinitz, Prussia, Germany (now Lubliniec, Poland) – 27 January 1972, New Rochelle, New York, USA). For an illuminating account of the lives of Richard Courant and David Hilbert, see the book of Constance Reid: *Hilbert–Courant*, Springer, New York, 1986.

[2] Boris Grigorievich Galerkin (4 March 1871, Polotsk, Russia (now in Belarus) – 12 June 1945, Moscow, USSR) studied mathematics and engineering at the St Petersburg Technological Institute. During his studies he supported himself by private tutoring and working as a designer. His ideas on the approximate solution of differential equations were published in 1915. From 1940 until his death, Galerkin was head of the Institute of Mechanics of the Soviet Academy of Sciences.

ified later in the chapter. On the other hand, as we shall see in Section 14.5, the Galerkin principle is more generally applicable and does not require symmetry of the boundary value problem.

To make these observations rigorous, we recall from Chapter 11 the concept of **Sobolev space**.

Definition 14.1 *For a positive integer k, we define the* **Sobolev space** $\mathrm{H}^k(a,b)$ *as the set of real-valued functions v defined on $[a,b]$ such that v and all of its derivatives of order up to and including $k-1$ are absolutely continuous on $[a,b]$ and*

$$v^{(k)} = \frac{\mathrm{d}^k v}{\mathrm{d}x^k} \in \mathrm{L}^2(a,b)\,.$$

Here $\mathrm{L}^2(a,b)$ denotes the set of all functions defined on (a,b) such that

$$\|v\|_2 = \|v\|_{\mathrm{L}^2(a,b)} = \left(\int_a^b |v(x)|^2 \mathrm{d}x\right)^{1/2}$$

is finite. We equip $\mathrm{H}^k(a,b)$ with the **Sobolev norm**

$$\|v\|_{\mathrm{H}^k(a,b)} = \left(\sum_{m=0}^{k} \|v^{(m)}\|^2_{\mathrm{L}^2(a,b)}\right)^{1/2},$$

where $v^{(0)} = v$.

The Sobolev spaces $\mathrm{H}^1(a,b)$ and $\mathrm{H}^2(a,b)$ corresponding, respectively, to $k=1$ and $k=2$ will be particularly relevant in this chapter. The next definition introduces variants of the space $\mathrm{H}^1(a,b)$ required for the imposition of the boundary conditions (14.2).

Definition 14.2 (*i*) *Given that A and B are real numbers, $\mathrm{H}^1_{\mathrm{E}}(a,b)$ will denote the set of all functions $v \in \mathrm{H}^1(a,b)$ such that $v(a) = A$ and $v(b) = B$.*

(*ii*) *$\mathrm{H}^1_0(a,b)$ will signify the set of all functions $v \in \mathrm{H}^1(a,b)$ such that $v(a) = 0$ and $v(b) = 0$.*

In the next section we shall state, using Sobolev spaces, the Rayleigh–Ritz and Galerkin principles associated with the boundary value problem (14.1), (14.2), and explore their relationship.

14.2 Rayleigh–Ritz and Galerkin principles

The Rayleigh–Ritz principle relies on converting the boundary value problem (14.1), (14.2) into a variational problem involving the minimisation of a certain quadratic functional over a function space.

Let us define the quadratic functional $\mathcal{J}\colon \mathrm{H}^1_{\mathrm{E}}(a,b) \to \mathbb{R}$ by

$$\mathcal{J}(w) = \tfrac{1}{2}\int_a^b [p(x)(w')^2 + r(x)w^2]\mathrm{d}x - \int_a^b f(x)w(x)\mathrm{d}x$$

where $w \in \mathrm{H}^1_{\mathrm{E}}(a,b)$, and consider the following *variational problem*:

$$\text{(RR)}\quad \text{find } u \in \mathrm{H}^1_{\mathrm{E}}(a,b) \text{ such that } \mathcal{J}(u) = \min_{w\in\mathrm{H}^1_{\mathrm{E}}(a,b)} \mathcal{J}(w)\,,$$

which we shall henceforth refer to as the **Rayleigh–Ritz principle**. For the sake of notational simplicity we define

$$\mathcal{A}(w,v) = \int_a^b [p(x)\, w'(x)v'(x) + r(x)\, w(x)v(x)]\mathrm{d}x$$

and recall from Chapter 9 the definition of inner product on $\mathrm{L}^2(a,b)$:

$$\langle w, v\rangle \;=\; \int_a^b w(x)v(x)\mathrm{d}x\,.$$

Using these, we can rewrite $\mathcal{J}(w)$ as follows:

$$\mathcal{J}(w) = \tfrac{1}{2}\mathcal{A}(w,w) - \langle f, w\rangle\,, \qquad w \in \mathrm{H}^1_{\mathrm{E}}(a,b)\,. \tag{14.4}$$

The mapping $\mathcal{A}\colon \mathrm{H}^1(a,b)\times\mathrm{H}^1(a,b) \to \mathbb{R}$ is a **bilinear functional** in the following sense:

❶ $\mathcal{A}(\lambda_1 w_1 + \lambda_2 w_2, v) = \lambda_1\mathcal{A}(w_1,v) + \lambda_2\mathcal{A}(w_2,v)$
for all $\lambda_1,\lambda_2 \in \mathbb{R}$ and all $w_1, w_2, v \in \mathrm{H}^1(a,b)$;

❷ $\mathcal{A}(w, \mu_1 v_1 + \mu_2 v_2) = \mu_1\mathcal{A}(w,v_1) + \mu_2\mathcal{A}(w,v_2)$
for all $\mu_1,\mu_2 \in \mathbb{R}$ and all $w, v_1, v_2 \in \mathrm{H}^1(a,b)$.

We note, in addition, that the bilinear functional $\mathcal{A}(\,\cdot\,,\,\cdot\,)$ is **symmetric**, in that

$$\mathcal{A}(w,v) = \mathcal{A}(v,w) \qquad \forall\, w, v \in \mathrm{H}^1(a,b)\,. \tag{14.5}$$

Our next result provides an equivalent characterisation of the Rayleigh–Ritz principle; it relies on the fact that the bilinear functional $\mathcal{A}(\,\cdot\,,\,\cdot\,)$ is symmetric in the sense of (14.5).

Theorem 14.1 *A function u in $\mathrm{H}^1_{\mathrm{E}}(a,b)$ minimises $\mathcal{J}(\,\cdot\,)$ over $\mathrm{H}^1_{\mathrm{E}}(a,b)$ if, and only if,*

$$\text{(G)} \qquad \mathcal{A}(u,v) = \langle f, v\rangle \qquad \forall\, v \in \mathrm{H}^1_0(a,b). \tag{14.6}$$

This identity will be referred to as the **Galerkin principle**.

Proof of theorem Suppose that $u \in \mathrm{H}^1_{\mathrm{E}}(a,b)$ minimises $\mathcal{J}(\,\cdot\,)$ over $\mathrm{H}^1_{\mathrm{E}}(a,b)$; that is, $\mathcal{J}(u) \le \mathcal{J}(w)$ for all $w \in \mathrm{H}^1_{\mathrm{E}}(a,b)$. Noting that $w = u + \lambda v$ belongs to $\mathrm{H}^1_{\mathrm{E}}(a,b)$ for all $\lambda \in \mathbb{R}$ and all $v \in \mathrm{H}^1_0(a,b)$, we deduce that

$$\begin{aligned} \mathcal{J}(u) &\le \mathcal{J}(u+\lambda v) = \tfrac{1}{2}\mathcal{A}(u+\lambda v, u+\lambda v) - \langle f, u+\lambda v\rangle \\ &= \mathcal{J}(u) + \lambda[\mathcal{A}(u,v) - \langle f,v\rangle] + \tfrac{1}{2}\lambda^2 \mathcal{A}(v,v) \end{aligned} \tag{14.7}$$

for all $v \in \mathrm{H}^1_0(a,b)$ and all $\lambda \in \mathbb{R}$. Here, in the transition from the first line to the second we made use of the fact that $\mathcal{A}(u,v) = \mathcal{A}(v,u)$ for all v in $\mathrm{H}^1_0(a,b)$, which follows from (14.5). Now, (14.7) implies that

$$-\tfrac{1}{2}\lambda^2 \mathcal{A}(v,v) \le \lambda[\mathcal{A}(u,v) - \langle f,v\rangle]$$

for all $v \in \mathrm{H}^1_0(a,b)$ and all $\lambda \in \mathbb{R}$. Let us suppose that $\lambda > 0$, divide both sides of the last inequality by λ and pass to the limit $\lambda \to 0$ to deduce that

$$0 \le \mathcal{A}(u,v) - \langle f,v\rangle \qquad \forall\, v \in \mathrm{H}^1_0(a,b)\,. \tag{14.8}$$

On replacing v by $-v$ in (14.8), we have that also

$$0 \ge \mathcal{A}(u,v) - \langle f,v\rangle \qquad \forall\, v \in \mathrm{H}^1_0(a,b)\,. \tag{14.9}$$

We conclude from (14.8) and (14.9) that

$$\mathcal{A}(u,v) = \langle f,v\rangle \qquad \forall\, v \in \mathrm{H}^1_0(a,b)\,, \tag{14.10}$$

as required.

Conversely, if $u \in \mathrm{H}^1_{\mathrm{E}}(a,b)$ is such that $\mathcal{A}(u,v) = \langle f,v\rangle$ for all v in $\mathrm{H}^1_0(a,b)$, then

$$\mathcal{J}(u+\lambda v) = \mathcal{J}(u) + \lambda\,[\mathcal{A}(u,v) - \langle f,v\rangle] + \tfrac{1}{2}\lambda^2\mathcal{A}(v,v) \ge \mathcal{J}(u)$$

for all $v \in \mathrm{H}^1_0(a,b)$ and all $\lambda \in \mathbb{R}$; therefore, u minimises $\mathcal{J}(\,\cdot\,)$ over $\mathrm{H}^1_{\mathrm{E}}(a,b)$. □

Thus we have shown that, as long as $\mathcal{A}(\,\cdot\,,\,\cdot\,)$ is a symmetric bilinear functional, $u \in \mathrm{H}^1_{\mathrm{E}}(a,b)$ satisfies the Rayleigh–Ritz principle if, and only if, it satisfies the Galerkin principle.[1] Our next task is to explain the

[1] In the language of the calculus of variations, (G) is the Euler–Lagrange equation for the minimisation problem (RR).

relationship between (RR) and (G) on the one-hand and (14.1), (14.2) on the other. Since in the case of a symmetric bilinear functional $\mathcal{A}(\,\cdot\,,\,\cdot\,)$ the principles (RR) and (G) are equivalent, it is sufficient to clarify the connection between (G), for example, and the boundary value problem (14.1), (14.2).

We begin with the following definition.

Definition 14.3 *If a function $u \in \mathrm{H}^1_{\mathrm{E}}(a,b)$ satisfies the Galerkin principle (14.6), it is called a* **weak solution** *to the boundary value problem (14.1), (14.2), and the Galerkin principle is referred to as the* **weak formulation** *of the boundary value problem (14.1), (14.2).*

Let us justify this terminology. Suppose that $u \in \mathrm{H}^2(a,b) \cap \mathrm{H}^1_{\mathrm{E}}(a,b)$ is a solution to the boundary value problem (14.1), (14.2). Then,

$$-\frac{\mathrm{d}}{\mathrm{d}x}\left(p(x)\frac{\mathrm{d}u}{\mathrm{d}x}\right) + r(x)u = f(x)\,, \tag{14.11}$$

for almost every $x \in (a,b)$ (see the discussion prior to Example 11.1 for a definition of **almost every**). Multiplying this equality by an arbitrary function $v \in \mathrm{H}^1_0(a,b)$, and integrating over (a,b), we conclude that

$$-\int_a^b \frac{\mathrm{d}}{\mathrm{d}x}\left(p(x)\frac{\mathrm{d}u}{\mathrm{d}x}\right) v\,\mathrm{d}x + \int_a^b r(x)uv\,\mathrm{d}x = \int_a^b f(x)v(x)\,\mathrm{d}x\,.$$

On integration by parts in the first term on the left-hand side,

$$-\int_a^b \frac{\mathrm{d}}{\mathrm{d}x}\left(p(x)\frac{\mathrm{d}u}{\mathrm{d}x}\right) v\,\mathrm{d}x = \left[p(x)\frac{\mathrm{d}u}{\mathrm{d}x}v\right]_{x=a}^{b} + \int_a^b p(x)\frac{\mathrm{d}u}{\mathrm{d}x}\frac{\mathrm{d}v}{\mathrm{d}x}\,\mathrm{d}x\,.$$

Since, by hypothesis, $v(a) = 0$ and $v(b) = 0$, it follows that

$$\int_a^b p(x)\frac{\mathrm{d}u}{\mathrm{d}x}\frac{\mathrm{d}v}{\mathrm{d}x}\,\mathrm{d}x + \int_a^b r(x)uv\,\mathrm{d}x = \int_a^b f(x)v(x)\,\mathrm{d}x$$

for all $v \in \mathrm{H}^1_0(a,b)$. Thus, we have shown the following result.

Theorem 14.2 *If $u \in \mathrm{H}^2(a,b) \cap \mathrm{H}^1_{\mathrm{E}}(a,b)$ is a solution to the boundary value problem (14.1), (14.2), then u is a weak solution to this problem; that is,*

$$\mathcal{A}(u,v) = \langle f, v\rangle \qquad \forall\, v \in \mathrm{H}^1_0(a,b)\,. \tag{14.12}$$

The converse implication, namely that any weak solution $u \in \mathrm{H}^1_{\mathrm{E}}(a,b)$ of (14.1), (14.2) belongs to $\mathrm{H}^2(a,b) \cap \mathrm{H}^1_{\mathrm{E}}(a,b)$ and solves (14.1), (14.2) in the usual (pointwise) sense, is not true in general, unless the weak solution can be shown to be sufficiently smooth to belong to $\mathrm{H}^2(a,b)$. It is for this reason that any function $u \in \mathrm{H}^1_{\mathrm{E}}(a,b)$ satisfying (14.12) is called a *weak* solution of the original boundary value problem.

Thus, Theorem 14.1 shows that $u \in \mathrm{H}^1_{\mathrm{E}}(a,b)$ is a weak solution to (14.1), (14.2) if, and only if, it minimises $\mathcal{J}(\,\cdot\,)$ over $\mathrm{H}^1_{\mathrm{E}}(a,b)$. Next, we show that if a weak solution exists then it must be unique.

Theorem 14.3 *The boundary value problem (14.1), (14.2) possesses at most one weak solution in* $\mathrm{H}^1_{\mathrm{E}}(a,b)$.

Proof The proof is by contradiction. Suppose that $u \in \mathrm{H}^1_{\mathrm{E}}(a,b)$ and $\tilde{u} \in \mathrm{H}^1_{\mathrm{E}}(a,b)$ are two weak solutions to (14.1), (14.2). Then, $u - \tilde{u}$ belongs to $\mathrm{H}^1_0(a,b)$, and

$$\mathcal{A}(u-\tilde{u},v) = \mathcal{A}(u,v) - \mathcal{A}(\tilde{u},v) = \langle f,v\rangle - \langle f,v\rangle = 0$$

for all $v \in \mathrm{H}^1_0(a,b)$. In particular,

$$\mathcal{A}(u-\tilde{u},u-\tilde{u}) = 0\,.$$

However, since $p(x) \geq c_0 > 0$ and $r(x) \geq 0$ for all x in $[a,b]$,

$$\mathcal{A}(v,v) = \int_a^b [p(x)(v')^2 + r(x)v^2]\mathrm{d}x \geq c_0 \int_a^b |v'|^2\mathrm{d}x\,.$$

On choosing $v = u - \tilde{u}$, this implies that

$$0 = \mathcal{A}(u-\tilde{u},u-\tilde{u}) \geq c_0 \int_a^b |(u-\tilde{u})'|^2\mathrm{d}x\,.$$

Since the right-hand side in the last inequality is nonnegative, it follows that $(u-\tilde{u})'(x) = 0$ for almost every x in (a,b); as $u - \tilde{u}$ is absolutely continuous on $[a,b]$ and $(u-\tilde{u})(a) = (u-\tilde{u})(b) = 0$, we conclude that $u = \tilde{u}$, and hence we get the desired uniqueness of a weak solution. □

It turns out that under the present hypotheses on p, q and f the *existence* of a weak solution $u \in \mathrm{H}^1_{\mathrm{E}}(a,b)$ is also ensured, although the proof of this is less simple and is omitted here; the interested reader is referred to the literature listed in the Notes at the end of the chapter.

14.3 Formulation of the finite element method

In the previous section we showed that the weak solution to the boundary value problem (14.1), (14.2) minimises $\mathcal{J}(\,\cdot\,)$ over $\mathrm{H}^1_{\mathrm{E}}(a,b)$. The finite element method is based on constructing an approximate solution u^h to the problem by minimising $\mathcal{J}(\,\cdot\,)$ over a finite-dimensional subset S^h_{E} of $\mathrm{H}^1_{\mathrm{E}}(a,b)$, instead.

A simple way of constructing S^h_{E} is to choose any function $\psi \in \mathrm{H}^1_{\mathrm{E}}(a,b)$, for example,

$$\psi(x) = \frac{B-A}{b-a}(x-a) + A \tag{14.13}$$

and a finite set of linearly independent functions φ_j, $j = 1, \dots, n-1$, in $\mathrm{H}^1_0(a,b)$ for $n \geq 2$, and then define

$$S^h_{\mathrm{E}} \;=\; \{v^h \in \mathrm{H}^1_{\mathrm{E}}(a,b)\colon v^h(x) = \psi(x) + \sum_{i=1}^{n-1} v_i\varphi_i(x)\,, \quad \text{where } (v_1,\dots,v_{n-1})^{\mathrm{T}} \in \mathbb{R}^{n-1}\}\,.$$

We consider the following approximation of problem (RR):

$$(\mathrm{RR})^h \quad \text{find } u^h \in S^h_{\mathrm{E}} \text{ such that } \mathcal{J}(u^h) = \min_{w^h \in S^h_{\mathrm{E}}} \mathcal{J}(w^h)\,.$$

Our next result is a finite-dimensional analogue of Theorem 14.1.

Theorem 14.4 *A function $u^h \in S^h_{\mathrm{E}}$ minimises $\mathcal{J}(\,\cdot\,)$ over S^h_{E} if, and only if,*

$$(\mathrm{G})^h \qquad \mathcal{A}(u^h, v^h) = \langle f, v^h\rangle \qquad \forall\, v^h \in S^h_0. \tag{14.14}$$

Here,

$$S^h_0 \;=\; \{v^h \in \mathrm{H}^1_0(a,b)\colon v^h(x) = \sum_{i=1}^{n-1} v_i\varphi_i(x)\,, \quad \text{where } (v_1,\dots,v_{n-1})^{\mathrm{T}} \in \mathbb{R}^{n-1}\}\,.$$

The problem $(\mathrm{G})^h$ can be thought of as an approximation to the Galerkin principle (G), and is therefore referred to as the **Galerkin method**. For a similar reason, $(\mathrm{RR})^h$ is called the Rayleigh–Ritz method, or just **Ritz method**. Thus, in complete analogy with the equivalence of (RR) and (G) formulated in Theorem 14.1, Theorem 14.4 now expresses the equivalence of $(\mathrm{RR})^h$ and $(\mathrm{G})^h$, the approximations to (RR) and (G), respectively. Of course, as in the case of (RR) and (G), the equivalence of $(\mathrm{RR})^h$ and $(\mathrm{G})^h$ relies on the assumption that the bilinear functional $\mathcal{A}(\,\cdot\,,\,\cdot\,)$ is symmetric. The proof is identical to that of Theorem 14.1, and is left as an exercise.

Theorem 14.4 provides no information about the existence and uniqueness of u^h that minimises $\mathcal{J}(\,\cdot\,)$ over S^h_{E} (or, equivalently, of the existence and uniqueness of u^h that satisfies (14.14)). This question is settled by our next result.

Theorem 14.5 *There exists a unique function $u^h \in S^h_{\mathrm{E}}$ that minimises $\mathcal{J}(\,\cdot\,)$ over S^h_{E}; this u^h is called the* **Ritz approximation** *to u. Equivalently, there exists a unique function $u^h \in S^h_{\mathrm{E}}$ that satisfies (14.14); this u^h is called the* **Galerkin approximation** *to u. The Ritz and Galerkin approximations to u coincide.*

Proof We shall prove the second of these two equivalent statements: we shall show that there exists a unique $u^h \in S^h_{\mathrm{E}}$ that satisfies (14.14). The proof of uniqueness of $u^h \in S^h_{\mathrm{E}}$ is analogous to the proof of Theorem 14.3, with u, $\tilde{u}$, $\mathrm{H}^1_{\mathrm{E}}(a,b)$ and $\mathrm{H}^1_0(a,b)$, replaced by u^h, $\tilde{u}^h$, S^h_{E} and S^h_0, respectively. Since S^h_{E} is finite-dimensional, the uniqueness of u^h satisfying (14.14) implies its existence. For an alternative proof of the existence and uniqueness of $u^h \in S^h_{\mathrm{E}}$ see the discussion following Definition 14.4. □

Having shown the existence and uniqueness of u^h minimising $\mathcal{J}(\,\cdot\,)$ over S^h_{E} (or, equivalently, satisfying (14.14)), we adopt the following definition.

Definition 14.4 *The functions φ_i, $i = 1, 2, \ldots, n-1$, appearing in the definitions of S^h_{E} and S^h_0 are called the* **Galerkin basis functions**.

Since any function $v^h \in S^h_0$ can be represented as a linear combination of the Galerkin basis functions φ_i, $1 \le i \le n-1$, it is clear that (14.14) is equivalent to

$$\mathcal{A}(u^h, \varphi_i) = \langle f, \varphi_i \rangle\,, \quad 1 \le i \le n-1\,. \tag{14.15}$$

As u^h belongs to S^h_{E}, it can be expressed in terms of ψ and the Galerkin basis functions as

$$u^h(x) = \psi(x) + \sum_{j=1}^{n-1} u_j \varphi_j(x)\,,$$

where $u_j \in \mathbb{R}$, $j = 1, \ldots, n-1$, are to be determined. On substituting this expansion of u^h into (14.15), we arrive at the following system of simultaneous linear equations:

$$\sum_{j=1}^{n-1} M_{ij} u_j = b_i\,, \quad 1 \le i \le n-1\,, \tag{14.16}$$

where

$$M_{ij} = \mathcal{A}(\varphi_j, \varphi_i), \quad b_i = \langle f, \varphi_i \rangle - \mathcal{A}(\psi, \varphi_i). \tag{14.17}$$

The coefficients u_j, $1 \le j \le n-1$, in the representation of the approximate solution are thus obtained by solving the system of linear equations (14.16). The matrix M is, clearly, symmetric (since the bilinear form $\mathcal{A}(\,\cdot\,,\,\cdot\,)$ is symmetric by hypothesis) and positive definite, because

$$\boldsymbol{v}^{\mathrm{T}} M \boldsymbol{v} = \mathcal{A}(v, v) > 0,$$

where $\boldsymbol{v} = (v_1, v_2, \ldots, v_{n-1})^{\mathrm{T}} \in \mathbb{R}^{n-1}$ is any nonzero vector and $v = v_1\varphi_1 + \cdots + v_{n-1}\varphi_{n-1} \in S_0^h$. Hence M is nonsingular.

The Ritz and Galerkin methods can be used to compute an approximation u^h to u as a linear combination of *any* finite set of linearly independent functions φ_i, $1 \le i \le n-1$, in $\mathrm{H}_0^1(a, b)$. We obtain the **Ritz finite element method** and the **Galerkin finite element method**, respectively, when we select the approximating subspaces S_{E}^h and S_0^h in the Ritz or the Galerkin method to be spaces of spline functions (see Chapter 11). Here we only consider the simplest case of linear splines, and choose the basis functions φ_i, $1 \le i \le n-1$, to be the hat functions (11.4). We begin by fixing a set of points x_k, $k = 0, 1, \ldots, n$, $n \ge 2$, in the interval $[a, b]$ such that

$$a = x_0 < x_1 < \cdots < x_n = b. \tag{14.18}$$

The intervals $[x_{i-1}, x_i]$, $1 \le i \le n$, are referred to as **elements**; hence the name *finite element method.* In the theory of the finite element methods (14.18) is called a **subdivision** of the computational domain $[a, b]$, and the points x_k are called **mesh points**. The function φ_i is the piecewise linear function which takes the value 0 at all the mesh points except x_i, where it takes the value 1. Thus,

$$\varphi_i(x) = \begin{cases} (x - x_{i-1})/h_i & \text{if } x_{i-1} \le x \le x_i, \\ (x_{i+1} - x)/h_{i+1} & \text{if } x_i \le x \le x_{i+1}, \\ 0 & \text{otherwise}, \end{cases} \tag{14.19}$$

where $h_i = x_i - x_{i-1}$. The functions φ_i, $1 \le i \le n-1$, are called the (piecewise linear) **finite element basis functions** and the associated Galerkin approximation u^h is referred to as the (piecewise linear) **finite element approximation** of u. The closure of the interval (x_{i-1}, x_{i+1}) over which φ_i is nonzero is called the **support** of the function φ_i. The piecewise linear finite element basis function φ_i, $1 \le i \le n-1$, with support $[x_{i-1}, x_{i+1}]$, is depicted in Figure 14.1.

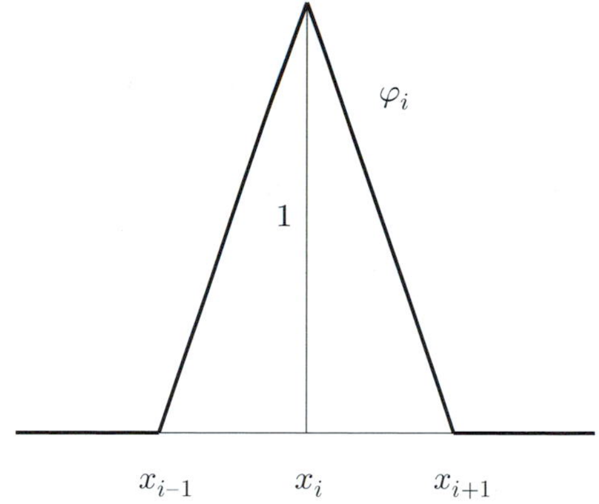

Fig. 14.1. A piecewise linear finite element basis function, φ_i, $1 \le i \le n-1$.

For the finite element method the important property of the basis functions φ_i, $1 \le i \le n-1$, is that they have *local* support, being nonzero only in one pair of adjacent intervals, $(x_{i-1}, x_i]$ and $[x_i, x_{i+1})$. This means that, in the matrix M,

$$M_{ij} = 0 \quad \text{if } |i-j| > 1\,.$$

The matrix M is, therefore, symmetric, positive definite and tridiagonal, and the associated system of linear equations can be solved very efficiently by the methods of Section 3.3, the most efficient algorithm being LU decomposition, without any use of symmetry. The fact that M is positive definite means that no interchanges are necessary.

The function ψ in (14.13), which is included in the definition of S^h_{E} to ensure that u^h satisfies the boundary conditions at $x = a$ and $x = b$, is easily rewritten as

$$\psi(x) = A\varphi_0(x) + B\varphi_n(x)\,,$$

which is also piecewise linear; clearly, $\psi(a) = A$ and $\psi(b) = B$. Here, φ_0 and φ_n are defined by setting, respectively, $i = 0$ and $i = n$ in (14.19) and restricting the resulting functions to the interval $[a, b] = [x_0, x_n]$; see (11.4). In (14.17) we see that the term $\mathcal{A}(\psi, \varphi_i)$ is nonzero only for $i = 1$ and $i = n-1$.

Before attempting to solve the system of linear equations we must, of course, first compute the elements of the matrix M, and the quantities on the right-hand side, b_i, $i = 1, \ldots, n-1$; see (14.16) and (14.17). The matrix elements are obtained from

$$M_{ij} = \mathcal{A}(\varphi_j, \varphi_i) = \int_a^b p(x)\varphi_j'(x)\varphi_i'(x)\mathrm{d}x + \int_a^b r(x)\varphi_j(x)\varphi_i(x)\mathrm{d}x\,,$$

with $1 \le i, j \le n-1$. We have written this as the sum of two terms, as the matrix M is often written in this way as the sum of two matrices which, for historical reasons, are often known as the **stiffness matrix** and the **mass matrix**, respectively. The terms M_{ij} are very simple; in fact in the first integral the derivatives φ_j' and φ_i' are piecewise constant functions over $[a, b]$.

It may be possible to compute these integrals analytically, but more generally some form of numerical quadrature will be necessary. It is then easy to show that if we use certain types of quadrature formulae we shall be led to the same system of equations as in the finite difference method of Section 13.5. Consider the particularly simple case where the mesh points are equally spaced, so that $x_j = a + jh$, $j = 0, 1, \ldots, n$, $h = (b-a)/n$. If we then approximate the integrals involved in the stiffness matrix by the midpoint rule (see Chapter 10), we obtain

$$\begin{aligned} \int_{x_{i-1}}^{x_i} p(x)\varphi_{i-1}'(x)\varphi_i'(x)\mathrm{d}x &= -(1/h^2)\int_{x_{i-1}}^{x_i} p(x)\mathrm{d}x \\ &\approx -p_{i-1/2}/h\,, \end{aligned}$$

where $p_{i-1/2} = p(x_i - h/2)$, and similarly for the other integrals involved. For the integrals in the mass matrix we use the trapezium rule, and then

$$\int_{x_{i-1}}^{x_i} r(x)\varphi_{i-1}(x)\varphi_i(x)\mathrm{d}x \approx 0\,,$$

since φ_i is zero at x_{i-1} and φ_{i-1} is zero at x_i. In the same way

$$\int_{x_{i-1}}^{x_i} r(x)[\varphi_i(x)]^2\mathrm{d}x \approx \tfrac{1}{2}hr_i\,,$$

where $r_i = r(x_i)$, since φ_i is zero at one end of the interval and unity at the other. The other part of the integral is, similarly,

$$\int_{x_i}^{x_{i+1}} r(x)[\varphi_i(x)]^2\mathrm{d}x \approx \tfrac{1}{2}hr_i\,. \tag{14.20}$$

Assuming that $f \in \mathrm{C}[a, b]$, approximating the integral on the right-hand side by the trapezium rule in the same way, and putting all the parts together, equation (14.14) now takes the approximate form

$$-\frac{p_{i-1/2}}{h}u_{i-1} + \frac{p_{i-1/2} + p_{i+1/2}}{h}u_i - \frac{p_{i+1/2}}{h}u_{i+1} + hr_iu_i = hf_i\,,$$

for $i = 1, 2, \ldots, n-1$, with the notational convention that $u_0 = A$ and $u_n = B$, and $f_i = f(x_i)$; clearly, this is the same as the finite difference equation (13.19). Of course, had we used a different set of basis functions φ_i, $1 \leq i \leq n-1$, or different numerical quadrature rules, the finite element and finite difference methods would have no longer been identical. Indeed, this example is just an illustration of the relation between the two methods; we should normally expect to compute the entries of the matrix M by using some more accurate quadrature method, such as a two-point Gauss formula.

In the next two sections we shall assess the accuracy of the finite element method. Our goal is to quantify the amount of reduction in the error $u - u^h$ as the mesh spacing h is reduced.

14.4 Error analysis of the finite element method

We begin with a fundamental result that underlies the error analysis of finite element methods.

Theorem 14.6 (Céa's Lemma) *Suppose that u is the function that minimises $\mathcal{J}(u)$ over $\mathrm{H}^1_{\mathrm{E}}(a, b)$ (or, equivalently, that u satisfies (14.6)), and that u^h is its Galerkin approximation obtained by minimising $\mathcal{J}(\,\cdot\,)$ over S^h_{E} (or, equivalently, that u^h satisfies (14.14)). Then,*

$$\mathcal{A}(u - u^h, v^h) = 0 \qquad \forall\, v^h \in S^h_0\,, \tag{14.21}$$

and

$$\mathcal{A}(u - u^h, u - u^h) = \min_{v^h \in S^h_{\mathrm{E}}} \mathcal{A}(u - v^h, u - v^h)\,. \tag{14.22}$$

The identity (14.21) is referred to as **Galerkin orthogonality**. The terminology stems from the fact that, since the bilinear functional $\mathcal{A}(\,\cdot\,,\,\cdot\,)$ is symmetric and $\mathcal{A}(v, v) > 0$ for all $v \in \mathrm{H}^1_0(a, b) \setminus \{0\}$, $\mathcal{A}(\,\cdot\,,\,\cdot\,)$ is an inner product in the linear space $\mathrm{H}^1_0(a, b)$. Therefore, by virtue of Definition 9.2, (14.21) means that $u - u^h$ is orthogonal to S^h_0 in $\mathrm{H}^1_0(a, b)$. A geometrical illustration of Galerkin orthogonality is shown in Figure 14.2. Given that ψ is a fixed element of $\mathrm{H}^1_{\mathrm{E}}(a, b)$, the mapping

$$R^h \colon u - \psi \in \mathrm{H}^1_0(a, b) \mapsto u^h - \psi \in S^h_0$$

which assigns a $u^h \in S^h_{\mathrm{E}}$ to $u \in \mathrm{H}^1_{\mathrm{E}}(a, b)$ (where u and u^h are as in Theorem 14.6) is called the **Ritz projector**.

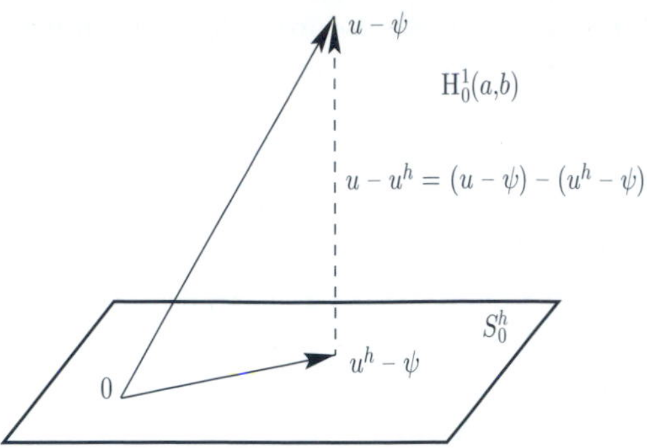

Fig. 14.2. Illustration of the Galerkin orthogonality property of the finite element method. $\mathcal{A}((u-\psi)-(u^h-\psi),v^h) = \mathcal{A}(u-u^h,v^h) = 0$ for all v^h in S^h_0. Here, $\psi(x) = A\varphi_0(x) + B\varphi_n(x)$, so that $u - \psi \in \mathrm{H}^1_0(a,b)$ and $u^h - \psi \in S^h_0$. The 0 in the figure denotes the zero element of the linear space S^h_0 (and, simultaneously, that of $\mathrm{H}^1_0(a,b)$), namely the function that is identically zero on the interval (a,b).

Proof of theorem By the definition of the Galerkin method $(\mathrm{G})^h$,

$$\mathcal{A}(u^h, v^h) = \langle f, v^h\rangle \qquad \forall\, v^h \in S^h_0 \,.$$

On the other hand, we deduce from (G) that

$$\mathcal{A}(u, v^h) = \langle f, v^h\rangle \qquad \forall\, v^h \in S^h_0 \,,$$

since $v^h \in S^h_0 \subset \mathrm{H}^1_0(a,b)$. The Galerkin orthogonality property (14.21) follows by subtraction.

Now suppose that v^h is any function in S^h_{E}; then,

$$\begin{aligned}\mathcal{A}(u-v^h,u-v^h) &= \mathcal{A}(u-u^h+u^h-v^h,u-u^h+u^h-v^h)\\ &= \mathcal{A}(u-u^h,u-u^h)+\mathcal{A}(u^h-v^h,u^h-v^h)\\ &\quad + 2\,\mathcal{A}(u-u^h,u^h-v^h)\\ &= \mathcal{A}(u-u^h,u-u^h)+\mathcal{A}(u^h-v^h,u^h-v^h)\,,\end{aligned}$$

by Galerkin orthogonality, given that $u^h - v^h \in S^h_0$. In the transition from the first line to the second, we made use of the fact that the bilinear functional $\mathcal{A}$ is symmetric. As the term $\mathcal{A}(u^h - v^h, u^h - v^h)$ is nonnegative, we deduce that

$$\mathcal{A}(u-u^h,u-u^h) \le \mathcal{A}(u-v^h,u-v^h) \qquad \forall\, v^h \in S^h_0 \,,$$

with equality when $v^h = u^h$; hence (14.22). □

Motivated by the minimisation property (14.22), we define the **energy norm** $\|\cdot\|_{\mathcal{A}}$ on $\mathrm{H}_0^1(a,b)$ *via*

$$\|v\|_{\mathcal{A}} = [\mathcal{A}(v,v)]^{1/2}\,. \tag{14.23}$$

Under our hypotheses on p and q, it is easy to see that $\|\cdot\|_{\mathcal{A}}$ satisfies all axioms of norm (see Chapter 2). The result we have just proved shows that u^h is the *best approximation* from S_{E}^h to the weak solution $u \in \mathrm{H}_{\mathrm{E}}^1(a,b)$ of our problem, when we measure the error of the approximation in the energy norm:

$$\|u-u^h\|_{\mathcal{A}} = \min_{v^h\in S_{\mathrm{E}}^h} \|u-v^h\|_{\mathcal{A}}\,. \tag{14.24}$$

A particularly relevant question is how the error $u-u^h$ depends on the spacing h of the subdivision of the computational domain $[a,b]$. We can obtain a bound on the error $u-u^h$, measured in the energy norm, by choosing a particular function $v^h \in S_{\mathrm{E}}^h$ in (14.24) whose closeness to u is easy to assess. For this purpose, we introduce the **finite element interpolant** $\mathcal{I}^h u \in S_{\mathrm{E}}^h$ of $u \in \mathrm{H}_{\mathrm{E}}^1(a,b)$ by

$$\mathcal{I}^h u(x) = \psi(x) + \sum_{i=1}^{n-1} u(x_i)\varphi_i(x)\,, \qquad x\in[a,b]\,.$$

Clearly,

$$\mathcal{I}^h u(x_j) = u(x_j)\,, \qquad j=0,1,\ldots,n\,,$$

which justifies our use of the word *interpolant*.

We then deduce from (14.24) that

$$\|u-u^h\|_{\mathcal{A}} \le \|u-\mathcal{I}^h u\|_{\mathcal{A}}\,; \tag{14.25}$$

hence, in order to quantify $\|u-u^h\|_{\mathcal{A}}$, we only need to estimate the size of $\|u-\mathcal{I}^h u\|_{\mathcal{A}}$. This leads us to the next theorem.

Theorem 14.7 *Suppose that $u \in \mathrm{H}^2(a,b)\cap\mathrm{H}_{\mathrm{E}}^1(a,b)$ and let $\mathcal{I}^h u$ be the finite element interpolant of u from S_{E}^h defined above; then, the following error bounds hold:*

$$\begin{aligned}
\|u-\mathcal{I}^h u\|_{\mathrm{L}^2(x_{i-1},x_i)} &\le \left(\frac{h_i}{\pi}\right)^2 \|u''\|_{\mathrm{L}^2(x_{i-1},x_i)}\,,\\
\|u'-(\mathcal{I}^h u)'\|_{\mathrm{L}^2(x_{i-1},x_i)} &\le \frac{h_i}{\pi}\|u''\|_{\mathrm{L}^2(x_{i-1},x_i)}\,,
\end{aligned}$$

for $i=1,2,\ldots,n$, where $h_i = x_i - x_{i-1}$.

Proof Consider an element $[x_{i-1}, x_i]$, $1 \le i \le n$, and define $\zeta(x) = u(x) - \mathcal{I}^h u(x)$ for $x \in [x_{i-1}, x_i]$. Then, $\zeta \in \mathrm{H}^2(x_{i-1}, x_i)$ and $\zeta(x_{i-1}) = \zeta(x_i) = 0$. Therefore ζ can be expanded into a convergent Fourier sine-series,

$$\zeta(x) = \sum_{k=1}^{\infty} a_k \sin \frac{k\pi(x - x_{i-1})}{h_i}, \qquad x \in [x_{i-1}, x_i].$$

Here, convergence is to be understood in the norm $\|\cdot\|_{\mathrm{L}^2(x_{i-1},x_i)}$. Hence,

$$\begin{aligned}
\int_{x_{i-1}}^{x_i} [\zeta(x)]^2 \mathrm{d}x &= \int_{x_{i-1}}^{x_i} \zeta(x)\zeta(x)\mathrm{d}x \\
&= \sum_{k,\ell=1}^{\infty} a_k a_\ell \int_{x_{i-1}}^{x_i} \sin \frac{k\pi(x - x_{i-1})}{h_i} \sin \frac{\ell\pi(x - x_{i-1})}{h_i} \mathrm{d}x \\
&= h_i \sum_{k,\ell=1}^{\infty} a_k a_\ell \int_0^1 \sin k\pi t \, \sin \ell\pi t \, \mathrm{d}t \\
&= \frac{h_i}{2} \sum_{k,\ell=1}^{\infty} a_k a_\ell \, \delta_{k\ell} \\
&= \frac{h_i}{2} \sum_{k=1}^{\infty} |a_k|^2,
\end{aligned}$$

where $\delta_{k\ell}$ is the Kronecker delta. Differentiating the Fourier sine series of ζ twice, we find that the Fourier coefficients of ζ' are $(k\pi/h_i)a_k$, while those of ζ'' are $-(k\pi/h_i)^2 a_k$. Thus, proceeding in the same way as above,

$$\begin{aligned}
\int_{x_{i-1}}^{x_i} [\zeta'(x)]^2 \mathrm{d}x &= \frac{h_i}{2} \sum_{k=1}^{\infty} \left(\frac{k\pi}{h_i}\right)^2 |a_k|^2, \\
\int_{x_{i-1}}^{x_i} [\zeta''(x)]^2 \mathrm{d}x &= \frac{h_i}{2} \sum_{k=1}^{\infty} \left(\frac{k\pi}{h_i}\right)^4 |a_k|^2.
\end{aligned}$$

Because $k^4 \ge k^2 \ge 1$, it follows that

$$\begin{aligned}
\int_{x_{i-1}}^{x_i} [\zeta(x)]^2 \mathrm{d}x &\le \left(\frac{h_i}{\pi}\right)^4 \int_{x_{i-1}}^{x_i} [\zeta''(x)]^2 \mathrm{d}x, \\
\int_{x_{i-1}}^{x_i} [\zeta'(x)]^2 \mathrm{d}x &\le \left(\frac{h_i}{\pi}\right)^2 \int_{x_{i-1}}^{x_i} [\zeta''(x)]^2 \mathrm{d}x.
\end{aligned}$$

However, $\zeta''(x) = u''(x) - (\mathcal{I}^h u)''(x) = u''(x)$ for $x \in (x_{i-1}, x_i)$, and hence the desired bounds on the interpolation error. □

Now, substituting the bounds from Theorem 14.7 into the definition of the norm $\|u - \mathcal{I}^h u\|_{\mathcal{A}}$, we arrive at the following estimate of the interpolation error in the energy norm.

Corollary 14.1 *Suppose that* $u \in \mathrm{H}^2(a,b) \cap \mathrm{H}^1_{\mathrm{E}}(a,b)$. *Then,*

$$\|u - \mathcal{I}^h u\|_{\mathcal{A}}^2 \le \sum_{i=1}^{n} \left\{ \left(\frac{h_i}{\pi}\right)^2 P_i + \left(\frac{h_i}{\pi}\right)^4 R_i \right\} \|u''\|^2_{\mathrm{L}^2(x_{i-1},x_i)},$$

where $P_i = \max_{x\in[x_{i-1},x_i]} p(x)$ *and* $R_i = \max_{x\in[x_{i-1},x_i]} r(x)$.

Proof Let us observe that

$$\begin{aligned}
\|v\|_{\mathcal{A}}^2 &= \mathcal{A}(v,v) \\
&= \int_a^b \left[p(x)|v'(x)|^2 + r(x)|v(x)|^2\right] \mathrm{d}x \\
&= \sum_{i=1}^{n} \int_{x_{i-1}}^{x_i} \left[p(x)|v'(x)|^2 + r(x)|v(x)|^2\right] \mathrm{d}x \\
&\le \sum_{i=1}^{n} \left\{ P_i \|v'\|^2_{\mathrm{L}^2(x_{i-1},x_i)} + R_i \|v\|^2_{\mathrm{L}^2(x_{i-1},x_i)} \right\}.
\end{aligned}$$

On letting $v = u - \mathcal{I}^h u$ and applying the preceding theorem on the right-hand side of the last inequality, with v' and v replaced by $u' - (\mathcal{I}^h u)'$ and $u - \mathcal{I}^h u$, respectively, the result follows. □

Inserting this estimate into (14.25) leads to the desired bound on the error between the analytical solution u and its finite element approximation u^h in the energy norm.

Corollary 14.2 *Suppose that* $u \in \mathrm{H}^2(a,b) \cap \mathrm{H}^1_{\mathrm{E}}(a,b)$. *Then,*

$$\|u - u^h\|_{\mathcal{A}}^2 \le \sum_{i=1}^{n} \left\{ \left(\frac{h_i}{\pi}\right)^2 P_i + \left(\frac{h_i}{\pi}\right)^4 R_i \right\} \|u''\|^2_{\mathrm{L}^2(x_{i-1},x_i)},$$

where $P_i = \max_{x\in[x_{i-1},x_i]} p(x)$ *and* $R_i = \max_{x\in[x_{i-1},x_i]} r(x)$. *Further,*

$$\|u - u^h\|_{\mathcal{A}} \le \frac{h}{\pi} \left\{ P + \left(\frac{h}{\pi}\right)^2 R \right\}^{1/2} \|u''\|_{\mathrm{L}^2(a,b)}, \tag{14.26}$$

where $P = \max_{x\in[a,b]} p(x)$, $R = \max_{x\in[a,b]} r(x)$, *and* $h = \max_{1\le i\le n} h_i$.

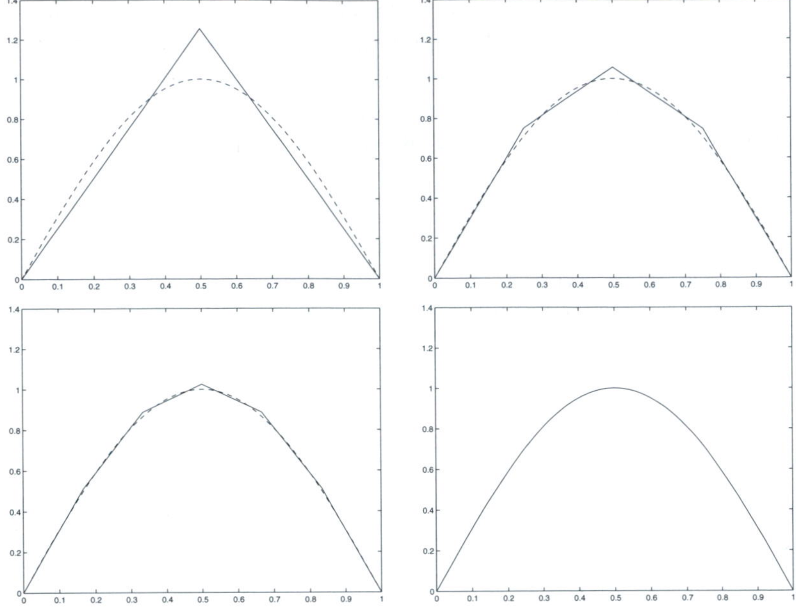

Fig. 14.3. Graph of the finite element approximation u^h to the analytical solution u of the boundary value problem (14.27) on a uniform subdivision of $[0, 1]$ of spacing $h = 1/n$, with $n = 2$ (top left), $n = 4$ (top right), $n = 6$ (bottom left), and $n = 100$ (bottom right). In each of the four subfigures, the dashed curve is the graph of the analytical solution $u(x) = \sin(\pi x)$. In the last figure the approximation error is so small that u and u^h are indistinguishable.

In order to illustrate the performance of the finite element method, we consider the following example:

$$-u''+r(x)u = f(x)\,, \qquad x \in (0,1)\,, \qquad u(0) = 0\,, \quad u(1) = 0\,. \tag{14.27}$$

If $r(x) \equiv 1$ and $f(x) = (1 + \pi^2)\sin(\pi x)$, the unique solution to this problem is $u(x) = \sin(\pi x)$. Let us pretend that we do not know the analytical solution u, and solve the boundary value problem numerically, using the finite element method on a subdivision of $[0, 1]$ of uniform spacing $h = 1/n$, for various values of n. The integrals $\langle f, \varphi_i \rangle$ involved in the definition of b_i in (14.17) have been approximated, on each of the elements $[x_{i-1}, x_i]$, $1 \le i \le n$, by means of the trapezium rule. The resulting approximations u^h, for $n = 2, 4, 6, 100$, are shown in Figure 14.3.

We see from Figure 14.3 that, as the spacing h of the subdivision is reduced, the finite element solution u^h approximates the analytical solution $u(x) = \sin(\pi x)$ with increasing accuracy. Indeed, the results corresponding to $n = 2$ and $n = 4$ in Figure 14.3 indicate that as the number of intervals in the subdivision is doubled (*i.e.*, h is halved), the maximum error between $u(x)$ and $u^h(x)$ is reduced by a factor of about 4. This reduction in the error cannot be explained by Corollary 14.2 which merely implies that halving h should lead to a reduction in $\|u - u^h\|_{\mathcal{A}}$ by a factor no less than 2. If you would like to learn more about the source of the observed enhancement of accuracy, consult Exercise 5 at the end of the chapter.

14.5 *A posteriori* error analysis by duality

The bound on the error between the analytical solution u and its finite element approximation u^h formulated in Corollary 14.2 shows that, in the limit of $h \to 0$, the error $\|u - u^h\|_{\mathcal{A}}$ will tend to zero as $\mathcal{O}(h)$. This is a useful result from the theoretical point of view: it reassures us that the unknown analytical solution may be approximated arbitrarily well by making h sufficiently small. On the other hand, asymptotic error bounds of this kind are not particularly helpful for the purpose of precisely quantifying the size of the error between u and u^h for a given, *fixed*, mesh size $h > 0$: as u is unknown, it is difficult to tell just how large the right-hand side of (14.26) really is.

The aim of the present section is, therefore, to derive a computable bound on the error, and to demonstrate how such a bound may be implemented into an adaptive mesh-refinement algorithm, capable of reducing the error $u - u^h$ below a certain prescribed tolerance in an automated manner, without human intervention. The approach is based on seeking a bound on $u - u^h$ in terms of the computed solution u^h rather than in terms of norms of the unknown analytical solution u. A bound on the error in terms of u^h is referred to as an ***a posteriori*** error bound, due to the fact that it becomes *computable* only *after* the numerical solution u^h has been obtained.

In order to illuminate the key ideas while avoiding technical difficulties, we shall consider the two-point boundary value problem

$$-(p(x)u')' + q(x)u' + r(x)u = f(x)\,, \qquad a < x < b\,, \tag{14.28}$$

$$u(a) = A\,, \quad u(b) = B\,, \tag{14.29}$$

where $p, q \in \mathrm{C}^1[a,b]$, $r \in \mathrm{C}[a,b]$ and $f \in \mathrm{L}^2(a,b)$. We shall assume, as at the beginning of the chapter, that $p(x) \geq c_0 > 0$, $x \in [a, b]$; however, instead of supposing that $r(x) \geq 0$, we shall now demand that

$$r(x) - \frac{1}{2}q'(x) \geq c_1 , \quad x \in [a, b] , \tag{14.30}$$

where c_1 is assumed to be a positive constant.[1]

Letting

$$\mathcal{A}(w, v) = \int_a^b [p(x)w'(x)v'(x) + q(x)w'(x)v(x) + r(x)w(x)v(x)]\mathrm{d}x ,$$

the weak formulation of (14.28), (14.29) is as follows:

$$\text{find } u \in \mathrm{H}^1_{\mathrm{E}}(a, b) \text{ such that } \mathcal{A}(u, v) = \langle f, v\rangle \qquad \forall v \in \mathrm{H}^1_0(a, b) . \tag{14.31}$$

Here, the bilinear functional $\mathcal{A}(\,\cdot\,,\,\cdot\,)$ is *not* symmetric, unless $q(x) \equiv 0$: indeed, $\mathcal{A}(w, v) = \mathcal{A}(v, w)$ for all $v, w \in \mathrm{H}^1(a, b)$ if, and only if, $q \equiv 0$. Hence, in general, the boundary value problem (14.28), (14.29) cannot be assigned a Ritz principle. On the other hand, the Galerkin principle (weak formulation) (14.31) is perfectly meaningful for any choice of q.

The Galerkin finite element approximation of (14.31) is constructed by introducing a (possibly nonuniform) subdivision of the interval $[a, b]$ defined by the points

$$a = x_0 < x_1 < \cdots < x_{n-1} < x_n = b$$

and considering the finite element space $S^h_{\mathrm{E}} \subset \mathrm{H}^1_{\mathrm{E}}(a, b)$ consisting of all continuous piecewise linear functions v^h on this subdivision that satisfy the boundary conditions $v^h(a) = A$ and $v^h(b) = B$. The Galerkin finite element approximation of the boundary value problem is

$$\text{find } u^h \in S^h_{\mathrm{E}} \text{ such that } \mathcal{A}(u^h, v^h) = \langle f, v^h\rangle \qquad \forall v^h \in S^h_0 . \tag{14.32}$$

We let $h_i = x_i - x_{i-1}$, $i = 1, \ldots, n$, and put $h = \max_i h_i$.

We wish to derive an *a posteriori* bound on the error in the $\|\cdot\|_{\mathrm{L}^2(a,b)}$ norm; that is, our aim is to quantify the size of $\|u - u^h\|_{\mathrm{L}^2(a,b)}$ in terms of the mesh parameter h and the computed solution u^h (rather than in terms of the analytical solution u as was the case in the *a priori* error analysis developed in the previous section). For this purpose, we

[1] At the expense of slight technical complications in the subsequent discussion, the requirement that $c_1 > 0$ can be relaxed to $c_1 > -\lambda_1$, where λ_1 is the smallest (positive) eigenvalue for the Sturm–Liouville eigenvalue problem $-(p(x)w')' = \lambda w$ for $x \in (a, b)$, $w(a) = 0$, $w(b) = 0$.

consider the auxiliary boundary value problem

$$-(p(x)z')' - (q(x)z)' + r(x)z = (u - u^h)(x), \quad a < x < b, \tag{14.33}$$

$$z(a) = 0, \ z(b) = 0, \tag{14.34}$$

called the **dual problem** (or adjoint problem).

We begin our error analysis by noting that the definition of the dual problem and straightforward integration by parts yield (recalling that $(u - u^h)(a) = 0$, $(u - u^h)(b) = 0$)

$$\begin{aligned}\|u - u^h\|^2_{\mathrm{L}^2(a,b)} &= \langle u - u^h, u - u^h\rangle \\ &= \langle u - u^h, -(pz')' - (qz)' + rz\rangle \\ &= \mathcal{A}(u - u^h, z)\,.\end{aligned}$$

On the other hand, (14.31) and (14.32) imply the Galerkin orthogonality property

$$\mathcal{A}(u - u^h, z^h) = 0 \qquad \forall\, z^h \in S^h_0\,.$$

In particular, by choosing

$$z^h = \mathcal{I}^h z \in S^h_0\,,$$

the continuous piecewise linear interpolant of the function $z \in \mathrm{H}^1_0(a, b)$, associated with the subdivision $a = x_0 < x_1 < \cdots < x_{n-1} < x_n = b$, we have that

$$\mathcal{A}(u - u^h, \mathcal{I}^h z) = 0\,.$$

Thus,

$$\begin{aligned}\|u - u^h\|^2_{\mathrm{L}^2(a,b)} &= \mathcal{A}(u - u^h, z - \mathcal{I}^h z) \\ &= \mathcal{A}(u, z - \mathcal{I}^h z) - \mathcal{A}(u^h, z - \mathcal{I}^h z) \\ &= \langle f, z - \mathcal{I}^h z\rangle - \mathcal{A}(u^h, z - \mathcal{I}^h z)\,, \end{aligned} \tag{14.35}$$

where the last transition follows from (14.31) with $v = z - \mathcal{I}^h z$.

We observe that the right-hand side no longer involves the unknown analytical solution u. Furthermore,

$$\begin{aligned}\mathcal{A}(u^h, z - \mathcal{I}^h z) &= \sum_{i=1}^{n} \int_{x_{i-1}}^{x_i} p(x)(u^h)'(x)\,(z - \mathcal{I}^h z)'(x)\,\mathrm{d}x \\ &\quad + \sum_{i=1}^{n} \int_{x_{i-1}}^{x_i} q(x)\,(u^h)'(x)\,(z - \mathcal{I}^h z)(x)\,\mathrm{d}x \\ &\quad + \sum_{i=1}^{n} \int_{x_{i-1}}^{x_i} r(x)\,u^h(x)\,(z - \mathcal{I}^h z)(x)\,\mathrm{d}x\,.\end{aligned}$$

Integrating by parts in each of the n integrals in the first sum on the right-hand side, noting that $(z - \mathcal{I}^h z)(x_i) = 0$, $i = 0, \ldots, n$, we deduce that

$$\begin{aligned}&\mathcal{A}(u^h, z - \mathcal{I}^h z)\\ &= \sum_{i=1}^{n} \int_{x_{i-1}}^{x_i} \left[-(p(x)(u^h)')' + q(x)(u^h)' + r(x)u^h\right] (z - \mathcal{I}^h z)(x)\,\mathrm{d}x\,.\end{aligned}$$

Furthermore,

$$\langle f, z - \mathcal{I}^h z\rangle = \sum_{i=1}^{n} \int_{x_{i-1}}^{x_i} f(x)\,(z - \mathcal{I}^h z)(x)\,\mathrm{d}x\,.$$

Substituting these two identities into (14.35), we deduce that

$$\|u - u^h\|^2_{\mathrm{L}^2(a,b)} = \sum_{i=1}^{n} \int_{x_{i-1}}^{x_i} R(u^h)(x)\,(z - \mathcal{I}^h z)(x)\,\mathrm{d}x\,, \tag{14.36}$$

where, for $1 \le i \le n$, and $x \in (x_{i-1}, x_i)$,

$$R(u^h)(x) = f(x) - \left[-(p(x)(u_h)')' + q(x)(u^h)' + r(x)u^h\right].$$

The function $R(u^h)$ is called **the finite element residual**; it measures the extent to which u^h fails to satisfy the differential equation

$$-(p(x)u')' + q(x)u' + r(x)u = f(x)$$

on the union of the intervals (x_{i-1}, x_i), $i = 1, \ldots, n$. Now, applying the Cauchy–Schwarz inequality on the right-hand side of (14.36) yields

$$\|u - u^h\|^2_{\mathrm{L}^2(a,b)} \le \sum_{i=1}^{n} \|R(u^h)\|_{\mathrm{L}^2(x_{i-1},x_i)}\,\|z - \mathcal{I}^h z\|_{\mathrm{L}^2(x_{i-1},x_i)}\,.$$

Recalling from Theorem 14.7 that

$$\|z - \mathcal{I}^h z\|_{\mathrm{L}^2(x_{i-1},x_i)} \le \left(\frac{h_i}{\pi}\right)^2 \|z''\|_{\mathrm{L}^2(x_{i-1},x_i)}\,, \quad i = 1, 2, \ldots, n\,,$$

we deduce that

$$\|u - u^h\|^2_{\mathrm{L}^2(a,b)} \le \frac{1}{\pi^2} \sum_{i=1}^{n} h_i^2 \|R(u^h)\|_{\mathrm{L}^2(x_{i-1},x_i)}\,\|z''\|_{\mathrm{L}^2(x_{i-1},x_i)}\,,$$

and consequently, using the Cauchy–Schwarz inequality for finite sums,

$$\sum_{i=1}^{n} a_i b_i \le \left(\sum_{i=1}^{n} |a_i|^2\right)^{1/2} \left(\sum_{i=1}^{n} |b_i|^2\right)^{1/2}$$

with

$$a_i = h_i^2 \|R(u^h)\|_{L^2(x_{i-1},x_i)} \qquad \text{and} \qquad b_i \;=\; \|z''\|_{L^2(x_{i-1},x_i)}\,,$$

we find that

$$\|u - u^h\|^2_{L^2(a,b)} \le \frac{1}{\pi^2} \left(\sum_{i=1}^{n} h_i^4 \|R(u^h)\|^2_{L^2(x_{i-1},x_i)} \right)^{1/2} \|z''\|_{L^2(0,1)}. \tag{14.37}$$

The rest of the discussion is aimed at eliminating $\|z''\|_{L^2(a,b)}$ from the right-hand side of (14.37). The desired *a posteriori* bound on the error $\|u - u^h\|_{L^2(a,b)}$ in terms of $R(u^h)$ will then follow.

Lemma 14.1 *Suppose that z is the solution of the dual problem (14.33), (14.34). Then, there exists a positive constant K, dependent only on p, q and r, such that*

$$\|z''\|_{L^2(a,b)} \le K \|u - u^h\|_{L^2(a,b)}\,.$$

Proof As

$$-pz'' - p'z' - qz' - q'z + rz = u - u^h\,,$$

it follows that

$$pz'' = u^h - u - (p' + q)\,z' + (r - q')\,z\,,$$

and therefore, recalling that $p(x) \ge c_0 > 0$ for $x \in [a,b]$,

$$\begin{aligned} c_0\|z''\|_{L^2(a,b)} \quad &\le \quad \|u - u^h\|_{L^2(a,b)} + \|p' + q\|_\infty\, \|z'\|_{L^2(a,b)} \\ &\quad\; + \|r - q'\|_\infty\, \|z\|_{L^2(a,b)}\,, \end{aligned} \tag{14.38}$$

where we used the notation $\|w\|_\infty = \max_{x\in[a,b]} |w(x)|$.

We shall show that both $\|z'\|_{L^2(a,b)}$ and $\|z\|_{L^2(a,b)}$ can be bounded in terms of $\|u - u^h\|_{L^2(a,b)}$ and then, using (14.38), we shall deduce that the same is true of $\|z''\|_{L^2(a,b)}$. Let us observe that, by (14.33),

$$\langle -(pz')' - (qz)' + rz, z\rangle = \langle u - u^h, z\rangle\,. \tag{14.39}$$

Integrating by parts in the terms involving p and q and noting that $z(0) = 0$ and $z(1) = 0$ yields

$$\begin{aligned} \langle -(pz')' - (qz)' + rz, z\rangle &= \langle pz', z'\rangle + \langle qz, z'\rangle + \langle rz, z\rangle \\ &\ge c_0\|z'\|^2_{L^2(a,b)} + \frac{1}{2}\int_a^b q(x)[z^2(x)]'\mathrm{d}x + \int_a^b r(x)[z(x)]^2\mathrm{d}x\,. \end{aligned}$$

Integrating by parts, again, in the second term on the right gives

$$\begin{aligned}\langle -(pz')' - (qz)' + rz, z\rangle \;&\geq\; c_0\|z'\|^2_{\mathrm{L}^2(a,b)} - \frac{1}{2}\int_a^b q'(x)[z^2(x)]\mathrm{d}x \\ &\quad + \int_a^b r(x)[z(x)]^2\mathrm{d}x\,.\end{aligned}$$

Hence, from (14.39),

$$c_0\|z'\|^2_{\mathrm{L}^2(a,b)} + \int_a^b \left(r(x) - \frac{1}{2}q'(x)\right)[z(x)]^2\mathrm{d}x \leq \langle u - u^h, z\rangle\,,$$

and thereby, noting (14.30) and using the Cauchy–Schwarz inequality on the right-hand side,

$$\begin{aligned}\min\{c_0, c_1\}\left(\|z'\|^2_{\mathrm{L}^2(a,b)} + \|z\|^2_{\mathrm{L}^2(a,b)}\right) &\leq \langle u - u^h, z\rangle \\ &\leq \|u - u^h\|_{\mathrm{L}^2(a,b)}\,\|z\|_{\mathrm{L}^2(a,b)}\,. \qquad (14.40)\end{aligned}$$

Therefore, also

$$\min\{c_0, c_1\}\|z\|^2_{\mathrm{H}^1(a,b)} \leq \|u - u^h\|_{\mathrm{L}^2(a,b)}\,\|z\|_{\mathrm{H}^1(a,b)}\,,$$

which means that

$$\begin{aligned}\left(\|z'\|^2_{\mathrm{L}^2(a,b)} + \|z\|^2_{\mathrm{L}^2(a,b)}\right)^{1/2} \;&=\; \|z\|_{\mathrm{H}^1(a,b)} \\ &\leq\; \frac{1}{\min\{c_0, c_1\}}\|u - u^h\|_{\mathrm{L}^2(a,b)}\,. \qquad (14.41)\end{aligned}$$

Now we substitute (14.41) into (14.38) to deduce that

$$\|z''\|_{\mathrm{L}^2(a,b)} \leq K\|u - u^h\|_{\mathrm{L}^2(a,b)}\,, \qquad (14.42)$$

where

$$K = \frac{1}{c_0}\left(1 + \frac{1}{\min\{c_0, c_1\}}\left(\|p' + q\|^2_\infty + \|r - q'\|^2_\infty\right)^{1/2}\right).$$

□

It is important to observe here that K involves only known quantities: the coefficients in the differential equation under consideration. Therefore K can be computed, or at least bounded above, without difficulties. On inserting (14.42) into (14.37), we arrive at our final result, the computable *a posteriori* error bound,

$$\|u - u^h\|_{\mathrm{L}^2(a,b)} \leq K_0\left(\sum_{i=1}^n h_i^4\|R(u^h)\|^2_{\mathrm{L}^2(x_{i-1},x_i)}\right)^{1/2}, \qquad (14.43)$$

where $K_0 = K/\pi^2$.

Next we shall describe the construction of an adaptive mesh refinement algorithm based on the *a posteriori* error bound (14.43).

Suppose that TOL is a prescribed tolerance and that our aim is to compute a finite element approximation u^h to the unknown solution u so that

$$\|u - u^h\|_{\mathrm{L}^2(a,b)} \le \mathtt{TOL}\,. \tag{14.44}$$

We shall use the *a posteriori* error bound (14.43) to achieve this goal by systematically refining the subdivision, and computing a succession of numerical solutions u^h on this sequence of subdivisions, until the inequality

$$K_0 \left(\sum_{i=1}^{n} h_i^4 \|R(u^h)\|^2_{\mathrm{L}^2(x_{i-1},x_i)} \right)^{1/2} \le \mathtt{TOL} \tag{14.45}$$

is satisfied. Clearly, if u^h satisfies (14.45), then, by virtue of (14.43), it also satisfies (14.44).

In order for the inequality (14.45) to hold it is sufficient to ensure that, on each interval $[x_{i-1}, x_i]$, $i = 1, 2, \ldots, n$, we have

$$h_i^4 \|R(u^h)\|^2_{\mathrm{L}^2(x_{i-1},x_i)} \le \frac{1}{n} \left(\frac{\mathtt{TOL}}{K_0} \right)^2 . \tag{14.46}$$

Thus, a sufficient condition for (14.44) is that (14.46) holds for all $i = 1, 2, \ldots, n$.

The mesh adaptation algorithm, therefore, proceeds as follows:

`Step 1.` Choose an initial subdivision

$$\mathcal{T}_0\colon\quad a = x_0^{(0)} < x_1^{(0)} < \cdots < x_{n_0-1}^{(0)} < x_{n_0}^{(0)} = b$$

of the interval $[a, b]$, with $h_i^{(0)} = x_i^{(0)} - x_{i-1}^{(0)}$, for $i = 1, 2, \ldots, n_0$; let $h^{(0)} = \max_i h_i^{(0)}$, and consider the associated finite element space $S_{\mathrm{E}}^{h^{(0)}}$ (of dimension $n_0 - 1$);

`Step 2.` Compute the corresponding solution $u^{h^{(0)}} \in S_{\mathrm{E}}^{h^{(0)}}$;

`Step 3.` Given a computed solution $u^{h^{(m)}} \in S_{\mathrm{E}}^{h^{(m)}}$ for some $m \ge 0$, defined on a subdivision $\mathcal{T}_m$, STOP if

$$K_0 \left(\sum_{i=1}^{n_m} \left(h_i^{(m)}\right)^4 \|R(u^{h^{(m)}})\|^2_{\mathrm{L}^2(x_{i-1}^{(m)},x_i^{(m)})} \right)^{1/2} \le \mathtt{TOL}\,; \tag{14.47}$$

`Step 4.` If not, then halve those elements $[x_{i-1}^{(m)}, x_i^{(m)}]$ in $\mathcal{T}_m$, with i in the set $\{1, 2, \ldots, n_m\}$, for which

$$\left(h_i^{(m)}\right)^4 \|R(u^{h^{(m)}})\|^2_{\mathrm{L}^2(x_{i-1}^m, x_i^m)} > \frac{1}{n_m}\left(\frac{\mathtt{TOL}}{K_0}\right)^2, \qquad (14.48)$$

denote by $\mathcal{T}_{m+1}$ the resulting subdivision of $[a, b]$ with n_{m+1} elements $[x_{i-1}^{(m+1)}, x_i^{(m+1)}]$ of respective lengths

$$h_i^{(m+1)} = x_i^{(m+1)} - x_{i-1}^{(m+1)}, \qquad i = 1, \ldots, n_{m+1},$$

and consider the associated finite element space $S_{\mathrm{E}}^{h^{(m+1)}}$ of dimension $n_{m+1} - 1$;

`Step 5.` Compute the finite element approximation $u^{h^{(m+1)}} \in S_{\mathrm{E}}^{h^{(m+1)}}$, increase m by 1 and return to `Step 3`.

The inequality (14.47) is called the **stopping criterion** for the mesh adaptation algorithm, and (14.48) is referred to as the **refinement criterion**. According to the *a posteriori* error bound (14.43), when the adaptive algorithm terminates, the error $\|u - u^h\|_{\mathrm{L}^2(a,b)}$ is guaranteed not to exceed the prescribed tolerance `TOL`.

We conclude the body of this chapter with a numerical experiment which illustrates the performance of the adaptive algorithm.

Example 14.1 *Let us consider the second-order ordinary differential equation*

$$-(p(x)u')' + q(x)u' + r(x)u = f(x), \qquad x \in (0, 1), \qquad (14.49)$$

subject to the boundary conditions

$$u(0) = 0, \qquad u(1) = 0. \qquad (14.50)$$

Suppose, for example, that

$$p(x) \equiv 1, \qquad q(x) \equiv 20, \qquad r(x) \equiv 10 \qquad \text{and} \qquad f(x) \equiv 1.$$

In this case, the analytical solution, u, can be expressed in closed form:

$$u(x) = C_1\,\mathrm{e}^{\lambda_1 x} + C_2\,\mathrm{e}^{\lambda_2 x} + \frac{1}{10},$$

where λ_1 and λ_2 are the two roots of the characteristic polynomial of the differential equation, $-\lambda^2 + 20\lambda + 10 = 0$, *i.e.*,

$$\lambda_1 = 10 + \sqrt{110}, \qquad \lambda_2 = 10 - \sqrt{110},$$

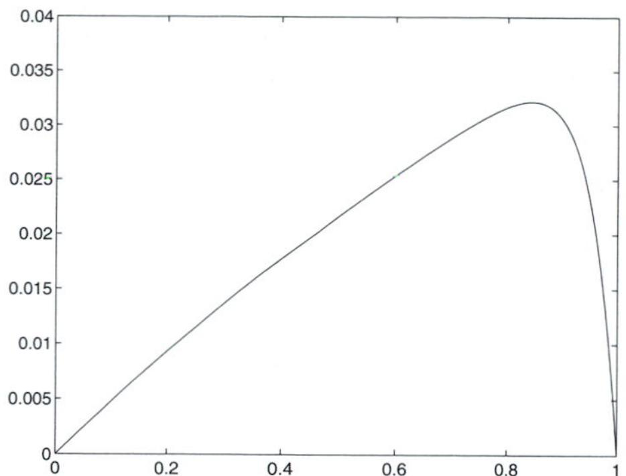

Fig. 14.4. Analytical solution of the boundary value problem (14.49), (14.50), with $p(x) \equiv 1$, $q(x) \equiv 20$, $r(x) \equiv 10$ and $f(x) \equiv 1$.

and C_1 and C_2 are constants chosen so as to ensure that $u(0) = 0$ and $u(1) = 0$; hence,

$$C_1 = \frac{\mathrm{e}^{\lambda_2} - 1}{10\,(\mathrm{e}^{\lambda_1} - \mathrm{e}^{\lambda_2})}\,, \qquad C_2 = \frac{1 - \mathrm{e}^{\lambda_1}}{10\,(\mathrm{e}^{\lambda_1} - \mathrm{e}^{\lambda_2})}\,.$$

The function u is shown in Figure 14.4.

Now, let us imagine for a moment that u is unknown, and let us compute a numerical approximation u^h to u, using the adaptive finite element algorithm described above, so that $\|u - u^h\|_{\mathrm{L}^2(0,1)} \leq \mathtt{TOL}$, where $\mathtt{TOL} = 10^{-4}$. The computation begins on a coarse subdivision of the interval $[0, 1]$ containing only 10 elements. This is then successively refined using the refinement criterion (14.48) until the stopping criterion (14.47) is satisfied; the resulting subdivisions are shown in Figure 14.5. In this example, the constant K_0 appearing in (14.43) and (14.45)–(14.48) is $(1 + \sqrt{500})/\pi^2$ (≈ 2.367).

Since we are in the fortunate (but highly idealised) position that, in addition to the numerical solution u^h, the analytical solution u is also available, we can assess the sharpness of our *a posteriori* error bound (14.43) by comparing the error $\|u - u^h\|_{\mathrm{L}^2(0,1)}$ appearing on the left-hand side of (14.43) with the computable *a posteriori* error bound on the right-hand side of (14.43). Figure 14.6 shows that the *a posteriori* bound consistently overestimates the error $\|u - u^h\|_{\mathrm{L}^2(0,1)}$ by about two orders of magnitude. By comparing the slopes of the two curves in Figure 14.6, we also see that the error and the *a posteriori* error bound decay at approximately the same rate as the number of mesh points increases in the course of mesh adaptation. ◇

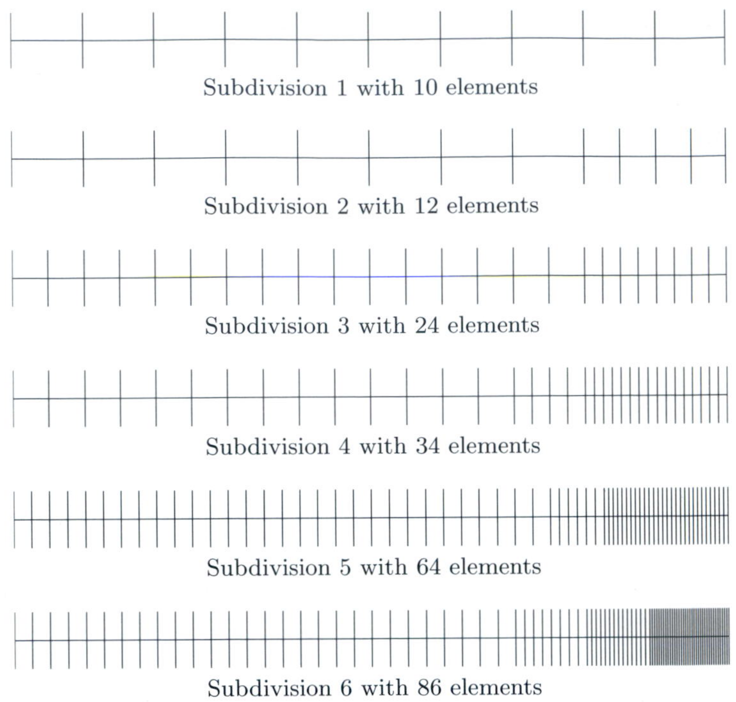

Fig. 14.5. Sequence of subdivisions of the interval $[0, 1]$ designed by the adaptive algorithm with $\texttt{TOL} = 10^{-4}$.

14.6 Notes

For further details concerning the mathematical theory and the implementation of the finite element method we refer to the following books.

- D. Braess, *Finite Elements*, Cambridge University Press, Cambridge, 2001.
- S. Brenner and L.R. Scott, *The Mathematical Theory of Finite Element Methods*, Second Edition, Springer, New York, 2002.
- C. Johnson, *Numerical Solution of Partial Differential Equations by the Finite Element Method*, Cambridge University Press, Cambridge, 1996.

For recent results on the theory of *a posteriori* error estimation for finite element approximations of differential equations, based on duality arguments, the interested reader may wish to consult the following review articles.

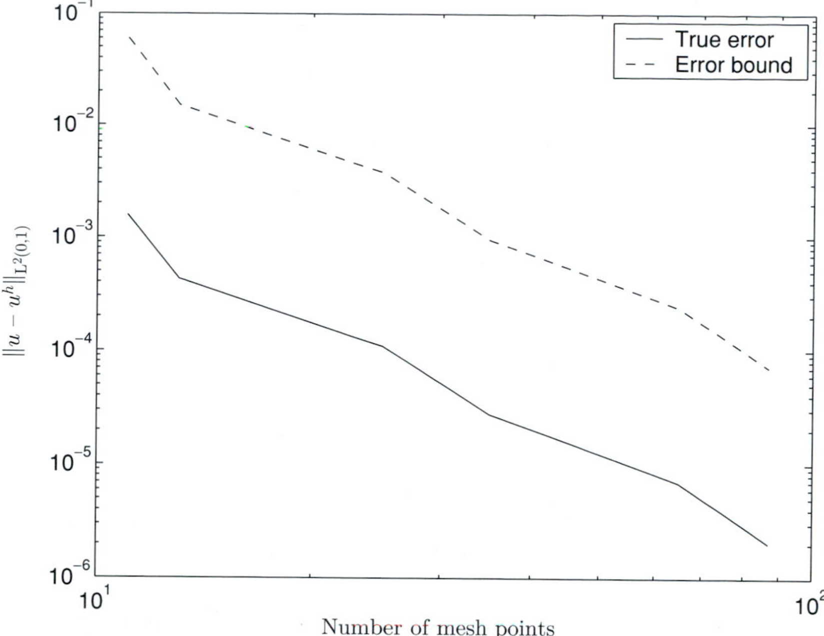

Fig. 14.6. Comparison of the true error $\|u - u^h\|_{L^2(0,1)}$ (solid curve) with the *a posteriori* error bound delivered by the adaptive algorithm (dashed curve) with $\mathtt{TOL} = 10^{-4}$.

- K. ERIKSON, D. ESTEP, P. HANSBO, AND C. JOHNSON, Introduction to adaptive methods for differential equations, in *Acta Numerica* **4** (A. Iserles, ed.), Cambridge University Press, Cambridge, 105–158, 1995.
- R. BECKER AND R. RANNACHER, An optimal control approach to a-posteriori error estimation in finite element methods, in *Acta Numerica* **10** (A. Iserles, ed.), Cambridge University Press, Cambridge, 1–102, 2001.
- M.B. GILES AND E. SÜLI, Adjoint methods for PDEs: superconvergence and adaptivity by duality, in *Acta Numerica* **11** (A. Iserles, ed.), Cambridge University Press, Cambridge, 145–236, 2002.

A detailed and general survey of the subject of *a posteriori* error estimation can be found in

- M. AINSWORTH AND J.T. ODEN, *A posteriori Error Estimation in Finite Element Analysis*, John Wiley & Sons, New York, 2000.

In this chapter we were concerned with the *a priori* error analysis of the piecewise linear finite element method in the energy norm, and its *a posteriori* error analysis in the L^2 norm. Using similar techniques, one can establish an *a priori* error bound in the L^2 norm and an *a posteriori* error bound in the energy norm. For extensions of the theory considered here to higher-order piecewise polynomial finite element approximations and generalisations to partial differential equations, the reader is referred to the books listed above.

Exercises

14.1 Given that (a,b) is an open interval of the real line, let

$$\mathrm{H}^1_{\mathrm{E}_0}(a,b) = \{v \in \mathrm{H}^1(a,b)\colon v(a) = 0\}\,.$$

(i) By writing

$$v(x) = \int_a^x v'(\xi)\mathrm{d}\xi\,,$$

for $v \in \mathrm{H}^1_{\mathrm{E}_0}(a,b)$ and $x \in [a,b]$, show the following (**Poincaré–Friedrichs**) inequality:

$$\|v\|^2_{\mathrm{L}^2(a,b)} \le \frac{1}{2}(b-a)^2\|v'\|^2_{\mathrm{L}^2(a,b)} \qquad \forall v \in \mathrm{H}^1_{\mathrm{E}_0}(a,b)\,.$$

(ii) By writing

$$[v(x)]^2 = \int_a^x \frac{\mathrm{d}}{\mathrm{d}\xi}[v(\xi)]^2\,\mathrm{d}\xi = 2\int_a^x v(\xi)v'(\xi)\mathrm{d}\xi$$

for $v \in \mathrm{H}^1_{\mathrm{E}_0}(a,b)$ and $x \in [a,b]$, show the following (**Agmon's**) inequality:

$$\max_{x\in[a,b]} |v(x)|^2 \le 2\,\|v\|_{\mathrm{L}^2(a,b)}\|v'\|_{\mathrm{L}^2(a,b)} \quad \forall v \in \mathrm{H}^1_{\mathrm{E}_0}(a,b)\,.$$

14.2 Given that $f \in \mathrm{L}^2(0,1)$, state the weak formulation of each of the following boundary value problems on the interval $(0,1)$:

(a) $-u'' + u = f(x)$, $u(0) = 0$, $u(1) = 0$;
(b) $-u'' + u = f(x)$, $u(0) = 0$, $u'(1) = 1$;
(c) $-u'' + u = f(x)$, $u(0) = 0$, $u(1) + u'(1) = 2$.

In each case, show that there exists at most one weak solution.

14.3 Give a proof of Theorem 14.4.

14.4 Prove Corollary 14.2.

14.5 Consider the boundary value problem

$$-p_0 u'' + r_0 u = f(x)\,, \quad u(0) = 0\,,\ u(1) = 0\,,$$

on the interval $[0,1]$, where p_0 and r_0 are positive constants and $f \in \mathrm{C}^4[0,1]$. Using equally spaced points

$$x_i = ih\,, \qquad i = 0,1,\ldots,n\,, \quad \text{with } h = 1/n,\ n \geq 2\,,$$

and the standard piecewise linear finite element basis functions (hat functions) φ_i, $i = 1,2,\ldots,n-1$, show that the finite element equations for $u_i = u^h(x_i)$ become

$$-p_0(u_{i-1} - 2u_i + u_{i+1})/h^2 + r_0(u_{i-1} + 4\,u_i + u_{i+1})/6 = \frac{1}{h}\langle f, \varphi_i\rangle$$

for $i = 1,2,\ldots,n-1$, with $u_0 = 0$ and $u_n = 0$. By expanding in Taylor series, show that

$$\frac{1}{h}\langle f, \varphi_i\rangle = f(x_i) + \tfrac{1}{12}h^2 f''(x_i) + \mathcal{O}(h^4)\,.$$

Interpreting this set of difference equations as a finite difference approximation to the boundary value problem, as in Chapter 13, show that the corresponding truncation error T_i satisfies

$$T_i = \tfrac{1}{12}h^2 r_0 u''(x_i) + \mathcal{O}(h^4)\,, \qquad i = 1,\ldots,n-1\,,$$

and use the method of Exercise 13.2 to show that

$$\max_{0\leq i\leq n} |u(x_i) - u^h(x_i)| \leq Mh^2\,,$$

where M is a positive constant.

14.6 In the notation of Exercise 5 suppose that all the integrals involved in the calculation are approximated by the trapezium rule. Show that the system of equations becomes identical to that obtained from the central difference approximation in Chapter 13, and deduce that

$$\max_{0\leq i\leq n} |u(x_i) - u^h(x_i)| \leq Mh^2\,,$$

where M is a positive constant.

14.7 Consider the differential equation

$$-\left(p(x)u'\right)' + r(x)u = f(x)\,, \qquad a < x < b\,,$$

with p, r and f as at the beginning of the chapter, subject to the boundary conditions

$$-p(a)u'(a) + \alpha u(a) = A\,, \qquad p(b)u'(b) + \beta u(b) = B\,,$$

where α and β are positive real numbers, and A and B are real numbers. Show that the weak formulation of the boundary value problem is

$$\text{find } u \in \mathrm{H}^1(a,b) \text{ such that } \mathcal{A}(u,v) = \ell(v) \text{ for all } v \in \mathrm{H}^1(a,b)\,,$$

where

$$\begin{aligned} \mathcal{A}(u,v) &= \int_a^b [p(x)u'(x)v'(x) + r(x)u(x)v(x)]\mathrm{d}x \\ &\quad + \alpha u(a)v(a) + \beta u(b)v(b)\,, \end{aligned}$$

and

$$\ell(v) = \langle f, v\rangle + Av(a) + Bv(b)\,.$$

Construct a finite element approximation of the boundary value problem based on this weak formulation using piecewise linear finite element basis functions on the subdivision

$$a = x_0 < x_1 < \cdots < x_{n-1} < x_n = b$$

of the interval $[a,b]$. Show that the finite element method gives rise to a set of $n+1$ simultaneous linear equations with $n+1$ unknowns $u_i = u^h(x_i)$, $i = 0, 1, \ldots, n$. Show that this linear system has a unique solution.

Comment on the structure of the matrix $M \in \mathbb{R}^{(n+1)\times(n+1)}$ of the linear system: (a) Is M symmetric? (b) Is M positive definite? (c) Is M tridiagonal?

14.8 Given that α is a nonnegative real number, consider the differential equation

$$-u'' + u = f(x) \quad \text{for } x \in (0,1)\,,$$

subject to the boundary conditions

$$u(0) = 0\,, \qquad \alpha u(1) + u'(1) = 0\,.$$

State the weak formulation of the problem. Using continuous piecewise linear basis functions on a uniform subdivision of $[0,1]$ into elements of size $h = 1/n$, $n \geq 2$, write down the finite element approximation to this problem and show that this has a unique solution u^h. Expand u^h in terms of the standard piecewise linear finite element basis functions (hat functions) φ_i,

$i = 1, 2, \ldots, n$, by writing

$$u^h(x) = \sum_{i=1}^{n} U_i \varphi_i(x)$$

to obtain a system of linear equations for the vector of unknowns $(U_1, \ldots, U_n)^{\mathrm{T}}$.

Suppose that $\alpha = 0$, $f(x) \equiv 1$ and $h = 1/3$. Solve the resulting system of linear equations and compare the corresponding numerical solution $u^h(x)$ with the exact solution $u(x)$ of the boundary value problem.

14.9 Consider the differential equation

$$-(p(x)u')' + r(x)u = f(x)\,, \quad x \in (0,1)\,,$$

subject to the boundary conditions $u(0) = 0$, $u(1) = 0$, where $p(x) \geq c_0 > 0$, $r(x) \geq 0$ for all x in the closed interval $[0, 1]$, with $p \in \mathrm{C}^1[0,1]$, $r \in \mathrm{C}[0,1]$ and $f \in \mathrm{L}^2(0,1)$. Given that u^h denotes the continuous piecewise linear finite element approximation to u on a uniform subdivision of $[0, 1]$ into elements of size $h = 1/n$, $n \geq 2$, show that

$$\|u - u^h\|_{\mathrm{H}^1(0,1)} \leq C_1 h \|u''\|_{\mathrm{L}^2(0,1)}\,,$$

where C_1 is a positive constant that you should specify. Show further that there exists a positive constant C such that

$$\|u - u^h\|_{\mathrm{H}^1(0,1)} \leq C h \|f\|_{\mathrm{L}^2(0,1)}\,.$$

Calculate the right-hand sides in these inequalities in the case when

$$p(x) \equiv 1\,, \qquad r(x) \equiv 0\,, \qquad f(x) \equiv 1\,,$$

for $x \in [0, 1]$, and $h = 10^{-3}$.

14.10 Consider the two-point boundary value problem

$$-u'' + u = f(x)\,, \quad x \in (0,1)\,, \qquad u(0) = 0\,,\ u(1) = 0\,,$$

with $f \in \mathrm{C}^2[0,1]$. State the piecewise linear finite element approximation to this problem on a nonuniform subdivision

$$0 = x_0 < x_1 < \cdots < x_n = 1\,, \qquad n \geq 2\,,$$

with $h_i = x_i - x_{i-1}$, assuming that, for a continuous piecewise linear function v^h,

$$\int_0^1 f(x)v^h(x)\mathrm{d}x$$

has been approximated by applying the trapezium rule on each element $[x_{i-1}, x_i]$.

Verify that the following *a posteriori* bound holds for the error between u and its finite element approximation u^h:

$$\|u - u^h\|_{\mathrm{L}^2(0,1)} \le K_0 \left(\sum_{i=1}^{n} h_i^4 \|R(u^h)\|^2_{\mathrm{L}^2(x_{i-1},x_i)} \right)^{1/2}$$

$$+ K_1 \max_{1\le i\le n} h_i^2 \left(\max_{x\in[x_{i-1},x_i]} |f''(x)|^2 + 4 \max_{x\in[x_{i-1},x_i]} |f'(x)| \right)^{1/2},$$

where $R(u^h) = f(x) - (-(u^h)''(x) + u^h(x))$ for $x \in (x_{i-1}, x_i)$, $i = 1, \dots, n$, and K_0, K_1 are constants which you should specify.

How would you use this bound to compute u to within a specified tolerance TOL?

Appendix A

An overview of results from real analysis

In this Appendix we gather a number of results from real analysis which are assumed at various places in the text. Some of these will be familiar from any course on the subject, and no proofs are given; a small number may be less familiar, and we give proofs of these for completeness.

Theorem A.1 (The Intermediate Value Theorem) *Suppose that f is a real-valued function, defined and continuous on the closed interval $[a, b]$ of $\mathbb{R}$. Then, f is a bounded function on the interval $[a, b]$ and, if y is any number such that*

$$\inf_{x\in[a,b]} f(x) \leq y \leq \sup_{x\in[a,b]} f(x),$$

then there is a number $\xi \in [a, b]$ such that $f(\xi) = y$. In particular, the infimum and the supremum of f are achieved, and can be replaced by $\min_{x\in[a,b]}$ and $\max_{x\in[a,b]}$, respectively.

The next result, known as Rolle's Theorem, was published in an obscure book in 1691 by the French mathematician Michel Rolle (1652–1719) who invented the notation $\sqrt[n]{x}$ for the nth root of x.

Theorem A.2 (Rolle's Theorem) *Suppose that f is a real-valued function, defined and continuous on the closed interval $[a, b]$ of $\mathbb{R}$, differentiable in the open interval (a, b), and such that $f(a) = f(b)$. Then, there exists a number $\xi \in (a, b)$ such that $f'(\xi) = 0$.*

It is often important in our applications that the point $\xi \in (a, b)$, *i.e.*, $a < \xi < b$. For instance it may happen that $f'(a) = f'(b) = 0$, as well as $f(a) = f(b)$; Theorem A.2 then states that, in addition to the endpoints of the interval $[a, b]$, there is also an interior point $\xi \in (a, b)$ at which the derivative vanishes.

Theorem A. 3 (The Mean Value Theorem) *Suppose that f is a real-valued function, defined and continuous on the closed interval $[a, b]$ of $\mathbb{R}$, and f is differentiable in the open interval (a, b). Then, there exists a number $\xi \in (a, b)$ such that*

$$f(b) - f(a) = f'(\xi)\,(b-a)\,.$$

Theorem A.4 (Taylor's Theorem) *Suppose that n is a nonnegative integer, and f is a real-valued function, defined and continuous on the closed interval $[a, b]$ of $\mathbb{R}$, such that the derivatives of f of order up to and including n are defined and continuous on the closed interval $[a, b]$. Suppose further that $f^{(n)}$ is differentiable on the open interval (a, b). Then, for each value of x in $[a, b]$, there exists a number $\xi = \xi(x)$ in the open interval (a, b) such that*

$$\begin{aligned} f(x) &= f(a) + (x-a)f'(a) + \cdots + \frac{(x-a)^n}{n!} f^{(n)}(a) \\ &\quad + \frac{(x-a)^{n+1}}{(n+1)!} f^{(n+1)}(\xi)\,. \end{aligned}$$

Theorem A.5 (Taylor's Theorem with integral remainder) *Let n be a nonnegative integer and suppose that f is a real-valued function, defined and continuous on the closed interval $[a, b]$ of $\mathbb{R}$, such that the derivatives of f of order up to and including n are defined and continuous on $[a, b]$, $f^{(n)}$ is differentiable on the open interval (a, b), and $f^{(n+1)}$ is integrable on (a, b). Then, for each $x \in [a, b]$,*

$$\begin{aligned} f(x) &= f(a) + (x-a)f'(a) + \cdots + \frac{(x-a)^n}{n!} f^{(n)}(a) \\ &\quad + \int_a^x \frac{(x-t)^n}{n!} f^{(n+1)}(t)\mathrm{d}t\,. \end{aligned}$$

Proof As this version of the theorem may be rather less familiar we include a proof.

The theorem is trivially true for $n = 0$. Suppose that the theorem is true for some nonnegative integer, say $n = k$. Then, provided that $f^{(k+1)}$ is differentiable on (a, b) and $f^{(k+2)}$ is integrable on (a, b), integration by parts shows that

$$\int_a^x \frac{(x-t)^{k+1}}{(k+1)!} f^{(k+2)}(t)\mathrm{d}t = -\frac{(x-a)^{k+1}}{(k+1)!} f^{(k+1)}(a) - \int_a^x -\frac{(x-t)^k}{k!} f^{(k+1)}(t)\mathrm{d}t\,;$$

use of the theorem when $n = k$ now shows that it is also true for $n = k+1$. The proof by induction is then complete. □

Theorem A.6 (The Integral Mean Value Theorem) *Suppose that f is a real-valued function, defined and continuous on a closed interval $[a,b]$ of $\mathbb{R}$, and let g be a function, defined, nonnegative and integrable on (a,b). Then, there exists a number $\xi \in (a,b)$ such that*

$$\int_a^b f(x)g(x)\,\mathrm{d}x = f(\xi)\int_a^b g(x)\,\mathrm{d}x\,.$$

Proof Since f is continuous on $[a,b]$, it is bounded on $[a,b]$, say

$$m \le f(x) \le M, \qquad x \in [a,b]\,.$$

Then, as $g(x) \ge 0$ for all $x \in (a,b)$, we have that

$$mg(x) \le f(x)g(x) \le Mg(x)\,, \qquad x \in (a,b)\,.$$

Integrating these inequalities gives

$$m\int_a^b g(x)\,\mathrm{d}x \le \int_a^b f(x)g(x)\,\mathrm{d}x \le M\int_a^b g(x)\,\mathrm{d}x\,.$$

If $\int_a^b g(x)\mathrm{d}x = 0$, then the result trivially follows. If, on the other hand, $\int_a^b g(x)\mathrm{d}x > 0$, then

$$m \le \frac{\int_a^b f(x)g(x)\,\mathrm{d}x}{\int_a^b g(x)\,\mathrm{d}x} \le M.$$

The existence of the required value of $\xi \in (a,b)$ now follows from the Intermediate Value Theorem. □

Theorem A.6 obviously also holds provided that $g(x) \le 0$ on (a,b); it is only important that g has constant sign on (a,b). Note also that we do not require that g is continuous, only that it is integrable. For example, Theorem A.6 will hold if f is a continuous function defined on $[0,1]$ and $g(x) = x^{-1/2}$, $x \in (0,1)$.

Bibliography

Abramowitz, M. and Stegun, I.A. (1972). *Handbook of Mathematical Functions with Formulas, Graphs, and Mathematical Tables*, Ninth printing (Dover, New York).

Ahlberg, J.H., Nilson, E.N. and Walsh, J.L. (1967). *The Theory of Splines and Their Applications*, Mathematics in science and engineering (Academic Press, New York).

Ainsworth, M. and Oden, J.T. (2000). *A posteriori Error Estimation in Finite Element Analysis* (John Wiley & Sons, New York).

Ascher, U.M., Mattheij, R.M.M. and Russell, R.D. (1995). *Numerical Solution of Boundary Value Problems for Ordinary Differential Equations,* Corrected reprint of the 1988 original, (SIAM, Philadelphia).

Axelson, O. (1996). *Iterative Solution Methods* (Cambridge University Press, Cambridge).

Bashforth, F. (1883). *An Attempt to Test the Theories of Capillary Action by Comparing the Theoretical and Measured Forms of Drops of Fluid. With an Explanation of the Method of Integration in Constructing Tables Which Give the Theoretical Form of Such Drops, by J.C. Adams* (Cambridge University Press, Cambridge).

Bauer, F.L. and Fike, C.T. (1960). Norms and exclusion theorems, *Num. Math.* **2**, 137–141.

Becker, R. and Rannacher, R. (2001). An optimal control approach to a-posteriori error estimation in finite element methods, *Acta Numerica* **10** 1–102, ed. A. Iserles, (Cambridge University Press, Cambridge).

Bernstein, S.N. (1912/13). Démonstration du théorème de Weierstrass fondée sur le calcul des probabilités, *Comm. Soc. Math. Kharkow* **13**, 1–2.

Björk, Å. (1996). *Numerical Methods for Least Squares Problems* (SIAM, Philadelphia).

Blyth, T.S. and Robertson E.F. (1998). *Basic Linear Algebra* (Springer, London).

Braess, D. (2001). *Finite Elements* (Cambridge University Press, Cambridge).

Brenner, S. and Scott, L.R. (2002). *The Mathematical Theory of Finite Element Methods*, Second Edition (Springer, New York).

Butcher, J.C. (1987). *The Numerical Analysis of Ordinary Differential Equa-*

tions. Runge–Kutta and General Linear Methods (John Wiley & Sons, Chichester).

Chabert, J.-L. (1999). *A History of Algorithms from the Pebble to the Microchip* (Springer, New York).

Cheney, E.W. (1966). *Introduction to Approximation Theory* (McGraw-Hill, New York).

Ciarlet, P.G. (1989). *Introduction to Numerical Linear Algebra and Optimisation* (Cambridge University Press, Cambridge).

Courant, R. (1943). Variational methods for the solution of problems in equilibrium and vibrations, *Bull. Amer. Math. Soc.* **49**, 1–23.

Cramer, G. (1750). *Introduction à l'analyse des lignes courbes algébriques* (Chez les Frères Cramer & Cl. Philibert, Genève).

Dahlquist, G. (1963). A special stability problem for linear multistep methods, *BIT* **3**, 27–43.

Davis, P.J. and Rabinowitz, P. (1984). *Methods of Numerical Integration*, Second Edition (Academic Press, Orlando, FL).

De Boor, C. (2001). *A Practical Guide to Splines*, Revised Edition (Springer, New York).

Dedieu, J.-P. and Shub, M. (2000). Multihomogeneous Newton methods, *Math. Comput.* **69**, 1071–1098.

Dekker, K. and Verver, J.G. (1984). *Stability of Runge–Kutta Methods for Stiff Nonlinear Differential Equations* (North-Holland, Amsterdam).

Douglass, S.A. (1996). *Introduction to Mathematical Analysis* (Addison–Wesley, Reading, MA).

Drazin, P.G. (1992). *Nonlinear Systems* (Cambridge University Press, Cambridge).

Dwyer, P.S. (1944). A matrix presentation of least squares and correlation theory with matrix justification of improved methods of solutions, *Ann. Math. Stat.* **15** 82–89.

Engels, H. (1980). *Numerical Quadrature and Cubature*, Computational Mathematics and Applications, (Academic Press, London).

Erikson, K., Estep, D., Hansbo, P. and Johnson, C. (1995). Introduction to adaptive methods for differential equations, in *Acta Numerica* **4**, 105–158, ed. A. Iserles (Cambridge University Press, Cambridge).

Freund, G. (1971). *Orthogonal Polynomials* (Pergamon Press, Oxford, New York).

Fujino, S. and Fischer J. (1998). Über S.A. Gerschgorin (1901–1933), *GAMM Mitt. Ges. Angew. Math. Mech.* **1**, 15-19.

Gauss, C.F. (1809). *Theoria motus corporum coelestium in sectionibus conicis solem ambientium.* (F. Perthes und I.H. Besser, Hamburg).

Gauss, C.F. (1809). Methodus nova integralium valores per approximationem inveniendi, in C.F. Gauss, *Werke*, **3**, 163–196 (Dietrich, Göttingen, 1863).

Gautschi, W. (1996). Orthogonal polynomials: Applications and computation, in *Acta Numerica* **5**, ed. A. Iserles (Cambridge University Press, Cambridge).

Gautschi, W. (1997). *Numerical Analysis: an Introduction* (Birkhäuser, Boston, MA).

Gautschi, W., Golub, G.H. and Opfer, G. (1999). *Applications and Computation of Orthogonal Polynomials* (Birkhäuser, Basel).

Giles, M.B. and Süli, E. (2002). Adjoint methods for PDEs: superconvergence and adaptivity by duality, *Acta Numerica* **11** 145–236, ed. A. Iserles, (Cambridge University Press, Cambridge).

Goldstine, H. (1977). *History of Numerical Analysis from the Sixteenth through the Nineteenth Century* (Springer, New York).

Golub, G.H. and Van Loan, C.F. (1996). *Matrix Computations,* Third Edition, (Johns Hopkins University Press, Baltimore).

Haar, A. (1918). Die Minkowskische Geometrie und die Annäherung an stetige Funktionen, *Math. Ann.* **78**, 294–311.

Hairer, E., Nørsett, S.P. and Wanner, G. (1993). *Solving Ordinary Differential Equations I: Nonstiff Problems,* Second Edition (Springer, Berlin).

Hairer, E. and Wanner, G. (1991). *Solving Ordinary Differential Equations II: Stiff and Differential-Algebraic Problems* (Springer, Berlin).

Hamilton, A.G. (1990). *Linear Algebra* (Cambridge University Press, Cambridge).

Henrici, P. (1962). *Discrete Variable Methods in Ordinary Differential Equations* (John Wiley & Sons, New York).

Higham, N.J. (1996). *Accuracy and Stability of Numerical Algorithms* (SIAM, Philadelphia).

Hildebrand, F.B. (1956). *Introduction to Numerical Analysis* (McGraw–Hill, New York).

Holladay, J.C. (1957). Smoothest curve approximation, *Math. Comput.* **11**, 233–243.

Horn, R.A. and Johnson, C.R. (1992). *Matrix Analysis* (Cambridge University Press, Cambridge).

Householder, A.S. (1964). *The Theory of Matrices in Numerical Analysis* (Blaisdell, New York).

Householder, A.S. (1970). *The Numerical Treatment of a Single Nonlinear Equation* (McGraw–Hill, New York).

Iserles, A. (1996). *A First Course in the Numerical Analysis of Differential Equations* (Cambridge University Press, Cambridge).

Jenkins, M.A. and Traub, J.F. (1970). A three-stage algorithm for real polynomials using quadratic iterations, *SIAM J. Numer. Anal.* **7**, 545–566.

Johnson, C. (1996). *Numerical Solution of Partial Differential Equations by the Finite Element Method* (Cambridge University Press, Cambridge).

Kaluza, R. (1996). *Through the Eyes of a Reporter: the Life of Stefan Banach* (Birkhäuser, Boston, MA).

Kantorovich, L.V., (1952). Functional analysis and applied mathematics. *Uspekhi Mat. Nauk* **3**, 89–185, 1948; English transl., Rep. 1509, National Bureau of Standards, Washington, DC.

Kantorovich, L.V. and Akilov, G.P. (1982). *Functional Analysis*, Second Edition, (Pergamon Press, Oxford, New York).

Keller, H.B. (1990). *Numerical Solution of Two-Point Boundary-Value Problems*, Fourth Printing, (SIAM, Philadelphia).

Keller, H.B. (1992). *Numerical Methods for Two-Point Boundary-Value Prob-*

lems, Reprint of the 1968 original published by Blaisdell (Dover, New York).

Kepler, J. (1615). *Nova stereometria doliorum vinariorum. Accessit stereometriae Archimedae supplementem*, in Johannes Keppler, *Gesammelte Werke* (Hrsg. von Franz Hammer, Bd IX *Mathematische Schriften*, 5–133, München, 1955).

Krommer, A.R. and Ueberhuber, C.W. (1998). *Computational Integration* (SIAM, Philadelphia).

Krylov, V.I. (1962). *Approximate Calculation of Integrals* (Macmillan, New York).

Lambert, J.D. (1991). *Numerical Methods for Ordinary Differential Systems* (John Wiley & Sons, Chichester).

Maclaurin, C. (1742). *A treatise of fluxions.* In two books. (printed by T.W. and T. Ruddimans, Edinburgh).

Mandelbrot, B. (1977). *Fractals: Form, Chance, and Dimension* (W.H. Freeman, San Francisco).

Mandelbrot, B. (1983). *The Fractal Geometry of Nature* (W.H. Freeman, New York).

Milnor, J.W. (1997). *Topology from the Differentiable Viewpoint* (Princeton University Press, Princeton, NJ).

Moulton, F.R. (1926). *New methods in exterior ballistics* (University of Chicago Press, Chicago).

Névai, P. (1979). *Orthogonal Polynomials* (American Mathematical Society, Providence, RI).

Ortega, J.M. and Rheinboldt, W.C. (2000). *Iterative Solution of Nonlinear Equations in Several Variables*, Reprint of the 1970 original (SIAM, Philadelphia).

Pan, V. (1997). Solving a polynomial equation: some history and recent progress, *SIAM Rev.* **39**, 187–220.

Parlett, B. (1980). *The Symmetric Eigenvalue Problem* (Prentice–Hall, Englewood Cliffs, NJ).

Pinkus, A. (2000). Weierstrass and approximation theory, *J. Approx. Th.* **107**, 1–66.

Powell, M.J.D. (1996). *Approximation Theory and Methods* (Cambridge University Press, Cambridge).

Pryce, J.D. (1993). *Numerical Solution for Sturm–Liouville Problems* (Clarendon Press, Oxford).

Rabinowitz, P. (1978). *A First Course in Numerical Analysis* (McGraw–Hill, New York).

Ralston, A. and Rabinowitz, P. (1978). *A First Course in Numerical Analysis*, Second Edition (McGraw–Hill, New York).

Reid, C. (1986). *Hilbert–Courant* (Springer, New York).

Robbins, H. (1955). A remark on Stirling's formula, *Amer. Math. Monthly* **62**, 26–29.

Romberg, W. (1955). Vereinfachte numerischen Integration, *Norske Vid. Selsk. Forh., Trondheim* **28**, 30–36.

Rudin, W. (1976). *Principles of Mathematical Analysis*, Third Edition, (McGraw–Hill, New York).

Schumaker, L.L. (1981). *Spline Functions: Basic Theory* (John Wiley & Sons, New York).

Shen, K., Crossley, J. and Lun, A.W.-C. (1999). *Chiu chang suan shu. The Nine Chapters on the Mathematical Art: Companion and Commentary* (Oxford University Press, Oxford).

Smale, S. (1986). Newton's method estimates from data at one point, in *The Merging of Disciplines: New Directions in Pure, Applied and Computational Mathematics.*, ed. R. Ewing, K. Gross, C. Martin (Springer, New York).

Stirling, J. (1730). *Methodus differentialis: sive tractatus de summatione et interpolatione serierum infinitarum* (G. Strahan, London).

Stuart, A.M. and Humphries, A.R. (1999). *Dynamical Systems and Numerical Analysis* (Cambridge University Press, Cambridge).

Sturm, J.C.F. (1835). Mémoire sur la résolution des équations numériques, *Mémoires présentés par divers Savants étrangers à l'Académie royale des sciences, section Sc. math. phys.* **6**, 273–318.

Sturm, J.C.F. (1836). Sur les équations différentielles linéaires du second ordre, *J. Math. Pures Appl.* **1**, 106–186.

Szegő, G. (1959). *Orthogonal Polynomials* (American Mathematical Society, Providence, RI).

Thomas, L.H. (1949). *Elliptic Problems in Linear Difference Equations over a Network,* Watson Sci. Comput. Lab. Rept. (Columbia University, New York).

Turing, A.M. (1948). Rounding-off errors in matrix processes, *Quart. J. Mech. Appl. Math.* **1**, 287–308.

Trefethen, L.N. and Bau, D. III (1997). *Numerical Linear Algebra* (SIAM, Philadelphia).

Van Dalen, D. (1999). *Mystic, Geometer, and Intuitionist. The Life of L.E.J. Brouwer: the Dawning Revolution* (Clarendon Press, Oxford).

Varga, R.S. (1962). *Matrix Iterative Analysis* (Prentice–Hall, Englewood Cliffs, NJ).

Wilkinson, J.H. (1961). Error analysis of direct methods of matrix inversion, *J. Assoc. Comput. Math.* **8**, 281–330.

Wilkinson, J.H. (1988). *The Algebraic Eigenvalue Problem* (Clarendon Press, Oxford University Press, New York).

Yosida, K. (1971). *Functional Analysis*, Third Edition (Springer, Berlin).

Ypma, T. (1995). Historical development of the Newton–Raphson method, *SIAM Rev.* **37**, 531–551.

Index

$(\cdot)_+$, 303
1-norm, 59, 66
2-norm, 59, 66, 225, 252, 255
  best approximation, 256
∞-norm, 59, 65, 225
  best approximation, 228
$\sim$, 43
$\succeq$, 99

A posteriori error analysis
  adaptivity, 410
  dual problem, 405
$A \setminus B$, 64
Absolutely continuous function, 295
Adaptive finite element algorithm
  refinement criterion, 410
  stopping criterion, 410
Agmon's inequality, 415
Asymptotic convergence, 16
  convergence rate, 13, 21

$B_\varepsilon(\boldsymbol{\xi})$, 104
Backward Differentiation Formulae, 349
Band matrix, 98
Bauer–Fike Theorem, 173, 174
Bernoulli numbers, 214
Bernstein polynomials, 227
Bessel's inequality, 266
Best approximation
  in 2-norm, 256
  in ∞-norm, 228
Bidiagonal matrix, 164
Bilinear functional, 388
  symmetric, 388
Binet–Cauchy Theorem, 51
Bisection method, 28
Boundary value problems, 361
  central difference approximation, 363
  derivative boundary condition, 367
  eigenvalue problem, 373
  error bound, 365, 368
  finite element approximation, 391
  global error, 364
  Maximum Principle, 365, 369, 372
  self-adjoint problem, 370
  truncation error, 364, 368, 371
  weak formulation, 390
  weak solution, 390
Brouwer's Fixed Point Theorem, 4, 125

$\mathrm{C}[a,b]$, 225
$\mathrm{C}^k[a,b]$, 293
$\mathbb{C}^n$, 62
$\mathbb{C}^{n\times n}$, 64, 145
$\mathbb{C}^n_*$, 64
Céa's Lemma, 397
Cauchy sequence, 105
Cauchy–Schwarz inequality, 59, 254
Central difference, 362
Characteristic polynomial, 136, 137
Chebyshev polynomials, 241, 263
Cholesky factorisation, 91
Closed ball, 63
Closed set, 105
Cofactor, 40
Comparison functions, 366, 372
Completeness, 105
Composite integration formulae, 209
Condition number, 58, 70
  ill-conditioned matrix, 70
  ill-conditioned problem, 68
Consistent linear multistep method, 337
Consistent one-step method, 321
Continuous function, 106
Contraction, 6

Contraction Mapping Theorem, 7, 110
Convergence
  asymptotic, 16
  asymptotic rate, 13
  linear, 12
  of linear multistep method, 340
  of one-step method, 322
  quadratic, 16, 22, 119
  sublinear, 13
  superlinear, 13
Cramer's rule, 41
Cubic splines, 298

Dahlquist's Theorems
  Barrier Theorem, 340
  Equivalence Theorem, 340
  Second Barrier Theorem, 348
de la Vallée Poussin's Theorem, 232
det(A), 40
Determinant, 40
Diagonal dominance, 96, 117, 367, 371
Differential equations
  boundary value problems, 361, 385
  initial value problems, 310

Eigenfunctions, 373
Eigenvalues, 133, 373
  characteristic polynomial, 136, 137
  definition, 66
  Jacobi's method, 137, 149
  QR algorithm, 162
  Rayleigh quotient, 170
  tridiagonal matrix, 156
Eigenvectors, 136
  definition, 66
  inverse iteration, 166
  Jacobi's method, 144
  orthogonal, 136
Energy norm, 399
Euler's method, 317, 323
  global error, 318
  truncation error, 318
Euler–Maclaurin formula, 211

Finite element method, 385
  a posteriori error analysis, 402
  a priori error analysis, 397
  adaptive algorithm, 409
  basis functions, 394
  Galerkin method, 394
  Galerkin principle, 386
  interpolant, 399
  Rayleigh–Ritz principle, 386
  residual, 406
  Ritz method, 394
  subdivision, 394
Fixed point
  definition, 4, 108
  simple iteration, 6
  simultaneous iteration, 108
  stable, 12
  unstable, 12
Frobenius norm, 141

Galerkin approximation, 393
Galerkin basis functions, 393
Galerkin finite element method, 394
Galerkin method, 392, 393
Galerkin orthogonality, 397, 405
Galerkin principle, 389
Gauss quadrature, 277
  composite, 285
  convergence, 283
  error estimate, 282
  quadrature points, 279
  quadrature weights, 279
Gaussian elimination, 44
  pivoting, 52
Gerschgorin discs, 145
Gerschgorin similarity transformation, 149
Gerschgorin theorems, 145
Global convergence, 29
  Newton's method, 31, 123
Global error
  boundary value problem, 364
  Euler's method, 318
  initial value problem, 317
Gram–Schmidt orthogonalisation, 261

$\mathrm{H}^{k+1}(a,b)$, 296
Hat function, 297, 394
Hermite cubic spline, 300
Hermite interpolation, 187, 277
  error, 190
Hilbert matrix, 72, 259
Hölder's inequality, 61
Householder matrix, 150
Householder reflector, 151
Householder's method, 155

Implicit methods
  linear multistep methods, 330
  one-step methods, 324
  Runge–Kutta methods, 351
Improved Euler method, 328
Infinity norm, 59, 65, 225
  best approximation, 228
Initial value problems, 310
  linear multistep methods, 329
  one-step methods, 317
Inner product, 252, 388
  inner product space, 253

orthogonality, 253
weight function, 255
Integral Mean Value Theorem, 421
Integration, 200
composite Simpson's rule, 210
composite trapezium rule, 209
Euler–Maclaurin formula, 211
Gauss quadrature, 277
Lobatto rule, 287
midpoint rule, 286
Newton–Cotes quadrature, 201
quadrature points, 202, 279
quadrature weights, 202
Radau rule, 287
Richardson extrapolation, 216
Romberg integration, 217
Simpson's rule, 203
trapezium rule, 202
Interchanges, 52
interchange matrix, 53
Interlace Theorem, 157
Intermediate Value Theorem, 419
Interpolation, 179, 244, 292
at Chebyshev points, 244
cubic spline, 298
Hermite, 187, 292
Lagrange, 180, 292
linear spline, 293
Interpolation points, 182
Inverse
of a lower triangular matrix, 47
of a matrix, 40
Inverse iteration, 166
Iteration, simple, 2

Jacobi's method, 149
classical, 140
convergence, 142
eigenvalues, 137
eigenvectors, 144
serial, 143
Jacobian matrix, 113, 346

Kronecker delta, 267

$L^2_w(a,b)$, 225
Lagrange interpolation, 180, 201
error, 183
Laguerre polynomials, 290
Least squares solution of linear equations, 74
Lebesgue integral, 256
Legendre polynomials, 263
Linear convergence, 12
Linear multistep methods, 329
A-stable, 348
absolutely stable, 348
characteristic polynomials, 332
consistency, 337
error constant, 338
explicit, 330
implicit, 330
order of accuracy, 338
region of absolute stability, 348
Root Condition, 332, 335
Simpson's rule method, 330
truncation error, 337
zero-stability, 331
Lipschitz condition, 7, 11, 109, 318
Lipschitz constant, 109
$\ln = \log_e$, 5, 315
Lobatto quadrature, 287
Logistic equation, 30
Lower triangular matrix, 46
inverse, 47
LU factorisation
existence, 50
of matrix, 48
with pivoting, 53

M-matrix, 101
Mass matrix, 396
Matrix
band, 98
bidiagonal, 164
condition number, 58, 70
diagonally dominant, 96
Hilbert, 72
lower triangular, 46
M-matrix, 101
monotone, 99
orthogonal, 138
permutation, 53
positive definite, 87, 88, 97
principal submatrix, 50
strictly diagonally dominant, 96
symmetric, 87
tridiagonal, *see* Tridiagonal matrix
unit lower triangular, 46
upper triangular, 47
Matrix factorisation
Cholesky, 90
LU, 48
QR, 76, 78, 163
Matrix norm, 58
1-norm, 66
2-norm, 66
∞-norm, 65
Frobenius norm, 141
subordinate norm, 64
Maximum norm, *see* Infinity norm
Maximum Principle, 365, 369, 372
comparison function, 366, 372

Mean Value Theorem, 8, 10, 11, 26, 113, 420
Midpoint rule, 286
Minimax approximation, 230
Minkowski's inequality, 62
Modified Euler method, 328
Monic polynomial, 243
Monotone matrix, 99
  tridiagonal, 100
Moore–Penrose generalised inverse, 70, 81

Natural cubic spline, 298
Near-minimax polynomial, 245, 270
Neighbourhood, 105
Newton's method, 19, 21, 116
  convergence, 23, 116
  global behaviour, 31, 123
  simultaneous equations, 118
Newton–Cotes quadrature, 201
  convergence, 208
  error estimate, 204
  Simpson's rule, 203
  trapezium rule, 202
Norm, 58, 224
  1-norm, 59, 66
  2-norm, 59, 66, 225, 252, 255
  ∞-norm, 59, 65, 225
  energy norm, 399
  Frobenius norm, 141
  induced norm, 254
  normed linear space, 58, 224
  p-norm, 60
  Sobolev norm, 387
  vector and matrix norm, 58
Normal equations, 76

One-step methods, 317
  consistent, 321
  convergence, 322
  Euler's method, 317, 323
  general form, 317
  implicit methods, 324
  improved Euler method, 328
  modified Euler method, 328
  order of accuracy, 323
  Runge–Kutta methods, 323, 325
  trapezium rule method, 324
  truncation error, 317
Open ball, 63
Open set, 104
Operation count, 92
Order of accuracy
  linear multistep methods, 338
  one-step methods, 323
Orthogonal, inner product space, 252
Orthogonal eigenvectors, 136
Orthogonal matrix, 138
Orthogonal polynomials, 259, 260, 277
  construction, 260
  zeros, 269, 279
Orthogonal transformation
  eigenvalues, 137
  invariance of sum of squares, 140
  plane rotation, 138
Orthonormal polynomials, 265
Oscillation Theorem, 232, 233, 243
  critical point, 233

$\mathcal{P}_n$, 180
Permutation matrices, 53
Piecewise polynomials, 292
Pivoting, 52, 92, 95
Plane rotations, 138, 163
Poincaré–Friedrichs inequality, 414
Positive definite matrix, 87, 97
  properties, 88
Principal submatrix, 50

QR algorithm, 162
  shift, 164
QR factorisation, 76, 78, 163
Quadratic convergence, 16, 22, 119
Quadrature, *see* Integration

$\mathbb{R}^{m\times n}$, $\mathbb{R}^{n\times n}$, 40
$\mathbb{R}^{n\times n}_{\mathrm{sym}}$, 87
$\mathbb{R}^n_*$, 64
Radau quadrature, 287
Rayleigh quotient, 170
Rayleigh–Ritz principle, 388
Relaxation, 19
  convergence, 20, 117
  simultaneous equations, 116
Richardson extrapolation, 216
Ritz approximation, 393
Ritz method, 392, 393
Ritz projector, 398
Rolle's Theorem, 184, 191, 419
Romberg integration, 217
Row operations, 46
Runge phenomenon, 208
Runge–Kutta methods, 323, 325
  algebraically stable, 354
  Butcher tableau, 352
  classical fourth order, 328
  diagonally implicit (DIRK), 353
  implicit, 349
  improved Euler method, 328
  modified Euler method, 328

Secant method, 25
  convergence, 26
Self-adjoint problem, 370

Set of measure zero, 295
Shift, QR algorithm, 165
Simple iteration, 2
  convergence, 11
  divergence, 15
  global behaviour, 29
Simpson's rule, 203
  composite, 210
  error estimate, 205
Simultaneous iteration, 106
  convergence, 110, 113
Simultaneous nonlinear equations, 104
  Newton's method, 118
Simultaneous relaxation, 116
Singular value
  decomposition, 82
  definition, 67
Sobolev norm, 387
Sobolev space, 296, 387
Solution of linear equations, 44, 55
  computational work, 56
  least squares, 74
  sensitivity, 71
Spline, 292, 394
  cubic, 298
  end conditions, 298
  Hermite cubic, 300
    error bound, 301
  interpolating cubic, 298
  knot, 292
  linear, 293
    basis functions, 297
    error bound, 293
    optimum property, 296
  natural cubic, 298
    construction, 299
    end conditions, 298
    optimum property, 300
Stability polynomial, 347
Stable fixed point, 12
Stiff linear ODE system, 345
Stiffness matrix, 396
Strictly diagonally dominant matrix, 96
Sturm sequence, 158
Sturm–Liouville problem, 373
Subdivision, 394
Sublinear convergence, 13
Subordinate matrix norm, 64
Superlinear convergence, 13
Support, 394
Symmetric bilinear functional, 388
Symmetric matrix, 87

Taylor's Theorem, 420
  several variables, 422
  with integral remainder, 420
Thomas algorithm, 95, 363
Trace, 136
Trapezium rule, 202
  composite, 209
Tridiagonal matrix, 93, 363, 367, 371, 373, 395
  eigenvalues, 156
  factorisation, 94
  monotone, 100
  reduction of real symmetric matrix, 150
Truncation error
  Euler's method, 318
  linear multistep method, 337
  one-step method, 317

Unit lower triangular matrix, 46
Upper triangular matrix, 47

Variational problem, 385
Vector norm, 58
  1-norm, 59
  2-norm, 59
  ∞-norm, 59
  p-norm, 60

Weak formulation, 390
Weak solution, 390
Weierstrass Theorem, 227, 283
Weight function, 255, 260, 277

Young's inequality, 61

Zero-stability, 331
  Root Condition, 335